INTERNATIONAL CENTRE FOR MECHANICAL SCIENCES

COURSES AND LECTURES No. 281

# THE CONSTITUTIVE LAW IN THERMOPLASTICITY

EDITED BY
Th. LEHMANN
RUHR-UNIVERSITY
BOCHUM

SPRINGER-VERLAG WIEN GMBH

This volume contains 111 illustrations.


Originally published by Springer Verlag Wien-New York in 1984

ISBN 978-3-211-81796-4 ISBN 978-3-7091-2636-3 (eBook)
DOI 10.1007/978-3-7091-2636-3

## CONTENTS

# THE CONSTITUTIVE LAW IN THERMOPLASTICITY
# AN INTRODUCTION

**THEODOR LEHMANN**
Ruhr-University Bochum
D–4630 Bochum

Within this course we are dealing mainly with non-isothermic inelastic deformations of such solid bodies which can be considered phenomenologically as classical continua. This means, we assume that the deformations of the considered bodies are completely derivable from the description of the motion of the material points in a well defined space of observation (space-fixed coordinate system). We shall see, however, that the description of the deformations can also be based merely on the consideration of the deformations of a comoving body-fixed coordinate system.

The restriction to classical continua is, of course, significant. Certain problems concerning deformations of structured media cannot be treated within this frame. This is, for instance, true for certain problems in rock mechanics. Such problems require an extended concept for the description of deformations taking into account generalized continua, so-called (simple or generalized) Cosserat continua.

Inelastic deformations of solid bodies represent coupled thermo-mechanical processes even if the body deforms quasistatically and isothermically. This is due to dissipation of mechanical work involved in inelastic deformations transforming mechanical work into heat. Moreover, inelastic deformations are accompanied by changes of the internal structure of the material, in general, which also represent coupled thermo-mechanical processes. Therefore each theoretical approach for the description of inelastic deformations has to be based on the fundamental thermodynamical laws.

Concerning the formulation of these fundamental laws of thermodynamics, however, we observe some significant differences. Sometimes it is said [1] we know as many different approaches for the theory of irreversible processes as authors in this field exist (see e.g. [1 to 15]). You will recognize this fact also within this course. It is, however, just intended to emphasize some different points of view with respect to the thermodynamics of irreversible processes. Therefore we shall treat this subject in different series of lectures.

One main difference concerns the question whether or not the thermodynamical state of a material element can be described uniquely by a set of thermodynamical state variables. When yes, we can deal with uniquely defined state functions. Otherwise we have to base our considerations on functionals representing the process history. This does not exclude that in certain cases both methods lead to the same result.

There exists, however, still another principal difficulty in the thermodynamical considerations of inelastic deformations of solid bodies. It arises from the different levels of description of these processes on the microscale or on the (phenomenological) macroscale, respectively. On the microscale, we can distinguish exactly between reversible processes representing a sequence of equilibrium states and irreversible processes characterized essentially by non-equilibrium states. On the macroscale, however, the situation is more complex. Besides reversible sequences of equilibrium states and rate dependent dissipative processes we obtain on the macroscale also sequences of equilibrium states which are nevertheless connected with irreversible (dissipative) transition processes. On the other hand we also find non-dissipative processes on the macroscale which are not governed by state equations. The first is true, for instance, for rate-independent irreversible plastic deformations, which are due to dissipative processes on the microscale with very short relaxation time. They appear on the macroscale like a sequence of equilibrium states. The second kind concerns certain hardening and softening processes connected with plastic deformations. These processes representing sequences of equilibrium states are connected with creation, migration, and annealing of lattice defects. The mechanical energy involved in these lattice defects is stored and not dissipated.

The process of storing and restoring, however, is not governed by state equations on the macroscale in contradiction to the corresponding processes on the microscale. This is due to the stochastic nature of the microprocesses which are influenced by randomly distributed material properties like grain boundaries etc. depending on the previous deformation and temperature history.

For these reasons the usual formulations of the classical thermodynamical theory of irreversible processes require certain modifications in applying them to a phenomenological theory of thermoplasticity. This concerns not only the question of reversibility of mechanical energy. We know also different kinds of internal processes which are not necessarily connected with external mechanical work. Radiation, for instance, may produce lattice defects, too. On the other hand certain annealing processes may lead to dissipation of internal energy without dissipation of mechanical work. All these phenomena have to be covered by a capable phenomenological theory of thermo-mechanical processes of solid bodies. We cannot expect, however, that such a theory is exact in the strict sense. This may be demonstrated again by a look on certain phenomena on the microscale level.

When we want to describe the kinematic state of a crystalline body or any discretely ordered body else within the frame of a continuum theory we may introduce continuous functions determining the disturbed state of the lattice (deformations of the lattice and distribution of the lattice defects). In the general case this requires the introduction of second rank and third rank tensor fields [16, 17]. Within the frame of a phenomenological theory supposing a classical continuum (or even a simple Cosserat continuum [18]) we can only use tensors of even rank as state variables. This means that we cannot map exactly the general kinematical state of a crystalline body into the state of a classical or simple Cosserat continuum, respectively. Therefore within this frame any phenomenological theory can only represent an approximation. It is principally connected with a loss of informations which can be gained at the microscale. The situation is - in a certain sense - comparable with the attempt to treat a threedimensional body within the frame of a two-dimensional or onedimensional theory. Nevertheless we can try to construct certain bridges between the description at the microscale level and a phenomenological theory at macroscale level. This has be done by different authors in different manners (e.g. [18 to 24]). Such attempts are, however, generally limited by the mentioned fact and therefore restricted to some basic phenomena.

In this situation we are left with the task to formulate a phenomenological theory of thermoplasticity which is at least self-consistent within the supposed frame. We demand, however, that such a theory does not contradict the facts significantly which we can learn from considerations of states and processes on the microscale. Therefore we shall spend some lectures to this subject.

The phenomenological theory of thermoplasticity which we have in mind shall not be restricted to small deformations since in many technical applications the considered deformations become large. In large deformations, however, the description of the total deformations and the decomposition of the deformations and of the corresponding mechanical work into a (strictly) reversible part and into the remaining part implicates some important questions. This problem has found increasing interest just recently [25 to 28]. We shall treat these questions within different lectures from different points of view depending on the respective model used for the material behaviour.

The centre of our series of lectures is formed by considerations concerning the definition of constitutive laws for material elements undergoing non-isothermic elastic-plastic or elastic-viscoplastic deformations, respectively. These considerations have to be based on thermodynamical reflections. It is obvious that different points of view concerning the thermodynamics of irreversible processes lead to different frames for the formulation of constitutive laws in thermoplasticity. It is not our aim to present an unified theory. It is rather intended to emphasize the differences following from the different concepts. However, we can not offer a complete review on all known or even possible concepts. Therefore this series of lectures has to be understood as an introduction into problems connected with the formulation of constitutive laws in thermoplasticity. It shall be stimulating for your own research work in this field.

In this context it may be emphasized that it becomes more and more important to combine theoretical and experimental work in this field.

Simple fundamental experiments represent the base for the formulation of a suitable theory of thermoplasticity. The capability of such a constituded theory has to be verified again in experiments with growing complexity which may lead to reformulations of the theory. In this way theory and experiment have to be mutual stimulating. However, we cannot expect to obtain a theory finally which covers the whole field of possible processes. In many cases we may even content ourselves with a rough approximation. It depends on the subject under consideration. Complex loading and temperature histories, for instance, or bifurcation problems (see e.g. [29 to 34]) demand for a more sophisticated theory than simple processes with proportional loading. In order to demonstrate this fact a series of lectures on bifurcation and localization problems is included in this course.

With respect to applications of the theory of thermoplasticity we have to think about the transfer of the constitutive law from the material (infinite) element to finite elements of structures. It is known that variational formulations of the fundamental laws represent a very helpful tool for this task. In thermoplasticity, however, the knowledge of variational principles is still very limited (see e.g. [35 to 37]). The last series of lectures deals with this subject including some applications, for instance, in limit analysis and limit design.

As already stated the course on constitutive law in thermoplasticity given here can only represent a certain introduction into the problems in this field. We do hope it is stimulating to further research work.

## REFERENCES


1. Hutter, K., The foundations of thermodynamics, its basic postulates and implications, *Acta Mech.* 27, 1977
2. Muschik, W., Thermodynamische Theorien, Überblick und Vergleich, *ZAMM* 61, 1981
3. Müller, I., *Thermodynamik*, Bertelsmann Universitätsverlag, Düsseldorf, 1973
4. Meixner, J. and Reik, H.G., Thermodynamik der irreversiblen Prozesse, *Handbuch der Physik*, Flügge, S., ed., Vol. III/2, 413, Springer-Verlag Berlin/Göttingen/Heidelberg, 1959
5. de Groot, S.R. and Mazur, P., *Non-equilibrium thermodynamics*, North-Holland Publ. Comp., Amsterdam, 1962
6. Gyarmati, I., *Non-equilibrium thermodynamics*, Springer-Verlag, Berlin/Heidelberg/New York 1970
7. Kluitenberg, G.A., Thermodynamical theory of elasticity and plasticity, *Physica* 28, 217, 1962
8. Meixner, J., Processes in simple thermodynamic materials, *Arch. Rat. Mech. Anal.* 33, 33, 1969

9. Coleman, B.D. and Gurtin, M.E., Thermodynamics with internal state variables, *Journ. Chem. Phys.* 47, 597, 1967

10. Day, W.A., *The thermodynamics of simple materials with fading memory*, Springer-Verlag Berlin/Heidelberg/New York 1972

11. Coleman, B.D. and Owen, D.R., A mathematical foundation of thermodynamics, *Arch. Rat. Mech. Anal.* 54, 1, 1974

12. Coleman, B.D. and Owen, D.R., On thermodynamics and elastic-plastic materials, *Arch. Rat. Mech. Anal.* 59, 25, 1975

13. Truesdell, A.C., *Rational thermodynamics*, McGraw-Hill, New York 1969

14. Noll, J.W., A new mathematical theory of simple materials, *Arch. Rat. Mech. Anal.* 48, 1, 1972

15. Green, A.E. and Naghdi, P.M., On continuum thermodynamics, *Arch. Rat. Mech. Anal.* 48, 352, 1972

16. Anthony, K.H., a) Die Reduktion von nichteuklidischen Objekten in eine euklidische Form und physikalische Deutung der Reduktion durch Eigenspannungszustände in Kristallen, *Arch. Rat. Mech. Anal.* 37, 161, 1970

b) Die Theorie der Disklinationen, *Arch. Rat. Mech. Anal.* 39, 43, 1970

c) Die Theorie der nichtmetrischen Spannungen in Kristallen, *Arch. Rat. Mech. Anal.* 40, 50, 1971

17. Kröner, E., Allgemeine Kontinuumstheorie der Versetzungen und Eigenspannungen, *Arch. Rat. Mech. Anal.* 4, 273, 1960

18. Lippmann, H., Eine Cosserat-Theorie des plastischen Fließens, *Acta Mech.* 8, 255, 1969

19. Kröner, E., Dislocation: A new concept in the continuum theory of plasticity, *Journ Math. Phys.* 42, 28, 1963

20. Teodosiu, E. and Sidoroff, F., A theory of finite elastovisco-plasticity of single crystals, *Int. J. Eng. Sci.* 14, 165, 1976

21. Eimer, Cz., Plasticity of polycrystal, *Arch. Mech.* 29, 687, 1977

22. Weng, G.J. and Phillips, A., An investigation of yield surfaces based on dislocation mechanics, *Int. J. Eng. Sci.* 15, I) 45, 1977 II) 61, 1977

23. Weng, G.J. and Phillips, A., The stress fields of continuous distribution of dislocations and of their movement in a poly-crystalline aggregate, *Int. J. Sol. Struct.* 14, 535, 1978

24. Werne, R.W. and Kelly, J.M., A dislocation theory of isotropic polycrystalline plasticity, *Int. J. Eng. Sci.* 16, 951, 1978

25. Nemat-Nasser, S., Decomposition of strain measures and their rates in finite deformation elastoplasticity, *Int. J. Sol. Struct.* 15, 155, 1979

26. Lee, E.H. and McMeeking, R.M., Concerning elastic and plastic components of deformation, *Int. J. Sol. Struct.* 16, 715, 1980

27. Lee, E.H., Some comments on elastic-plastic analysis, *Int. J. Sol. Struct.* 17, 859, 1981

28. Lehmann, Th., Some remarks on the decomposition of deformations and mechanical work, *Int. J. Eng. Sci.* 20, 281, 1982

29. Rice, J.R., The localization of plastic deformation, *Proc. 14. IUTAM Cong. Delft*, 207, 1976

30. Needleman, A., Non-normality and bifurcation in plane strain tension and compression, *J. Mech. Phys. Solids* 27, 231, 1979

31. Bruhns, O., and Raniecki, B., Bounds to bifurcation stresses in solids with non-associated plastic flow law at finite strain, *J. Mech. Phys. Solids* 29, 153, 1981

32. Zander, G., Zur Bestimmung von Verzweigungslasten dünnwandiger Kreiszylinder unter kombinierter Längs- und Torsionslast, Diss. Ruhr-Universität Bochum, *Mitt. Inst. für Mechanik, Ruhr-Univ. Bochum*, Nr. 27, 1981

33. Lehmann, Th., On the concept of stress-strain relation in plasticity, *Acta Mech.* 42, 263, 1982

34. Lehmann, Th., Some theoretical considerations and experimental results concerning elastic-plastic stress-strain relations, to be published in *Ing.-Arch.*

35. Biot, M.A., Variational-Lagrangian irreversible thermodynamics of nonlinear thermorheology, *Quart. Appl. Math.* 34, 213, 1976

36. Mroz, Z. and Raniecki, B., Uniqueness and variational principles in thermoplasticity, *Int. J. Eng. Sci.* 14, 211, 1976

37. Besseling, J., Finite element methods, in: *Trends in solid mechanics.* Besseling and v.d.Heijden, Eds., Delft Univ. Press, 53, 1979

# THERMODYNAMIC THEORIES OF THERMOELASTICITY AND SPECIAL CASES OF THERMOPLASTICITY

Ingo Müller
FB 9 – Hermann-Föttinger-Institut
TU Berlin

Course given at the International Centre for Mechanical Sciences (CISM)
Udine, Academic Year 1982. The Laplace Session. July 19 - 30, 1982.

Contents

# 1. CONSTITUTIVE RELATIONS OF NON-LINEAR THERMOELASTICITY

## 1.1. General Field Equations

1.1.1. Objective of Thermodynamics of Solids. The objective of thermodynamics of solids is the determination of the fields of

$$\begin{aligned} &\text{density} \quad \rho(X_A, t) \\ &\text{motion} \quad x_i(X_A, t) \\ &\text{temperature} \quad T(X_A, t) \end{aligned} \tag{1.1}$$

in all particles of the body and at all times. $X_A$ characterizes a particle by its position in reference configuration.

The density $\rho$ is a measure for the inertia of the particle and the temperature T indicates how hot the particle is. The motion $x_i$ determines the position of the particle in the configuration at time $t$.

1.1.2. Equations of Balance. In order to determine the fields $\rho$, $x_i$ and T we must have field equations and it is customary in thermodynamics to base such field equations upon the equations of balance of mechanics and thermodynamics, viz.

$$\text{balance of mass} \quad \frac{d\rho}{dt} + \rho\frac{\partial v_i}{\partial x_i} = 0\,, \qquad (1.2)_1$$

balance of momentum $$\rho \frac{dv_i}{dt} - \frac{\partial t_{ij}}{\partial x_j} = \rho f_i \,, \qquad (1.2)_2$$

balance of (internal) energy $$\rho \frac{d\varepsilon}{dt} + \frac{\partial q_i}{\partial x_i} = t_{ij} \frac{\partial v_i}{\partial x_j} + \rho r \qquad (1.2)_3$$

$v_i$ is the velocity $\dfrac{\partial x_i(X_A , t)}{\partial t}$ , $f_i$ is the specific body force and $r$ is the specific supply of internal energy, usually by radiation.

It is obvious that these equations in their present form are not field equations for the determination of density, motion and temperature, because they contain the new fields

stress tensor $t_{ij}$

specific internal energy $\varepsilon$ , and

heat flux $q_i$ .

Therefore the equations of balance must be supplemented by constitutive equations.

1.1.3. Constitutive Equations. Experience shows that stress, heat flux, and internal energy are related to density, motion and temperature in a materially dependent manner by constitutive relations.

In simple bodies the constitutive quantities $t_{ij}$ , $q_i$ and $\varepsilon$ of one particle and time may depend upon the histories of $\rho$ , $x_i$ and $T$ of that particle and on the histories of motion and temperature in the immediate neighbourhood. Thus we write fairly general constitutive relations in the form

$$t_{ij}(X_A, t) = \underset{s=0}{\overset{\infty}{\mathscr{F}_{ij}}} [\rho(X_A, t-s), x_i(X_A, t-s), F_{iA}(X_A, t-s), T(X_A, t-s), g_A(X_A, t-s)] ,$$

$$q_i(X_A, t) = \underset{s=0}{\overset{\infty}{\mathscr{I}_i}} [\text{———————————————}]$$

$$\varepsilon(X_A, t) = \underset{s=0}{\overset{\infty}{\mathscr{N}}} [\text{———————————————}] \tag{1.3}$$

where $F_{iA} = \dfrac{\partial x_i(X_A, t)}{\partial X_A}$ and $g_A = \dfrac{\partial T}{\partial X_A}$ are deformation gradient and temperature gradient respectively.

1.1.4. Thermodynamic Processes. If the constitutive functionals $\mathscr{F}_{ij}, \mathscr{I}_i, \mathscr{N}$ were known we should be able to eliminate $t_{ij}$, $q_i$ and $\varepsilon$ between the equations of balance (1.2) and the constitutive relations (1.3) and obtain a set of explicit field equations provided that $f_i$ and $r$ are given functions. Every solution of these field equations is called a thermodynamic process.

1.1.5. Restrictive Principles. While solutions of such functional field equations would be difficult to obtain, their determination would be a problem for the mathematician, if indeed the constitutive functionals were explicitly known. The situation is worse though, because there is no material for which we know the complete form of the functionals. Therefore the thermodynamicist attempts to find restrictions on the functionals with the goal of reducing their generality to, maybe, simple functions of only a few variables which may then be measured or determined from atomistic models by use of the methods of statistical mechanics.

This search for restrictions upon the constitutive relations and their eventual determination is the subject of the constitutive theory.

The tools of the constitutive theory are universal physical principles which have been abstracted from long experience with bodies of arbitrary material. The most important ones among such principles are

the principle of material frame indifference, and

the entropy principle.

Also, sometimes the constitutive functions of a body are a priori known to have certain properties which - although not universal - are shared by a large number of bodies. Examples are

isotropy, and

incompressibility,

## 1.2. Thermodynamic Constitutive Theory I. Material Frame Indifference.

1.2.1. A Modicum of Kinematics. The polar decomposition theorem states that the deformation gradient $F_{iA}$ has two unique decompositions, viz.

$$F_{iA} = R_{iB}\, U_{BA} \qquad \text{and} \qquad F_{iA} = V_{ij}\, R_{jA} \,, \tag{1.4}$$

where $R_{iB}$ is an orthogonal matrix and $U_{BA}$ and $V_{ij}$ are symmetric positive definite matrices. $\underset{\sim}{R}$ characterizes the rotational part of $\underset{\sim}{F}$ whereas $\underset{\sim}{U}$ and $\underset{\sim}{V}$ are measures for the change of lengths of a particle. The right and left Cauchy-Green tensors

$$\underset{\sim}{C} = \underset{\sim}{F}^T \underset{\sim}{F} = \underset{\sim}{U}^2 \qquad \text{and} \qquad \underset{\sim}{B} = \underset{\sim}{F}\,\underset{\sim}{F}^T = \underset{\sim}{V}^2 \tag{1.5}$$

are also useful measures for length changes during the deformation.

From the definition of $\underset{\sim}{F}$ we obtain a relation between $\dot{F}$ and the velocity gradient, namely

$$\dot{F}_{iA} = \frac{\partial v_i}{\partial x_j} F_{jA} \qquad \text{or} \qquad \frac{\partial v_i}{\partial x_j} = \dot{F}_{iA} \overset{-1}{F}_{Aj} \tag{1.6}$$

The determinant of F is denoted by J and from (1.6) it is easy to derive the equation

$$\dot{J} = J \frac{\partial v_i}{\partial x_i} . \tag{1.7}$$

It follows that the mass balance $(1.2)_1$ can be integrated to give

$$\rho = \frac{\rho_0}{J} , \tag{1.8}$$

where $\rho_0$ is the constant density in the reference configuration.

By (1.8) $\rho$ and $\underset{\sim}{F}$ are algebraically related and therefore $\rho$ may be dropped as a variable in the constitutive relations (1.3).

1.2.2. Objective Scalars, Vectors and Tensors. Objectivity of quantities refers to Euclidean transformations of frame, i.e. transitions from one rectangular cartesian frame to another one in arbitrary translational and rotational motion. The corresponding coordinate transformation reads

$$x_i^* = \mathcal{O}_{ij}(t) x_j + b_i(t) , \tag{1.9}$$

where $\mathcal{O}_{ij}$ is an orthogonal time-dependent matrix and $b_i$ a time dependent vector.

Quantities $s, V_i$ and $T_{ij}$ are called objective scalars, vectors and tensors respectively, if they transform according to the rules

$$s^* = s , \quad V_i^* = \mathcal{O}_{ij} V_j , \quad T_{ij}^* = \mathcal{O}_{ik} \mathcal{O}_{j\ell} T_{k\ell} . \tag{1.10}$$

Examples for objective scalars are density $\rho$ and temperature $T$, while the heat flux $q_i$ and the stress $T_{ij}$ are an objective vector and tensor respectively.

From (1.9) it follows that the deformation gradient transforms according to $F^*_{iA} = \mathcal{O}_{ij} F_{jA}$ and hence the right and left Cauchy-Green tensor obey the rules

$$C^*_{AB} = C_{AB} \qquad \text{and} \qquad B^*_{ij} = \mathcal{O}_{ik}\, \mathcal{O}_{j\ell}\, B_{k\ell} \tag{1.11}$$

so that the left Cauchy-Green tensor is an objective tensor.

1.2.3. Formulation of the Principle of Material Frame Indifference. The principle of material frame indifference states that the constitutive functionals $\mathcal{F}_{ij}$, $\mathcal{I}_i$ and $\mathcal{N}$ are the same ones for observers in different Euclidean frames. This means for the constitutive relations (1.3) that

$$\begin{aligned}
\underset{\sim}{t} &= \underset{\sim}{\mathcal{F}}[\overset{t}{\underset{\sim}{x}}(s), \overset{t}{\underset{\sim}{F}}(s), \overset{t}{T}(s), \overset{t}{\underset{\sim}{g}}(s)] = \mathcal{O}^T(t)\underset{\sim}{\mathcal{F}}[\overset{t}{\mathcal{O}}(s)\overset{t}{x}(s) + \overset{t}{b}(s), \overset{t}{\mathcal{O}}(s)\overset{t}{F}(s), \overset{t}{T}(s), \overset{t}{g}(s)]\mathcal{O}(t) \\
\underset{\sim}{q} &= \underset{\sim}{\mathcal{I}}[\text{———}] = \underset{\sim}{\mathcal{O}}^T(t)\,\underset{\sim}{\mathcal{I}}[\text{———}] \\
\varepsilon &= \mathcal{N}[\text{———}] = \mathcal{N}[\text{———}]
\end{aligned} \tag{1.12}$$

must hold for all $\mathcal{O}(t)$ and $b(t)$.

1.2.4. Consequences of Material Frame Indifference. First one can choose $\overset{t}{\underset{\sim}{\mathcal{O}}}(s) = \underset{\sim}{1}$ and $\overset{t}{\underset{\sim}{b}}(s) = -\overset{t}{\underset{\sim}{\chi}}(s)$ and obtain from (1.12) that neither one of the constitutive quantities can depend on $\overset{t}{x}(s)$. Next one chooses $\overset{t}{\underset{\sim}{\mathcal{O}}}(s) = \overset{t}{\underset{\sim}{R}}{}^T(s)$, , where $\underset{\sim}{R}$, according to (1.4), is the rotational part of $\underset{\sim}{F}$. In this case (1.12) implies that the constitutive relations must have the forms

$$\underset{\sim}{t} = \underset{\sim}{\mathcal{F}}[\overset{t}{\underset{\sim}{F}}(s), \overset{t}{T}(s), \overset{t}{\underset{\sim}{g}}(s)] = \underset{\sim}{R}(t)\underset{\sim}{\mathcal{F}}[\overset{t}{\underset{\sim}{U}}(s), \overset{t}{T}(s), \overset{t}{\underset{\sim}{g}}(s)]\underset{\sim}{R}^T(t) \tag{1.13$_1$}$$

$$\underset{\sim}{q} = \underset{\sim}{\mathcal{I}}[\text{———}] = \underset{\sim}{R}(t)\underset{\sim}{\mathcal{I}}[\text{———}] \tag{1.13$_2$}$$

$$\varepsilon = \mathcal{N}[\overset{t}{\underset{\sim}{F}}(s), \overset{t}{T}(s), \overset{t}{\underset{\sim}{g}}(s)] = \mathcal{N}[\overset{t}{\underset{\sim}{U}}(s), \overset{t}{T}(s), \overset{t}{\underset{\sim}{g}}(s)]. \qquad (1.13)_3$$

It follows that the internal energy is independent of $\underset{\sim}{R}$, while stress and heat flux depend explicitly only on the present value of $\underset{\sim}{R}$, but are independent of its history.

The equations (1.13) are called the reduced forms of the constitutive equations. They assume a simpler form, if we introduce the material stress $t_{AB}$ and the material heat flux $q_A$ by the definitions

$$t_{AB} = J \overset{-1}{F}_{Ai} t_{ij} (F^T)^{-1}_{jB} \qquad \text{and} \qquad q_A = J \overset{-1}{F}_{Ai} q_i \, . \qquad (1.14)$$

If this is introduced into (1.13) the rotation tensors cancel out and when we also replace U by C, which is possible by (1.5), we obtain

$$\begin{aligned} t_{AB} &= t_{AB}[\overset{t}{\underset{\sim}{C}}(s), \overset{t}{T}(s), \overset{t}{\underset{\sim}{g}}(s)], \\ q_A &= q_A[\text{———————}], \\ \varepsilon &= \varepsilon[\text{———————}]. \end{aligned} \qquad (1.15)$$

For convenience in notation I have dropped here and often later the distinction between a functional and its value.

For an extensive treatment of the ideas presented in this Section and the following one see [1], [2] or [3].

## 1.3. Thermodynamic Constitutive Theory II, Material Isotropy

1.3.1. Dependence of Constitutive Functionals upon Reference Configuration. If $\kappa$ and $\hat{\kappa}$ are two reference configurations in which a particle lies at $X_A$ and $\hat{X}_A$ respectively, we have

$$\hat{X}_A = \hat{X}_A(X_B) \qquad \text{with} \qquad P_{AB} = \frac{\partial \hat{X}_A}{\partial X_B} \tag{1.16}$$

The deformation gradients and temperature gradients referred to the two reference configurations are different, viz.

$$\underset{\kappa}{F} = \underset{\hat{\kappa}}{F}\,P \qquad \text{and} \qquad \underset{\kappa}{g} = \underset{\hat{\kappa}}{g}\,P\,. \tag{1.17}$$

But of course, the constitutive quantities must be unaffected by the change of reference configuration so that the constitutive functionals must make up for the dependence of $\underset{\sim}{F}$ and $\underset{\sim}{g}$ on the reference configuration. If C and $\mathcal{L}$ stand for any one of the constitutive quantities and functions respectively we must therefore have

$$C = \underset{\kappa}{\mathcal{L}}[\underset{\kappa}{F}(s), T(s), \underset{\kappa}{g}(s)] = \underset{\hat{\kappa}}{\mathcal{L}}[\underset{\kappa}{F}(s)\,P^{-1}, T(s), \underset{\kappa}{g}(s)\,P^{-1}]\,. \tag{1.18}$$

When $\underset{\kappa}{\mathcal{L}}$ is known, $\underset{\hat{\kappa}}{\mathcal{L}}$ can be determined from (1.18).

1.3.2. Isotropy Condition. When a particle starts out in two different reference configurations $\kappa$ and $\hat{\kappa}$ and is subject to the same histories $\overset{t}{\underset{\sim}{F}}(s)$ , $\overset{t}{T}(s)$ and $\overset{t}{\underset{\sim}{g}}(s)$ , it will in general have different values of $t, q$ and $\varepsilon$ at time $t$. Should it happen, however, that the constitutive quantities assume the same values in the two cases, we say that the particle is isotropic with respect to the transformation $\underset{\sim}{P}$ from $\kappa$ to $\hat{\kappa}$. In that case we have

$$\underset{\kappa}{\mathcal{L}}[\overset{t}{\underset{\sim}{F}}(s), \overset{t}{T}(s), \overset{t}{\underset{\sim}{g}}(s)] = \underset{\hat{\kappa}}{\mathcal{L}}[\overset{t}{\underset{\sim}{F}}(s), \overset{t}{T}(s), \overset{t}{\underset{\sim}{g}}(s)]\,, \tag{1.19}$$

or, by (1.18)

$$\underset{\kappa}{\mathcal{L}}[\overset{t}{\underset{\sim}{F}}(s), \overset{t}{T}(s), \overset{t}{\underset{\sim}{g}}(s)] = \underset{\kappa}{\mathcal{L}}[\overset{t}{\underset{\sim}{F}}(s)\underset{\sim}{P}, \overset{t}{T}(s), \overset{t}{\underset{\sim}{g}}(s)\underset{\sim}{P}]. \tag{1.20}$$

This equation is called the isotropy condition. It holds for all $\underset{\sim}{P}$ for which the particle is isotropic. Thus the isotropy condition represents a restriction for the constitutive functionals.

1.3.3. Isotropic Bodies. There is no solid body for which the isotropy condition holds for non-orthogonal transformations $\underset{\sim}{P}$ . Often in a solid it holds for a subgroup of the orthogonal group and we say the body is isotropic with respect to that subgroup. But the most restrictive conditions result, if (1.20) holds for all orthogonal transformations Q. In that case we say that the body is isotropic as such.

If Q stands for an arbitrary orthogonal transformation, we thus have for an isotropic body

$$\begin{aligned} t_{ij} &= t_{ij}[\overset{t}{\underset{\sim}{F}}(s), \overset{t}{T}(s), \overset{t}{\underset{\sim}{g}}(s)] = t_{ij}[\overset{t}{\underset{\sim}{F}}(s)\underset{\sim}{Q}, \overset{t}{T}(s), \overset{t}{\underset{\sim}{g}}(s)\underset{\sim}{Q}] \\ q_i &= q_i[\text{———}] = q_i[\text{———}] \\ \varepsilon &= \varepsilon[\text{———}] = \varepsilon[\text{———}] \end{aligned} \tag{1.21}$$

for all Q. We combine this statement with the requirement (1.12) of material frame indifference in which we may choose $\underset{\sim}{\mathcal{O}}^t(s) = \underset{\sim}{Q}$. It follows that the constitutive functionals of an isotropic body must obey the conditions

$$\begin{aligned} Q_{i\kappa} t_{\kappa\ell} Q^T_{\ell j} &= t_{ij}[\underset{\sim}{Q}(\overset{t}{\underset{\sim}{F}}(s)\underset{\sim}{Q}^T, \overset{t}{T}(s), \underset{\sim}{Q}\overset{t}{\underset{\sim}{g}}(s)] = Q_{i\kappa} t_{\kappa\ell}[\overset{t}{\underset{\sim}{F}}(s), \overset{t}{T}(s), \overset{t}{\underset{\sim}{g}}(s)]Q^T_{\ell j} \\ Q_{i\kappa} q_\kappa &= q_i[\text{———}] = Q_{i\kappa} q_\kappa[\text{———}] \\ \varepsilon &= \varepsilon[\text{———}] = \varepsilon[\text{———}] \end{aligned} \tag{1.22}$$

for all $Q$. Functionals of this type are called isotropic functionals of $\overset{t}{\underset{\sim}{F}}(s)$, $\overset{t}{T}(s)$, and $\overset{t}{\underset{\sim}{g}}(s)$, because they are the constitutive functionals of isotropic bodies.

The equations (1.22) indicate that the internal energy remains unchanged, when $\underset{\sim}{F}$ is replaced by $\underset{\sim}{Q}\,\underset{\sim}{F}\,\underset{\sim}{Q}^T$ and $\overset{t}{\underset{\sim}{g}}(s)$ by $Q\overset{t}{\underset{\sim}{g}}(s)$. The heat flux $\underset{\sim}{q}$ changes to $\underset{\sim}{Q}\underset{\sim}{q}$ upon this replacement, while the stress changes to $\underset{\sim}{Q}\,\underset{\sim}{t}\,\underset{\sim}{Q}^t$. This can be seen to imply the conditions

$$Q_{AC}\, t_{CD}\, Q^T_{DB} = t_{AB}[\underset{\sim}{Q}\underset{\sim}{C}(s)\,Q^T, T(s), \underset{\sim}{Q}\underset{\sim}{g}(s)] = Q_{AC}\, t_{CD}[\underset{\sim}{C}(s), T(s), g(s)]\,Q^T_{DB}$$

$$Q_{AC}\, q_C = q_A\,[\rule{4cm}{0.4pt}] = Q_{AC}\, q_C\,[\rule{3cm}{0.4pt}] \qquad (1.23)$$

$$\varepsilon = \varepsilon\,[\rule{4cm}{0.4pt}] = \varepsilon\,[\rule{3cm}{0.4pt}]$$

for the constitutive equation (1.15), provided of course that the body is isotropic.

## 1.4. Thermodynamic Constitutive Theory III. Limited Memory and Representations.

1.4.1. <u>Range of Memory and List of Variables.</u> The memory of a body reaches back into the past by a certain length of time $\tau$. If the interesting process in the body varies slowly within that time, the remembered part of the histories $\overset{t}{\underset{\sim}{F}}(s)$, $\overset{t}{T}(s)$ and $\overset{t}{\underset{\sim}{g}}(s)$ may be represented by the present values $\underset{\sim}{F}(t)$, $T(t)$, $\underset{\sim}{g}(t)$ and, maybe by a few time derivatives of $\underset{\sim}{F}$, $T$ and $\underset{\sim}{g}$ at time $t$. In that case the constitutive functionals reduce to constitutive functions.

The most important case is the one of a thermoelastic solid, in which $\underset{\sim}{F}(t)$, $T(t)$, and $\underset{\sim}{g}(t)$ suffice to determine stress, heat flux and internal energy. Another frequently considered case is that of a viscoelastic solid, in which $\underset{\sim}{F}(t)$, $\dot{\underset{\sim}{F}}(t)$, $T(t)$ and $\underset{\sim}{g}(t)$ are the independent variables.

We shall later consider the case of thermoelastic bodies with a memory for temperature, so that $\dot{T}$ must be taken into account among the variables. The constitutive equations read

$$\begin{aligned} t_{AB} &= t_{AB}(\underset{\sim}{C}, T, \dot{T}, \underset{\sim}{g}), \\ q_A &= q_A \;(\text{———}), \\ \varepsilon &= \varepsilon \;\;(\text{———}), \end{aligned} \tag{1.24}$$

and the constitutive functions must be isotropic. We shall concentrate on this case, because it is a simple extension of thermoelasticity and because it provides two interesting conclusions which cannot be derived for properly thermoelastic bodies.

1.4.2. Representation of Isotropic Functions. By (1.23) the internal energy $\varepsilon$ can only depend on scalar combinations of the variables, in particular on the magnitude $g$ of the temperature gradient and on the principal invariants $I_1$, $I_2$, $I_3$ of the right Cauchy-Green tensor. $I_1$ is the trace of $\underset{\sim}{C}$, and $I_3$ is det $C$, while $I_2$ is composed of the traces of $\underset{\sim}{C}$ and $\underset{\sim}{C}^2$: $I_2 = -\frac{1}{2}(C^2)_{ii} + \frac{1}{2}C_{ii}^2$ In addition to $g$ and $I_1$, $I_2$, $I_3$ there are two more independent scalar combinations of the components of $\underset{\sim}{g}$ and $\underset{\sim}{C}$, viz. $g_A C_{AB} g_B$ and $g_A C^2_{AB} g_B$.

Again, by (1.23), the heat flux $q_A$ must be a linear combination of $g_A$, $C_{AB} g_B$ and $C^2_{AB} g_B$ with coefficients that depend upon the scalar combinations of $\underset{\sim}{g}$ and $\underset{\sim}{C}$ and upon scalar variables.

Finally the stress $t_{AB}$ must be a linear combination of the six tensors

$$\delta_{AB}, \; C_{AB}, \; C^2_{AB}, \; g_A g_B, \; g_C C_{C(AgB)}, \; g_C C^2_{C(AgB)}$$

with coefficients that may be dependent on scalars.

In summary we write the constitutive relations for an isotropic thermoelastic solid as

$$t_{AB} = a\,\delta_{AB} + b\,C_{AB} + c\,C^2_{AB} + d g_A g_B + e g_C C_{C(AgB)} + f g_C C^2_{C(AgB)},$$

$$q_A = \alpha g_A + \beta C_{AB} g_B + \gamma C^2_{AB} g_B, \tag{1.25}$$

$$\varepsilon = \varepsilon(T, \dot{T}, g, I_1, I_2, I_3, g_A C_{AB} g_B, g_A C^2_{AB} g_B),$$

where all coefficients a through f and $\alpha$ through $\gamma$ may depend on the same variables as does $\varepsilon$ according to $(1.25)_3$.

It is not altogether easy to show briefly that (1.25) is the general representation of isotropic functions of $C, T, \dot{T}, g$, and that this representation is not reducible, but this can be proved. An important role in such proofs is played by the Hamilton Cayley theorem which for any matrix states

$$\underset{\sim}{M}^3 - \overset{M}{I}_1 \underset{\sim}{M}^2 + \overset{M}{I}_2 \underset{\sim}{M} - \overset{M}{I}_3 \underset{\sim}{1} = 0 \tag{1.26}$$

where the $\overset{M}{I}$'s now are the principal invariants of $\underset{\sim}{M}$.

## 1.5. Thermodynamic Constitutive Theory IV. Incompressibility

1.5.1. <u>The Treatment of Incompressibility.</u> For conciseness in the following treatment of incompressibility we restrict the attention to the stress as the only constitutive quantity and leave out the thermal variables T and $\underset{\sim}{g}$. In an isotropic solid we then have by $(1.25)_1$

$$t_{AB} = a\,\delta_{AB} + b\,C_{AB} + c\,C^2_{AB}$$

or going back to the actual stress tensor $t_{ij}$ by (1.14)

$$t_{ij} = \bar{a}\,\delta_{ij} + \bar{b}B_{ij} + \bar{c}\,(B^2)_{ij}\,, \tag{1.27}$$

where $\bar{a}, \bar{b}$ and $\bar{c}$ may be functions of $I_1, I_2, I_3$.

No material is really fully incompressible, rather we must view incompressibility as the limiting case of decreasing compressibility of a body. In order to describe decreasing compressibility and eventually incompressibility as a material property it is useful to make a shift of dependent and independent variables (see [4]): We shall consider the trace $t_{\ell\ell}$ of the stress tensor $t_{ij}$ as a variable and $I_3 = \det C$ as a constitutive quantity. Incompressibility then results as the limiting case where the constitutive quantities - including $I_3$ - are independent of $t_{\ell\ell}$.

With the indicated shift of variables we may replace the six constitutive equations (1.27) by the five equations

$$t_{ij} - \frac{1}{3}\,t_{\ell\ell}\,\delta_{ij} = \hat{b}(B_{ij} - \frac{1}{3}\,B_{\ell\ell}\,\delta_{ij}) + \hat{c}\,(B^2_{ij} - \frac{1}{3}B^2_{\ell\ell}\,\delta_{ij}) \tag{1.28}$$

and in addition by the one equation

$$I_3 = I_3(I_1\,, I_2, t_{\ell\ell})\,; \tag{1.29}$$

$\hat{b}$ and $\hat{c}$ may in general depend on $I_1, I_2$ and $t_{\ell\ell}$. But in an incompressible body $\hat{b}, \hat{c}$ and $I_3$ are independent of $t_{\ell\ell}$. In fact, according to [4], thermodynmaics implies that $I_3$ is also independent of $I_1\,, I_2$ so that we must have

$$I_3 = 1 \tag{1.30}$$

in an incompressible body.

1.5.2. Stress- Strain Relations for Rubber; a Mooney-Rivlin Material. Rubber may be considered as an incompressible isotropic thermoelastic material. There is no evidence that the stress depends upon the temperature gradient so that we may use (1.28) as the constitutive equation for $t_{ij} - \frac{1}{3} t_{\ell\ell}\delta_{ij}$ :

$$t_{ij} - \frac{1}{3} t_{\ell\ell}\delta_{ij} = \hat{b}(B_{ij} - \frac{1}{3} B_{\ell\ell}\delta_{ij}) + \hat{C}(B^2_{ij} - \frac{1}{3} B^2_{\ell\ell}\delta_{ij}) . \tag{1.31}$$

Upon use of the Hamilton-Cayley theorem (1.26) for the elimination of $(B^2)_{ij}$ we obtain

$$t_{ij} - \frac{1}{3} t_{\ell\ell}\delta_{ij} = s_1 (B_{ij} - \frac{1}{3} B_{\ell\ell}\delta_{ij}) + s_{-1} (B^{-1}_{ij} - \frac{1}{3} B^{-1}_{\ell\ell}\delta_{ij}) , \tag{1.32}$$

where $s_1$ and $s_{-1}$ may depend upon $I_1$ and $I_2$ . This is the most useful form of the constitutive relation for rubber, because considerable information has been collected about the coefficients $s_1$ and $s_{-1}$ . The experiments of Rivlin and Saunders [5] have shown that $s_1$ is independent of $I_1$ and $I_2$ , and that $s_{-1}$ , while independent of $I_1$ , increases linearly with $I_2$ $s_{-1}$ is negative valued and its average magnitude in the range $3 < I_2 < 30$ is about one tenth of the value of $s_1$ .

Even though $s_{-1}$ is not a constant, it is true that the analysis based upon (1.32) with constant coefficients reflects the qualitative features of non-linear elasticity correctly. And of course such an analysis is very much simpler than for non-constant coefficients. Therefore it is customary to use (1.32) with constant values $s_1$ and $s_{-1}$ and we say that it characterizes a Mooney-Rivlin material.

1.5.3. A Rubber Slab under Normal Stresses. When the deformation gradient of a rubber slab has the form

$$F_{iA} = \begin{bmatrix} \lambda_1 & & 0 \\ & \lambda_2 & \\ 0 & & \lambda_3 \end{bmatrix} \tag{1.33}$$

the stress must be normal to the sides as shown in Figure 1.1. Indeed, by (1.32) we must have

$$t_{ij} - \frac{1}{3} t_{\ell\ell}\,\delta_{ij} = s_1 \begin{bmatrix} \frac{2}{3}\lambda_1^2 - \frac{1}{3}(\lambda_2^2 + \lambda_3^2) & 0 & 0 \\ 0 & \frac{2}{3}\lambda_2^2 - \frac{1}{3}(\lambda_1^2 + \lambda_3^2) & 0 \\ 0 & 0 & \frac{2}{3}\lambda_3^2 - \frac{1}{3}(\lambda_1^2 + \lambda_2^2) \end{bmatrix} +$$

$$+ s_{-1} \begin{bmatrix} \frac{2}{3}\frac{1}{\lambda_1^2} - \frac{1}{3}(\frac{1}{\lambda_2^2} + \frac{1}{\lambda_3^2}) & 0 & 0 \\ 0 & \frac{2}{3}\frac{1}{\lambda_2^2} - \frac{1}{3}(\frac{1}{\lambda_1^2} + \frac{1}{\lambda_3^2}) & 0 \\ 0 & 0 & \frac{2}{3}\frac{1}{\lambda_3^2} - \frac{1}{3}(\frac{1}{\lambda_2^2} + \frac{1}{\lambda_1^2}) \end{bmatrix} \tag{1.34}$$

where, because of incompressibility $\lambda_1 \lambda_2 \lambda_3 = 1$ holds (see (1.30)).

For biaxial loading with $t_{33} = 0$ we obtain by elimination of $t_{\ell\ell}$

$$t_{11} = s_1(\lambda_1^2 - \lambda_3^2)(1 - \frac{s_{-1}}{s_1}\lambda_2^2), \tag{1.35}$$

$$t_{22} = s_1(\lambda_2^2 - \lambda_3^2)(1 - \frac{s_{-1}}{s_1}\lambda_1^2)$$

and finally for **uniaxial** loading with $t_{22} = t_{33} = 0$, and $\lambda_2 = \lambda_3 = \frac{1}{\sqrt{\lambda_1}}$

$$t_{11} = s_1(\lambda_1^2 - \frac{1}{\lambda_1})(1 - \frac{s_{-1}}{s_1}\frac{1}{\lambda_1}) . \tag{1.36}$$

When we consider rubber as a Mooney-Rivlin material, $s_1$ and $s_{-1}$ are independent of the deformation. However, experiments show that both are linear homogenous functions of temperature, because $t_{11}$ grows with increasing temperature as shown by Figure 1.2.

$t_{11}$ in (1.36) is the force on the rubber slab in the 1-direction referred to an area that has decreased in the deformation by lateral contraction. The force referred to the original area is equal to $S = t_{11}\lambda_2\lambda_3 = {}^{t_{11}}/_{\lambda_1}$ . It follows that S is given by

$$S = \sigma T(\lambda_1 - \frac{1}{\lambda_1^2}) , \tag{1.37}$$

where for simplicity the small quotient $s_{-1}/s_1$ has been neglected. $\sigma$ is a constant factor. Figure 1.3 represents S as a function of $\lambda_1$ for different temperatures.

Even though $s_{-1}/s_1$ may be as big as 0.1 we shall later neglect it, unless the presence of this ratio is important for its qualitative effect on the conclusions.

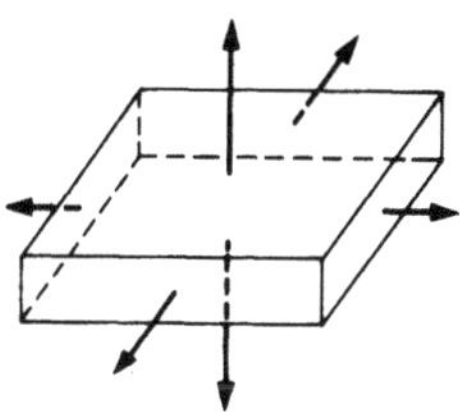

Figure 1.1

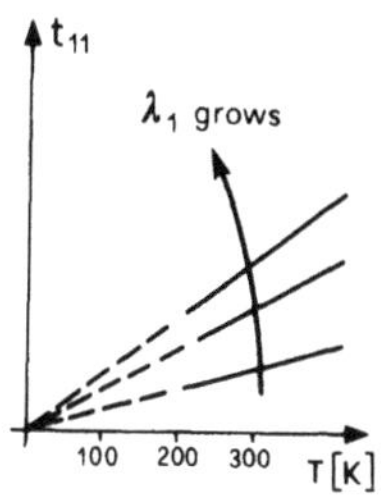


Figure 1.2

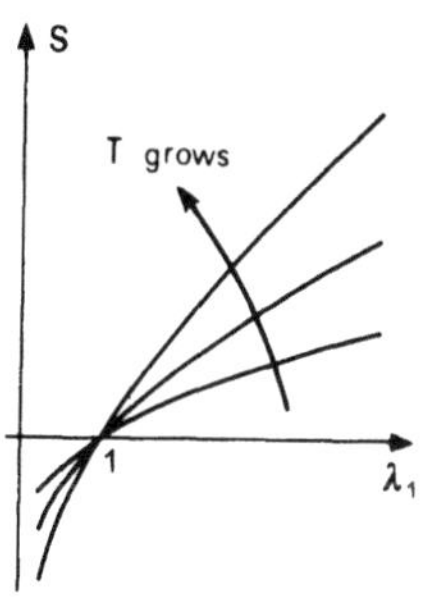


Figure 1.3

# 2. ENTROPY

## 2.1. Purpose of the Entropy Principle

The entropy principle is abstracted from our basic experience that all processes have a particular direction, and that they approach equilibrium, i.e. a stationary and generally uniform state.

The entropy principle is used to find restrictions upon the constitutive equations and in its classical form, due to Clausius and Carathéodory it has provided a link between the thermal and caloric equations of state which is most powerful in helping the experimentalist to determine the internal energy.

In non-equilibrium thermodynamics the proper form of the entropy principle is still a matter of debate and different propositions can be discussed. But whatever the form, here too the purpose of the principle is the derivation of restrictions upon constitutive functions.

In this section I start oout presenting the classical arguments of Clausius on fluids. But then I proceed to consider thermoelastic bodies with memory presenting the thermostatics of Carathéodory for those. Next comes the consideration of a thermoelastic body with a (short) memory for temperature, i.e. the constitutive relations are those of equation (1.24).

For the present purpose this set of equations describes a model material which I use to illustrate the procedure of three competing thermodynamic theories, viz.

linear irreversible thermodynamics

thermodynamics based on Clausius-Duhem inequality (Coleman, Noll)

rational thermodynamics with Lagrange multipliers.

The experimentally verifiable results agree for all three theories and, indeed, they concern thermostatics and agree with that theory. Where the results differ, there is insufficient empirical knowledge, that would favour one proposition and throw out the other, because thermodynamic experimentation in non-equilibrium is difficult.

## 2.2. Thermodynamics to Thermostatics

### 2.2.1. Thermostatic Processes and Slow Thermostatic Processes.

Thermostatics neglects acceleration and satisfies the momentum balance by considering only uniform fields of density, deformation gradient and temperature. Thus a thermostatic process consists of the functions

$$\rho(t)\,,\quad F_{iA}(t)\,,\quad T(t) \tag{2.1}$$

and the equations of balance $(1.2)_{1,3}$ reduce to the form

$$\frac{\dot{\rho}}{\rho} = -\frac{\dot{J}}{J}$$

$$\varepsilon = q' + \frac{1}{2\rho_0} t_{AB}\,\dot{C}_{AB} \qquad \text{with} \qquad \begin{aligned} \varepsilon(t) &= n\,[\overset{t}{C}(s)\,,\overset{t}{T}(s)] \\ t_{AB}(t) &= f_{AB}[\text{————}]\,, \end{aligned}$$

where $q' = -\frac{1}{m}\oint_F q_i\, e_i\, ds$ is the heating. $q'$ represents the influx of heat through the boundary F of the body referred to its mass. In (2.2) I have written the working term by use of the material stress. This is appropriate, since by material frame indifference $\varepsilon$ and $t_{AB}$ depend only on the six independent components of $\underset{\sim}{C} = \underset{\sim}{F}^T \underset{\sim}{F}$ rather than on the nine components of $\underset{\sim}{F}$. Radiation is ignored here.

Note that $\frac{1}{m} \oint t_{ij} v_i e_j \, ds = \frac{1}{\rho} t_{ij} (\dot{\underset{\sim}{F}} \underset{\sim}{F}^{-1})_{ij} = \frac{1}{2\rho_0} t_{AB} \dot{C}_{AB}$ is the specific power exerted on the body. Therefore we may write

$$\dot{\varepsilon} = q' + \ell \tag{2.3}$$

or upon integration between $t_1$, and $t_2$ with $q_{12} = \int_{t_1}^{t_2} q' dt$ and $a_{12} = \int_{t_1}^{t_2} \ell dt$

$$\varepsilon_2 - \varepsilon_1 = q_{12} + a_{12} \,. \tag{2.4}$$

$q_{12}$ is the heat exchanged and $a_{12}$ is the specific work done to the body.

In particular, if a process is so slow that it does not change during the span of the memory of the body, $\varepsilon$ and $t_{AB}$ at time $t$ depend only on $C(t)$ and $T(t)$. Thus $\varepsilon$ and $t_{AB}$ are no different in such a slow thermostatic process than they are in equilibrium, which is a uniform and time-independent process. Therefore we write $\varepsilon = \varepsilon|_E(\underset{\sim}{C}, T)$ and $t_{AB} = t_{AB}|_E(\underset{\sim}{C}, T)$ and the balance of internal energy reads

$$\dot{\varepsilon}|_E(\underset{\sim}{C}, T) = q' + \frac{1}{2\rho_0} t_{AB}|_E \dot{C}_{AB} \,. \tag{2.5}$$

In a slow thermostatic process of a fluid the stress $t_{AB|E}$ is equal to $-p|_E \delta_{AB}$, where $p|_E$ is the pressure. Also the internal energy and the pressure depend only on $\rho$ and $T$. Thus in a fluid we have, instead of (2.5)

$$\dot{\varepsilon}|_E(\rho, T) = q' - p|_E(\rho, T)\left(\frac{1}{\rho}\right)^{\bullet} . \tag{2.6}$$

Slow thermostatic processes are also called reversible, because it follows from (2.5) and (2.6) that the heating $q'$ is reversed into its negative, when the process $C(t)$, $T(t)$ or $\rho(t)$, $T(t)$ changes its temporal direction.

Equations (2.5) and (2.6) represent but one equation for the slow thermostatic process $\underset{\sim}{F}(t)$, $T(t)$ or $\rho(t)$, $T(t)$ in a solid or a fluid respectively. And that equation also contains $q'$.

2.2.2. Solvable Problems in Thermostatics of Fluids. A slow thermostatic process in a fluid can be represented by a curve in a two dimensional diagram like the $(T, {}^1\!/_\rho)$ -diagram, or the $(p|_E, {}^1\!/_\rho)$ -diagram. In latter diagram the work done on the fluid, viz. $a_{12} = \int_1^2 - p|_E \, d({}^1\!/_\rho)$ is given by the area under the curve, and the heat exchanged is equal to $q_{12} = \varepsilon_2 - \varepsilon_1 + \int_1^2 p|_E \, d({}^1\!/_\rho)$. These quantities can be calculated once the process $\rho(t)$, $T(t)$ is given; or alternatively, if $q'$ is given as well as $\rho(t)$ (say) we may calculate $T(t)$. In both cases, of course, we must know the constitutive relations $p = p|_E(\rho, T)$ and $\varepsilon = \varepsilon|_E(\rho, T)$, which are called the thermal and caloric equations of state in thermostatics.

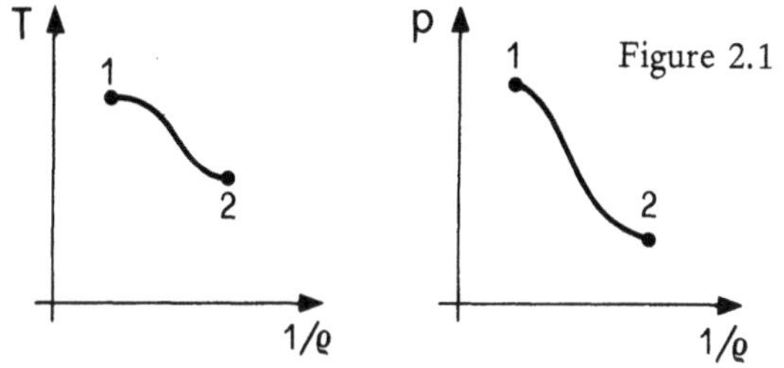

Figure 2.1

For instance in an ideal gas with

$$p|_E = \rho \frac{R}{M} T \qquad \text{and} \qquad \varepsilon|_E = z \frac{R}{M} T + \alpha \tag{2.7}$$

the work done to the body and the heat provided to it in an isothermal process are

$$a_{12} = -\frac{R}{M} T \ln \frac{\rho_1}{\rho_2} \quad \text{and} \quad q_{12} = \frac{R}{M} T \ln \frac{\rho_1}{\rho_2} . \tag{2.8}$$

Or else, if the heating $q'$ is zero, so that the process is adiabatic, we have

$$p|_E (1/\rho)^{\frac{z+1}{z}} = \text{const}, \quad a_{12} = z \frac{R}{M}(T_2 - T_1), \quad q_{12} = 0 . \tag{2.9}$$

2.2.3. Efficiency of a Cycle. A cycle has the same initial and final values of $\varepsilon$ and hence, by (2.4), the heat and work exchanged by the body in a cycle add up to zero

$$q_0 + a_0 = 0 . \tag{2.10}$$

Thus the heat exchanged with the body is converted into work. Experience shows that the exchanged heat is composed of positive and negative parts denoted by $q_+$ and $q_-$, and the quality of a cycle for the conversion of heat to work is estimated by its efficiency

$$e = \frac{|a_0|}{q_+} = 1 + \frac{q_-}{q_+} . \tag{2.11}$$

A Carnot cycle exchanges heat only at two temperatures $T_+$ and $T_- < T_+$ (say). In particular in a slow thermostatic cycle performed with an ideal gas we can use (2.8) and (2.9) and a minimum of calculation to obtain

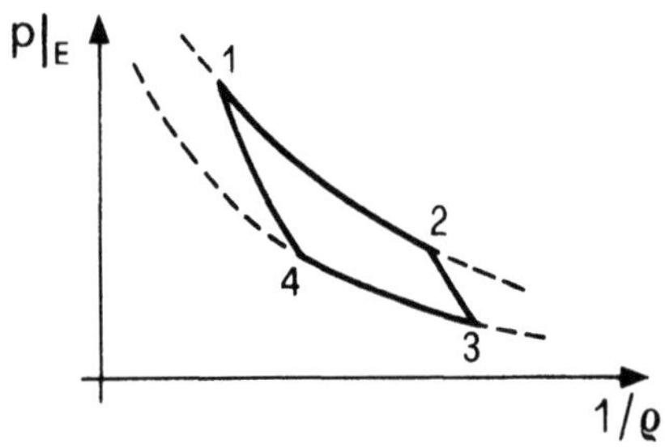

Figure 2.2

$$e = 1 - \frac{T_-}{T_+} \tag{2.12}$$

## 2.3. The Classical Work by Clausius [6]

2.3.1. Second Law of Thermodynamics. Clausius has drawn conclusions from the basic experience that

"Heat cannot pass from a colder to a warmer body all by itself".

He has also given an alternative statement of this observation which reads

"A transition of heat from a colder to a warmer body cannot occur without compensation".

Because of the important implications of this sentence it is often called the Second Law of Thermodynamics, the energy balance being the First Law.

2.3.2. Universal Efficiency of a Carnot Cycle. The second law implies what was already stated, namely that $q_0$ in a cycle consists of positive and negative parts $q_+$ and $q_-$. Indeed, if $q_-$ were zero we could withdraw heat from a cold reservoir and convert it fully into work. That work could than be converted back into heat by stirring a hot gas (say), thus providing a net transition of heat from cold to warm which contradicts the second law.

Also by similar reasoning it is obvious that in a Carnot cycle, that produces work, $q_+$ is provided at $T^+$ and $q_-$ is withdrawn at $T_-$.

Clausius proceeds to consider two Carnot cycles I and II working with different fluids and in different ranges of density but between the same temperatures $T_+$ and $T_-$. The work produced by one cycle is absorbed by the other one.

At first slow thermostatic, i.e. reversible Carnot processes are considered. The second law could be violated, if cycle I delivered less heat at $T_-$ than cycle II absorbs there. But it is also impossible that cycle I delivers more heat at $T_-$ than cycle II absorbs there, because by reversing the cycles one could get back to the previous impossible situation. Therefore $q_+$ and $q_-$ must have the same values in both cycles and it follows that

$$\left.\frac{|a_0|}{q_+}\right|_I = \left.\frac{|a_0|}{q_+}\right|_{II} \qquad \text{or by (2.11)} \qquad e|_I = e|_{II} \, . \tag{2.13}$$

Clausius concluded that the efficiency of a slow thermostatic Carnot cycle is universal, i.e. independent of material, and since, by (2.12), we know it for an ideal gas, we know it for all fluids:

$$e = 1 - \frac{T_-}{T_+} \, . \tag{2.14}$$

2.3.3. Absolute Temperature as an Integrating Factor. Combination of (2.11) and (2.14) gives

$$\frac{q_+}{T_+} + \frac{q_-}{T_-} = 0 \tag{2.15}$$

and this relation can easily be generalized into

$$\oint \frac{q'}{T} \, dt = 0 \tag{2.16}$$

for arbitrary slow thermostatic cycles by the realization that each such cycle can be decomposed into a sequence of Carnot cycles as shown in Figure 2.3.

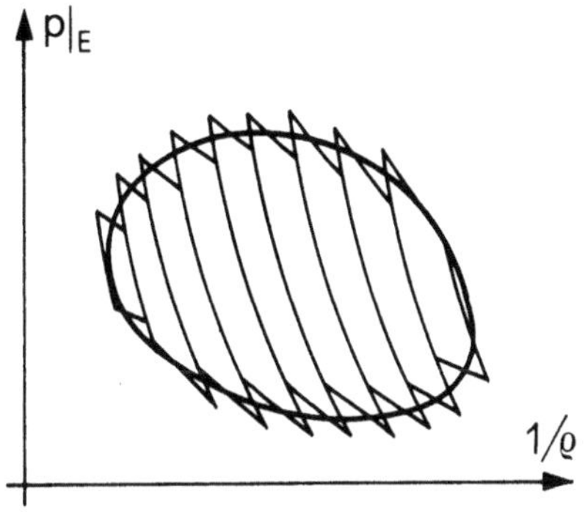


Figure 2.3

By use of (2.6) we may write (2.16) in the form

$$\oint \frac{1}{T}(d\varepsilon|_E + p|_E\, d(1/\rho)) = 0 .$$

This implies that $\frac{1}{T}(d\varepsilon|_E + p|_E\, d(1/\rho))$ is an exact differential and we say that $\frac{1}{T}$ is an integrating factor of

$$d\varepsilon|_E + p|_E\, d(1/\rho) .$$

Note that it is not the existence of an integrating factor that is important here, because any differential form of two variables has integrating factors. What is important, is the fact that one of these integrating factors is independent of $\rho$ or in fact, that it is the inverse of the absolute temperature.

2.3.4. Entropy. The integral function of the exact differential $\frac{1}{T}(d\varepsilon|_E + p|_E\, d(1/\rho))$ is called the entropy and is denoted by $\eta|_E$ so that we have

$$d\eta|_E = \frac{1}{T}(d\varepsilon|_E + p|_E\, d(1/\rho)) . \tag{2.17}$$

We may also write

$$d\eta|_E = \frac{q' dt}{T} \tag{2.18}$$

and this relation provides a suggestive interpretation of entropy, because it relates the change of entropy of a fluid to the heat it exchanges with the environment in a slow thermostatic process. In particular the entropy is constant in a slow thermostatic adiabatic process.

2.3.5. An Integrability Condition. Equation (2.17) is nowadays usually called Gibbs equation although it first appears in Clausius' papers. Since $\varepsilon|_E$ is a function of $\rho$ and T we may write

$$d\eta|_E = \frac{1}{T}\left(\left(\frac{\partial \varepsilon|_E}{\partial T}\right) dT + \left(\frac{\partial \varepsilon|_E}{\partial 1/\rho} + p|_E\right) d\left(1/\rho\right)\right) \tag{2.19}$$

and this implies the equation

$$\frac{\partial \varepsilon|_E}{\partial 1/\rho} = T\,\frac{\partial p|_E}{\partial T} - p|_E \tag{2.20}$$

as an integrability condition for $\eta|_E$. Equation (2.20) provides a relationtion between the thermal equation of state $p|_E(\rho, T)$ and the caloric equation of state $\varepsilon|_E(\rho, T)$.

Equation (2.20) is the first, and at the same time the most powerful example for a restriction that the existence of entropy imposes upon constitutive functions. It reduces significantly the necessary efforts for the determination of the caloric equation of state $\varepsilon|_E(\rho, T)$. Specifically it allows one to determine $\varepsilon|_E$ to within an additive constant by measurements of pressure and measurements of the specific heat $\frac{\partial \varepsilon|_E}{\partial T}$ at one density.

2.3.6. The Growth of Entropy. After having derived (2.20) Clausius returned to the argument about the two Carnot cycles I and II of Section 2.3.2. He modified that argument by assuming that cycle II is not performing a slow thermostatic i.e. reversible process. In that case it is impossible to reach the conclusions (2.13) and (2.16), rather one derives the inequalities

$$\left.\frac{|a_0|}{q_+}\right|_I > \left.\frac{|a_0|}{q_-}\right|_{II} \quad \text{or by (2.11)} \quad e|_I > e|_{II}\,. \tag{2.21}$$

Since cycle I is still reversible, (2.12) still holds and therefore (2.21) implies for the rapid or irreversible cycle

$$\frac{q_+}{T_+} - \frac{q_-}{T_-} < 0 \tag{2.22}$$

or by extension to an arbitrary irreversible cycle

$$\oint \frac{q' dt}{T} < 0 . \tag{2.23}$$

When a cycle is only irreversible on the part (1-2) of the process but reversible on the remainder, we may use (2.18) on the reversible part and rewrite (2.23) as

$$\overset{1}{\eta}|_E - \overset{2}{\eta}|_E \geqslant \int_{t_2}^{t_1} \frac{q' dt}{T} \tag{2.24}$$

In words this can be expressed by saying that the change of entropy of an irreversible process is greater than the heat supplied to it divided by the temperature where it is applied.

If the two states 1 and 2 are close enough in time we may write (2.24) as

$$\dot{\eta}|_E \geqslant \frac{q'}{T} \tag{2.25}$$

2.4. The Classical Work of Carathéodory [7].

2.4.1. <u>Energy Balance for an Adiabatic Thermostatic Process.</u> Carathéodory has given an alternative approach to entropy, an approach which at the same time introduces the absolute temperature in a manner that is not linked to its occurrence in the ideal gas laws.

We may say that at the outset we only have an empirical temperature $\vartheta$. For a thermostatic process the equation $(2.2)_2$ may then be written as

$$\dot{\varepsilon} = q' + \frac{1}{2\rho_0} t_{AB} \dot{C}_{AB} \qquad \text{with} \qquad \begin{aligned} \varepsilon &= \mathcal{N}[\overset{t}{C}(s), \overset{t}{\vartheta}(s)] \\ t_{AB} &= t_{AB}[\text{———}] . \end{aligned}$$

Carathéodory considers only adiabatic processes so that $q' = 0$.+)
For abbreviation we introduce

$$F_\alpha = G_{AB}\,, \quad p_\alpha = -\frac{1}{2\rho_0} t_{AB} \qquad (\alpha = 1,2,\dots n) \tag{2.26}$$

and obtain

$$\dot{\varepsilon} + \sum_{\alpha=1}^{n} p_\alpha \dot{F}_\alpha = 0 \qquad \text{with} \qquad \begin{aligned} \varepsilon &= \mathcal{N}[\overset{t}{F_\alpha}(s)\,,\overset{t}{\vartheta}(s)] \\ p_\alpha &= \mathcal{P}[\text{———}] \end{aligned} \tag{2.27}$$

for a thermostatic adiabatic process and

$$\dot{\varepsilon}|_E + \sum_{\alpha=1}^{n} p_\alpha|_E \dot{F}_\alpha = 0 \qquad \text{with} \qquad \begin{aligned} \varepsilon &= \varepsilon|_E\,(F_\alpha, \vartheta) \\ p_\alpha &= p_\alpha|_E(\text{—}) \end{aligned} \tag{2.28}$$

in a slow such process. In general $\alpha$ will run from 1 through 6, because $C_{AB}$ has 6 independent components, but in some bodies, e.g. isotropic ones, $t_{AB}$ has a special form which reduces the number of $C_{AB}$'s that occur in the energy balance.

2.4.2. Carathéodory's axiom and lemma assure the existence of integrating factors for the differential form $\dot{\varepsilon}|_E + \sum_{\alpha=1}^{n} p_\alpha|_E \dot{F}_\alpha$. The axiom reads:

"In each neighbourhood of an arbitrary initial state
there exist states that cannot be reached by an adiabatic process"

We might say that this axiom replaces Clausius' statement about the impossibility of heat passing from cold to warm by itself. And while the Carathéodory axiom is less suggestive than Clausius' second law at first

+) While the bulk of Carathéodory's work refers to adiabatic slow thermostatic processes, the remainder - reported in Section 2.4.5 - considers adiabatic processes in general not necessarily thermostatic ones. Here, however, for simplicity I stick to those.

sight, it can still be motivated well by simple **intuitive** considerations of the following type: Let us look at a special case of an adiabatic process that starts with $\vartheta^i, F^i_\alpha$ and ends with $\vartheta^f, F^f_\alpha$ so that $F^f_\alpha = F^i_\alpha$ i.e.the process carries the body back to its original form. $\vartheta^f$ must satisfy the energy balance

$$\varepsilon|_E(\vartheta^f, F^i_\alpha) - \varepsilon|_E(\vartheta^i, F^i_\alpha) = -\int_t^{t_f} p_\alpha \dot{F}_\alpha \, dt$$

from which it is obvious that different values of $\vartheta^f$ can be reached depending on how the process is conducted. But whatever the value of $\vartheta^f$ may be, we feel sure that it cannot be smaller than $\vartheta^i$, because dissipation will heat up the body. Thus in this example all those states $(\vartheta^i, F^i_\alpha)$ are inaccessible from $(\vartheta^f, F^i_\alpha)$ in an adiabatic process for which $\vartheta^f < \vartheta^i$.

The axiom becomes relevant to thermodynamics by Carathéodory's lemma which reads:

"Let $\dot{x}_0 - \sum_{\alpha=1}^{s} X_\alpha \dot{x}_\alpha = 0$ be a Pfaffian equation in which the $X_\alpha$ 's are finite, continuous and differentiable functions of $x_0$ and $x_\alpha$. If one knows that in each neighbourhood of an arbitrary point in the space spanned by $x_0, x_\alpha$ there are points that cannot be reached along curves which satisfy the Pfaffian equation, the Pfaffian form $\dot{x}_0 - \sum_{\alpha=1}^{s} X_\alpha \dot{x}_\alpha$ must have an integrating factor."

It is obvious how this lemma applies to thermostatics, because the energy balance (2.28) for slow thermostatic adiabatic processes is a

Pfaffian equation and Carathéodory's axiom ensures that "in each neighbourhood of an arbitrary point in the space spanned by $\vartheta, F_\alpha$ there are points that cannot be reached along curves which satisfy the Pfaffian equation". Therefore the Pfaffian form $\dot{\varepsilon} - \sum_{\alpha=1}^{n} p_\alpha \dot{F}_\alpha$ has integrating factors and we write

$$\dot{h}_\Lambda|_E = \Lambda(\dot{\varepsilon}|_E(\vartheta, F_\alpha) + \sum_{\alpha=1}^{n} p_\alpha|_E(\vartheta, F_\alpha)\dot{F}_\alpha) , \tag{2.29}$$

where $h_\Lambda|_E$ is the potential function corresponding to the integrating factor $\Lambda$. Obviously we may consider $\Lambda$ as a function of $\vartheta$ and $F_\alpha$.

The existence of integrating factors as such is not in itself useful for finding restrictions on the constitutive functions $\varepsilon|_E(\vartheta, F_\alpha)$ and $p_\alpha|_E(\vartheta, F_\alpha)$. For one thing, if there is one integrating factor, there are many. But Carathéodory's argument continues and he proceeds to show that <u>one</u> integrating factor is a universal function of temperature alone. Thus he comes to the concept of entropy and absolute temperature.

2.4.3. <u>Entropy and Absolute Temperature.</u> Carathéodory considers two different bodies I and II which he brings into contact and he defines equilibrium of the compound body as the state where the temperatures of the two bodies are equal while $F_\alpha^{I}$ and $F_\alpha^{II}$ are arbitrary and independent of each other. He can then prove that the integrating factors $\Lambda^{I}$ and $\Lambda^{II}$ may be written as

$$\Lambda^{I} = \frac{\alpha^{I}(h_{\Lambda^{I}})}{T(\vartheta)} \quad \text{and} \quad \Lambda^{II} = \frac{\alpha^{II}(h_{\Lambda^{II}})}{T(\vartheta)} \tag{2.30}$$

Thus $\Lambda$ is only a function of $h_\Lambda$ and of $\vartheta$ and the $\vartheta$-dependence of the two bodies is governed by the same function $T(\vartheta)$ i.e. $T(\vartheta)$ is a universal function.

By use of (2.30) we may write (2.29) as

$$\frac{1}{\alpha(h_\Lambda)}\dot{h}_\Lambda|_E = \frac{1}{T(\vartheta)}(\dot{\varepsilon}|_E + \sum_{\alpha=1}^{n} p_\alpha|_E \dot{F}_\alpha) \tag{2.31}$$

or with $\eta|_E = \int_{h_\Lambda 0}^{h_\Lambda|_E} \frac{dx}{\alpha(x)}$

$$\dot{\eta}|_E = \frac{1}{T(\vartheta)}(\dot{\varepsilon}|_E - \sum_{\alpha=1}^{n} p_\alpha|_E \dot{F}_\alpha). \tag{2.32}$$

We conclude that among all integrating factors of $\dot{\varepsilon}|_E + \sum_{\alpha=1}^{n} p_\alpha|_E \dot{F}_\alpha$ there is one which is a universal function of temperature alone. The corresponding potential function, which has been denoted by $\eta|_E$ is called the entropy.

Equation (2.32) implies the integrability condition

$$\frac{d\ell n T}{d\vartheta} = \frac{\partial p\alpha|E / \partial\vartheta}{p_\alpha|_E + \partial \varepsilon|_E / \partial F_\alpha} \tag{2.33}$$

by which the combination on the right hand side is a universal function of $\vartheta$ alone. But (2.33) may be exploited further to establish the absolute temperature as a rational measure for temperature. The argument runs as follows: (2.33) permits the calculation of T to within a multiplicative constant, if only the right hand side of that equation is calculated as a function of $\vartheta$ for any material. If this is done it turns out that $T(\vartheta)$ is a monotonically increasing function of $\vartheta$. Therefore $T(\vartheta)$ can be used as a measure of temperature and it is called the absolute temperature. The multiplicative constant is determined by fixing the absolute temperature scale so that the temperature difference between melting ice and boiling water is equal to 100 degrees.

With the definitions (2.26) we can write (2.32) as

$$\dot{\eta}|_E = \frac{1}{T}(\dot{\varepsilon}|_E - \frac{1}{2\rho_0} t_{AB}|_E \dot{C}_{AB}) , \tag{2.34}$$

where $\eta|_E$ , $\varepsilon|_E$ and $t_{AB}|_E$ are all functions of T and $C_{AB}$ . This is the Gibbs equation for solids. It implies the integrability conditions

$$\frac{\partial \varepsilon|_E}{\partial C_{AB}} = -\frac{1}{2\rho_0}(T\frac{\partial t_{AB}}{\partial T} - t_{AB}) \quad \text{and} \quad \frac{\partial t_{CD}|_E}{\partial C_{AB}} = \frac{\partial t_{AB}|_E}{\partial C_{CD}} \tag{2.35}$$

Equation $(2.35)_1$ is the counterpart in solids to equation (2.20) which held for fluids. $(2.35)_2$ is a symmetry relation which the stress of an elastic body must satisfy.

2.4.4. Additivity of Entropy. Equation (2.32), if applied to a compound body of two bodies I and II in equilibrium reads

$$\dot{\eta}|_E = \frac{1}{T}(\dot{\varepsilon}|_E + \sum_{\alpha=1}^{n} p_\alpha^I|_E \dot{F}_\alpha^I + \sum_{\beta=1}^{m} p_\beta^{II}|_E \dot{F}_\beta^{II}$$

and with $\varepsilon|_E = \varepsilon^I|_E + \varepsilon^{II}|_E$ we may write

$$\dot{\eta}|_E = \frac{1}{T}(\dot{\varepsilon}^I|_E + \sum_{\alpha=1}^{n} p_\alpha^I|_E \dot{F}_\alpha^I) + \frac{1}{T}(\dot{\varepsilon}^{II}|_E + \sum_{\beta=1}^{m} p_\beta^{II}|_E \dot{F}_\beta^{II})$$

or, by (2.32)

$$\dot{\eta}|_E = \dot{\eta}^I|_E + \dot{\eta}^{II}|_E . \tag{2.36}$$

Thus we conclude that the entropy of the compound body is the sum of the entropies of the partial bodies.

Carathéodory, after having derived this property comments it as follows:

> "The additivity of entropy has induced many physicists to consider entropy as a physical quantity which, similar to the mass, belongs to each body and depends on the state of this body."

2.4.5. Growth of Entropy. It is obvious from (2.32) that the entropy is constant in an adiabatic slow thermostatic process, because in such a process $\dot{\varepsilon}|_E + \sum_{\alpha=1}^{n} p_\alpha|_E \dot{F}_\alpha$ vanishes. In an arbitrary adiabatic process we have

$$\dot{\eta}|_E > 0 \tag{2.37}$$

by Carathéodory's axiom, i.e. the entropy grows.

For proof of (2.37) we consider all possible thermostatic adiabatic processes - i.e. rapid and slow ones - which start at $\vartheta^i, F^i_\alpha$ and end at $\vartheta^f, F^f_\alpha$. According to the balance of internal energy $\varepsilon|_E(\vartheta^f, F^f_\alpha)$ is given by the equation

$$\varepsilon|_E(\vartheta^f, F^f_\alpha) - \varepsilon|_E(\vartheta^i, F^i_\alpha) = -\int_{t_i}^{t_f} \sum_{\alpha=1}^{n} p_\alpha \dot{F}_\alpha \, dt$$

and it will differ for different processes, since the power term $\int \sum_{\alpha=1}^{n} p_\alpha \dot{F}_\alpha \, dt$ is different. We assume that all possible values of the power lie in a connected interval. Therefore the possible values $\varepsilon|_E(\vartheta^f, F^f_\alpha)$ also lie in a connected interval and the same holds true for all possible vaDues of $\vartheta^f$ and $\eta|_E(\vartheta^f, F^f_\alpha)$. While $\varepsilon|_E(\vartheta^i, F^i_\alpha)$ and $\vartheta^i$ may or may not lie in those intervals of possible final energies and temperatures, $\eta|_E(\vartheta^i, F^i_\alpha)$ will <u>certainly</u> lie within the interval of final entropies, since among all adiabatic processes there is the slow thermostatic adiabatic process in which $\eta|_E$ does not change. In fact Carathéodory's axiom implies that $\eta|_E(\vartheta^i, F^i_\alpha)$ lies at one end of the interval of final entropies. Indeed, if $\eta|_E(\vartheta^i, F^i_\alpha)$ were in an interior point of that interval, we could - by a suitable rapid adiabatic process - first reach any value of $\eta|_E$ in the neighbourhood of $\eta|_E(\vartheta^i, F^i_\alpha)$ and then - by a slow adiabatic process - go over to any desired set of values $F_\alpha$. Thus it would be possible to reach any point $\vartheta^f, F^f_\alpha$ in the neighbourhood of $\vartheta^i, F^i_\alpha$ by an adiabatic process and contradict Carathéodory's axiom. Therefore $\eta|_E(\vartheta^i, F^i_\alpha)$ must lie in one of the end points of the interval of values $\eta|_E(\vartheta^f, F^f_\alpha)$ so that $\eta|_E$ can only either grow or decrease in an adiabatic process. One experiment with any body ensures that $\eta|_E$ in fact grows. This proves (2.37).

## 2.5. Linear Irreversible Thermodynamics

2.5.1. <u>Principle of Local Equilibrium.</u> Linear irreversible thermodynamics (see [8] or [9]) offers a rather straightforward extension of the ideas of Clausius and Carathéodory by assuming that the Gibbs equations (2.17) or (2.34), which were derived for slow thermostatic processes, also hold in thermodynamics. $\eta|_E$, $\varepsilon|_E$ and $t_{AB}|_E$ at a time t and at position $x_i$ are thus all considered as functions of T and $C_{AB}$ at that time and place. Rather obviously this may be a good assumption for thermodynamic processes that do not vary too rapidly and in which all fields differ but little across a particle of the body. The assumption is called the principle of local equilibrium. We proceed to investigate its consequences for a thermoelastic body.

2.5.2. Entropy Inequality. We combine the Gibbs equation

$$\dot{\eta}|_E = \frac{1}{T}(\dot{\varepsilon}|_E - \frac{1}{2\rho_0} t_{AB}|_E \dot{G}_{AB}) \tag{2.38}$$

with the equation of balance of internal energy $(1.2)_3$ where, because of local equilibrium $\varepsilon$ is replaced by $\varepsilon|_E$ and $t_{AB}$ by $t_{AB}|_E$. We eliminate $\dot{\varepsilon}|_E$ between the two equations and write the result inthe form

$$\rho\eta|_E + \frac{\partial}{\partial x_i}(\frac{q_i}{T}) = -\frac{1}{T^2} q_i \frac{\partial T}{\partial x_i} . \tag{2.39}$$

This equation has the form of an equation of balance for entropy and

$$\phi_i = \frac{q_i}{T} \qquad \text{and} \qquad \sigma = -\frac{1}{T^2} q_i \frac{\partial T}{\partial x_i} \tag{2.40}$$

are therefore interpreted as the flux of entropy and the entropy production. Integrated over the body and over times from $t_1$ to $t_2$ (2.39) assumes the form

$$\int_V \rho\eta|_E \, dV\Big|_{t_2} - \int_V \rho\eta|_E \, dV\Big|_{t_2} = -\int_{t_1}^{t_2} (\oint \frac{q_i da_i}{T}) dt - \int_{t_1}^{t_2} (\int_V \frac{q_i}{T^2} \frac{\partial T}{\partial x_i} dV) dt \tag{2.41}$$

In this form the entropy balance compares easily with the result (2.24) of Clausius and it is seen to be a straightforward generalization of that result to the case of a non-uniform temperature field: In particular, the entropy of the body cannot now be written as $\eta|_E V$, because different particles have different specific entropies. Also, the heat added to the body can no longer be $q'dt$ divided by the temperature of the body, because the body has different temperatures in different particles. Therefore in (2.41) the heat added at a particular particle is divided by the temperature of that particle.

This being understood, the comparison of (2.41) and (2.24) requires that

$$\int_{t_1}^{t_2} \left( \int_V \frac{q_i}{T^2} \frac{\partial T}{\partial x_i} dV \right) dt \geqslant 0 \tag{2.42}$$

holds and the easiest way to satisfy that condition is by requiring that $\sigma$ in (2.40) be non-negative. Thus there results the local entropy inequality

$$\rho \dot{\eta}|_E + \frac{\partial}{\partial x_i}\left(\frac{q_i}{T}\right) \geqslant 0 \qquad \text{or} \qquad q_i \frac{\partial T}{\partial x_i} \leqslant 0 . \tag{2.43}$$

2.5.3. Phenomenological Equation for $q_i$. The entropy inequality $(2.43)_2$ suggests a constitutive equation for $q_i$ of the form

$$q_i = - \kappa_{ik} \frac{\partial T}{\partial x_k} \tag{2.44}$$

where $\kappa_{ik}$, the tensor of heat conductivity, must be a positive definite matrix. In the case of a thermoealstic solid this is the only conclusion from the entropy inequality, it restricts the constitutive equation fcr the heat flux.

## 2.6. Clausius-Duhem Inequality and its Exploitation

2.6.1. Clausius-Duhem Inequality. Heat cannot only be added to a body at its surface, but also in the interior by absorption of radiation. Thus if r is the specific heat supply by radiation, (2.25) should contain a term r/T on its right hand side and its obvious generalization to non-uniform processes should therefore read

$$\frac{d}{dt}\int_V \rho\eta \, dV + \oint_F \frac{q_i}{T}\, da_i - \int_V \frac{r}{T}\, dV \geqslant 0 , \tag{2.45}$$

or in local form

$$\rho\dot{\eta} + \frac{\partial}{\partial x_i}\left(\frac{q_i}{T}\right) - \frac{\rho r}{T} \geqslant 0 . \tag{2.46}$$

This is the Clausius-Duhem inequality which forms the basis of most of the modern work in thermodynamics after Coleman and Noll [10] have suggested a method for its exploitation. The inequality must be interpreted as an equation of balance of entropy where the entropy flux has been assumed to be $q_i/T$ just like in linear irreversible thermodynamics, and the heat supply has been assumed to be $\rho r/T$. However, contrary to linear irreversible thermodynamics, in the exploitation of the Clausius-Duhem inequality, one does not normally restrict $\eta$ to be the entropy of a slow thermostatic process. I.e. the principle of local equilibrium is not accepted here; rather the entropy may depend on history of $\underset{\sim}{F}^t(s)$, $\overset{t}{T}(s)$, and possibly $\underset{\sim}{\overset{t}{g}}(s)$ just as stress, heat flux and internal energy do in general. This is why we now write $\eta$ instead of $\eta|_E$ for the specific entropy.

2.6.2. <u>Reduction of the Clausius-Duhem Inequality.</u> The Clausius-Duhem inequality is postulated to hold for all thermodynamic processes, i.e. solutions $\rho(X_A, t)$, $x_i(X_A, t)$ and $T(X_A, t)$ of the field equations that are obtained from the equations of balance (1.2) and from constitutive equations (see Section 1.1.4.). The momentum balance $(1.2)_2$ and the energy balance $(1.2)_3$ do not restrict the fields, because whatever the fields are, the equations $(1.2)_{2,3}$ may be satisfied by a suitable body force $f_i$ and radiation supply $r$. The mass balance $(1.2)_1$ is usually satisfied by eliminating $\rho$ from the field equations by use of (1.8).

Thus, if $r$ is eliminated between (2.46) and $(1.2)_3$, the resulting inequality

$$-\rho(\varepsilon - T\eta)^{\cdot} - \rho\eta\dot{T} - \frac{1}{T} q_i \frac{\partial T}{\partial x_i} + \frac{\rho}{2\rho_0} t_{AB}\dot{C}_{AB} \geqslant 0 \qquad (2.47)$$

must hold for all fields $x_i(X_A, t)$ and $T(X_A, t)$.

2.6.3. <u>Exploitation of the Clausius-Duhem Inequality for a Thermoelastic Body.</u> In order to illustrate the method of exploitation of the Clausius-Duhem inequality (2.47) in a simple manner, which nevertheless is indicative of the greater generality of the present theory compared to linear irreversible thermodynamics, we consider a thermoelastic body with a memory for past temperatures. The constitutive equations are given by (1.24), viz.

$$\begin{aligned} t_{AB} &= t_{AB}(C_{AB}, T, \dot{T}, T_{,B}) \\ q_A &= q_A(\text{———}) \\ \varepsilon &= \varepsilon\ (\text{———}) \\ \eta &= \eta\ (\text{———}), \end{aligned}$$

where we assume, in the absence of information to the contrary, that all variables may be present on all constitutive relations. This assumption is known as the principle of equipresence.

If the equations (2.48) are introduced into (2.47) and when all indicated differentiations are performed we obtain the equation

$$(-\rho\frac{\partial\psi}{\partial C_{AB}}+\frac{\rho}{2\rho_0}t_{AB})\dot{C}_{AB}-\rho(\frac{\partial\psi}{\partial T}+\eta)\dot{T}-\rho\frac{\partial\psi}{\partial\dot{T}}\ddot{T}-\rho\frac{\partial\psi}{\partial T_{,B}}\dot{T}_{,B}-\frac{q_i}{T}\frac{\partial T}{\partial x_i}\geqslant 0\,. \tag{2.49}$$

where the specific free energy $\psi = \varepsilon - T\eta$ has been introduced. The left hand side of this inequality is explicitly linear in $\dot{C}_{AB}, \ddot{T}, \dot{T}_{,B}$. But (2.49) must hold for all fields $x_i(X_A, t)$ and $T(X_A, t)$, hence in particular for arbitrary values of $\dot{C}_{AB}$ and $\ddot{T}$. Therefore the inequality could be violated unless the factors of $\dot{C}_{AB}$ and $\dot{T}$, $\dot{T}_{,B}$ were zero. Hence follows

$$\begin{aligned}\frac{\partial\eta}{\partial C_{AB}} &= \frac{1}{T}(\frac{\partial\varepsilon}{\partial C_{AB}}-\frac{1}{2\rho_0}t_{AB})\,,\\ \frac{\partial\eta}{\partial\dot{T}} &= \frac{1}{T}\frac{\partial\varepsilon}{\partial T}\,,\\ \frac{\partial\eta}{\partial T_{,B}} &= \frac{1}{T}\frac{\partial\varepsilon}{\partial T_{,B}}\,.\end{aligned} \tag{2.50}$$

There remains the residual inequality

$$-(\frac{\partial\varepsilon}{\partial T}-T\frac{\partial\eta}{\partial T})\dot{T}-\frac{1}{\rho T}q_i\frac{\partial T}{\partial x_i}\geqslant 0\,. \tag{2.51}$$

From (2.50) we get as integrability conditions for $\eta$ that the stress $t_{AB}$ is independent of $\dot{T}$ and $T_{,B}$.

Inspection of the residual inequality shows that its left hand side, which we denote by $\Sigma$, has its minimum, namely zero, in equilibrium, where $\dot{T}$ and $\frac{\partial T}{\partial x_i}$ both vanish. Necessary conditions for this are

$$\frac{\partial\Sigma}{\partial\dot{T}}\Big|_E = 0\,, \quad \frac{\partial\Sigma}{\partial T_{,B}}\Big|_E = 0 \quad \text{and} \quad \begin{bmatrix} \frac{\partial^2\Sigma}{\partial\dot{T}^2}\Big|_E & \frac{\partial^2\Sigma}{\partial\dot{T}\partial T_{,B}}\Big|_E \\ \frac{\partial^2\Sigma}{\partial\dot{T}\partial T_{,B}}\Big|_E & \frac{\partial^2\Sigma}{\partial T_{,A}\partial T_{,B}}\Big|_E \end{bmatrix} \ \text{non-negative definite} \quad (2.52)$$

where the index E denotes equilibrium. $(2.52)_{1,2}$ require

$$\frac{\partial\eta|_E}{\partial T} = \frac{1}{T}\frac{\partial\varepsilon|_E}{\partial T} \quad \text{and} \quad q_i|_E = 0\,, \quad (2.53)$$

while $(2.52)_3$ implies the conditions

$$\frac{\partial\varepsilon}{\partial\dot{T}}\Big|_E \leqslant 0 \quad \text{and} \quad \frac{\partial q_i}{\partial T_{,k}}\Big|_E - \text{negative definite.} \quad (2.54)$$

2.6.4. Summary of Results. The equations $(2.53)_1$ and $(2.50)_1$ in equilibrium may be combined to give

$$d\eta|_E = \frac{1}{T}\left(d\varepsilon|_E - \frac{1}{2\rho_0}t_{BC}|_E\, dC_{BC}\right) \quad (2.55)$$

which is equivalent to the Gibbs equation (2.34) of thermostatics. It implies integrability conditions for $\eta|_E$ , viz.

$$\frac{\partial\varepsilon|_E}{\partial C_{AB}} = -\frac{1}{2\rho_0}\left(T\frac{\partial t_{AB}|_E}{\partial T} - t_{AB}|_E\right) \quad \text{and} \quad \frac{\partial t_{CD}|_E}{\partial C_{AB}} = \frac{\partial t_{AB}|_E}{\partial C_{CD}}. \quad (2.56)$$

These are identical to the conditions (2.35) of thermostatics. In addition we now get results on the dependence of the constitutive quantities on the non-equilibrium variables $\dot{T}$ and $T_{,A}$. In particular, since $\psi = \varepsilon - T\eta$, by $(2.50)_{2,3}$, is independent of $\dot{T}$ and $T_{,A}$, so is $t_{AB}$ by $(2.50)_1$. And while $\varepsilon$ and $\eta$ may depend on $\dot{T}$ and $T_{,A}$ they must do so in a manner that leaves $\psi$ independent of these variables. We shall discuss the dependence of $\varepsilon$ on $\dot{T}$ later.

The heat flux $q_i$ is restricted by the conditions $(2.53)_2$ and $(2.54)_2$ In a theory that is linear in $T,_A$ and $\dot{T}$ these imply

$$q_i = a_i \dot{T} - \kappa_{ik} \frac{\partial T}{\partial x_k} , \qquad (2.57)$$

where $a_i$ is a constant vector and $\kappa_{ik}$ a positive-definite matrix. (2.57) is a strange result in view of our intuitive expectation that $q_i$ should vanish, when the temperature gradient vanishes. However, the Clausius-Duhem inequality does not exclude the term $a_i\dot{T}$ . Of course, if the body is isotropic, that term is absent and (2.57) reduces to the result (2.44) of linear irreversible thermodynamics.

## 2.7. Rational Thermodynamics with Lagrange Multipliers.

2.7.1. Entropy Principle. As we have seen, the Clausius-Duhem inequality (2.46) was motivated from the results of thermostatics, but its scope was extended beyond thermostatics by allowing $\eta$, $\varepsilon$ and $t_{AB}$ to depend on non-equilibrium variables. However, if that extension is made, it would seem inconsistent to leave the heating terms in the form which they have in thermostatics. Therefore we shall now investigate the possibility that the entropy flux is no longer equal to heat flux divided by temperature. We shall instead let the entropy flux be given by a constitutive equation in its own right. We shall consider processes that are free from the supplies $f_i$ and $r$ and we assume that under such circumstances there is no supply of entropy either. Thus the entropy principle inequality reads (see [11])

$$\rho\dot{\eta} + \frac{\partial \phi_i}{\partial x_i} \geqslant 0 . \qquad (2.58)$$

Just as in the exploitation of the Clausius-Duhem inequality we assume that the entropy inequality (2.58) must hold for all thermodynamic processes, i.e. solutions of the field equations based on the equations of balance (1.2).

But now there must be a second part to the entropy principle to compensate for the fact that now the entropy flux is no longer given. We assume that the normal component of the entropy flux is continuous at a wall where the temperature is continuous.

$$[\phi_i e_i] = 0 \quad , \quad \text{if} \qquad [T] = 0 . \tag{2.59}$$

For motivation I remark that a gradient of temperature is the prime cause for a production of entropy. Thus, if there is no temperature gradient across a wall it is reasonable to assume that there is no entropy production in the wall and that implies $(2.59)_1$. Also of course (2.59) is satisfied, if $\phi_i = q_i/T$ holds, because the heat flux usually has a continuous normal component.

2.7.2. Exploitation of the Entropy Inequality for a Thermoelastic Body [12]. The general entropy principle (2.58), (2.59) will be used to derive restrictions on the constitutive equations

$$\begin{aligned} t_{AB} &= t_{AB}(C_{BC}\, , T\, , \dot{T}\, , T_{,A}) \\ q_A &= q_A(\text{———————}) \\ \varepsilon &= \varepsilon\ (\text{———————}) \\ \eta &= \eta\ (\text{———————}) \\ \phi_A &= \phi_A(\text{———————}) \end{aligned} \tag{2.60}$$

for a thermoelastic body with a memory for temperature. This is the same type of body which has been considered by thermostatics, linear irreversible thermodynamics and by the Coleman-Noll theory that is based upon the Clausius-Duhem inequality. The purpose of this multiple treatment of the same material is to provide an appreciation of the relative merits of those theories.

The key for the exploitation of the entropy inequality (2.58) is the requirement that it must not hold for all fields $T(X_A, t)$ and $T(X_A, t)$ but only for thermodynamic processes, i.e. solutions of the field equations. We can get rid of this constraint by the use of Lagrange multipliers, because Liu has shown in [13], that the following inequality must hold for all fields

$$\rho\dot{\eta} + \frac{\partial\phi_i}{\partial x_i} - \Lambda^{v_i}(\rho\dot{v}_i - \frac{\partial t_{ij}}{\partial x_j}) - \tag{2.61}$$

$$- \Lambda^{\varepsilon}(\rho\dot{\varepsilon} - \frac{\partial q_i}{\partial x_i} - t_{ij}\frac{\partial v_i}{\partial x_j}) \geqslant 0 .$$

The $\Lambda$'s are called Lagrange multipliers and, according to Liu's proof, they may be functions of the variables in the constitutive equations (2.60).

When the constitutive equations are introduced into the inequality and all indicated differentiations are performed, it turns out that the left hand side is explicitly linear in

$$\dot{v}_i, \dot{C}_{AB}, \ddot{T}, \dot{T}_{,B}, G_{AB,C}, T_{,AB}. \tag{2.62}$$

Now, since (2.61) must hold for all fields $x_i(X_A, t)$ and $T(X_A, t)$, it must hold in particular for arbitrary values of the derivatives (2.62) at one point and time. Therefore these derivatives must not contribute to the inequality (2.61) and this requirement leads to the following conditions:

$$\Lambda^{v_i} = 0$$

$$\frac{\partial \eta}{\partial C_{BC}} - \Lambda^{\varepsilon} \frac{\partial \varepsilon}{\partial C_{BC}} + \frac{\Lambda^{\varepsilon}}{2\rho_0} t_{BC} = 0 \ ,$$

$$\frac{\partial \eta}{\partial \dot{T}} - \Lambda^{\varepsilon} \frac{\partial \varepsilon}{\partial \dot{T}} = 0 \ ,$$

$$\frac{\partial \eta}{\partial T_{,A}} - \Lambda^{\varepsilon} \frac{\partial \varepsilon}{\partial T_{,A}} + \frac{1}{\rho_0}\left(\frac{\partial \phi_A}{\partial \dot{T}} - \Lambda^{\varepsilon} \frac{\partial q_A}{\partial \dot{T}}\right) = 0 \ ,$$

$$\frac{\partial \phi_{i\rangle}}{\partial C_{E\langle B}} - \Lambda^{\varepsilon} \frac{\partial q_{A\rangle}}{\partial C_{E\langle B}} = 0 \ ,$$

$$\frac{\partial \phi_{\langle A}}{\partial T_{,B\rangle}} - \Lambda^{\varepsilon} \frac{\partial q_{1A}}{\partial T_{,B}} = 0 \ .$$

There remains a residual inequality of the form

$$\left(\frac{\partial \eta}{\partial T} - \Lambda^{\varepsilon} \frac{\partial \varepsilon}{\partial T}\right) T + \frac{1}{\rho_0}\left(\frac{\partial \phi_A}{\partial T} - \Lambda^{\varepsilon} \frac{\partial q_A}{\partial T}\right) T_{,A} \geqslant 0 \ . \qquad (2.63)_7$$

These are the consequences from the entropy inequality. We expect them to be restrictive for the constitutive functions $t_{AB}$, $q_A$ and $\varepsilon$. But before such restrictions can be derived, we must eliminate the

Lagrange multiplier $\Lambda^\varepsilon$ and the entropy and its flux from (2.63). This is the difficult part of rational thermodynamics. What is missing here in comparison to linear irreversible thermodynamics and to the Coleman Noll theory is a simple relation between the entropy flux and the heat flux. For an isotropic body it can be shown that

$$\phi_i = \Lambda^\varepsilon q_i \tag{2.64}$$

holds and that $\Lambda^\varepsilon$ can only be a function of T and $\dot{T}$. The general proof of (2.64) is somewhat complex even for an isotropic body. But the ideas underlying the proof can easily be appreciated by looking at the special case, where the heat flux $q_A$ is given by $\alpha T_{,A}$ instead of by the full representation (1.25) and where similarly the entropy flux $\phi_A$ is given by $\varphi T_{,A}$. $\alpha$ and $\varphi$ are taken to be functions of $T, \dot{T}, g$ and the principal invariants $I_1, I_2, I_3$ of $\underset{\sim}{C}$. Insertion of these simplified relations into $(2.63)_6$ gives

$$(\varphi - \Lambda^\varepsilon \alpha)\,\delta_{AB} + 2\left(\frac{\partial \varphi}{\partial g} - \Lambda^\varepsilon \frac{\partial \alpha}{\partial g}\right) T_{,A}\, T_{,B} = 0 \,,$$

whence follows that $\varphi = \Lambda^\varepsilon \alpha$ holds and that $\Lambda^\varepsilon$ is independent of $g$. Thus $\phi_A = \Lambda^\varepsilon q_A$ and if this is inserted into $(2.63)_5$ it follows that $\Lambda^\varepsilon$ is independent of C. Therefore $\phi_A = \Lambda^\varepsilon(T, \dot{T}) q_A$ holds and we have proved (2.64) for this case.

2.7.3. <u>Consequence of the Entropy Principle for a Non-Isotropic Body of Fourier Type.</u> We shall restrict the attention to non-isotropic bodies where

$$q_A = \alpha_{AB}\, T_{,B} \qquad \text{and} \qquad \phi_A = \varphi_{AB}\, T_{,B} \tag{2.65}$$

holds and where $\underset{\sim}{\alpha}$ and $\underset{\sim}{\varphi}$ may depend on $T, \dot{T}, g$ and the density $\rho$. We say that such bodies are of Fourier type, because the heat flux obeys Fourier's law of heat conduction. Here too we can show that

$$\phi_A = \Lambda^\varepsilon(T, \dot{T}) q_A \qquad \text{or} \qquad \phi_i = \Lambda^\varepsilon(T, \dot{T}) q_i \tag{2.66}$$

holds much along the same lines as in an isotropic body.

Once we know that (2.66) is true, we can use it to eliminate $\phi_A$ from the results (2.63) and obtain

$$\frac{\partial \eta}{\partial C_{BC}} = \Lambda^\varepsilon \frac{\partial \varepsilon}{\partial C_{BC}} - \frac{\Lambda^\varepsilon}{2\rho_0} t_{BC} , \tag{2.67}$$

$$\frac{\partial \eta}{\partial \dot{T}} = \Lambda^\varepsilon \frac{\partial \varepsilon}{\partial \dot{T}} ,$$

$$\frac{\partial \eta}{\partial T_{,A}} = \Lambda^\varepsilon \frac{\partial \varepsilon}{\partial T_{,A}} - \frac{1}{\rho_0} \frac{\partial \Lambda^\varepsilon}{\partial \dot{T}} q_A ,$$

$$\left(\frac{\partial \eta}{\partial T} - \Lambda^\varepsilon \frac{\partial \varepsilon}{\partial T}\right) \dot{T} + \frac{1}{\rho_0} \frac{\partial \Lambda^\varepsilon}{\partial T} q_A T_{,A} \geqslant 0 .$$

The integrability conditions for $\eta$ implied by (2.67) are

$$\frac{\partial q_A}{\partial T_{,B}} - \frac{\partial q_B}{\partial T_{,A}} = 0 ,$$

$$\frac{\partial t_{BC}}{\partial C_{AD}} - \frac{\partial t_{AD}}{\partial C_{BC}} = 0 ,$$

$$\frac{\partial \ell n \Lambda^{\varepsilon}}{\partial \dot{T}}(2\rho_0 \frac{\partial \varepsilon}{\partial C_{BC}} - t_{BC}) = \frac{\partial t_{BC}}{\partial \dot{T}}$$

$$\frac{\partial \ell n \Lambda^{\varepsilon}}{\partial \dot{T}} \frac{\partial q_A}{\partial C_{BC}} = \frac{1}{2} \frac{\partial t_{BC}}{\partial T_{,A}}$$

$$\frac{\partial}{\partial \dot{T}}(\ell n \frac{\partial \Lambda^{\varepsilon}}{\partial \dot{T}}) q_A = \rho_0 \frac{\partial \varepsilon}{\partial T_{,A}} - \frac{\partial q_A}{\partial \dot{T}}.$$

From $(2.69)_1$ and $(2.65)_1$ we conclude that

$$\alpha_{AB} = \alpha_{BA} \tag{2.70}$$

holds. Thus it comes out that the tensor of heat conductivity is symmetric. This is a result which is postulated in phenomenological thermodynamics as an Onsager relation in linear irreversible thermodynamics. Here it comes out of the theory itself.

Equation $(2.69)_2$ implies a symmetry of the stress of which the equilibrium conterpart was also derived in thermostatics, see $(2.35)_2$ and in linear irreversible thermodynamics, see $(2.56)_2$.

Most restrictive are the conditons $(2.69)_{3,4,5}$ which imply that the combinations of constitutive quantities

$$\frac{\dfrac{\partial t_{BC}}{\partial \dot{T}}}{2\rho_0 \dfrac{\partial \varepsilon}{\partial C_{BC}} - t_{BC}}, \quad \frac{\dfrac{\partial t_{BC}}{\partial T_{,A}}}{\dfrac{\partial q_A}{\partial C_{BC}}}, \quad \frac{\rho_0 \dfrac{\partial \varepsilon}{\partial T_{,A}} - \dfrac{\partial q_A}{\partial \dot{T}}}{q_A} \tag{2.71}$$

are only functions of $T$ and $\dot{T}$, independent of $C_{AB}$ and $T_{,A}$. Moreover the first two of them are equal. We shall see in one of the next sections that one can argue that these combinations are even universal functions of $T$

and $\dot{T}$.

In the Coleman Noll theory $(2.69)_1$ is not present, and $(2.69)_{3,4}$ reduce to the simple statement, that $t_{AB}$ is independent of $\dot{T}$ and $T_{,A}$. This is due to the fact that in that theory $\Lambda^{\varepsilon}$ in effect was a priory set equal to $1/T$, a relation that comes out to be valid in the present theory only in equilibrium, see (2.73) below.

The residual inequality (2.68) is evaluated in the same manner as in linear irreversible thermodynamics by use of the fact that its left hand side has a minimum in equilibrium. This implies the condition $\frac{\partial \eta|_E}{\partial T} = \Lambda^{\varepsilon}|_E \frac{\partial \varepsilon|_E}{\partial T}$ which combined with $(2.67)_1$ gives

$$d\eta|_E = \Lambda^{\varepsilon}|_E \left(d\varepsilon|_E - \frac{1}{2\rho_0} t_{BC}|_E \, dC_{BC}\right) \tag{2.72}$$

Comparison of this equation with the Gibbs equation (2.34) of thermodynamics, linear irreversible thermodynamics and the Coleman Noll theory allows us to identify the equilibrium value of $\Lambda^{\varepsilon}$ with the reciprocal of the absolute temperature

$$\Lambda^{\varepsilon}|_E = \frac{1}{T} \,. \tag{2.73}$$

Further evaluation of the residual inequality provides the inequalities

$$\frac{\partial \varepsilon}{\partial \dot{T}}\Big|_E \leqslant - T^2 \frac{\partial \Lambda^{\varepsilon}}{\partial \dot{T}}\Big|_E \frac{\partial \varepsilon|_E}{\partial T} \quad \text{and} \quad \alpha_{AB} \ - \ \text{negative definite} \tag{2.74}$$

These are the counterparts of the conditions $(2.54)_{1,2}$ of the Coleman Noll theory and we conclude that $\partial\varepsilon/\partial\dot{T}$ may be positive in the present theory. We proceed to derive a consequence of this possibility.

2.7.4. Finite Speed of Heat Propagation. Let us consider the process of heat conduction in a body at rest in its reference configuration. The only relevant equation of balance in this case is the balance of internal energy $(1.2)_3$ of which the linear form in an isotropic body with $q_A = \alpha T,_A$ reads

$$\rho\left(\frac{\partial \varepsilon|_E}{\partial T}\right)\frac{\partial T}{\partial t} + \rho \frac{\partial \varepsilon}{\partial \dot{T}}\bigg|_E \frac{\partial^2 T}{\partial t^2} = -\alpha \frac{\partial^2 T}{\partial x_i \partial x_i} . \tag{2.75}$$

Since $\alpha$ is negative this is a hyperbolic equation, if $\partial \varepsilon / \partial \dot{T}|_E$ is positive. In that case (2.75) predicts a finite speed of propagation of disturbances in temperature. Note that, by $(2.74)_1$ $\partial \varepsilon / \partial \dot{T}|_E$ may be positive in this theory, while in the Coleman Noll theory (2.54) states that it must be negative.

There are other methods of obtaining a finite speed for temperature. All of them require that we somehow take the history of a body into account in the determination of its present state. A particularly convincing way for doing that is shown by the kinetic theory of gases which has motivated Cattaneo [14] and Müller [15] to formulate non-conventional constitutive theories in thermodynamics, see also [16].

2.7.5. The Coldness. Many results so far still contain the Lagrange multiplier $\Lambda^\varepsilon$ e.g. $(2.69)_{3,4,5}$. It is true that we were able to identify $\Lambda^\varepsilon|_E$ as the reciprocal of the absolute temperature, but we should also like to know $\Lambda^\varepsilon$ in non-equilibrium. This can be determined by use of the jump conditions (2.59) as follows.

Let us consider two different thermoelastic bodies I and II separated by a wall, where the temperature is continuous. By (2.59) we have

$$[\phi_i e_i] = 0 \qquad \text{or by (2.66)} \qquad [\Lambda^\varepsilon(T, \dot{T}) q_i e_i] = 0 \quad ,$$

or, since $q_i e_i$ is continuous under very general conditions

$$[\Lambda^\varepsilon(T, \dot{T})] = 0 .$$

More explicitly we may write this as

$$\Lambda^\varepsilon_I (T, \dot{T}) = \Lambda^\varepsilon_{II} (T, \dot{T}) \tag{2.76}$$

since the temperature is continuous. Thus $\Lambda^\varepsilon$ is the same function of T and $\dot{T}$ in the two bodies and we say that $\Lambda^\varepsilon(T, \dot{T})$ is a universal function; Müller [12]has called it the coldness.

Once the universal character of $\Lambda^\varepsilon(T, \dot{T})$ is established, the results $(2.69)_{3,4,5}$ assume a new dimension. Indeed they indicate that all quantities not only are functions of T and $\dot{T}$ alone, but that they are <u>universal</u> functions, the same functions in all thermoelastic bodies.

Therefore in principle we have a possibility to determine the form of the function $\Lambda^\varepsilon(T, \dot{T})$ without too much effort: We need only measure any one of the three expressions (2.71) as functions of T and $\dot{T}$ for just one material and determine $\Lambda^\varepsilon(T, \dot{T})$ by integration.

## 2.8. Criticism.

2.8.1. <u>Comparison of Theories.</u> We may say that equation (2.34) and the ensuing integrability conditions (2.35) form the single hard result of the thermostatics of Carathéodory. No doubt is cast upon this result by either one of the thermodynamic theories. In fact linear irreversible thermodynamics accepts (2.34) as a starting point, while both the Coleman Noll theory and rational thermodynamics derive that result.

The single addition which linear irreversible thermodynamics provides to thermostatics is the positive definiteness of the tensor of heat conductivity and that result is also obtained by the Coleman Noll theory and rational thermodynamics.

A big step toward thermodynamics proper occurs when the principle of local equilibrium is given up in the Coleman Noll theory; and there are more results too, some of them annulling the greater generality in the initial assumptions. Thus for instance, although stress and free energy were initially allowed to depend on $\dot{T}$ and $T,_A$ , the calculation of the Clausius Duhem inequality forbids that possibility.

One more step away from thermostatics and toward thermodynamics is taken in rational thermodynamics where the entropy flux is considered to be an independent constitutive quantity, not a priori related to the heat flux. In that theory stress and free energy may depend on the non-equilibrium variables $\dot{T}$ and $T,_A$ and there are several interconnections between those dependencies, viz. (2.69).

2.8.2. Criticism. Because of the difficulties of thermodynamic experimentation we cannot decide which one of the thermodynamic theories is general enough for the description of a particular material. Therefore the theories are judged on their appeal to logic or on personal preference.

Given our desire to have a local and instantaneous entropy inequality, it seems impossible to further generalize the inequality (2.58). But one may ask whether that degree of generality is needed or whether perhaps the Clausius Duhem inequality is good enough. One may also ask, whether the supplementary condition (2.59) is generally valid; if (2.59) is given up, the remarks in Section 2.7.5 become invalid, and there is no universal coldness.

One can also argue that, once we leave thermostatics and consider non-uniform and rapidly changing processes, it is not enough to throw $\dot{T}$ and $T_{,A}$ in to broaden the set of variables. It may not even be enough to throw in the whole history of $T$. Rather one might think that more spatial derivatives become essential to deal with non-local action. The kinetic theory of gases supports such doubts and suggests directions of research that are now beginning to be investigated (e.g. see [14] and [15]).

Unfortunately the addition of more and more variables makes a theory based on (2.58) or even on (2.46) cumbersome and often it is impossible to derive suggestive results.

2.8.3. Hope for Comfort. So where does this leave us with respect to thermoplasticity? Surely one could expect thermoplasticity to be more complex and more difficult to describe thermodynamically than the simple theory of thermoelasticity that we have just been through. However, there is hope for comfort, because there is the possibility that the description of thermoplasticity might not require a sophisticated theory of functional constitutive relations defined on unspeakble spaces or of intrinsic time scales activated by successive plastic deformations.

In fact there is the possibility that maybe the basic facts of thermoplasticity may be understood as a reflection of simple instabilities in the material properties that lead to unstable parts in the free energy of a body similar to those that occur in phase transitions. Thus it is conceivable that the simple formulae of thermostatics go a long way in describing thermoplasticity.

In the following section I shall explain these ideas in the simplest possible terms, and in the section after that I shall apply them to alloys with shape memory.

## 3. Phase Transition in Gases and Rubbers as Particular Cases of Thermoplasticity.

### 3.1. Influence of Entropy and Internal Energy upon Pressure and Load.

The mathematical treatment of fluid and elastic solids in slow thermostatic processes exhibits considerable differences. This is due to the fact that in a fluid the stress $t_{AB}$ is equal to $-p\delta_{AB}$ and that the pressure p and the internal energy $\varepsilon$ depend only on $\rho$ and T, while in a solid the stress may be an arbitrary (symmetric) tensor. It is true though that in the treatment of uniaxial stresses those differences become minimal; in fact they are reduced to differences in notation and sign of the power term.

Even so, however we shall treat fluids and solids separately: the formulae and sketches on the left hand side of the following pages refer to fluids while those on the right hand side refer to solids. All constitutive functions in this chapter and the next refer to slow thermostatic processes, but I drop the index E of Section 2, since no confusion can arise.

| Fluids | Solids |
|---|---|

The thermal and caloric equations of state in fluids are equations, which relate the pressure p and the internal energy $\varepsilon$ to the temperature T and to the density. In an elastic solid under uniaxial stress the thermal and caloric equations of state relate the stress S and the internal energy $\varepsilon$ to the temperature T and to the deformation gradient $\lambda_1$ or simply $\lambda$ (see Section 1.5.3. for the notations). The general form of these equations reads

$$\varepsilon = \varepsilon(\rho, T), \quad p = p(\rho, T) \qquad \Bigg\| \qquad \varepsilon = \varepsilon(\lambda, T), \quad S = S(\lambda, T) \tag{3.1}$$

In both cases the equations of state are not independent, because of the existence of the entropy for which we may also write an equation of state, viz.

$$\eta = \eta(\rho, T) \qquad \Bigg\| \qquad \eta = \eta(\lambda, T) \tag{3.2}$$

The equations (3.1), (3.2) are related to each other by the Gibbs equation (2.17) and (2.34) for a fluid and a solid respectively.

$$d\eta = \frac{1}{T}(d\varepsilon + p d(1/\rho)) \qquad \Bigg\| \qquad d\eta = \frac{1}{T}(d\varepsilon - \frac{1}{\rho_0} S d\lambda) \tag{3.3}$$

$$= \frac{1}{T} \left(\frac{\partial \varepsilon}{\partial T}\right)_\rho dT + \left(\frac{\partial \varepsilon}{\partial (1/\rho)} + p\right) d(1/\rho) \qquad \Bigg\| \qquad \frac{1}{T} \left(\frac{\partial \varepsilon}{\partial T}\right)_\lambda dT + \left(\left(\frac{\partial \varepsilon}{\partial \lambda}\right)_T - \frac{1}{\rho_0} S\right) d\lambda \tag{3.4}$$

An alternative form of the Gibbs equation results, when this equation is rewritten as a differential form for the free energy $\psi = \varepsilon - T\eta$

$$d(\varepsilon - T\eta) = -\eta dT - p d(1/\rho) \qquad \Bigg\| \qquad d(\varepsilon - T\eta) = -\eta dT + \frac{1}{\rho_0} S d\lambda \tag{3.5}$$

Inspection of these formulae shows that

$$p = -\left(\frac{\partial(\varepsilon - T\eta)}{\partial\, 1/\rho}\right)_T \qquad\qquad S = \rho_0\left(\frac{\partial(\varepsilon - T\eta)}{\partial\lambda}\right)_T \tag{3.6}$$

hold. We may express this by saying that the free energy is the "potential" for p or S . We can also see that p or S are composed of two parts of which the first one represents the $\rho$ – or $\lambda$ – dependence of the internal energy, while the second one results from the $\rho$ – or $\lambda$ – dependence of the entropy. Obviously the entropic contribution increases with growing temperature.

The integrability conditions for $\eta$ and $\varepsilon - T\eta$ which are implied by the Gibbs equations (3.4) and (3.5) read

$$\left(\frac{\partial\varepsilon}{\partial\, 1/\rho}\right)_T = T\left(\frac{\partial p}{\partial T}\right)_\rho - p \qquad\qquad \left(\frac{\partial\varepsilon}{\partial\lambda}\right)_T = \frac{1}{\rho_0}\left(T\frac{\partial S}{\partial T} - S\right) \tag{3.7}$$

$$\left(\frac{\partial\eta}{\partial\, 1/\rho}\right)_T = \left(\frac{\partial p}{\partial T}\right)_\rho \qquad\qquad \left(\frac{\partial\eta}{\partial\lambda}\right)_T = -\frac{1}{\rho_0}\left(\frac{\partial S}{\partial T}\right)_\lambda .$$

and these enable us to identify the energetic and entropic parts in p and S suggestively as properties of curves that are easily measured, namely the curves which represent p , or S as functions of T for a fixed density or for fixed $\lambda$. See Figures 3.1.

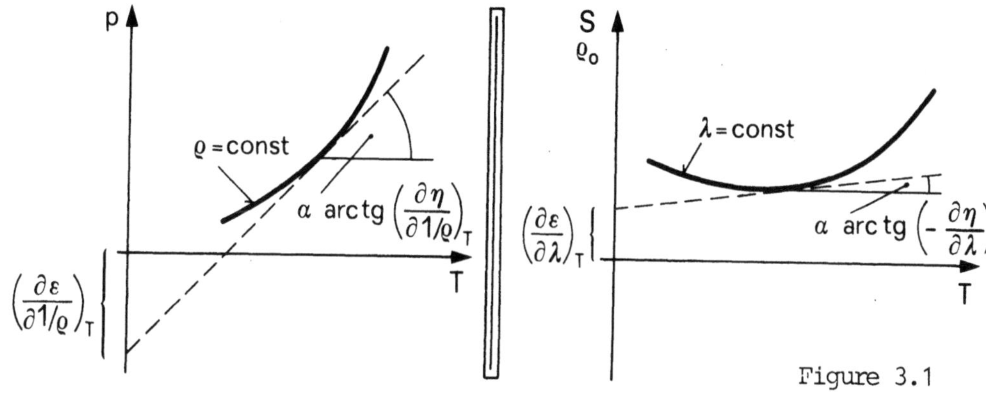


Figure 3.1

It follows that the tangent to the (p, T) -curve in the state $(\rho, T)$ determines the energetic and entropic part of the pressure in this state; specifically the point of intersection of the tangent and the p -axis gives the energetic part of the pressure and the slope of the tangent determines the entropic part of pressure. An increase of temperature increases the entropic contribution. An analogous interpretation can be given to the energetic and entropic parts of the stress in an elastic solid.

3.2. Examples for Entropy-Induced Pressure and Stress.

As an example for a fluid we consider an ideal gas and as an example for an elastic solid we consider rubber.

| Ideal Gas | Ideal Rubber |
|---|---|

Equations $(2.7)_1$ and (1.37) are the thermal equations of state in the two cases

$$p = \rho \frac{R}{M} T \qquad \Bigg\| \qquad S = \sigma T(\lambda - \frac{1}{\lambda^2}). \tag{3.9}$$

We say that $(3.9)_2$ characterizes ideal rubber.

The two equations (3.9) imply that the $(p, T)$ -curves and the $(S, T)$ -curves are straight lines through the origin as shown in Figure 3.2.

Figure 3.2

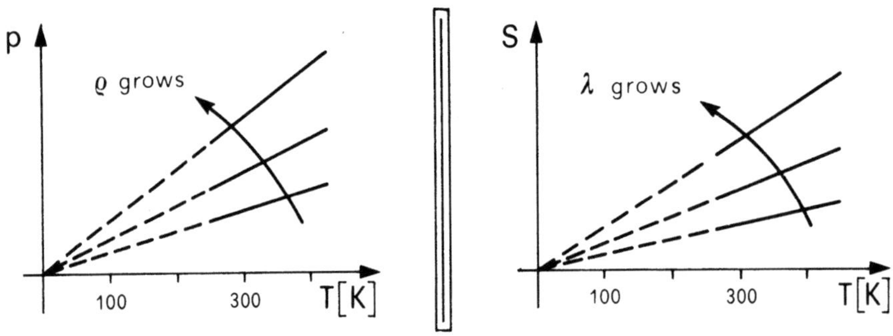


It follows that in ideal gases and in ideal rubber the pressure $p(T, \rho)$ and the stress $S(T, \lambda)$ are determined by the entropy alone and we say that the elasticity of these materials is "entropy-induced."

According to (3.9) the dependence of $p$ on $\rho$ or $S$ on $\lambda$ can be represented by isotherms of the form shown in Figure 3.3.

For small values of $\rho$ and $\lambda$ these isotherms do also describe the behaviour of real gases and rubbers.

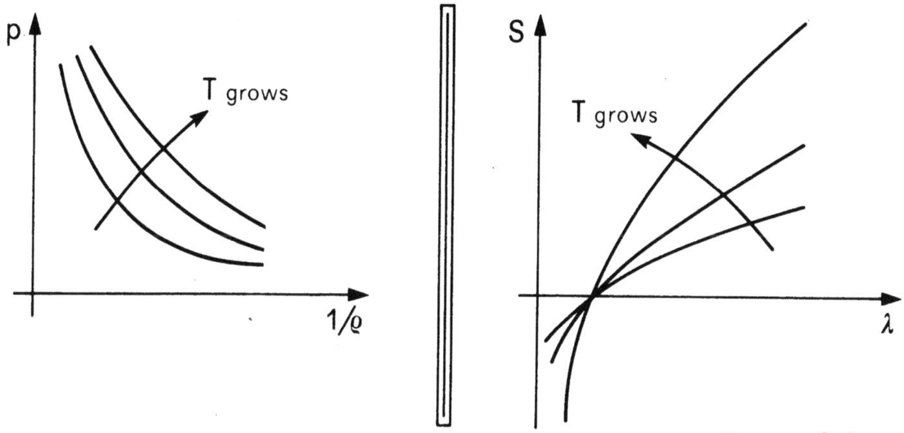


Figure 3.3

By (3.8) the entropy $\eta(T,\rho)$ and $\eta(T,\lambda)$ follows by integration

$$\left(\frac{\partial\eta}{\partial 1/\rho}\right)_T = \left(\frac{\partial p}{\partial T}\right)_\rho$$

$$= \rho\,\frac{R}{M}$$

$$\Longrightarrow \eta = \eta_0(T) + \frac{R}{M}\,\ell n\,\frac{\rho_0}{\rho}$$

$$\left(\frac{\partial\eta}{\partial\lambda}\right)_T = -\frac{1}{\rho_0}\left(\frac{\partial S}{\partial T}\right)_\lambda$$

$$= -\frac{\sigma}{\rho_0}\left(\lambda - \frac{1}{\lambda^2}\right) \qquad (3.10)$$

$$\Longrightarrow \eta = \eta_0(T) - \frac{\sigma}{\rho_0}\left(\frac{\lambda^2}{2} + \frac{1}{\lambda} - \frac{\lambda_0^2}{2} - \frac{1}{\lambda_0}\right)$$

We plot $-T(s - s_0(T))$ as a function of $1/\rho$ and $\lambda$ for gas and rubber respectively in order to get an idea how the entropy contributes to the $1/\rho$ -dependence or the $\lambda$ -dependence of the free energy, see (3.6).

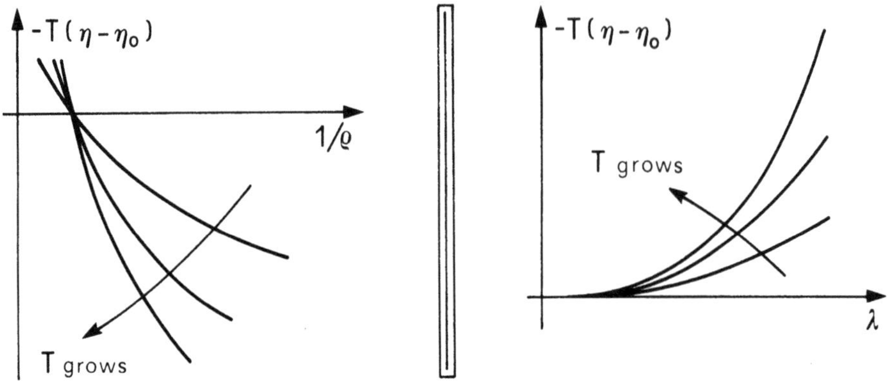


Figure 3.4

3.3. Examples for the Energetic Contribution to Pressure and Stress in a Real Gas and Crystallized Rubber.

| Real Gas | Crystallized Rubber |
|---|---|

The deviation of real gases and real rubbers from the ideal behaviour is best described in the form of their isotherms in the $(P, 1/\rho)$- and $(\varepsilon, \lambda)$-diagram, see Figure 3.5.

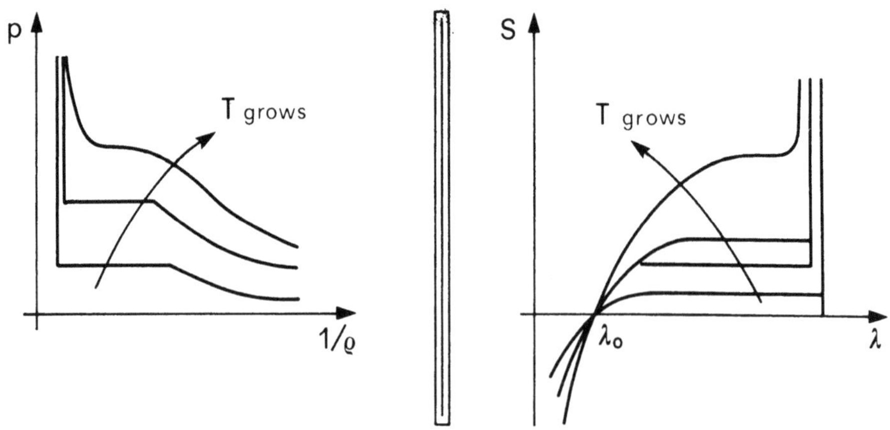


Figure 3.5

It is true that in a real gas for small values of $\rho$ the isotherms of a real gas coincide with those of an ideal gas, but for intermediate values of $\rho$ and high temperatures the isotherm falls below the one of an ideal gas and eventually for big values of $\rho$ it becomes very steep. At smaller temperatures there is even a horizontal part to the isotherms.

In rubber we observe a similar behaviour: For small strains $\lambda$ the isotherms are like those of ideal rubber but for intermediate strains and high temperature the slopes are smaller until for big values of $\lambda$ they become very steep. At smaller temperatures the isotherms have a horizontal part. But here something new occurs: In loading the horizontal part of the isotherm is higher than in unloading and therefore the isotherm exhibits a hysteresis. For small temperatures this type of behaviour even leads to the effect that after unloading there remains a residual deformation.

The occurrence of a residual deformation may be construed as an indication that the rubber has undergone a plastic deformation as it was pulled along the horizontal part of the isotherm. Also this is a true case of thermoplasticity, because there is a considerable release of heat in the deformation and the "yield stress' increases with increasing temperature. The curves of Figure 3.5 imply that the residual deformation is recovered, upon an increase of temperature, i.e. the body returns to the original strain

All this will be demonstrated in the lectures that accompany the presentation of this manuscript.

The deviation in the behaviour of real gases and rubbers from ideality are due to phase transitions. In the gas the transition is between liquid and vapour and in the rubber the transition is a crystallization. In both cases the cause of the phase transition is the interaction force between the molecules of the gas or rubber. In other words there is an interaction energy and we may therefore conjecture that the non-ideal behaviour may be described by assuming a $\rho-$, or $\lambda$ -dependence of $\varepsilon$ as well as of $\eta$.

This conjecture is confirmed by simple molecular considerations: Indeed for small densities the molecules of a gas are too far apart for the mutual interaction forces to be felt and therefore the interaction energy is equal to zero. For intermediate and large densities the molecules are close together, they attract each other and the interaction energy becomes smaller with growing density. In rubber the molecular model is very different. Rubber molecules consist of long chains of in-

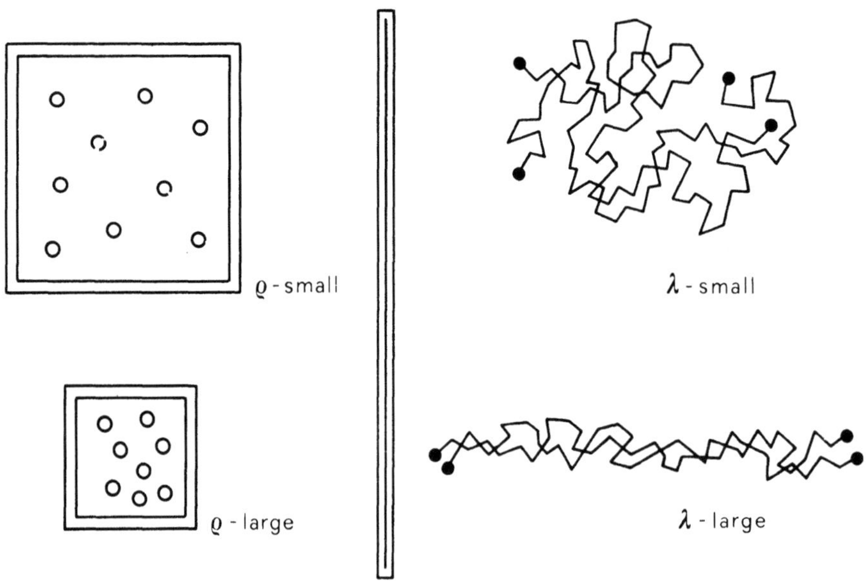


Figure 3.6

dividual rods which in the unstrained state are strongly entangled as shown in Figure 3.6. There are but few places where the rods are close enough to each other to feel their mutual attraction and this remains so for small strains. It follows that for small strains the interaction energy is almost equal to zero. For intermediate and large strains, however, the chains disentangle to accommodate the tensile stress and the lateral contraction presses them close together as shown in Figure 3.6. Thereby the rods are forced into interaction and the internal energy decreases. It becomes smaller with increasing strain. Both in the gas and in the rubber the decrease of the energy comes to an end when the molecules touch each other, because a strong molecular repulsion sets in and correspondingly there is a steep increase in energy.

These molecular considerations are reflected in the Figures 3.7 which represent the energy as a function of density and strain. The sudden increase at large density and strain is due to the molecular repulsion upon contact.

Figure 3.7

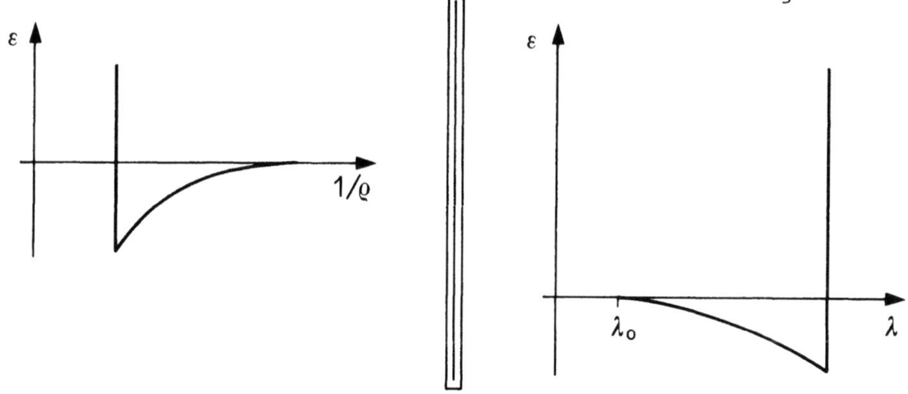

3.4. The Free Energy of Gases and Rubbers. Pressure and Stress.

We assume that the entropy is not influenced by the molecular interaction. Therefore the curves of free energy vs. density in a gas can be obtained by adding the $(\varepsilon, 1/\rho)$-curves and the $(-T\eta, 1/\rho)$-curves of Figures 3.4 and 3.7. Correspondingly, if we want to obtain the free energy as a function of T and $\rho$ in rubber we must add the $(\varepsilon, \lambda)$-and $(-T\eta, \lambda)$-curves of the Figures 3.4 and 3.7.

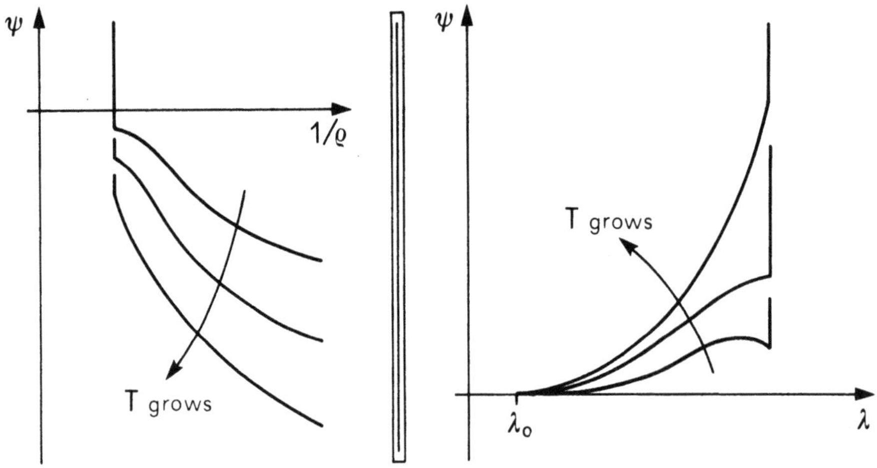


Figure 3.8

By differentiation of these curves we obtain the pressure as a function of $\rho$ and the stress as a function of $\lambda$. Figures 3.9 represent these functions by the drawn out and dashed parts of the curves.

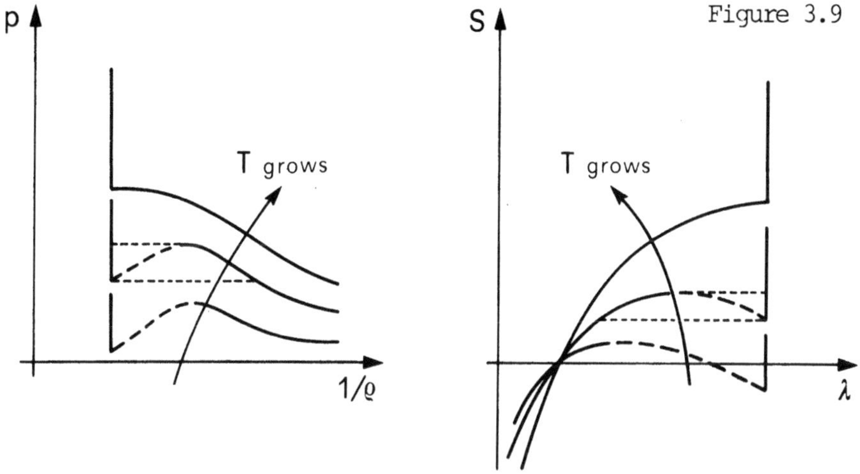


Figure 3.9

Comparison of these curves with the observed behaviour of Figure 3.5 shows rough qualitative agreement at high temperatures. However, at low temperatures there are no horizontal parts in the $(p, 1/\rho)$ -curves of Figure 3.9, nor are there hystereses in the $(S, \lambda)$ -curves of that Figure.

But then, the occurrence of hysteresis is implied by the above diagrams. This is best seen as follows: If the pressure or the stress are increased isothermally starting with small densities and strains, the increase ought to end when the maximum is reached. A further slight increase must lead to a horizontal jump over the dashed region of the isotherms. The jump ends on the vertical part of the isotherms. In unloading we stay on the vertical lines as long as they are there and then we jump horizontally below the dashed region until we hit the isotherm again. These jumps are indicated by the dotted lines on one isotherm of the $(S, \lambda)$ -diagram of Figure 3.9.

Thus we understand the hysteresis and now the question arises why there is no hysteresis in gases but only one horizontal line and also we should like to know where this horizontal line has to be drawn.

3.5. Phase Equilibrium

We see from Figure 3.8 that the curves of free energy vs. $\rho$ or $\lambda$ have parts with a negative curvature which in Figure 3.10 are dashed. The states on those parts are thermodynamically unstable and the body cannot assume them.

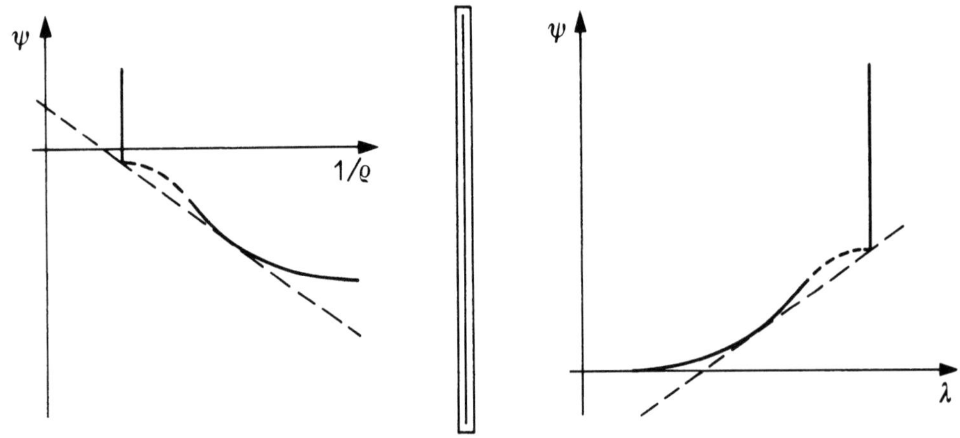


Figure 3.10

In both figures I have also drawn a tangent which spans the unstable part. The body has the possibility to bridge the range of instability by falling apart into two phases whose densities or strains are equal to the abscissae of the points of contact of tangent and curve. In-between the body changes its density or strain by changing the fraction of the phases. In this manner its average free energy and its average density or strain are found on points on the tangent. In this manner the body succeeds in reducing its free energy below the value which it would have if it were following the curved path.

When this construction with the tangent is introduced into the $(p, 1/\rho)$ -diagram or the $(S, \lambda)$-diagram, one obtains a horizontal isobar in the range of the phase separation, the so-called Maxwell line. This isobar is constructed, as indicated in Figure 3.11, by making the two shaded areas between the isobar and the calculated isotherms equal in

Figure 3.11

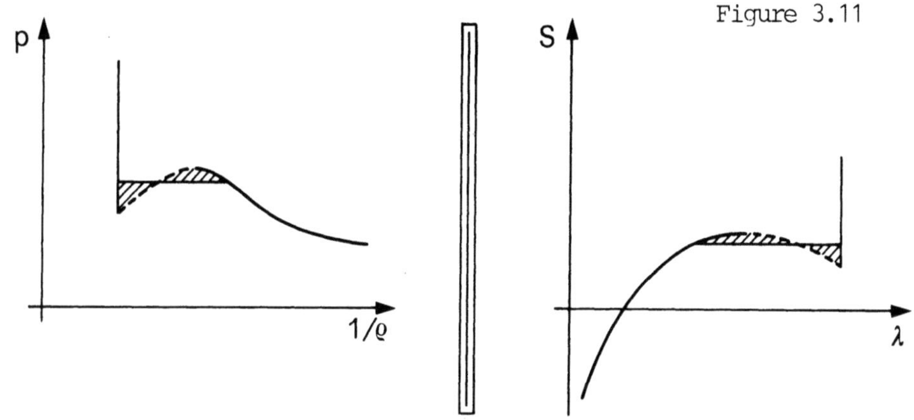


size. This follows from the observation that the tangents in Figure 3.10 have the equations

$$\psi_I + \frac{\partial\psi}{\partial 1/\rho}\left(\frac{1}{\rho_{II}} - \frac{1}{\rho_I}\right) = \psi_{II} \qquad\qquad \psi_I + \frac{\partial\psi}{\partial\lambda}(\lambda_{II} - \lambda_I) = \psi_{II}$$

$$\psi_I - \mathscr{P}\left(\frac{1}{\rho_{II}} - \frac{1}{\rho_I}\right) = \psi_{II} \qquad\qquad \psi_I + \frac{S}{\rho_0}(\lambda_{II} - \lambda_I) = \psi_{II} \qquad (3.11)$$

$$\psi_{II} - \psi_I = -p\left(\frac{1}{\rho_{II}} - \frac{1}{\rho_I}\right). \qquad\qquad \psi_{II} - \psi_I = \frac{S}{\rho_0}(\lambda_{II} - \lambda_I).$$

On the other hand by integration along the isotherms we obtain

$$\psi_{II} - \psi_I = -\int_{1/\rho_I}^{1/\rho_{II}} p\, d(1/\rho) \qquad\Bigg\|\qquad \psi_{II} - \psi_I = \int_{\lambda_I}^{\lambda_{II}} \frac{S}{\rho_0}\, d\lambda\,. \tag{3.12}$$

Comparison of (3.12) and $(3.11)_3$ shows that

$$\int_{1/\rho_I}^{1/\rho_{II}} p d(1/\rho) = p\left(\frac{1}{\rho_{II}} - \frac{1}{\rho_I}\right) \qquad\Bigg\|\qquad \int_{\lambda_I}^{\lambda_{II}} S d\lambda = S(\lambda_{II} - \lambda_I) \tag{3.13}$$

must hold and that is the mathematical expression for the construction of the Maxwell line.

In this manner we now understand why there is a single horizontal range in the isotherms of real gases instead of a hysteresis. In this range liquid and vapour coexist in phase equilibrium.

It remains to be explained, why in real rubber the isotherm does not follow the Maxwell line but traces out a hysteresis loop instead. The reason must be that steric hindrances within the solid body prevent the realization of phase equilibrium.

## 3.6. Evaluation

The above discussion has made use of very simple considerations with the purpose of obtaining free energy and load characteristics that are capable of describing the qualitative features of a phase transition. The synoptic treatment of gases and rubbers as representatives of fluids and solids has brought to light an important difference in the comportment of those two classes of material: In the fluid phase equilibrium was established in the transition range while in the solid the transition range was surrounded by a hysteresis loop.

While in both cases there is "yielding", i.e. large deformation upon a tiny increase of load as a result of a material instability the residual or "plastic" deformation occurs only in the solid; it accompanies the phase transition and results from the failure of the body to establish phase equilibrium.

In the next chapter in the treatment of memory alloys I shall expound the idea that yielding and plasticity is the consequence of material instability and constrained phase equilibrium both of which are most clearly exhibited by the form of the free energy as a function of deformation and temperature. This presentation is intended to advertise the view that similar reasons underly thermoplasticity in general, and that simple models might be constructed to describe it.

# 4. A Model for Shape Memory

## 4.1. Observed Stress-Strain Curves

This section presents a model which is capable of simulating the observed stress-strain curves of a memory alloy and their dependence on temperature. Figs. 4.1a through 4.1f show such curves schematically for increasing temperatures.

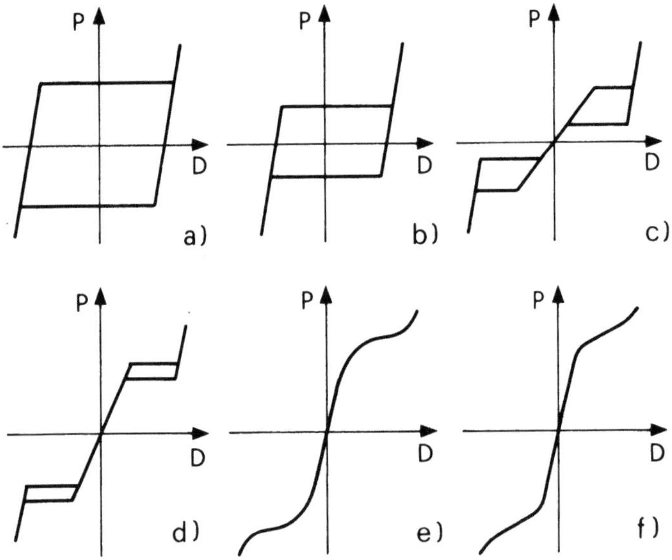


Figure 4.1

At low temperatures the behaviour is much like that of an elastic ideally plastic body: A virginal elastic line is followed by a yield line once a critical yield load is exceeded. The yield stops - and this of course is different from a plastic body - on a second elastic branch which may be traversed up and down. Unloading leaves a residual deformation. At a somewhat higher temperature this behaviour is qualitatively unchanged but the yield load decreases with increasing temperature. At intermediate temperatures there is still an elastic line through the origin and a yield load as well as a second elastic branch, but when unloading we fall below a recovery load, the body creeps back to the original elastic line and thence to the origin upon removal of the load. This behaviour is called pseudoelasticity, because the body is elastic in that it returns to the origin but its load-deformation curve exhibits a hysteresis which is absent in elasticity. At slightly higher temperatures the pseudoelastic behaviour is unchanged but both the yield load and the recovery load increase and the two grow closer together, so that the hysteresis becomes smaller. The initial elastic branch becomes steeper as temperature grows. Eventually at a critical temperature the hysteresis vanishes and from then on we have truly (non-linear) elastic behaviour.

It is obvious from the load-deformation curves of Figure 4.1 why the body is said to have memory: when we deform it at low temperature and then unload there will be a residual deformation according to Figures 4.1a or 4.1b. But when we now raise the temperature of the unloaded body to a value corresponding to one of the Figures 1c through 1f, there is only one possible deformation, viz. $D = 0$, so that the body creeps back to it.

The structure of the model is motivated by the observation that at high temperatures the alloy assumes an austenitic phase while at low temperatures it is martensitic and usually twinned. The curves of Fig. 4.1 are abstracted from the articles [17] and [18].

4.2. Model

The basic element of the model is a lattice particle - a small piece of the crystal lattice - which Fig. 4.2 shows in three different equilibria, denoted by A for austenite and by $M\pm$ for the martensitic twins. $M\pm$ can be thought of as sheared versions of A and, of course, there may be intermediate positions with smaller shear lengths $\Delta$. In all positions the particles have a different potential energy and the top part of Fig. 4.2 shows the postulated form of this energy. It has a metastable minimum for the austenitic particle and two lateral minima for the martensitic twins. In-between there are potential barriers.

The body itself is modelled as a stack of layers, each one consisting of lattice particles as shown in Fig. 4.3. In Fig. 4.3a all layers are austenitic while in 4.3b they are martensitic and alternate between $M_+$ and $M_-$. This model was first described in [19] and its properties were systematically developped in [20] and [21].

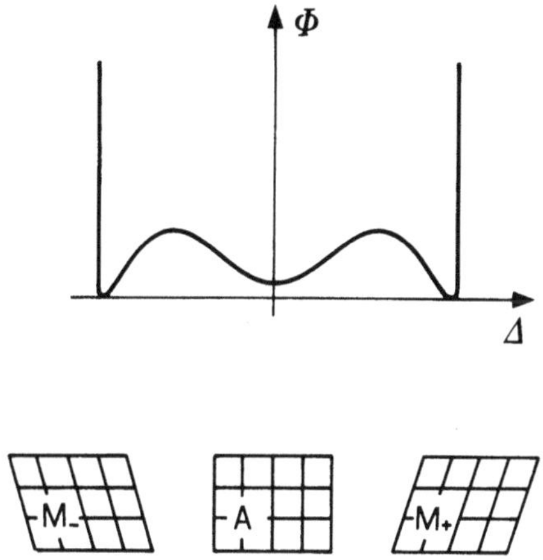


Figure 4.2

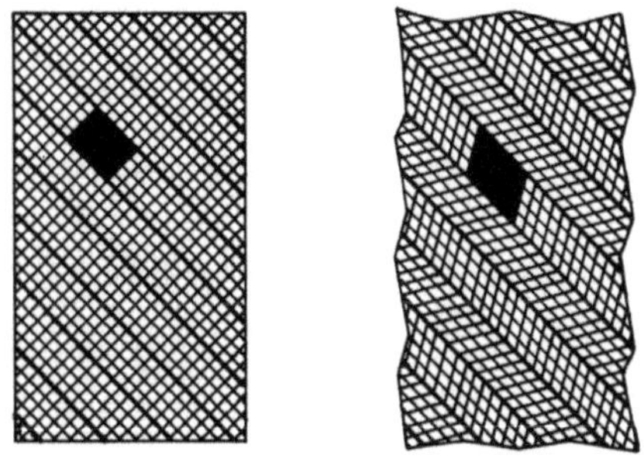

a) b)

Figure 4.3

4.3. Suggestive Interpretation of the Model.

As is required by the observations, the model predicts that the configurations of Figs. 4.3a and 4.3b prevail at high and low temperatures respectively in the virginal state of the unloaded body. This can be seen as follows: When the sample comes from the melt, it is hot and its lattice particles will have a mean kinetic energy kT far bigger than the potential barriers between $M_-$, A, and $M_+$. Therefore the particles will constantly change between phases, and the mean position will be in the middle, i.e. the body is austenitic. As the body cools, kT will eventually become so small that the particles cannot leave the lateral minima while they may still be able to jump out of the central minimum. As a result we expect the virginal body to consist of $M_+$ and $M_-$ in equal proportion at low temperatures.

We proceed to describe the behaviour of the model at low temperatures under a load as shown in Fig. 4.4: A small load will shear the lattice particles a little, the $M_-$ layers become steeper, and the $M_+$ layers become flatter, see Fig. 4.4a. Each of the N layers contributes the vertical component of its shear length $\Delta$ to the deformation so that we have

$$D = \frac{1}{\sqrt{2}} \sum_{i=1}^{N} \Delta_i \; . \tag{4.1}$$

Unloading allows the stack of layers to relax to its original configuration. But an increase of the load to the yield load will make the $M_-$ layers flip over to the $M_+$ position and the accompanying large increase in shear length will lead to a drastic increase in deformation. Unloading will now bring the particles into the equilibrium position $M_+$ and therefore a residual deformation remains. Figs. 4.4b and 4.4c sketch this behaviour.

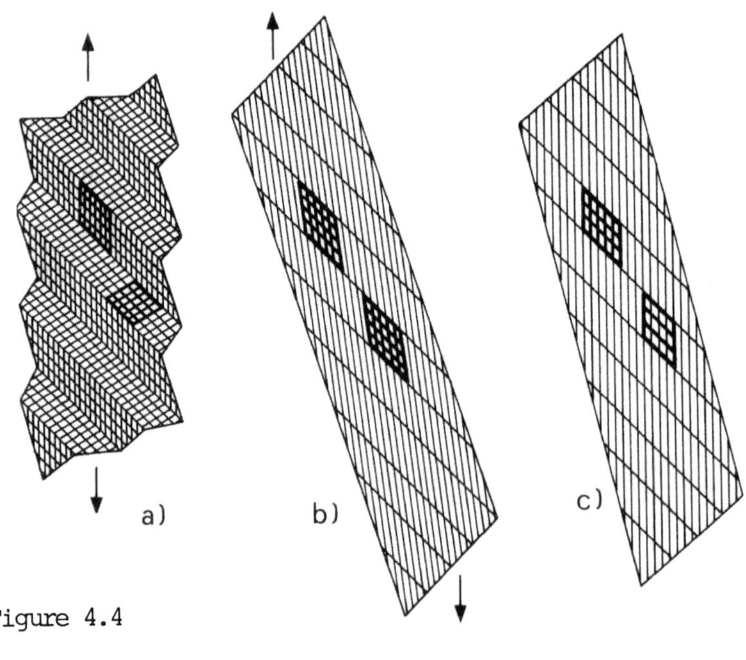


Figure 4.4

We conclude that the model can simulate all observed properties of of a memory alloy at low temperature, i.e. the virginal elastic curve, the yielding and the residual deformation after unloading.

In order to understand the decrease of the yield load at an increase of temperature we use energetic arguments to describe deformation and yield: Under a load the potential energy of Fig. 4.2 is deformed by the addition of the potental energy $-P\Delta$ of the load P as shown in Fig. 4.5. At zero temperature all lattice particles lie still in the lateral minima of these curves and the elastic deformation is composed of the small shifts to the right of these minima. The $M_-$ particles flip over to the $M_+$ configuration only when the load is so high that the left minimum is eliminated as shown in Fig. 4.5c. At higher temperatures the lattice particles fluctuate about the lateral minima with a mean kinetic energy kT which may be visualized by the height of the little pools in Figs. 4.5d and 4.5e. It is obvious that the barrier can now be overcome at the load $P_1$. Thus we conclude that the yield load decreases at higher temperatures.

4.4. Statistical Mechanics of the Model. Helmholtz Free Energy and Load-Deformation Curves at Intermediate and High Temperatures.

We have seen by suggestive arguments that the model can simulate the low temperature behaviour of memory alloys. At intermediate and high temperatures the properties of the model are less easy to explain, because they are not only determined by the energy of the model but also by its entropy. The formal ideas of statistical mechanics can help in this situation. The basic tenet here is that the probability $p_{\Delta i}$ for particle i to have the shear length $\Delta_i$ is proportional to the Boltzmann factor $\exp\{-\phi(\Delta_i)/kT\}$. The form of the partition function depends on the type of loading. Relevant formulae are summarized in Fig. 4.6.

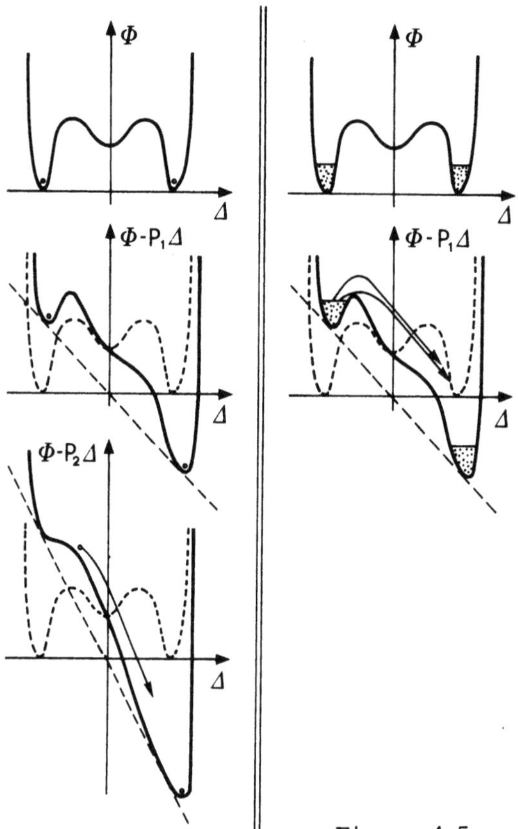


Figure 4.5

First we consider the body clamped to a fixed deformation D. The partition function Z given in (4.2) is a function of T and D, because the summation extends over sets of $\Delta_i$ that satisfy condition (4.1). From the partition function we obtain the Helmholtz free energy (4.3) and hence, by differentiation, the load p as a function of D and T.

CLAMPING

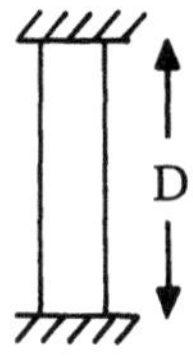


DEAD LOADING

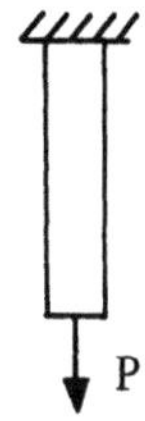


PARTITION FUNCTION

$$Z(D,T) = \sum_{\Delta_1=-J}^{J} \dots \sum_{\Delta_N=-J}^{J} e^{-\frac{\sum_{i=1} \phi(\Delta_i)}{kT}} \tag{4.2}$$

$$Z(P,T) = z \int_{-\frac{NJ}{\sqrt{2}}}^{\frac{NJ}{\sqrt{2}}} \sum_{\Delta_1=-J}^{J} \dots \sum_{\Delta_N=-J}^{J} e^{-\frac{\sum_{i=1}^{N} \phi(\Delta_i) - PD}{kT}} dD \tag{4.5}$$

$$= z \int_{-\frac{NJ}{\sqrt{2}}}^{\frac{NJ}{\sqrt{2}}} e^{\frac{PD}{kT}} Z(D,T)\, dD \tag{4.6}$$

FREE ENERGY

HELMHOLTZ $\psi = - kT \ln Z$ (4.3) GIBBS $G = - kT \ln Z$ (4.7)

STRESS STRAIN RELATION

$$P = \frac{\partial \psi}{\partial D} \tag{4.4}$$

$$D = -\frac{\partial G}{\partial P} . \tag{4.8}$$

In [20] the partition function Z(D,T) has been calculated approximately under assumptions that may be valid at intermediate and high temperatures. The result reads

$$\psi(D,T) - \psi(0,T) = - NkT[\beta J \frac{\sqrt{2}D}{NJ} + \ln(\frac{1}{J} \int e^{-\frac{\phi(\Delta)}{kT}} \cosh(\beta\Delta) d\Delta)] , \quad (4.9)$$

where $\beta$ is a function of D that is obtained by inversion of the equation

$$\frac{\sqrt{2}D}{ND} = - \operatorname{ctgh}(\beta J) - \frac{1}{\beta J} .$$

The introduction of the dimensionless variables

$$d = \frac{\sqrt{2}D}{NJ}, \ \delta = \frac{\Delta}{J}, \ \varphi = \frac{\phi}{\phi_s}, \ \Theta = \frac{kT}{\phi_s}, \ p = \frac{JP}{\sqrt{2}\phi_s}, \ b = \beta J, \ \psi = \frac{\psi - \psi(0,T)}{N\phi_s}, \ g = \frac{G - G(0,T)}{N\phi_s}$$

4.10

simplifies the formulae and we obtain with $H(b, \Theta) = \int_0^1 e^{-\frac{\varphi(\delta)}{\Theta}} \cos h(b\,\delta)d\delta$ :

$$\psi(d, \Theta) = -\Theta(b\,d + \ln H) ,$$

hence, by (4.4)

$$p(d, \Theta) = -\Theta(b + \frac{b^2 \sin h^2 b}{b^2 - \sin h^2 b}[d + \frac{\partial \ln H}{\partial b}]) . \tag{4.11}$$

An explicit calculation of $\psi$ and p is possible for the simplified box potential that is drawn in the lower part of Fig. 4.7. The other pictures of Fig. 4.7 represent the Helmholtz free energy $\psi(d, \Theta)$ and the load $p(d, \Theta)$.

Comparison with Fig. 4.1 shows that at high temperatures the load-deformation curve has the qualitative properties that are observed. At intermediate temperatures the curves have decreasing parts. Those parts are thermodynamically unstable and in a loading and unloading experiment they will be bridged, because a breakthrough will occur along the dotted lines of Fig. 4.7a. Thus at intermediate temperatures the model simulates the hystereses that are observed according to Fig. 4.1.

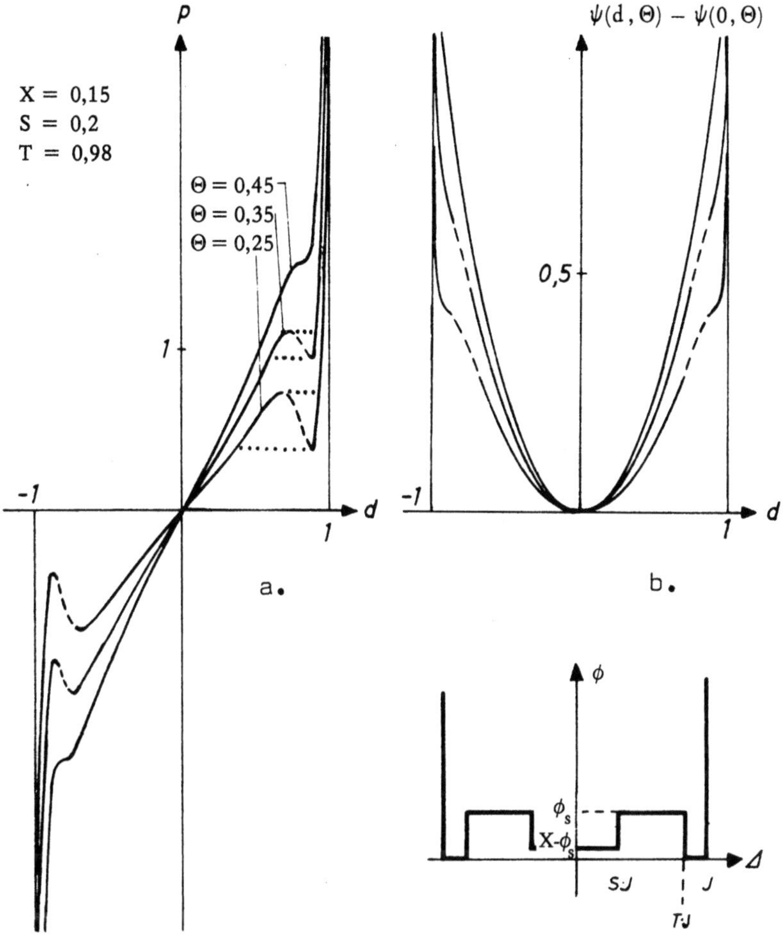


Figure 4.7

## 4.5. On Phase Equilibrium

The non-monotonic load-deformation curves of Fig. 4.7a correspond to the free energy curves of Fig. 4.7b, that have parts with negative curvature. Such parts can be bypassed by the body by moving along the subtangent that is represented by a dotted line in Fig. 4.8a. Here the body establishes phase equilibrium by falling apart into two phases, A and $M_+$, and it increases deformation by decreasing the fraction of A. In the load-deformation diagram 4.8b, phase equilibrium exists on the horizontal Maxwell-line whose height is equal to the slope of the tangent in Fig. 4.8a. The observations reported in Fig. 4.1 show that a memory alloy does not establish phase equilibrium. Were it otherwise the hystereses would not appear. Recall the corresponding discussion in Section 4.5 for rubber, where the hysteresis also appeared, because phase equilibrium was not established.

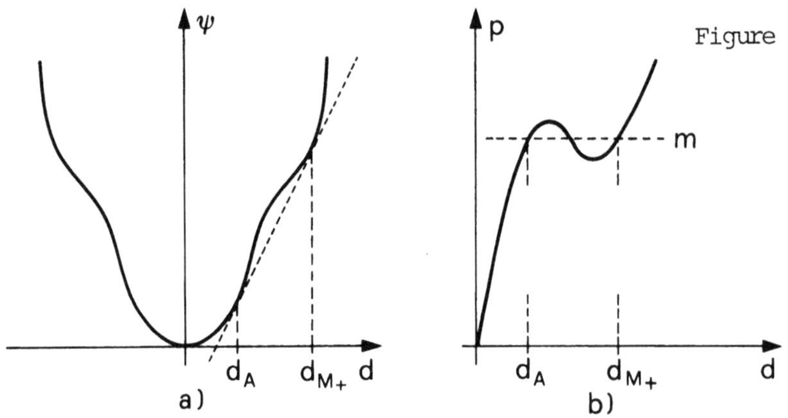

Figure 4.8

Another aspect of the constrained phase equilibrium results from the consideration of the body under a dead load. Equns. (4.5) and (4.6) in Fig. 4.6 give the partition function for this case, whence follows the Gibbs free energy as

$$g(p,\Theta) = -\,\Theta\frac{1}{N}\;\ln\Big(z\,\frac{NJ}{\sqrt{2}}\int_{-1}^{1}\big(e^{-\frac{\psi-pd}{\Theta}}\big)^{N}\,dd\Big)\,. \qquad (4.12)$$

When $\psi$ is taken from Fig. 4.7 for $\Theta = 0{,}25$, the exponent $\psi - p\cdot d$ in (4.12) is represented in Fig. 4.9a for different loads p. Since N is a large number, the minima of $\psi - p\,d$ correspond to sharp peaks in the integrand of (4.12) which is practically zero except where the peaks are. The left minimum in Fig. 4.9a corresponds to the austenitic phase and the right one to the $M_+$ phase. Thus equ. (4.12) implies that the depth of these minima gives the functions $g_A(p,T)$ and $g_{M_+}(p,T)$ for austenite and martensite respectively. Fig. 4.9b represents these functions. The range where the overlap occurs is the range of constrained phase equilibrium. Equilibrium would prevail, if the body always assumed the lowest minimum in Fig. 4.9a for every p. Experience shows, however, that the body, once it is in a particular phase, likes to remain there, even if it could decrease the Gibbs free energy by changing phase.

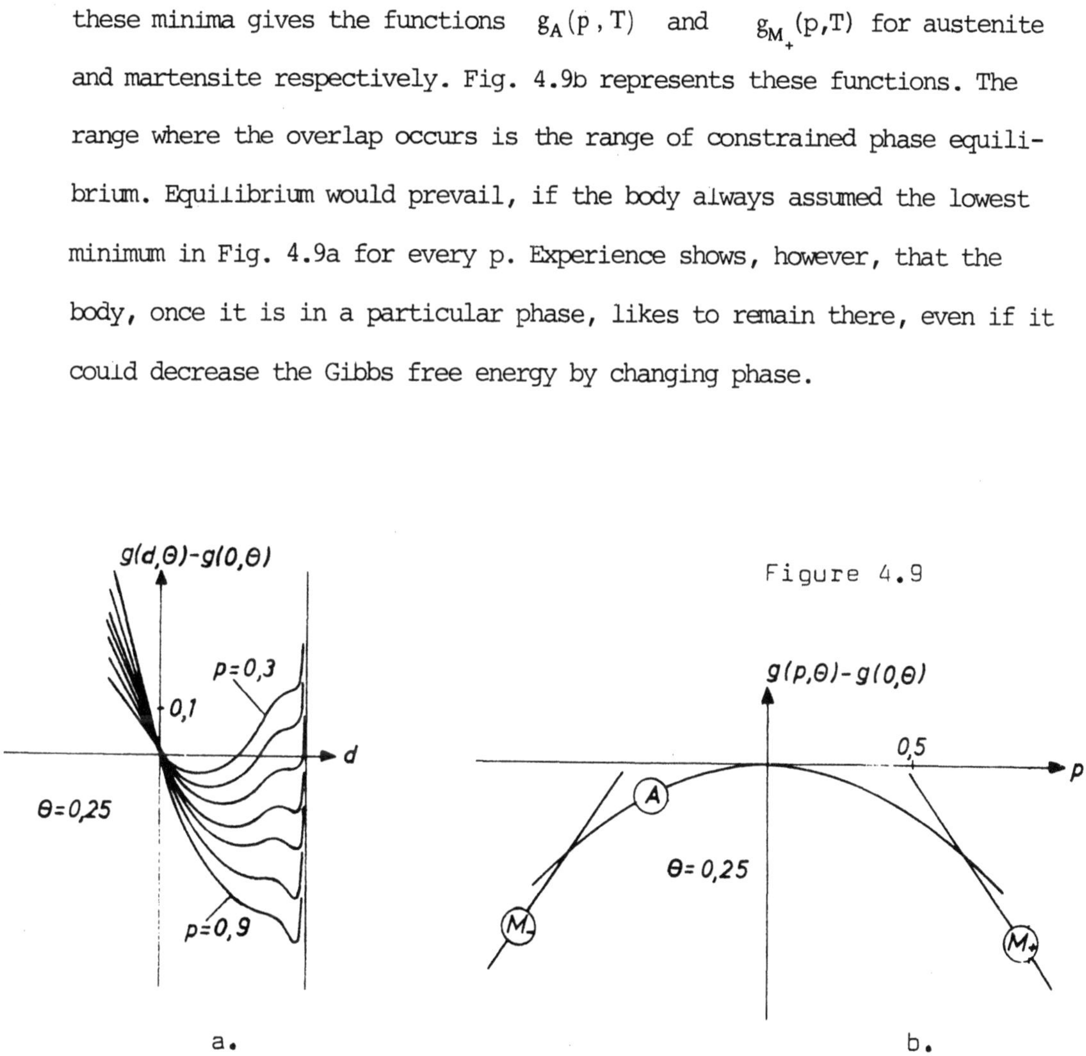


Figure 4.9

a.  b.

## 4.6. The Dynamic Theory

While the above static theory of the model offers some insight into the effects of the phase transition on the nature of the load-deformation-temperature behaviour of memory alloys, the complexity of the partition function has prevented the development of a uniform theory that is valid for all temperatures. Indeed, low temperatures were treated separately from intermediate and high temperatures, and the theory cannot say anything about the transition from low to intermediate temperatures.

The dynamic theory of the model knows no such inhibitions, it is currently being developed by Achenbach and Müller, and the following description presents a first account of its results. The basic tenets of the theory and results for a somewhat simpler model have been described in [22] .

We now denote by $M_+$, A, and $M_-$ the number of lattice particles in the left, central and right potential wells of Figs. 4.2 and 4.5. For the change of these numbers we write the rate laws of Fig. 4.10, where $\overset{-0}{p}$, $\overset{0+}{p}$ etc. are transition probabilities from $M_-$ to A, or from A to $M_+$ etc. The temperature changes according to equ. (4.16), because each transition between the wells converts potential energy into kinetic energy or vice versa. $\Delta\phi_L$ and $\Delta\phi_R$ in Fig. 4.10 denote these energetic differences. Also there is a flux of energy when the external temperature $T_E$ differs from that of the body. The transition probability over a barrier is proportional to the Boltzmann factor exp $\{-\phi_{B/kT}\}$, where $\phi_B$ is the height of the barrier. More specifically the transition probability $\overset{-0}{p}$ (say) is given by the formula

$$\overset{-0}{p} = a \frac{e^{-\frac{\phi(\Delta_L)-P\Delta_L}{kT}}}{\int_{-J}^{\Delta_L} e^{-\frac{\phi(\Delta)-P\Delta}{kT}}}$$

where $\Delta_L$ is the abscissa of the left maximum. Thus $\overset{-0}{p}$ is a function of P and T. $a$ depends on the frequency with which the particles run against the barrier and this frequency increases with temperature while it decreases with the width of the potential well.

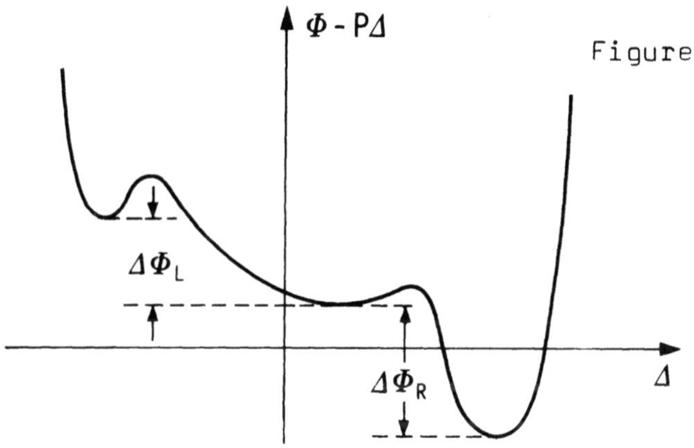


Figure 4.9

$$\dot{M}_- = + A\overset{0-}{p} \qquad - M_- \overset{-0}{p} \tag{4.13}$$

$$\dot{A} = - A\overset{0-}{p} - A\overset{0+}{p} + \quad + M_- \overset{-0}{p} + M_+ \overset{+0}{p} \tag{4.14}$$

$$\dot{M}_+ = \qquad A\overset{0+}{p} \qquad - M_+ \overset{+0}{p} \tag{4.15}$$

$$C\dot{T} = - \Delta\phi_L \dot{M}_- + \Delta\phi_R \dot{M}_+ - a(T_E - T) . \tag{4.16}$$

The rate laws (4.13) through (4.16) describe activated processes which means that barriers have to be overcome before the process can occur. There is always a non-zero transition probability and this leads to creep even at small loads and low temperatures. As long as the load is well below the yield load, the creep is too slow to be noticed. Its speed increases steadily as the load grows and when the load is big enough there is a sudden yield as a result of a chain reaction: When a lattice particle jumps into a deeper potential well, it converts potential energy into kinetic energy, i.e. heat. The resulting increase of temperature makes it easier for other particles to jump and this heats the body further so that more jumps can occur, etc. We call this a chain reaction, because of the formal analogy to the well-known activated processes in chemistry.

The set of equations (4.13) through (4.16) are ordinary differential equations which can be solved numerically for given loads p(t) and external temperatures $T_E(t)$. For a slowly alternating tensile and compressive load of a sawtooth type one obtains deformation d, which are represented by the load-deformation curves of Fig. 4.11 for increasing values of $T_E$. By comparison with the schematized representation of the observed load-deformation diagrams of Fig. 4.1 we conclude that the model is capable of simulating all observed properties of memory alloys in a qualitative manner. The figure also suggests how the model changes load-deformation curves from low to intermediate temperatures.

The adjustment of the parameters of the model to the experimental data is now in progress.

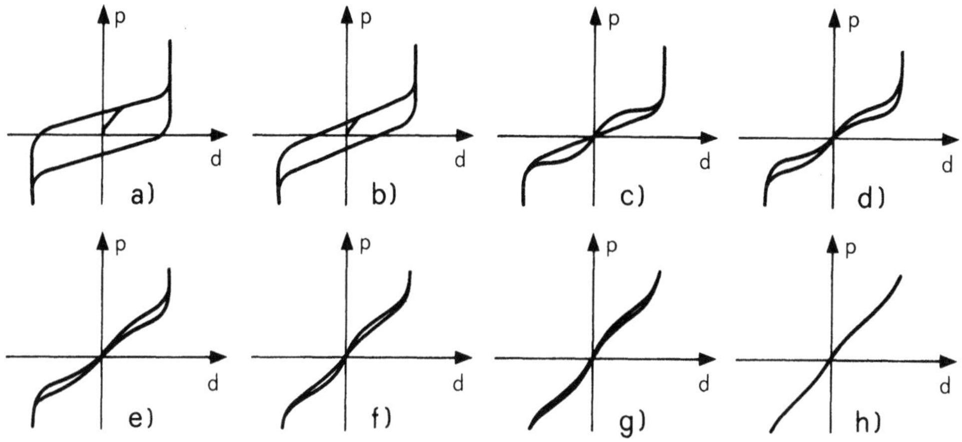


Figure 4.10

4.7. Discussion

If we look back to the Sections 3 and 4 we see that special cases of thermoplasticity have been described qualitatively and that there were two ingredients in those descriptions.

i.) A free energy as a function of deformation and temperature which contains unstable branches of negative curvature. See Figs. 4.7 and 4.8 for example.

ii.) Rate laws of the type (4.13) through (4.16) which govern an activated homogenous process that may represent creep or yield depending on its speed.

It seems to be a challenging task to develop similarly simple models for other types of thermoplasticity.

References

[1] Truesdell, C., Toupin, R.A., The Classical Field Theories Handbuch der Physik III/1, Springer Berlin, Göttingen, Heidelberg, 1967, 226

[2] Truesdell, C., Noll, W., The Non-linear Field Theories of Mechanics Handbuch der Physik III/3, Springer Berlin, Heidelberg, New York, 1965, 1

[3] Müller, I., Thermodynamik, Grundlagen der Materialtheorie Bertelsmann Universitätsverlag,Düsseldorf, 1973

[4] Alts, T., A new approach to a thermodynamic theory of fibre-reinforced thermo-elastic materials with thermo-kinematic constraints, in Mechanics of Structures Media, Ed. A.P.S. Selvadurai, Elsevier Publ. Co., Amsterdam, 1981, part A, 29

[5] Rivlin, R.S., Saunders, D.W., Large elastic deformations of isotropic materials VII. Experiments on the deformation of rubber. Phil. Trans. Roy. Soc. London, 243, 251, 1951

[6] Clausius, R., Mechanische Wärmetheorie, Vieweg und Sohn, Braunschweig, 1887

[7] Carathéodory, C., Untersuchungen über die Grundlagen der Thermodynamik, Math. Annalen, 67, 355, 1909

[8] Meixner, J., Reik, H.G., Die Thermodynamik der irreversiblen Prozesse in kontinuierlichen Medien mit inneren Umwandlungen, in Handbuch der Physik III/2, Springer Berlin, Heidelberg, New York, 1959

[9] de Groot, S.R., Mazur, J., Non-equilibrium thermodynamics, North Holland Publ. Co., Amsterdam, 1962

[10] Coleman, B.D., Noll, W., The thermodynamics of elastic materials with heat-conduction and viscosity, Arch. Rational Mech. Anal., 13, 167, 1963

[11] Müller, I., Die Kältefunktion, eine universelle Funktion in der Thermodynamik viskoser wärmeleitender Flüssigkeiten, Arch. Rational Mech. Anal., 40, 1, 1971

[12] Müller, I., The coldness, a universal function in thermoelastic bodies, Arch. Rational Mech. Anal., 41, 319, 1971

[13] Liu, I.-Shih, Method of Lagrange multipliers for exploitation of the entropy principle, Arch. Rational Mech. Anal., 46, 131, 1972

[14] Cattaneo, C., Sulla conduzione della calore, Atti del Sem. mat. e fis. Univ. di Modena, 3, 3, 1948

[15] Müller, I., Zum Paradoxon der Wärmeleitung, Z. f. Physik, 198, 329, 1967

[16] Ruggeri, T., Generators of hyperbolic heat equation in non-linear thermoelasticity, Rend. Sem. Mat. Padova (to appear)

[17] Nakanishi, N., Lattice softening and the origin of SME in shape memory effects in alloys, Ed. Perkins, Plenum Press, N.Y. and London, 147, 1975

[18] Rodriguez, C. and Brown, L.C. The Mechanical Properties of SME Alloys, in Shape Memory Effects in Alloys. Ed. J. Perkins, Plenum Press, N.Y. and London p. 29 , 1975

[19] Müller, I., Wilmanski, K., State functions for a pseudoelastic body, Proc. of Euromech 113 in Grenoble, Villars de Lans, (in press)

[20] Müller, I., Wilmanski, K., A model for phase transition in pseudo-elastic bodies, Il Nuovo Cimento, 57B, 283, 1980

[21] Müller, I., Wilmanski, K., Memory alloys - phenomenology and Ersatz-model, in Continuum Model of Discrete Systems, Brulin, O. and Hsieh, R.K.T., Eds., North Holland Publ. Co., 495, 1981

[22] Achenbach, M., Müller, I., Creep and yield in martensitic transformations, Ingenieurarchiv 53, p.73, 1983

# PHYSICAL THEORY OF PLASTICITY IN CRYSTALLINE SOLIDS

**VOLKMAR GEROLD**
**Max-Planck-Institut für Metallforschung,**
**Institut für Werkstoffwissenschaften, and**
**Institut für Metallkunde, Univ. Stuttgart**
**D–7000 STUTTGART 1, West-Germany**

Abstract

The basic atomistic mechanisms responsible for plastic deformation are described in this paper. The description is mostly restricted to the behaviour of metals because of their relative ease of deformation. The first chapter outlines the description of crystal lattices as far as it is necessary to understand the different deformation mechanisms. The second chapter describes the most important lattice defect responsible for plastic deformation in crystalline solids; the dislocation. The knowledge of dislocation theory enables one to understand the plastic behaviour of pure metal single crystals as well as polycrystals which is outlined in the third chapter. The last chapter deals with the plastic behaviour of crystalline solids at high temperatures where steady deformation at constant loading can occur which is known as creep. The sometimes drastic influence of alloy conditions on the plastic properties (e.e. solid solutions, precipitation, martensitic transformation) are not included in this paper.

# BASIC MECHANISMS OF PLASTICITY

## Description of Crystals

Introduction. To understand the basic micromechanisms of plasticity in solid bodies one has to consider the atomic structure of these bodies. As it is well known most solid material consists of many small crystals which have a periodic atomic arrangement. The centers of the atoms form a crystal lattice. Its structure depends on the material. Since the lecture will be restricted to the mechanical properties of metals and alloys only two types of crystal lattices will be discussed. These are the face centered cubic (fcc) and the body centered cubic (bcc) lattices. Fortunately, most of the pure metals crystalize in one of both lattices and many alloys show these structures, too.

The atomic structure of a lattice can be fully described by a smallest crystal unit, the so called elementary cell. Fig. 1 shows the elementary cells of both lattice types. The length of the cube edges is called the lattice constant a. The whole lattice is obtained by a multiple repetition of this cell. For example, each square face of the elementary cell is at the same time a face of the neighbouring cell, and so on. The fcc lattice is characterized by an additional center atom at each square face whereas the bcc structure has an additional atom in the center of the elementary cell. The fcc cell has four atoms, the bcc cell only two atoms per cell.

Lattice Directions and Lattice Planes. For further discussions atomic distances in various directions and the different directions play an important role. Distances and directions are described by a triple number which describes the coordinate fractions (in units of a) in three directions

a b c

Fig. 1. Hard ball models of cubic unit cells. From Reed-Hill[1]
a) face-centered cubic (fcc) cell
b) same cell as in (a) with a corner atom removed to show a close-packed (111) plane
c) body-centered cubic (bcc) cell.

parallel to three cube edges which form a carthesian coordinate system in this case. The trible number is set in brackets as |uvw|. Fig. 2 gives a two-dimensional example of a fcc lattice. The following distances are marked by arrows: |110|/2, |310|/2, |100|. The corresponding directions are usually given by the smallest set of three whole numbers. The most important directions have always low indicees. Crystallographically equivalent directions are described by <uvw>. Another important feature of crystal lattices is the existence of lattice planes. Each lattice plane passes through the centers of atom positions and all centers form a set of parallel lattice planes. Many different sets of lattice planes are possible. Three of them are marked by lines in Fig. 2. Each set is indexed by a tripel number (hkl) which is put in paranthesis. The corresponding indicees are also given in the diagram. As a rule for cubic lattices the indicees (hkl) of a lattice plane are the same as the direction indicees |uvw| for a direction normal to the planes. (Sometimes the indicees (hkl) have a common divisor. Thus the planes (100), (110) or (310) in the fcc lattice are usually indexed (200), (220) or (620)). The most important lattice planes have a relatively large plane spacing d. These planes have always low indicees (hkl). They have also a high density of atom centers in a plane. Crystallographic equivalent planes are described by {hkl}.

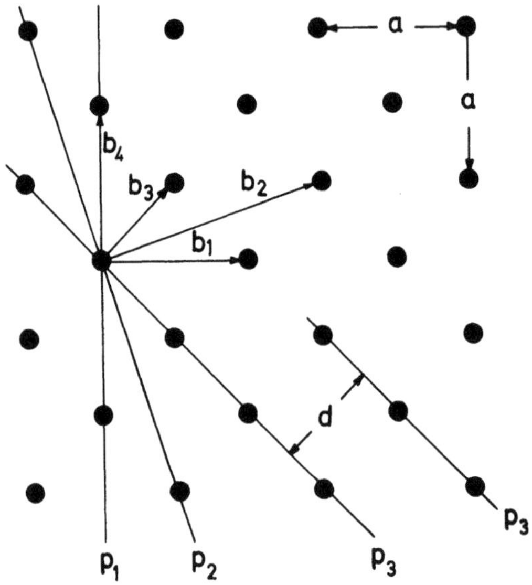


Fig. 2. Distances |uvw| and lattice planes (hkl) in the cube plane of a fcc crystal (only in two dimensions).
a = lattice constant.
Distances: $b_1 = |100|$, $b_2 = \frac{1}{2}|310|$, $b_3 = \frac{1}{2}|110|$, $b_4 = |010|$
Lattice planes: $p_1 = (100)$, $p_2 = (310)$, $p_3 = (110)$
d = spacing of the (110) planes

In the fcc lattice the planes (111) have the largest spacing; they are also "close-packed". If the atoms are modelled by hard spheres they form a close packed hexagonal structure in the (111) planes (Fig. 3). Some distances and directions in this plane are indexed. The next parallel plane has atomic locations either at positions B or C relative to the plotted plane A. If the stacking sequence is ABCABC ... or ACBACB ... it results a fcc lattice which is a close-packed lattice. If the sequence is ABAB or ACAC then a hexagonal lattice is obtained which is called hexagonal close-packed lattice (hcp).

The bcc crystal is not a close-packed structure. The plane (hkl) with largest spacing d is (110) in that case. Such a plane is plotted in Fig. 4

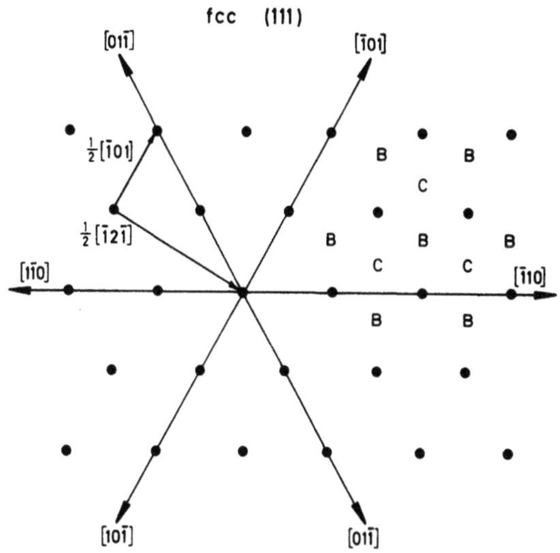


Fig. 3. Atomic structure of the (111) plane in the fcc lattice. All close-packed directions $\langle 1\bar{1}0\rangle$ are marked. Letters B and C mark positions of atoms above and below the drawing plane.

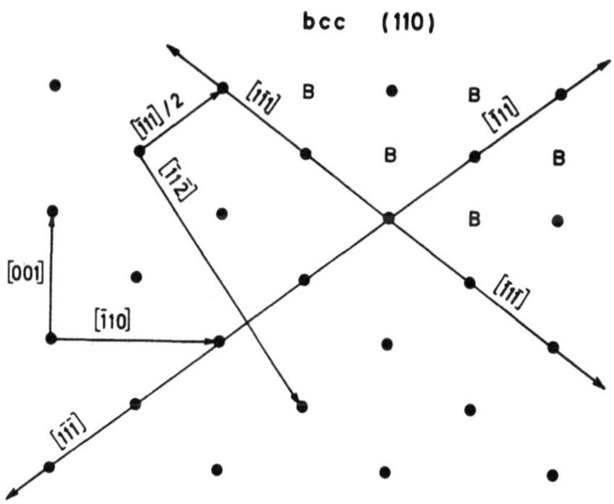


Fig. 4. Atomic structure of the (110) plane in the bcc lattice. All close-packed directions $\langle 1\bar{1}1\rangle$ are marked. The letter B marks positions of atoms above and below the drawing plane.

together with some distances. The next plane has the position B and the stacking is ABAB ... It is no close-packed stacking, of course.

Real crystals contain a variety of lattice defects. Depending on the dimensions of these defects one can distinguish point defects (vacancies impurity atoms, interstitial atoms), line defects (dislocations), two-dimensional defects (stacking faults, grain boundaries) and three-dimensional defects (precipitates, dispersoids). They all play either active or passive roles in plastic deformation.

## Deformation by Glide: Dislocations

There exist different mechanisms for plastic deformation but the glide mechanism is the most important one. On a microscopic scale dense-packed lattice planes slip one against each other caused by a shear stress acting on these planes. The slip or glide direction is always the direction of the shortest atomic distance b in this plane. One set of slip planes and one slip direction form a slip system. For example, the fcc lattice has four different slip planes {111} each having three possible slip directions <110> (see Fig. 3). This gives a total of 12 slip systems.

On a submicroscopic scale the basic deformation step is the slip by a unit atomic distance b. Without the existence of lattice defects this would need a very high shear stress of the order of G/20 where G is the shear modulus. In pure metals the actual shear stress is much lower namely ca. $10^{-5}$G. The unit slip does not occur in a whole slip plane at the same time. The slip occurs only along a line in the slip plane which passes the slip plane and eventually ends at the surface (Fig. 5). The line is the center of a lattice distortion called dislocation. This mechanism is described in the following chapters.

## Deformation by Diffusional Creep: Vacancies

At high temperatures ($T > 0.5\ T_s$) an additional mechanism is possible which is connected with so-called self-diffusion of atoms in a lattice. The latter occurs with the help of lattice sites which are empty. They are called vacancies and exist in thermal equilibrium. Their density increases with temperature. The thermal equilibrium is possible because

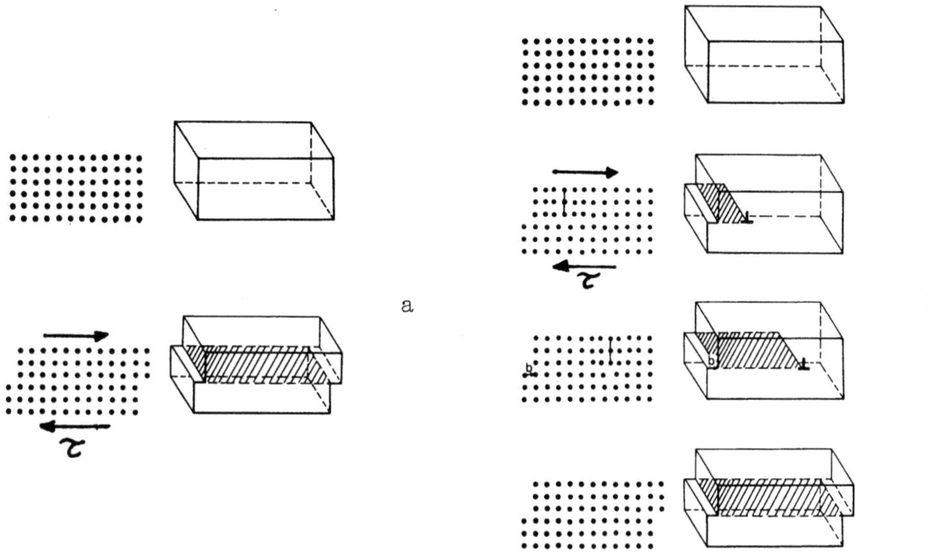


Fig. 5. The basic glide mechanism in a primitive cubic lattice.
a) Slip on a slip plane by one atomic distance b
b) Local slip on the same slip plane by the glide motion of a dislocation line

there exist enough sources and sinks for vacancies. One important source and sink are the grain boundaries. Under an axial stress the grain boundaries parallel and perpendicular to this axis act in a different way, as shown in Fig. 6. The boundary perpendicular to the axis act more often as a source for vacancies than as a sink. In order to act as source an atom has to be placed into the boundary which gives a local elongation in the stress axis. As a result a net flow of vacancies occurs from perpendicular grain boundaries to parallel ones and a flow of atoms exists in the opposite direction. The smaller the grain size the more important is this creep process (so-called Nabarro-Herring creep). This mechanism will be discussed in the last chapter.

Deformation by Twinning: Deformation Twins

This mechanism occurs in cases where deformation by slip is difficult: At low temperatures or in non-cubic crystal lattices. The twin formation or

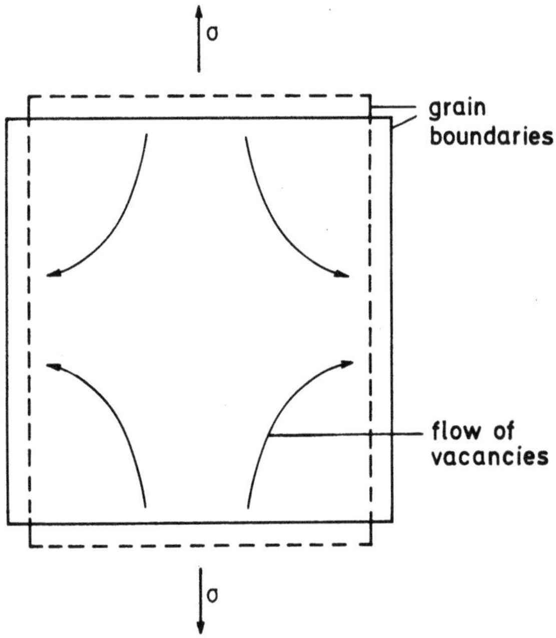


Fig. 6. Diffusional creep by flow of vacancies from a grain boundary perpendicular to the stress axis to a grain boundary parallel to this axis (schematic).

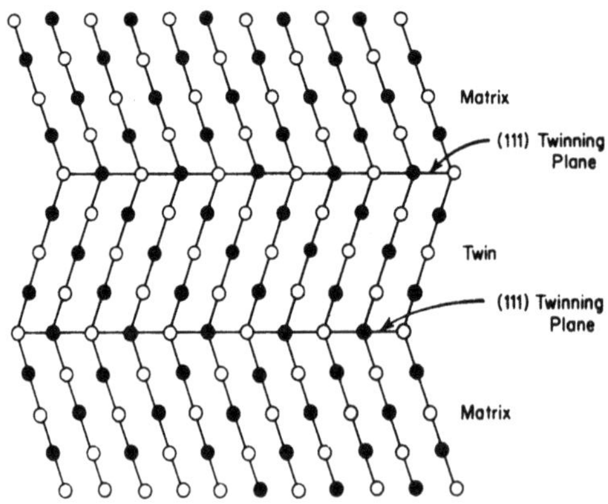


Fig. 7. Deformation twinning in the fcc lattice. Drawing plane $(1\bar{1}0)$. Full and open circles represent atoms on different position levels. From Reed-Hill[1]

twinning is a process which occurs simultaneously in small needle-shaped areas. Very often it is connected with noise (Famous example: tin cry). The principle is shown in Fig. 7. It will not be discussed further.

References

1. Reed-Hill, R.E., *Physical Metallurgical Principles*, Van Nostrand, Princeton, 1964.

# DISLOCATION THEORY

## Definition of Dislocations

Introduction. Dislocations are one-dimensional defects in crystals.[1-4] They occur in all kinds of crystals and are usually produced during the solidification or crystallization process. Usually, they form networks in these crystals. Under special solidification conditions it is possible to produce dislocation-free crystals. Besides these grown-in dislocations additional dislocations are produced in huge amounts during plastic deformation.

Dislocation Line and Burgers Vector. In order to define a dislocation we discuss an artifical configuration namely a glissile dislocation loop in a slip plane (Fig. 8). This loop is the boundary of a circular area where an elementary slip by one atomic distance has occurred. The direction and length of this slip is called the Burgers vector $\underline{b}$ of this dislocation loop. In this definition the sign of the direction is left open. In order to define the sign one has to give the dislocation line a direction. If this direction is reverted then the direction of the Burgers vector has to be reverted, too. According to this definition a dislocation is always a boundary between two areas in the lattice where a shear on an atomic scale has occurred on one of both areas. Therefore, a dislocation line cannot end in an otherwise perfect crystal. It can end only on a free or on an internal surface as a grain boundary. A dislocation line can react with others and form a dislocation network as will be discussed in a later paragraph.

A corresponding definition of the Burgers vector $\underline{b}$ is the so-called Burgers circuit around a dislocation line.

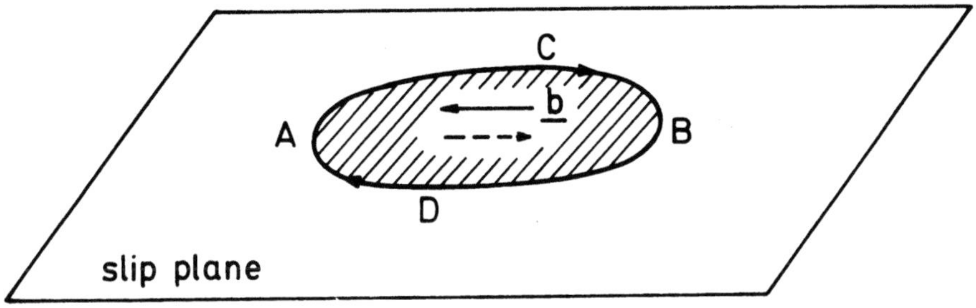


Fig. 8. A glissile dislocation loop on a slip plane. The area inside the loop has slipped by a shortest atomic distance b (Burgers vector of the dislocation loop).

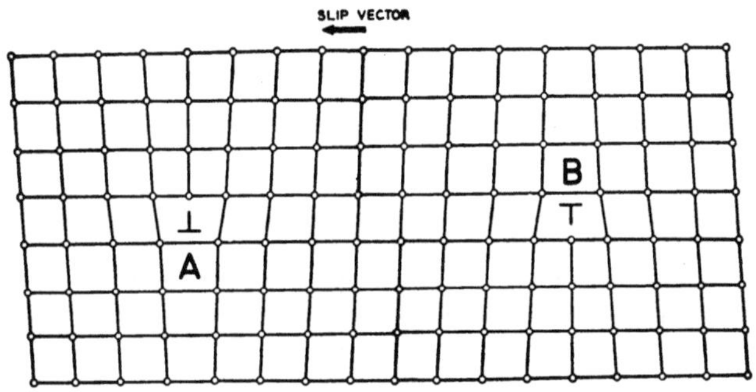


Fig. 9. Section A - B normal to the slip plane of Fig. 8. A primitive cubic lattice is assumed. At A and B two edge dislocations with opposite sign run perpendicular to the drawing plane. From Read[1]

Edge and Screw Dislocations. The angle between the line direction and the Burgers vector b varies along the dislocation loop. There are four special positions A to D where b is either parallel (C, D) or perpendicular (A, B) to the line direction. The lattice configurations at A and B are shown on Fig. 9. In these positions the dislocation loop forms so-called edge dislocations. According to the definition, the direction of the dislocation lines at A and B is opposite to each other and b is the same. Therefore, these are edge dislocations of opposite sign. The symbol for an edge dislocation is ⊥ where the horizontal line marks the slip plane and the vertical line the so-called "extra half plane".

At positions C and D the dislocation loop forms screw dislocations. The lattice configuration at position C is scetched on Fig. 10. The lattice planes perpendicular to the dislocation line form a left-handed screw plane (line direction and $\underline{b}$ are antiparallel). At position D there exists a right-handed screw dislocation (line direction and $\underline{b}$ parallel).

A dislocation is called a mixed dislocation if $\underline{b}$ is neither parallel nor perpendicular to the line direction. Then the Burgers vector $\underline{b}$ has components parallel and normal to the line which describe the screw and edge parts of the dislocation.

The Motion of Dislocations. Dislocations are able to change their position in the lattice due to the action of stresses. The most important motion is the glide of dislocations. This glide occurs on slip planes of the crystal and is the basic process during plastic deformation. The resulting macroscopic slip direction is parallel to the Burgers vector. As a condition for dislocation glide the motion must occur in the slip plane

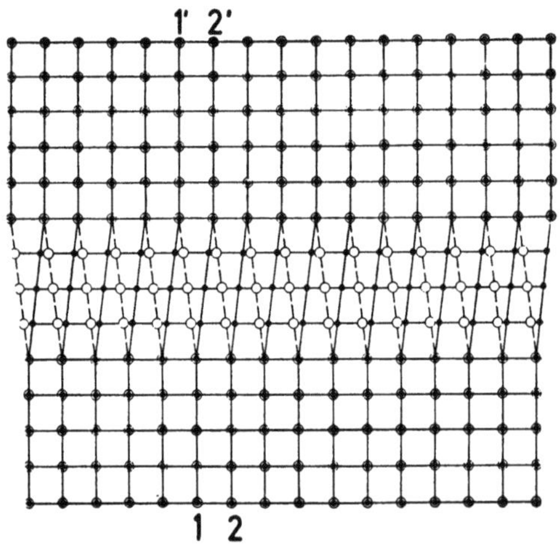


Fig. 10. Section parallel to the dislocation line at position C and normal to the slip plane of Fig. 8. The screw dislocation runs horizontally between the two atomic planes (dots and circles, respectively). Behind the dislocation the vertical lattice planes run from position 1 to position 1' whereas in front of the screw they run from 1 to 2'. This set of planes has changed to a left-handed screw plane. From Read[1]

which is parallel to the line direction and to the Burgers vector. For an edge dislocation there exists, therefore, only one slip plane. The screw, however, is able to glide in different planes because line direction and Burgers vector are parallel to each other. The screw can cross-slip (Fig. 11).

The motion of an edge dislocation normal to its slip plane can only occur if self diffusion occurs in the crystal. Such a motion of an edge is called dislocation climb (Fig. 12). During climb, lattice vacancies are permanently either nucleated or annihilated at the dislocation line. This process occurs only at elevated temperatures (recovery, creep).

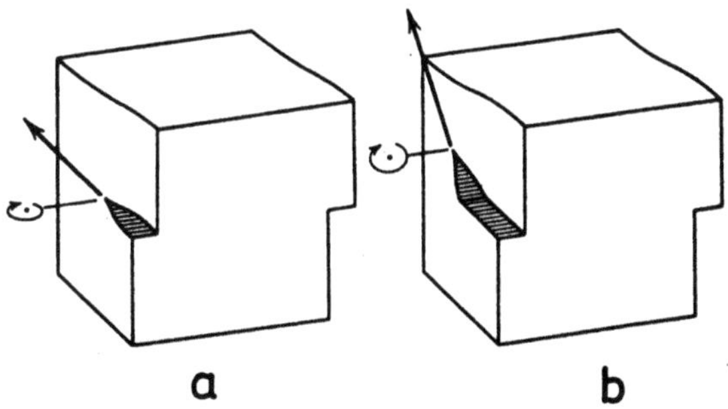


Fig. 11. Cross slip of a screw dislocation. On Fig. a the dislocation has slipped on the primary slip plane. On Fig. b it has started to cross slip in the cross-slip plane. After Hayden et al.[4]

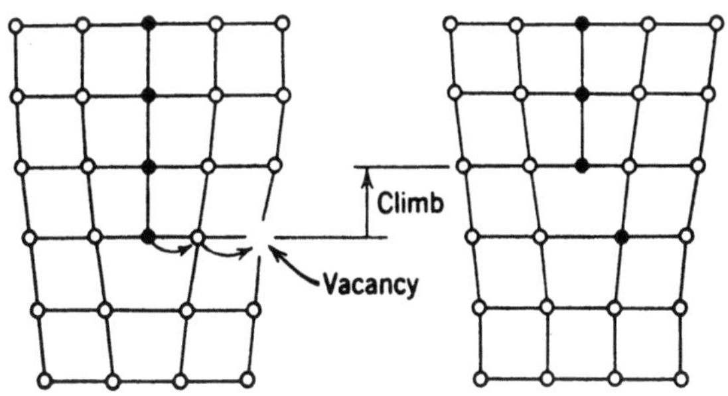


Fig. 12. Climb of an edge dislocation. From Hayden et al.[4]

## Properties of a Single Dislocation

The Stress Field Around a Dislocation. The lattice distortions around a dislocation line can be calculated based on the linear theory of elasticity. The following equations are valid for straight dislocations and assume elastic isotropy. They fail for distances $r \lesssim r_i \approx 2b$ from the position of the dislocation line.

Stress Field Around a Screw Dislocation. This stress field has cylindrical symmetry. In cylindric coordinates $(r, \psi, z)$ there exists only one component $\sigma_{\psi z}$ of the stress tensor if the dislocation is located on the z axis. For a right-handed screw it is

$$\sigma_{\psi z} = Gb/(2\pi r) \tag{1}$$

In Carthesian coordinates this becomes $(r^2 = x^2 + y^2)$:

$$\begin{aligned} \sigma_{yz} &= Gb(x/r)/(2\pi r) \\ \sigma_{zx} &= -Gb(y/r)/(2\pi r) \end{aligned} \tag{2}$$

For a left-handed screw the signs on the right-handed side of Eqs. (1) and (2) have to be reverted.

The stress field of a dislocation decreases with increasing distance r as 1/r which is a slowly decreasing field. This is the case for all kinds of dislocations. The long range stress fields interact with each other with the result that the motion of an individual dislocation is strongly influenced by the presence of the others.

Stress Field Around an Edge Dislocation. Contrary to the screw an edge dislocation has areas of elastic compression and dilatation. Therefore, the stress tensor consists of several components. If the dislocation is positioned on the z-axis and the Burgers vector in the x-direction (the additional half plane at positive y values, see Fig. 13) the components are $(D = Gb/2\pi(1-\nu))$

$$\sigma_{xx} = -Dy(3x^2+y^2)/r^4 \tag{3}$$

$$\sigma_{yy} = -Dy(x^2-y^2)/r^4 \tag{4}$$

$$\sigma_{zz} = \nu(\sigma_{xx}+\sigma_{yy}) \tag{5}$$

$$\sigma_{xy} = Dx(x^2-y^2)/r^4 \tag{6}$$

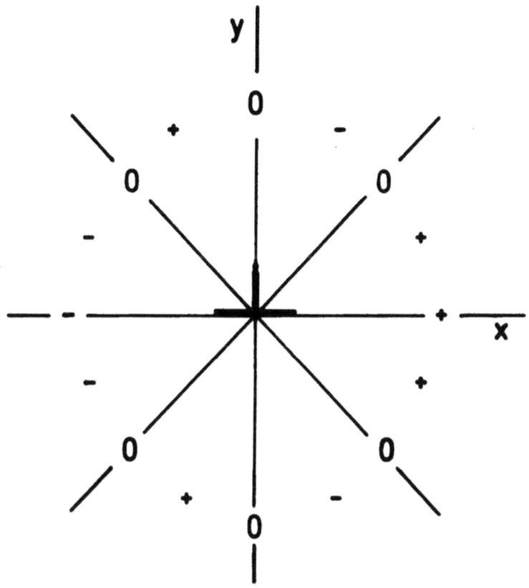


Fig. 13. The change of sign of the shear stress component $\sigma_{xy}$ from an edge dislocation.

For an edge dislocation of opposite sign the signs have to be reverted. For further application in plastic deformation the shear component $\sigma_{xy}$ is the most important one. Along a slip plane y = const the shear stress $\sigma_{xy}$ changes signs in the following way (Fig. 13). We start on the positive x axis where the sign is positive. If we rotate anticlockwise the value decreases and reaches zero on the $45^{\circ}$ axis. Then $\sigma_{xy}$ becomes negative. After an increase it decreases again and reaches zero on the positive y axis. Similar sequences occur in the other quadrants of the drawing, partly with opposite sign. More details will be discussed later on.

The Stress Field of a Mixed Dislocation. The stress field of a mixed dislocation consists of a screw part and an edge part depending on the components of b parallel and normal to the line direction.

The Line Energy. For a straight dislocation line the elastic energy stored in the stress field can be calculated. The energy is given per unit length of the dislocation, the so-called line energy $U_L$:

Screw dislocation

$$U_L = \frac{Gb^2}{4\pi} \ln(r_a/r_i) \tag{7}$$

Edge dislocation

$$U_L = \frac{Gb^2}{4\pi(1-\nu)} \ln(r_o/r_i) \tag{8}$$

The radii $r_i$ and $r_o$ are the so-called inner and outer cut-off radii needed to get a finite value: $r_i \simeq 2b$, $r_o \simeq 10^{-6}$ m.

As a rough estimate:

$$U_L \simeq Gb^2/2. \tag{9}$$

The line energy is proportional to $b^2$. Therefore, only such dislocations will occur with the lowest possible Burgers vector which corresponds to the shortest atomic distance in the lattice.

Forces Acting on Dislocations. In this chapter only such forces will be discussed which lead to dislocation glide. Since dislocation glide is connected with macroscopic shear in a slip system the origin of this force is the shear stress $\tau$ acting on the slip system at the position of the dislocation line. If the dislocation line L glides for a distance r, a local slip b occurs over an area A = Lr. The shear stress $\tau$ produces a work W:

$$\begin{aligned} W &= \tau A b \\ &= \tau b L r \\ &= F r \end{aligned} \tag{10}$$

This work can be described as the work of a force F acting on the dislocation line and shifting it by the distance r. By comparing the Eqs. (10) with each other a force F/L per unit length of the dislocation results:

$$F/L = \tau b \tag{11}$$

This force acts always in the slip plane and is normal to the dislocation line.

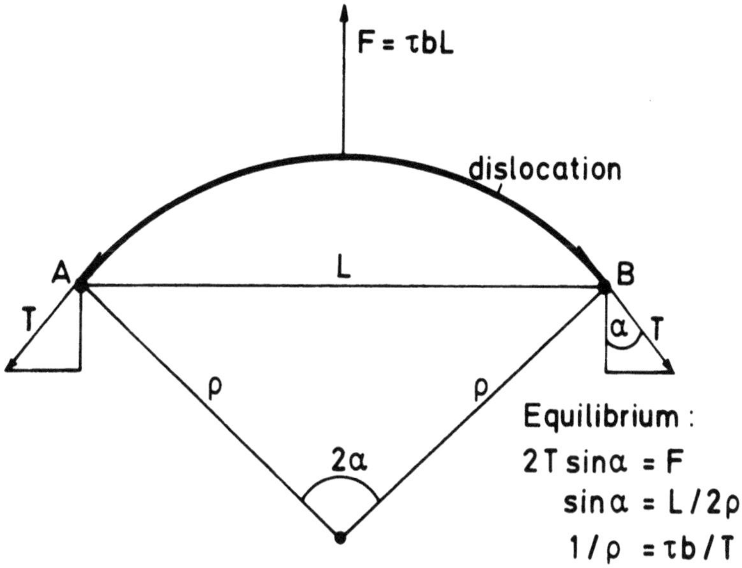


Fig. 14. Bending of a dislocation line which is fixed at points A and B.

Bending of a Dislocation: The Line Tension. If a dislocation line in a slip plane is fixed for some reason at two points A and B (Fig. 14) and if a shear stress $\tau$ acts as a force on the dislocation line it will start to bend between points A and B until an equilibrium will be reached between the applied force $F/L = \tau b$ and a reverse force which is introduced by the curvature of the dislocation line. This second force is caused by the line tension T (with the dimension of a force) of the dislocation line. This tension always tries to keep the dislocation line as short as possible.

From geometry it follows for the radius of curvature $\rho$:

$$1/\rho = \tau b/T \qquad (12)$$

Unfortunately, the line tension T depends in a complex way on the configuration of the dislocation line. As a rough approximation one uses

$$T \simeq Gb^2/2 \qquad (13)$$

which results in

$$1/\rho \simeq 2\tau/Gb \tag{14}$$

The curvature $1/\rho$ increases with $\tau$ and decreases with Gb which describes the rigidity of the dislocation line.

Extended Dislocations. So far only those properties have been discussed which result from the long range stress field of dislocations. Another important factor is the atomic structure of the core of the dislocation line, i. e. the structure at its center. This structure depends on the lattice type of the crystal and is quite different for the fcc and the bcc lattice.

Dislocation Splitting in the fcc Lattice. The reason for the extension or splitting can be discussed easily for an edge dislocation which is characterized by an "extra half plane" perpendicular to the Burgers vector b. For the fcc lattice b = <110>/2. A closer inspection of the "extra half plane" results in the fact that it is a double plane. (In Fig. 3 this plane corresponds to the zig-zag row of atoms which is perpendicular to the $|\bar{1}10|$-direction). The situation is sketched in Fig. 15a where the double layer nature of the lattice planes under discussion is marked by full and broken lines. The resulting elastic distortion of the lattlattice could be reduced if the dislocation would split into two so-called partial dislocations as shown in Fig. 15b. In trying to do this the dislocation creates a defect area between the partials. In the sketch always the wrong planes meet each other and create this faulted area S. It is called a stacking fault (SF) and is a well-known defect in fcc lattices. A similar splitting occurs also in screw dislocations.

A detailed analysis shows that the SF can be described by a wrong stacking sequence of the (111) planes parallel to the SF. It is connected with a specific energy the so-called stacking fault energy (SFE) $\gamma$. Its value varies widely with the material. A low SFE (as in Au, Ag, $\alpha$ brass) results in a wide dislocation extension ($d > 10b$) whereas a high SFE (as in Al) gives a negligible extension.

A high extension aggravates processes as cross slip or climb.

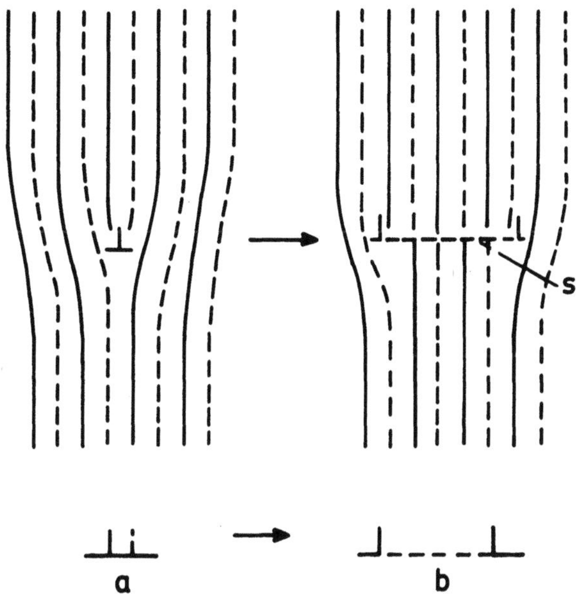


Fig. 15. Configuration of an edge dislocation in a fcc lattice. The lattice planes normal to b consists of double planes marked by full and dashed lines. resp.

a) Full dislocation
b) Split dislocation with a stacking fault between the partial dislocation.

Screw Dislocations in bcc Crystals. In the bcc lattice the Burgers vector is b = <111>/2. This direction has a three-fold symmetry. For a screw dislocation which is always parallel to b the core atoms show displacements (in addition to the elastic distortion) which have the same threefold symmetry. Because of this symmetry the screw dislocations have no more a preferred slip plane. They easily cross slip and, therefore macroscopically move in a plane which is parallel to b and which has the maximum applied shear stress. With decreasing temperatures it becomes more and more difficult to move these screw dislocations. This is the reason for the strong temperature dependence of the yield stress of bcc metals.

Interaction Between Dislocations

Forces Between Dislocations. Since dislocations have long range stress fields they create attractive or repelling forces between them. The discussion will be restricted to the interaction between parallel dislocations of the same slip system. The force $F_2$ on a dislocation $D_2$ will be discussed which is created by the stress field $\tau_1$ from a dislocation $D_1$ positioned along the z axis. The slip plane for $D_2$ is y = const. Both dislocations are either screws or edges. The general equation is

$$F_2/L = \tau_1 b_2 \tag{15}$$

Forces Between two Screw Dislocations. In this case $\tau_1$ is given by $\sigma_{yz}$, Eq. (2). If $b_1$ and $b_2$ are parallel to each other ($b_1 = b_2 = b$) the force is

$$F_2/L = Gb^2x/(2\pi r^2) \tag{16}$$

The dislocation $D_2$ is repelled from $D_1$. If the Burgers vectors are antiparallel (dislocations of opposite sign) the force has the reverse sign and $D_2$ is attracted by $D_1$. Since screw dislocations usually can cross slip they can meet each other and annihilate themselves.

Forces Between two Edge Dislocations. In this case $\tau_1$ is given by $\sigma_{xy}$, Eq. (6). For two edge dislocations with the same sign the force $F_2$ is given by (see also Fig. 13)

$$F_2/L = Dbx(x^2-y^2)/r^4. \tag{17}$$

For positions $|x|>|y|$ this force repells $D_2$ from $D_1$. For a slip plane $y \neq 0$ and for positions $|x| = |y|$ the force is zero, an unstable equilibrium is reached. For $|x|<|y|$ the force becomes attractive and $D_2$ reaches a stable equilibrium at the position $x = 0$ in the slip plane $y \neq 0$. This unusual behaviour of parallel dislocations with the same sign is typical for edge dislocations and results from the compressive and dilative regions which attract each other.

For antiparallel dislocations, the sign of $F_2$ is reverted which gives stable equilibrium at positions $|x| = |y|$ and an unstable one at $x = 0$

for $y \neq 0$. If $y = 0$ the dislocations $D_1$ and $D_2$ will annihilate each other. The stable equilibrium condition does occur very often and is called a dislocation dipole. More than two dislocations of opposite signs form a stable dislocation multipole.

Forces Between Mixed Dislocations. For two parallel dislocations having Burgers vectors $\underline{b}_1$ and $\underline{b}_2$ each can be separated into a screw and an edge component. Each component pair leads to a force which then superimpose. In this way the resulting force can be calculated for cases where $\underline{b}_1$ and $\underline{b}_2$ are not parallel or antiparallel to each other.

Dislocation Networks. If two dislocations attract each other they can meet and react with each other. In this way dislocation networks can be formed. Such networks have knots where three dislocations come together. There exists a rule for the Burgers vectors $\underline{b}_i$ of these dislocations: The sum of $\underline{b}_i$ for all dislocations with directions running into the knot is equal to the sum of $\underline{b}_j$ for all dislocations with directions running off the knot. Fig. 16 shows as an example the reaction between two dislocations $D_1$ and $D_2$ on a (110) plane of the bcc lattice (see also Fig. 4). $D_1$ is a pure edge dislocation with $\underline{b}_1 = |1\bar{1}1|/2$. $D_2$ is a mixed dislocation with $\underline{b}_2 = |111|/2$, which can be separated in an edge and a screw component. The edge component interacts with $D_1$ and leads to an attractive force. Both dislocations attract each other and will form a third dislocation $D_3$ with $\underline{b}_3 = \underline{b}_2 + \underline{b}_2 = |001|$. If $D_1$ and $D_2$ are fixed for some reason at points A and B the resulting configuration will be as shown in Fig. 16b.

Small Angle Grain Boundaries. Two grains having only slightly different orientations to each other are separated by a small angle boundary. The misorientation between the two grains can be described by a rotation axis and a rotating angle which is of the order of $1^{\circ}$ or less. Two ideal cases exist where the rotation axis is either in or normal to the boundary. The first case is a tilt boundary. It consists of equal-spaced edge dislocations running parallel to the rotation axis (Fig. 17a). The rotation angle is given by

$$\alpha = b/d \tag{18}$$

where d is the spacing of the dislocations.

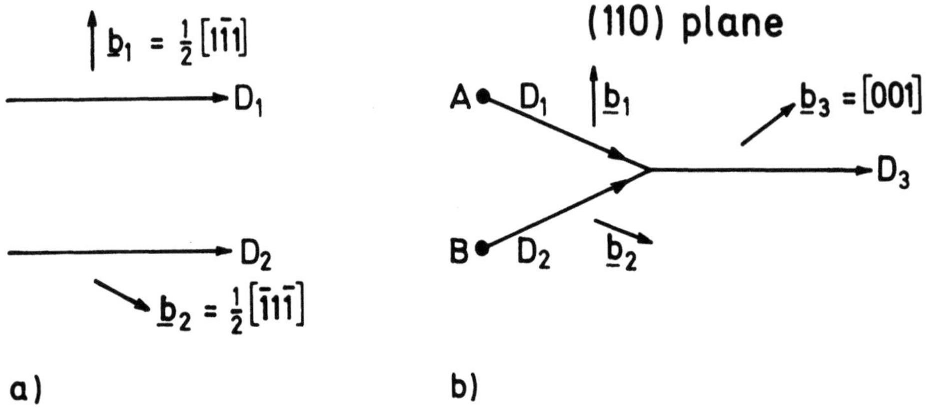


Fig. 16. Reaction between two dislocations $D_1$ and $D_2$ on a (110) plane of the bcc lattice. (a) Before the reaction, (b) after the reaction.

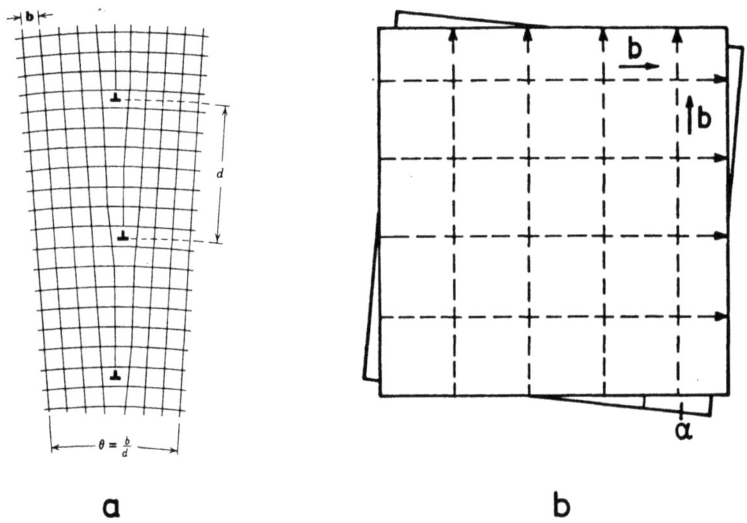


Fig. 17. Small angle grain boundaries

a) Tilt boundary (normal to the drawing plane). From Hayden et al.[4]
b) Twist boundary (in the drawing plane) formed by screw dislocations (dashed lines).

The other case is the twist boundary where the rotation axis is normal to the boundary. In that case the boundary consists of at least two sets of parallel screw dislocations as sketched on Fig. 17b. The twist angle $\alpha$ is also given by Eq. (18). A twist boundary in the fcc lattice parallel to a (111) plane consists of three sets of screw dislocations which interact and form a hexagonal network (Fig. 18).

Dislocation Multiplication: Frank Read Source. During plastic deformation huge amounts of dislocations are created from dislocation sources. Such sources are three-dimensional dislocation networks as the Frank-Read source sketched on Fig. 19. This source consists of a short piece (length L) of a glissile dislocation on a active slip plane which is part of a three-dimensional network with knots at positions A and B. An applied shear stress $\tau$ in the slip system bends the dislocation between A and B. When $\tau$ reaches the value 2T/bL the dislocation forms a half circle (position 2) with a bending radius $\rho$ = L/2 (Eq. 12). This is an unstable equilibrium. A small increase of $\tau$ leads to further motion of the dislocation which rotates around the fix points A and B to position 4. On the reverse side the dislocation will meet itself and react. As a result a dislocation loop (position 5) is split away which spreads over the slip plane. At the

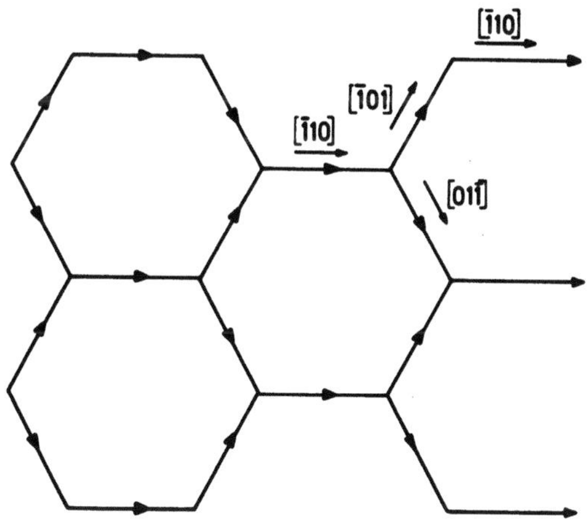


Fig. 18. Small angle twist boundary on a (111) plane of a fcc lattice.

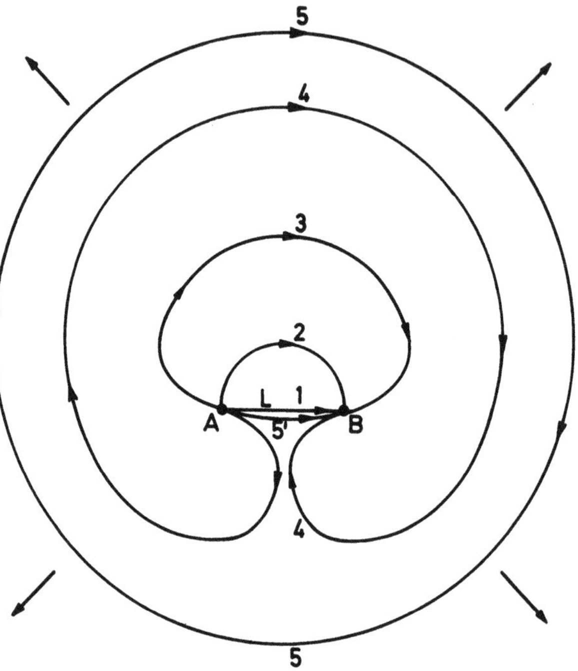


Fig. 19. Frank-Read source of length L on a slip plane. The points A and B are knots of a three-dimensional dislocation network.

same time, the short piece 5' of the dislocation is pushed between the points A and B by the applied shear stress $\tau$ and repeats the bending and splitting off of a dislocation loop many times. Finally the source ceases to nucleate loops because of the internal back stress created by all the preceding loops. In practical cases more than 100 dislocation loops can be produced by a single source.

Pile up of Dislocations. If the dislocations nucleated from a source meet an obstacle as, for example, a grain boundary they are piled up against the obstacle by the applied shear stress $\tau$. If there are n dislocations piled up (Fig. 20) they create a stress concentration $\tau_n$ at the obstacle which is given by

$$\tau_n = n\tau \qquad (19)$$

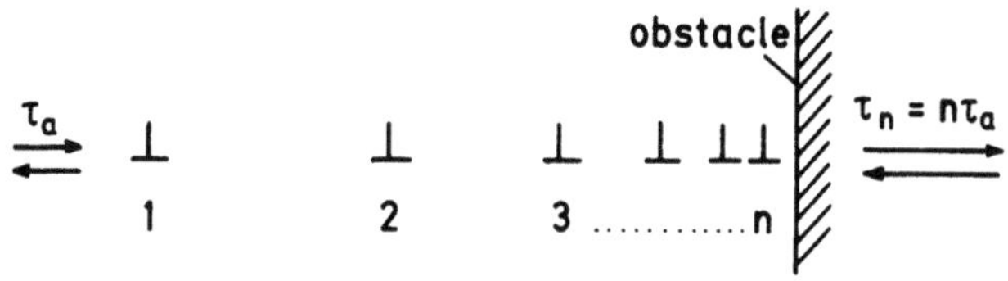


Fig. 20. Pile-up dislocations at an obstacle. The shear stress τ on the slip plane pushes the dislocations against the obstacle.

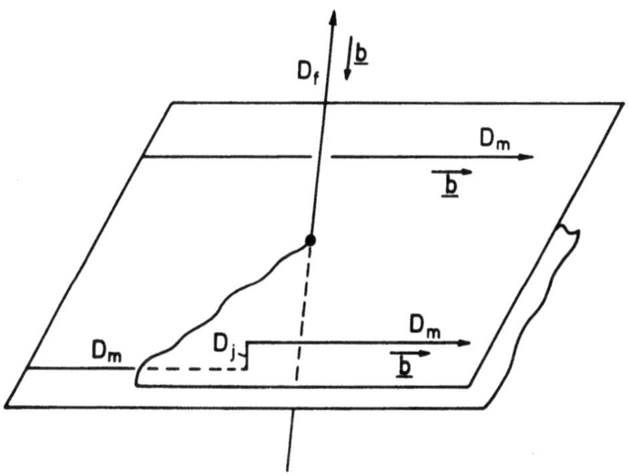


Fig. 21. Jog formation in a mobile screw dislocation $D_m$ by cutting a forest dislocation $D_f$ which is also a screw. The jog $D_j$ can glide only parallel to $D_m$ but has to climb when $D_m$ moves on the slip plane.

Such a stress concentration at a grain boundary leads either to plastic deformation in the neighbouring grain by activating dislocation sources or by failure of the grain boundary, i.e., grain boundary cracking.

Cutting Between Dislocations: Formation of Jogs. Another important interaction between dislocations is the cutting of so-called forest dislocations (the name for immobile dislocations running nearly perpendicular to an active slip plane) by glide dislocations. If a forest dislocation $D_f$ is a screw the active slip plane forms actually a screw plane. If a mobile dislocation $D_m$ moves on such a slip plane and passes the forest dislocation it afterwards moves on two consecutive planes as shown in Fig. 21. The two parts of the moving dislocation are connected by a jog $D_j$. Such jogs sometimes are not able to glide together with the disloca-tion because its motion in the given direction is only possible by climb. In that case it continously has to create either vacancies or interstitial atoms. It is well-known that such point defects are created during plastic deformation.

References

1. Read, W.T., *Dislocations in Crystals,* McGraw Hill, New York, 1953.

2. Weertman, J. and Weertman, J.R., *Elementary Dislocation Theory,* The Macmillan Comp., New York, 1964.

3. Hull, D., *Introduction to Dislocations,* Pergamon Press, Oxford, 1965.

4. Hayden, W., Moffat, W.G. and Wulff, J., *Mechanical Behaviour. Structure and Properties of Materials,* Vol. III, J. Wiley, New York, 1965.

# PLASTICITY OF PURE METALS

## Deformation of Single Crystals

Basic Relationships.[1,2] Plastic deformation occurs by microscopic shear movements along crystallographic planes in well defined crystallographic directions. One shear plane combined with one shear direction is called a slip system. The shear direction is then parallel to the Burgers vector of these dislocations. Therefore, the slip direction occurs always parallel to a close-packed direction which is <110> in fcc, and <111> in a bcc lattice. The shear or slip planes usually are also close-packed planes as {111} and {110}, respectively. On a microscopic scale, the deformation is quite inhomogeneous. On the surface distinct slip lines or more coarse slip bands can be observed (Fig. 22) depending on the nature of the crystal and on the amount of deformation.

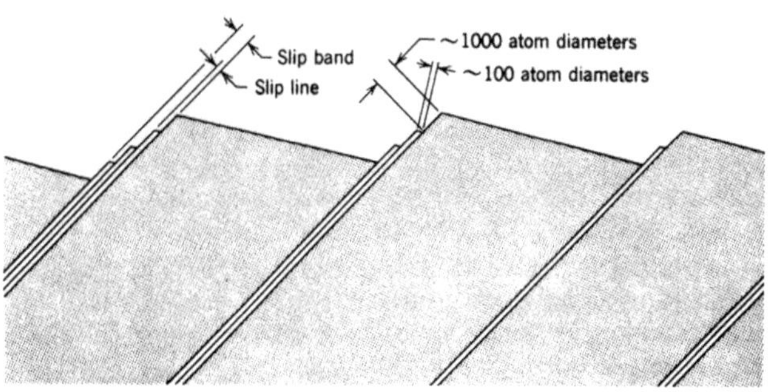


Fig. 22. Section through a deformed single crystal close to the surface (schematic). From Hayden et al.[1]

Each slip system is activated by an applied shear stress $\tau$ which acts on the slip plane in the shear direction. As a rule that slip system will be activated first which has the highest applied shear stress. (Fig. 23). This shear stress depends on the applied tensile stress $\sigma$ and the angles $\phi_o$ and $\lambda_o$ between the tensile axis and either the normal to the slip plane ($\phi_o$) or the slip direction ($\lambda_o$):

$$\tau = \sigma \cos \phi_o \cos \lambda_o \tag{20}$$

The product $\cos \phi_o \cos \lambda_o$ is called Schmid factor. Each slip system has a distinct Schmid factor. If one of these factors is much larger than the others the corresponding slip system will have a much larger shear stress $\tau$ than the others. If this $\tau$ reaches a critical value $\tau_o$ (critical resolved shear stress, CRSS) then plastic deformation will start (Schmid law for single slip orientation).

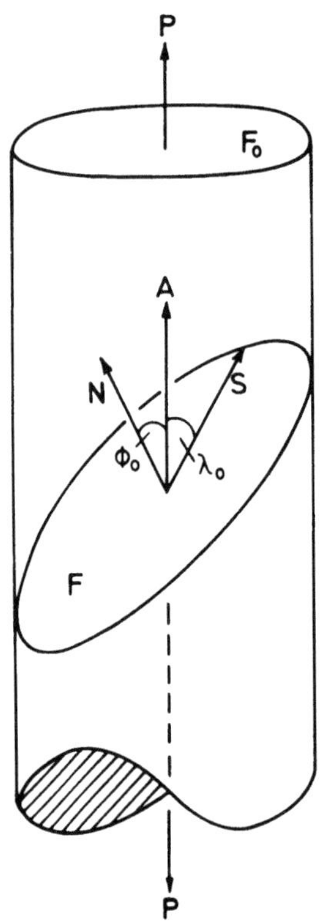


Fig. 23. The definition of the angles $\phi_o$ and $\lambda_o$.

A = axis of the single rod

N = normal to the slip plane

$\phi_o$ = angle between A and N

S = slip direction

$\lambda_o$ = angle between A and S

$F_o$ = cross section of the rod

F = $F_o/\cos \phi_o$ = slip plane area

P = tensile load

$P \cos \lambda_o$ = load component in the direction S

$\tau$ = $(P/F_o)\cos \phi_o \cos \lambda_o$ $\sigma \cos \phi_o \cos \lambda_o$.

Due to the shear process in a single slip system the orientation of the tensile axis will change with increasing deformation. Equations exist which correlate the applied shear stress $\tau$ and the shear strain $\gamma$ with the applied tensile stress $\sigma$ or load P and the elongation $\varepsilon$.

Face-Centered Cubic Crystals.[2] The slip systems are $\{111\} <1\bar{1}0>$. There exist altogether 12 slip systems in this lattice. Stress-strain curves depend on the purity of the crystal, its orientation, temperature and crystal size.

The Critical Resolved Shear Stress $\tau_o$. The CRSS depends mainly on purity and temperature. It depends also on orientation which reflects the fact that besides the so-called primary slip system also other slip systems are more or less involved. The crystallographic orientation of the tensile axis is described on a standard (001) projection of the crystal (Fig. 24). Usually the direction A of the axis is given in the standard orientation triangel which is marked by full lines. Typical single slip orientations are close to the center of the triangle. The primary slip system (111) $|\bar{1}01|$ is marked by P. During deformation the tensile axis is shifted into the direction of slip, $|\bar{1}01|$. For single slip orientations and high purity $\tau_o$ is of the order of 1 MPa at room temperature. It increases about 30% to lower temperatures (Fig. 25).

For single-slip orientations there exist three stages of deformation (Fig. 26):

Stage I Hardening. This so-called easy glide hardening has a very low hardening coefficient. Only dislocations of the primary slip system are observed which can slip over large distances (several mm). The maximum strain in this stage is $\gamma$ = 5 to 40%. Low stacking fault energy and low temperatures give larger maximum strains (Fig. 27).

Stage II Hardening. This so-called linear hardening stage has a hardening coefficient $\theta = d\tau/d\gamma$ of the order of G/300.G is the shear modulus. Secondary slip systems are activated and interact with the still dominating primary slip system. Therefore, slip distances decrease continuously with deformation, the dislocation density $\rho$ is heavily increasing. The following relationship is valid:

$$\tau = \alpha G b \sqrt{\rho} \tag{21}$$

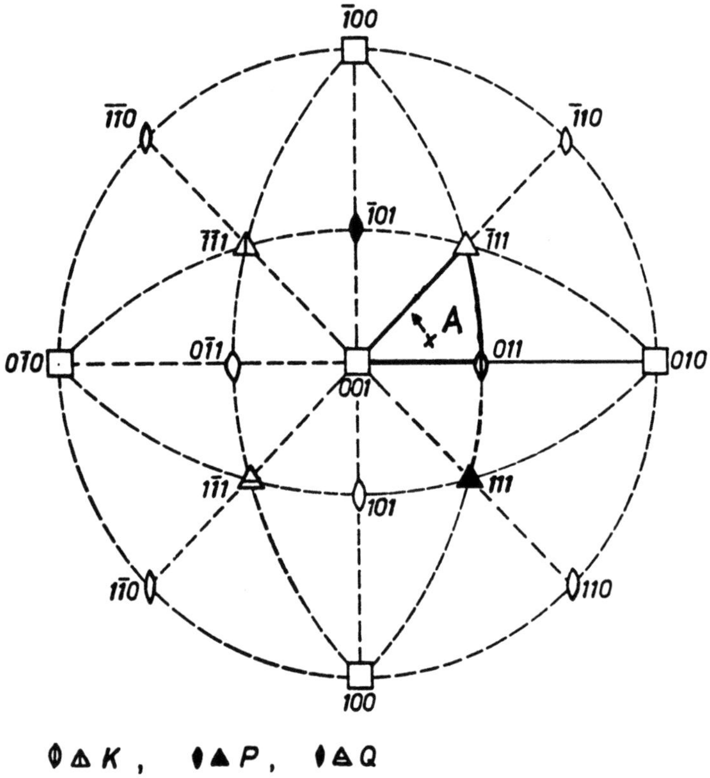


Fig. 24. Standard projection |001| of a cubic crystal.
Orientation triangle |001| - |$\bar{1}$11| - |011| with orientation A of the specimen axis.
P = primary slip system
K = conjugate slip system
Q = cross-slip system

where α is a constant between 0.2 and 0.5. The distribution of dislocations becomes more inhomogeneous and forms for example dipole walls in planes perpendicular to the primary Burgers vector.

<u>Stage III Hardening</u>. In this stage the hardening coefficient θ decreases due to dynamic recovery. The main mechanism for recovery is marked cross slip of primary screw dislocations. Since cross slip is a

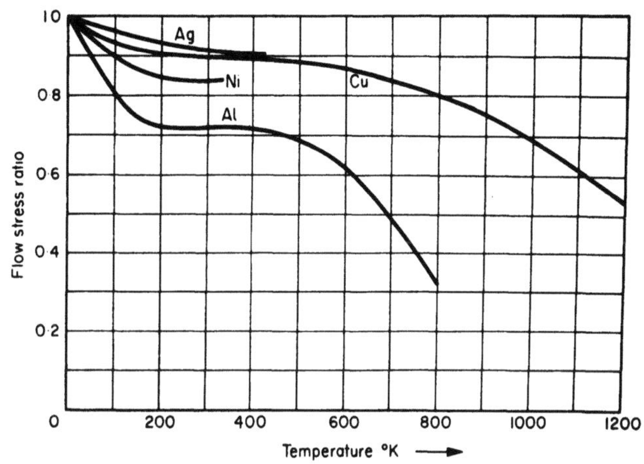


Fig. 25. Normalized CRSS of fcc crystal as function of the temperature T. After Mitchel[3] from Honeycombe[2]

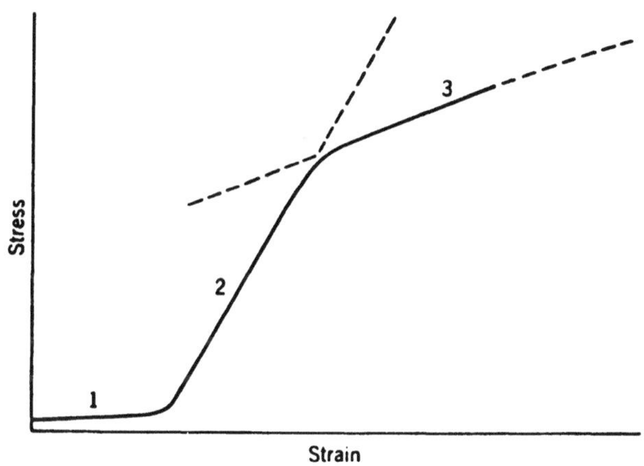


Fig. 26. Schematic shear stress-shear strain diagram of a fcc crystal in easy glide orientation. From McLean[12]

thermally activated process it occurs at larger $\tau$ for lower temperatures (Fig. 27). Low stacking fault energies also increase the stress $\tau$ needed for dynamic recovery.

For <u>multi-slip orientations</u> stage I is no more observed (Fig. 28). Such orientations are close to the borders $|001|-|\bar{1}11|$ and $|011|-|\bar{1}11|$ of the orientation triangle where the Schmid factors of at least two slip systems are the same.

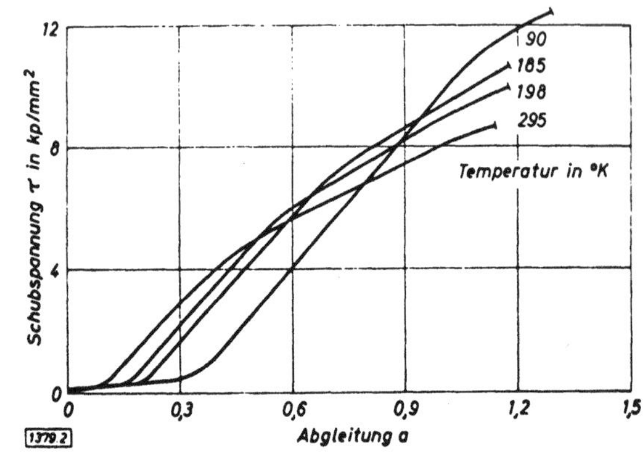


Fig. 27. Strain hardening curves of copper single crystals for various temperatures (in K). After Diehl and Berner[4]

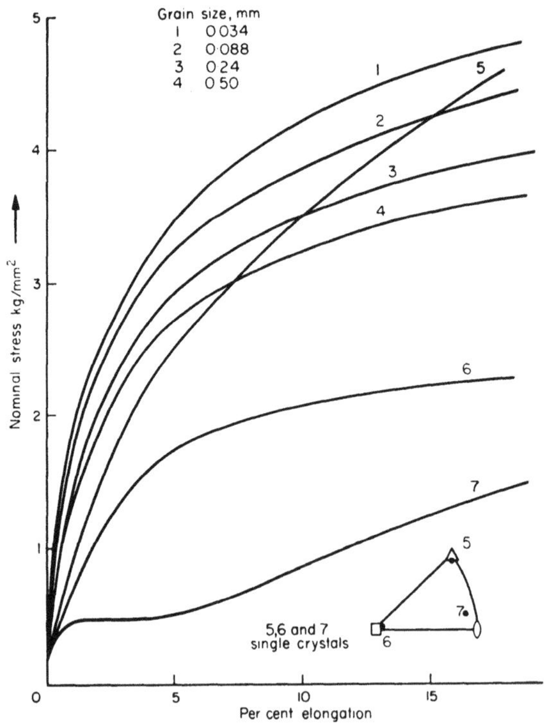


Fig. 28. Stress-strain curves of aluminium single crystals of various orientations and of polycrystals of various grain sizes. After Honeycombe[2]

Hardening Theories. Detailed theories try to explain the hardening coefficient for stages I and II.[5]

Body Centered Cubic Crystals.[6,7] The plastic deformation of bcc crystals is of more recent interest. Its deformation behaviour is much more complex. Compared to the fcc crystals they have a much steeper temperature dependence of the CRSS at low temperatures which reaches much higher values of the order of 300 MPa and more (Fig. 29). Only at medium temperatures ($T > T_c \approx 0.25\ T_s$) three stage stress-strain curves comparable to those of fcc crystals are found (Fig. 30). The reason is the strong temperature dependence of the mobility of screw dislocation. These dislocations are parallel to a <111> direction which has a threefold symmetry. Their dislocation core shows atomic displacements reflecting this symmetry. This is not the case for other types of dislocations in bcc crystals.

With decreasing temperature ($T < T_c$) the screw dislocations become less mobile than the edge dislocations and determine the

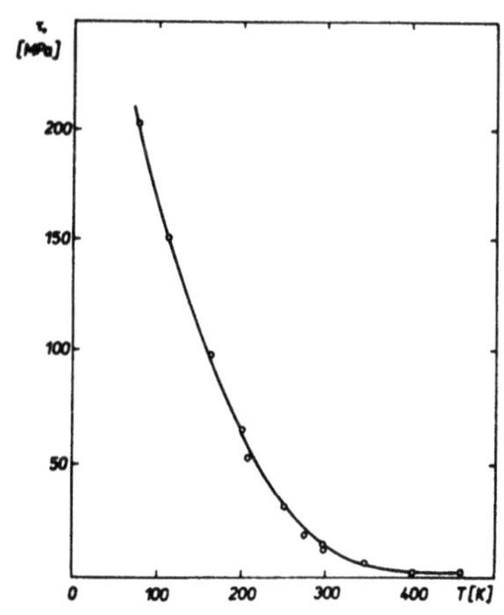


Fig. 29. The temperature dependence of the CRSS of α-iron single crystals. After Novác and Sesták[8]

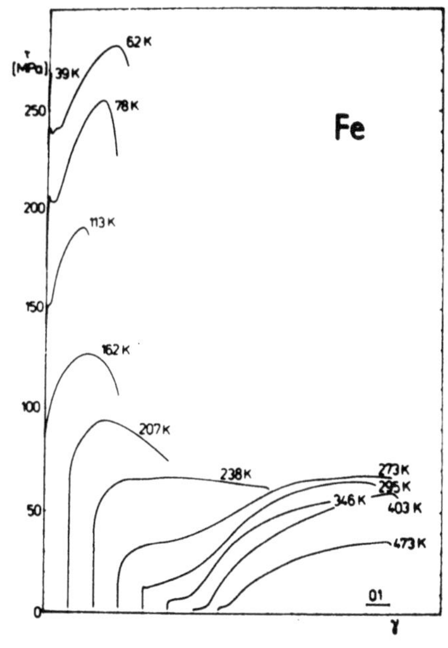


Fig. 30. Shear stress-shear strain curves of α-Fe single crystals at various temperatures. After Novák and Sesták[8]

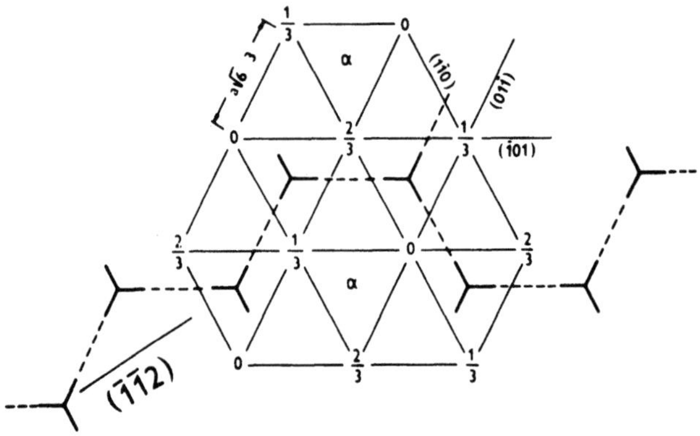


Fig. 31. A |111| projection of the bcc lattice showing the possible slip path of a right-handed screw dislocation. From Wüthrich[9]

The numbers mark the positions of atoms and their height above the drawing plane in units of the Burgers vector length $b = a\sqrt{3}/2$. The stable positions of the screw are in the centers of the triangles marked by α (with corners pointing downwards). Possible slip path parallel to planes ($\bar{1}\bar{1}2$) and ($\bar{1}01$). The slip on ($\bar{1}\bar{1}2$) from left to right is easier than in the reverse direction.

plastic behaviour of the crystal. They move by thermally activated cross slip on an atomistic scale (Fig. 31). Therefore, the screws have not well defined slip planes. They very often move parallel to planes of maximum resolved shear stress. No sharp slip lines are developed at the crystal surface.

Since <111> is also the direction for twinning shear, different stresses are necessary to deform a crystal either in tensile or in compression. If the applied shear stress counter acts a possible twinning deformation in the corresponding twin system then a higher stress is needed for plastic shear deformation than under reverse stress. Since the twinning plane is $\{\bar{1}\bar{1}2\}$ this difference occurs when the plane with maximum shear stress is $\{\bar{1}\bar{1}2\}$.

## Deformation of Polycrystals[10]

Introduction. In a polycrystal the deformation of individual grains is not independent from each other. Therefore, multiple slip must occur, at least in regions close to the grain boundaries. Also, grains have various orientations and slip will begin at different tensile stresses for differently oriented grains. The stress-strain curve resembles a stress-strain curve for edge oriented single crystals where also multiple slip occurs (Fig. 28).

The yield stress $\sigma_o$ can be related to the CRSS $\tau_o$ by a proportionality factor M called Taylor factor:

$$\sigma_o = M\tau_o \tag{22}$$

M has the approximate value 3 for fcc material. It results from a detailed theory where Taylor found at least 5 active slip systems necessary in a grain to satisfy the boundary conditions to the neighbouring grains.

Influence of Grain Size. The grain boundaries act as obstacles for dislocations and additionally increase the yield stress $\sigma_y$ with decreasing grain size D. This can be described by the Hall Petch relationship

$$\sigma_y = \sigma_o + k_y/\sqrt{D} \tag{23}$$

where $\sigma_o$ is the yield stress in the interior of the grains and $k_y$ is a constant. This type of grain size dependence is mostly found in alloys where a yield point phenomenon is observed (e.g., steel, Fig. 32). In other cases the constant $k_y$ is much smaller and the exponent for D deviates considerably from -1/2. In case of a yield phenomenon the yield spreads across the specimen as a Lüders band. In that case the following model seems to be applicable:

In a deforming grain the moving dislocations are piled up against the grain boundary to a neighbouring undeformed grain. The number n of dislocations in a pile up is proportional to $(\sigma-\sigma_o)$ D and the local concentration of shear stress, $\tau'$, at the boundary is proportional to $(\sigma-\sigma_o)$ n which gives ($\alpha$ = proportionality factor)

$$\tau' = (\sigma - \sigma_o)^2 D / (\alpha M^2 Gb) \tag{24}$$

When $\tau'$ reaches a critical value $\tau'_c$, the deformation is nucleated in the neighbouring grain and yield occurs. Thus, as a final result an expression for $k_y$ can be derived:

$$k_y = \sqrt{(\alpha M^2 \tau'_c Gb)}, \tag{25}$$

where M is the Taylor factor and $\alpha$ a constant of the order of 1. There exist also theories which explain the same law by other models which either need also pile-ups or use properties of the grain boundary. The value of $k_y$ is material dependent and can vary by one order of magnitude. For fine grained material ($D \simeq 2$ µm) the increase in yield stress can reach more than 300 MPa.

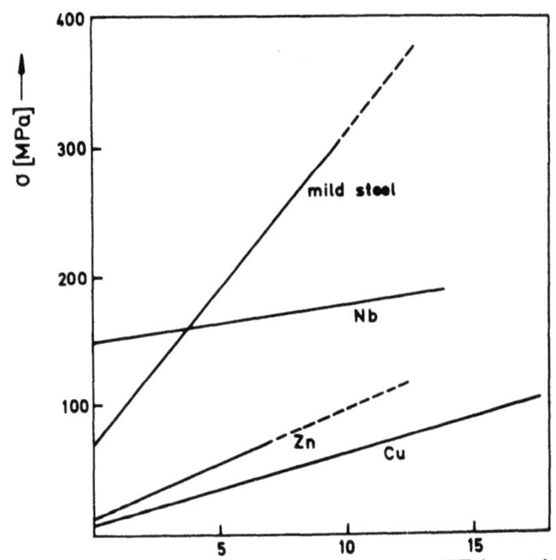


Fig. 32. Experimentally observed grain size dependence for various metals at room temperature. From Mecking[10]. Yield stress $\sigma$ versus $\sqrt{1/D}$. D = grain size.

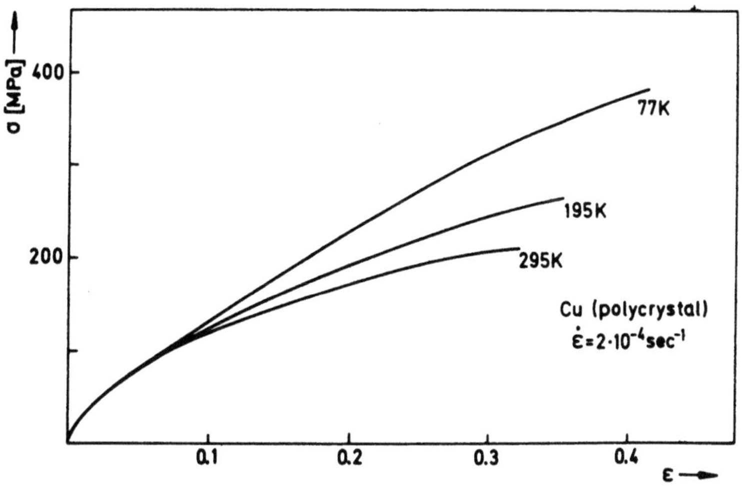


Fig. 33. Stress-strain curves of copper polycrystals of 450 μm grain size at different temperatures for a fixed strain rate. From Biermann[11]

The Stress-Strain Curves. Some analogies exist between stress-strain curves of edge oriented single crystals (e.g., <111> oriented) and polycrystalline specimens (Fig. 28). In both cases no stage I is observed. Stage II shows no longer linear hardening and continuously changes into stage III hardening (Fig. 33). The onset of stage III is temperature dependent whereas stage II (the beginning of the hardening curve) is temperature independent.

The stress-strain curve can be described phenomenologically by various equations. The most simple Ludwik equation

$$\sigma = k\ \varepsilon^n \qquad (26)$$

with n of the order of 0.5 shows deviations at small and large stresses on a $\log\sigma$ - $\log\varepsilon$ plot (Fig. 34). Neither the replacement of $\sigma$ by $(\sigma-\sigma_c)$ nor of $\varepsilon$ by $(\varepsilon-\varepsilon_c)$ gives a reasonable improvement.

A totally different equation due to Voce is

$$\sigma_s - \sigma = (\sigma_s-\sigma_o)\exp(-(\varepsilon-\varepsilon_o)/\varepsilon_c) \qquad (27)$$

where $\sigma_s$ is the saturation stress and $\varepsilon_o$ is the deformation at $\sigma_o$.

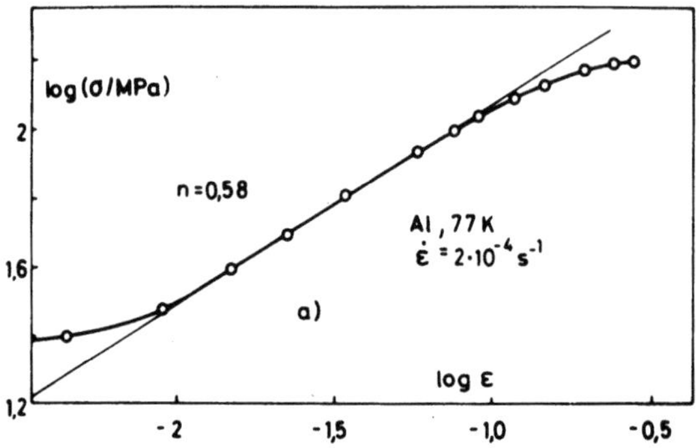


Fig. 34. Log σ vers. log ε plot of a stress-strain curve of an aluminium polycrystal at 77 K. From Mecking[10]. The curve deviates from a straight line for small and for large strains.

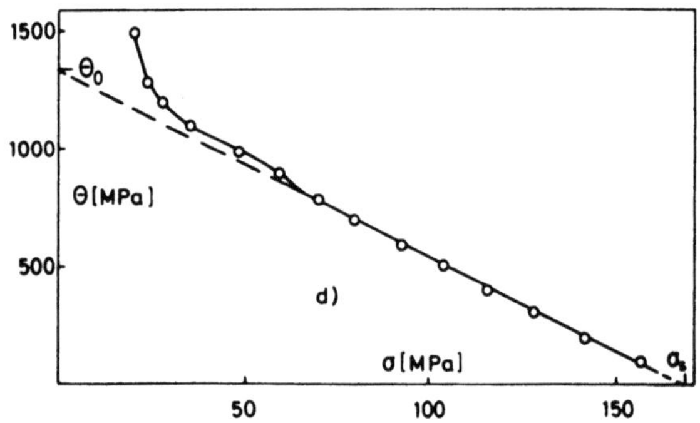


Fig. 35. Strain hardening coefficient Θ vers. flow stress. Same data as on Fig. 34. From Mecking[10]. At large stresses a straight line according to Eq. (28) is observed.

The derivative $\Theta = d\sigma/d\varepsilon$ results in

$$\Theta = (\sigma_s/\varepsilon_c)(1-\sigma/\sigma_s) \tag{28}$$

This equation fits a large range of flow stresses especially to higher stresses and $\sigma_s$ has a physical meaning. At $\sigma = \sigma_s$ the hardening has dropped to zero. A plot of $\Theta$ versus $\sigma$ results in a reasonable straight line which meets the stress axis at $\sigma_s$ (Fig. 35). For pure Al $\sigma_s \simeq 120$ MPa at room temperature and 170 MPa at 77 K. The constant $\Theta_o = \sigma_s/\varepsilon_c$ can be interpreted as approximate slope in stage II. Quantities as $\Theta_o$ and $\sigma_s$ are thought to be interpretable hy hardening theories. For example, at elevated temperatures $\sigma_s$ (for a distinct strain rate) should have the same value as the constant stress used in a creep test which shows the same steady state creep rate (see last chapter).

Development of Dislocation Structures. While dislocations are stored athermally in stage II and concentrate especially at grain boundaries, in stage III they become thermally activated (cross slip at low, climb at high temperatures) and rearrange (dynamic recovery). At large strains well defined cell structures or subgrains (the latter at higher temperatures) are observed. At large strains, depending on the type of deformation, preferred orientation of grains is obtained (rolling texture, wire drawing texture).

References


1. Hayden, H.W., Moffat, W.G. and Wulff, J., *Mechanical Behaviour.Structure and Properties of Materials*, Vol. III, J. Wiley, New York, 1965.

2. Honeycombe, R.W.K., *The Plastic Deformation of Metals*, Arnold, London, 1968.

3. Mitchell, T.E., *Progr. Mater. Res.*, 6, 119, 1964.

4. Diehl, J. and Berner, R., Temperaturabhängigkeit der Verfestigung von Kupfer-Einkristallen oberhalb 78 K, *Z. f. Metallkde.*, 51, 522, 1960.

5. Kronmüller, H., Theorie der plastischen Verformung, in *Moderne Probleme der Metallphysik*, Seeger, A., Ed., Springer-Verlag, Berlin-Heidelberg-New York, 1965, p. 126.

6. Sesták, B., Plasticity and Crystal Structure, in *Strength of Metals and Alloys*, Haasen, P., Gerold, V. and Kostorz, G., Eds., Pergamon Press, Oxford, 1980, p. 1461.

7. Sesták, B. and Seeger, A., Gleitung und Verfestigung in kubisch-raumzentrierten Metallen und Legierungen (I), (II), (III), *Z. f. Metallkde.*, 69, 195, 355 and 425, 1978.

8. Novák, V. and Sesták, B., Work hardening in single crystals of high-purity iron at intermediate temperatures, in *Proc. 4th Int. Conf. on the Strength of Metals and Alloys*, Vol. 1., Nancy, 1976, p. 156.

9. Wüthrich, C., Struktur der Kinken und Einschnürungen in Schrauben- und 71$^{\circ}$-Versetzungen in kubisch-raumzentrierten Metallen, *Phil. Mag.*, No. 2, 35, 325, 1977.

10. Mecking, H., Deformation of Polycrystals, in *Strength of Metals and Alloys*, Haasen, P., Gerold, V. and Kostorz, G., Eds., Pergamon Press, Oxford, 1980, p. 1573.

11. Biermann, M., Doctoral Thesis, RWTH Aachen, 1977. See Ref. 10.

12. McLean, D., *Mechanical Properties of Metals*, Wiley, New York, 1962.

## CREEP PROCESSES

### Introduction

Creep is usually discussed under constant stress conditions and is controlled by thermally activated slip processes of dislocations. These processes usually are connected with the interaction of the gliding dislocations with obstacles on the slip plane which have dimensions of the order of b (the Burgers vector). Examples for such obstacles are solute atoms (as Zn atoms in α brass) or immobile dislocations on other planes penetrating the slip plane (so-called forest dislocations). In addition, the nucleation of cross slip of screw dislocations and the climb of edges are also thermally activated.

Creep is mostly studied at stresses above the yield stress. Thus the material is already deforming during loading. At constant stress the material continues to deform in creep (Fig. 36). At low temperatures ($T < 0.5\ T_s$) it is hardening continuously during the creep and the creep rate decreases slowly to zero (low temperature creep). At higher temperatures recovery of the dislocation structure occurs in addition and creep is continuing until failure (high temperature creep).

### Low Temperature Creep[1,2]

A simple model for thermal activated slip leads to the equation

$$\dot{\varepsilon} = \rho_m b L \nu \qquad (29)$$

where $\dot{\varepsilon}$ is the creep rate, $\rho_m$ the density of mobile dislocations, L the

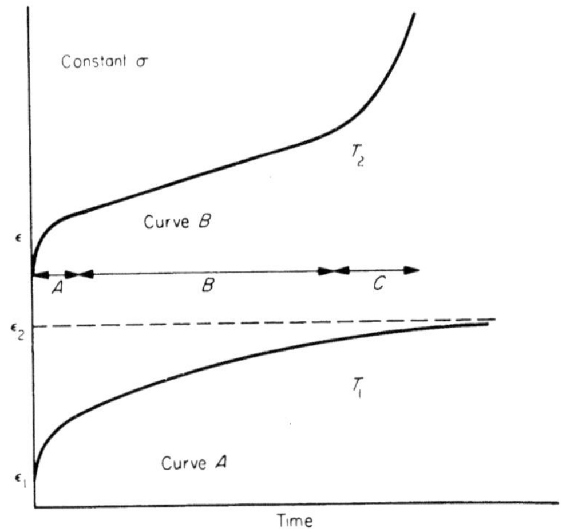


Fig. 36. Typical creep curves of a single phase material. Lower curve at $T_1 < 0.5\ T_s$, upper curve at $T_2 > 0.5\ T_s$.

average distance between obstacles and ν the frequency with which the obstacles are passed (Fig. 37). This frequency is assumed to be

$$\nu = \nu_o \exp(-U(\sigma)/kT) \tag{30}$$

with

$$U(\sigma) = U_o - (\sigma-\sigma_i)bLd \tag{31}$$

Here, $U_o$ is the energy needed to pass an obstacle. This energy is reduced by the second term in Eq. (31) which describes the work done by the applied stress σ. $\sigma_i$ is the long range internal stress and Ld the area a the dislocation sweeps during overcoming the obstacle (Fig. 37). The product bLd is usually called the activation volume v.

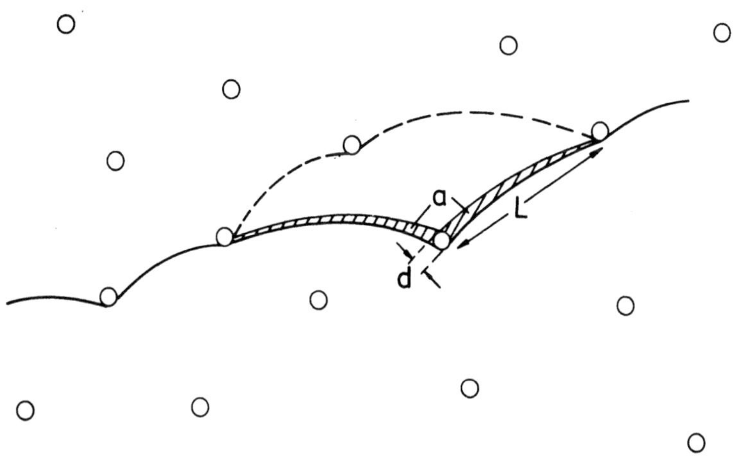


Fig. 37. Schematic sketch of a dislocation overcoming one obstcle in the slip plane. L = planar distance between obstacles, d = obstacle depth, a = Ld=activation area swept by the dislocation on overcoming the obstacle.

A further analysis is very difficult since during creep all quantities as $\rho_m$, L, $\sigma_i$ and $\nu$ will vary. In order to continue one has to simplify the problem. For low temperature creep the most important parameter seems to be the internal stress $\sigma_i$. If all other parameters are kept constant and if it is assumed that $\sigma_i$ increases with further deformation $\Delta\varepsilon$ as

$$\sigma_i = \sigma_{io} + \Theta \, \Delta \, \varepsilon \tag{32}$$

with $\Theta$ being the hardening coefficient and $\Delta\varepsilon$ the increase in strain during creep one gets the result

$$\Delta\varepsilon(t) = (kT/v\Theta) \, \ell n \, (1+t/t_o) \tag{33}$$

For $t \gg t_o$ this expression leads to a logarithmic creep law. It is observed experimentally for medium creep times. Therefore, low temperature creep is also called logarithmic creep.

## High Temperature Creep

The Typical Creep Curve. At $T > 0.5\ T_s$ three stages of creep are usually observed (Fig. 36). During primary creep (or transient creep) a dislocation substructure is developed which finally leads to secondary creep with a constant creep rate: the so-called stationary creep rate $\varepsilon_s$. Finally, in tertiary creep the sample fails. The most interesting feature is the occurrence of stationary creep where the dislocation structure does not change very much with deformation.

This typical creep behaviour is found in most single phase materials where creep is controlled by the interaction between dislocations. A different behaviour will result if the creep is controlled by the interaction between dislocations and solute atoms. In that case the dislocation can only move in dragging a cloud of solute atoms with them. This results in a quite different transient creep curve.

In two-phase (age-hardened) materials usually no steady state creep is observed. Instead, only a minimum creep rate results during deformation at constant stress and temperature.

In the following only the case of single phase material will be discussed which shows creep curves as in Fig. 36.

Steady State Creep in Pure Metals. The constant creep rate $\dot{\varepsilon}_s$ depends on stress and temperature. For low stresses ($\sigma \lesssim 10^{-3}G$) the following power law creep is found:

$$\dot{\varepsilon} = k\ \sigma^n \exp(-Q/kT). \qquad (34)$$

The exponent n is usually around 4 to 5. The activation energy Q is very accurately the same as for self diffusion[3] (Fig. 38). Therefore, self diffusion controls the creep process via dislocation climb. This leads to the phenomenological Dorn equation[5]

$$\dot{\varepsilon}_s = AD(Gb/kT)(\sigma/G)^n \qquad (35)$$

where A is a numerical constant and D is the selfdiffusion coefficient.

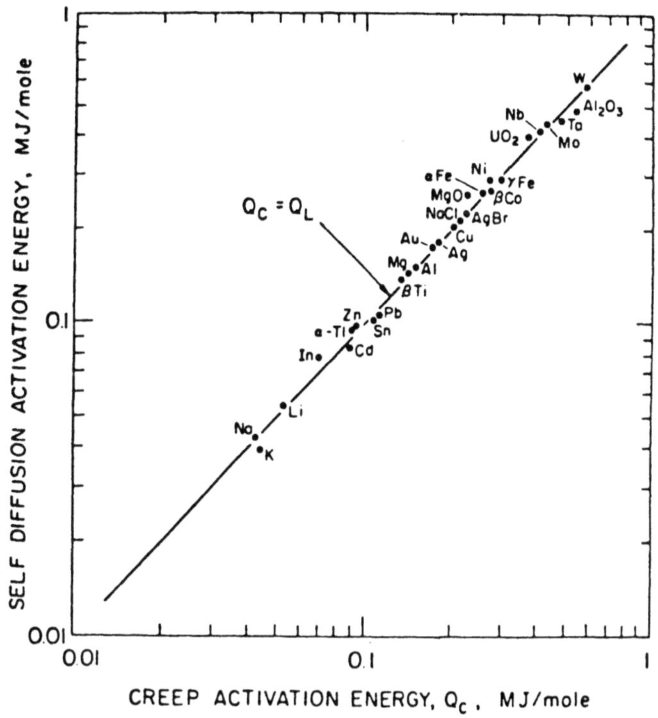


Fig. 38. Comparison of the self diffusion activation energy with the creep activation energy for various metals. After Sherby and Miller from Nix and Ilschner[4]. Both energies are identical.

<u>Dislocation Structures</u>. The creep substructure typical for steady state creep develops during primary creep (transient creep). It consists of a <u>cell structure</u> with well developed cell walls. The cell size is inversely proportional to the applied stress $\sigma$ (Fig. 39). The cell structure develops <u>back stresses</u> ($\sigma_i$) which result in backward creep after unloading (Fig. 40). This long range back stress has its origin in the cell walls where annihilation of dislocations occur via climbing. This is supposed to be the recovery mechanism controlling steady state creep.

<u>Model for Steady State Creep</u>.[4] A model has been proposed recently which assumes the existence of two processes which may control the creep

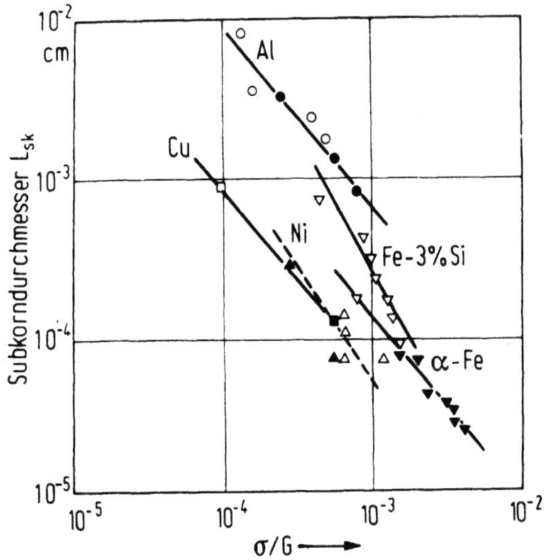


Fig. 39. Relationship between cell size L and normalized creep stress σ/G. After Ilschner.[2]

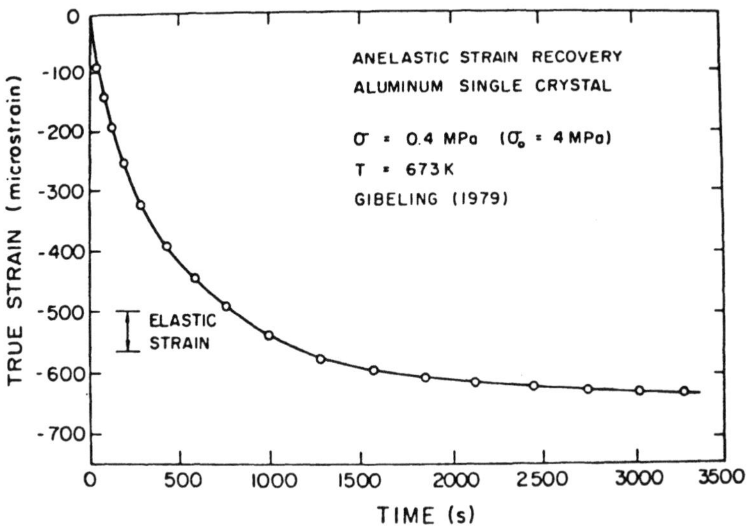


Fig. 40. Anelastic strain recovery (backflow) for an Al single crystal after unloading from $\sigma_o$ = 4 MPa to σ = 0.4 MPa at 673 K. From Nix and Ilschner.[4]

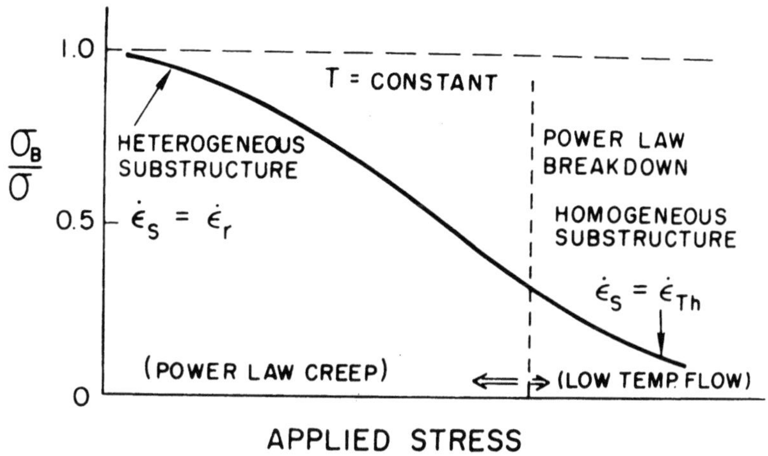


Fig. 41. Schematic sketch of the relationship between the long range internal backstress $\sigma_B$ (or $\sigma_i$) and the applied stress $\sigma$ at steady state creep. After Nix and Ilschner.[4]

rate (Fig. 41). The first process is the diffusion controlled recovery described above which may be dominant especially at low stress where power law creep is observed, Eq. (35). In this range, the magnitude of back stress ($\sigma_B$ or $\sigma_i$) is close to that of the applied stress. A cell structure is developed and the recovery creep rate $\dot{\varepsilon}_r$ controls $\dot{\varepsilon}_s$.

At higher stresses the power law breaks down. At the same time the back stress $\sigma_i$ becomes much smaller than the applied stress $\sigma$. Here, thermally activated dislocation glide similar to Eqs. (29) to (31) is assumed to take over the control of $\dot{\varepsilon}$. A more homogeneous dislocation substructure is observed.

The total steady state creep rate is given as

$$\dot{\varepsilon}_s = \alpha_1 \, D(Gb/kT)(\sigma/G)^4 + \alpha_2 \, (\sigma/G)^2 \exp(-U_o/kT) \; \{\exp \frac{v(\sigma-\sigma_i)}{kT} - 1\} \qquad (36)$$

with $\alpha_1 \simeq 3000$ and $\alpha_2 \;\; 0.5 \cdot 10^{12}$.

## Diffusional Creep

At high temperatures and low stress levels diffusional creep comes into play and may dominate because it depends only linearly on the applied stress $\sigma$. The creep is performed by vacancies which are nucleated and annihilated with different rates at different parts of the grain boundaries. The reason is the influence of the applied stress $\sigma$ on these rates. For grain boundaries perpendicular to the stress axis the nucleation rate is increased compared to the annihilation rate with the result that the mean vacancy concentration $c_v$ close to these boundaries becomes a slightly higher value compared to the concentration $c_{vo}$ close to the boundaries parallel to the stress axis:

$$c_v \simeq c_{vo} \; (1+V_a \sigma/kT) \tag{37}$$

which is of the order of 1.005 $c_{vo}$. In this equation $V_a$ is the atomic volume and k the Boltzmann constant. As a consequence a flow of vacancies occurs as shown on Fig. 6 which corresponds to a flow of atoms in the opposite direction. The resulting creep rate $\dot{\varepsilon}$ depends on the average grain diameter d and is given by

$$\dot{\varepsilon}_s = (14 \; D \; V_a/kT) \; \sigma/d^2, \tag{38}$$

where D is the bulk self diffusion coefficient. This is the so-called Nabarro-Herring creep. At lower temperatures the diffusion along the grain boundaries dominates which then is called Coble creep and has a different grain size dependence:

$$\dot{\varepsilon}_s = (45 \; \delta D_B \; V_a/kT) \; \sigma/d^3. \tag{39}$$

In this equation $D_B$ is the boundary diffusion coefficient and $\delta$ the effective width of the boundary (only the product $\delta \; D_B$ is experimentally available).

## Deformation Mechanism Maps

Which type of deformation dominates depends in a characteristic way on

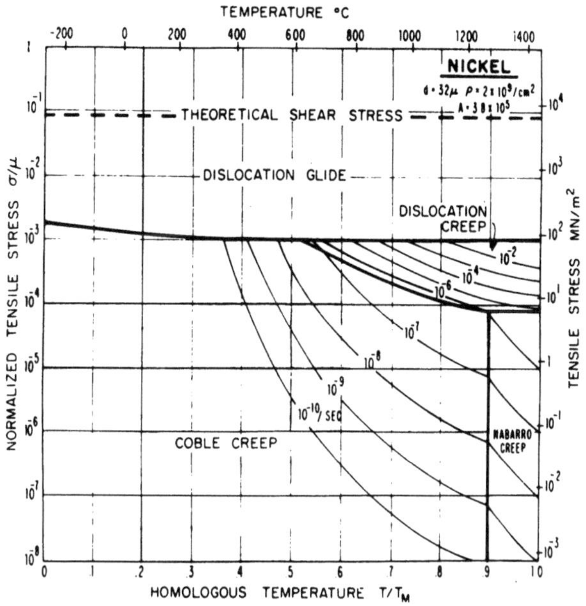


Fig. 42. Deformation mechanism map for pure polycrystalline nickel with a grain size of 32 μm. From Ashby.[6]

the applied stress and the temperature. This led Ashby to the publication of deformation mechanism maps for various materials. He plotted the normalized stress σ/G versus the normalized temperature $T/T_s$ where $T_s$ is the melting temperature. In these diagrams he was able to define areas where one deformation mechanism dominates. Fig. 42 shows the diagram for nickel as an example. Besides the different types of creep mechanisms also the normal dislocation glide is shown.

The diagram also shows the creep rates for the various mechanisms. In the lower left part at low stresses and low temperatures only elastic deformation occurs. With increasing temperature Coble creep starts to occur, first at very low creep rates which then increase with increasing stress and temperature.

At $T/T_s = 0.9$ the border to Nabarro-Herring creep is crossed. For stress levels σ/G above $10^{-4}$ the area for dislocation creep is found

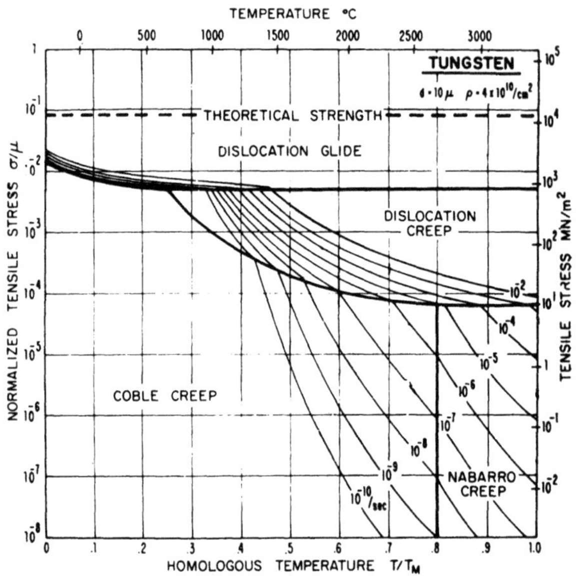


Fig. 43. Deformation mechanism map for pure tungsten with a grain size of 10 µm. From Ashby.[6]

which then changes to dislocation glide at levels $\sigma/G \simeq 10^{-3}$. The latter area is limited by the theoretical shear limit of an ideal crystal which is of the order of $\sigma/G \simeq 0.1$.

These diagrams vary from material to material as can be found by comparing Fig. 42 (nickel) with Fig. 43 (tungsten).

References

1. Honeycombe, R.W.K., *The Plastic Deformation of Metals*, Arnold, London, 1968.

2. Ilschner, B., *Hochtemperaturplastizität*, Springer-Verlag, Berlin, 1973.

3. Sherby, O.D. and Burke, P.M., Mechanical behavior of crystalline solids at elevated temperature, *Progr. Mater. Sci.*, 13, 325, 1968.

4. Nix, W.D. and Ilschner, B., Mechanisms Controlling Creep of Single Phase Metals and Alloys, in *Strength of Metals and Alloys*, Haasen, P. Gerold, V. and Kostorz, G., Eds. Pergamon Press, Oxford, 1980, 1503.

5. Mukherjee, A.K., Bird, J.E. and Dorn, J.E., *ASM Trans. Quart.*, 62, 155, 1969.

6. Ashby, M.F., A first report on deformation mechanism maps, *Acta Met.*, 20, 887, 1972.

# PLASTICITY AND NONEQUILIBRIUM THERMODYNAMICS

G.A. KLUITENBERG
Eindhoven University of Technology
Department of Mathematics and Computing Science
Eindhoven – The Netherlands

CONTENTS

ACKNOWLEDGEMENT

The author is indebted to Mrs. Elsina Baselmans-Weijers for typing and retyping the manuscript very carefully.

## 1. INTRODUCTION

In general mechanical processes in continuous media involve dissipation of heat. The exceptions to the rule are deformations of a purely elastic nature and the flow of ideal fluids.

Dissipation of heat (as a consequence of internal friction) is closely connected with the production of entropy. Hence, in general mechanical processes in continua are irreversible and it seems worthwhile to investigate these phenomena with the aid of methods based on nonequilibrium thermodynamics.

In the references 1-13 the possibility is discussed to introduce new tensorial variables in thermodynamics. It appears to be possible to derive all classical linear rheological equations (Hooke, Newton, Kelvin-Voigt, Maxwell, Jeffreys, Poynting-Thomson, Burgers). However, it also turns out to be possible to obtain significant results for nonlinear phenomena, as plasticity, without violating the basic concepts of the thermodynamics of irreversible processes.

It has been my purpose to write these notes as self contained as possible. For more thorough discussions of nonequilibrium thermodynamics the reader is referred to De Groot,[14] De Groot and Mazur[15] (in particular the first four chapters), Prigogine[16,17] and Meixner and Reik.[18] Furthermore, for the foundations of the physics of continua the reader may consult, for instance, Landau and Lifshitz,[19,20] Sedov,[21] Germain,[22] Carini[23] and, in particular for inelastic phenomena, Koiter,[24] Freudenthal and Geiringer[25] and Reiner.[26]

The basic ideas of the theory which is discussed in these notes are published twenty years ago in the journal Physica (see refs. 1-4). It should be emphasized that many authors have investigated the relation between irreversible thermodynamics and continuum mechanics. In particular the thermodynamics of relaxation phenomena has been explored in detail. We refer here to papers by Eckart,[27] Meixner,[28-33] Staverman and Schwarzl,[34,35] Polder,[36] Falk and Meixner[37] and König and Meixner.[38,39]

In these notes mainly those results obtained in refs. 1-13 are discussed which have a bearing on elastic-plastic media.

Some new results on transformations of internal variables are given in sections 7 and 8. Similar transformations, however, were discussed in refs. 40-42, which deal with dielectric and magnetic relaxation (see also ref. 43).

It is perhaps interesting to note that there is a close relation among the thermodynamic methods with the aid of which one may describe mechanical, dielectric and magnetic relaxation phenomena. Indeed, there is a striking similarity among the Poynting-Thomson equation for mechanical relaxation (see refs. 4 and 26), the Debye equation for dielectric relaxation[43] and Snoek's equation for magnetic relaxation.[43] An analogous remark holds for the Kelvin-Voigt equation[2] for mechanical relaxation and the De Groot-Mazur equations for dielectric and magnetic relaxation.[15,41]

The theory of internal thermodynamic variables is also closely related with network theory and the mathematical system theory.

In section 19 one finds some remarks which are in essence not new but which are usually not made in the context of a discussion of the possibilities to apply methods of nonequilibrium thermodynamics. Finally, some remarks on shape memory effects in section 27 were not published before.

I limited myself to the case of deformations and rotations which are small from a geometrical (kinematical) point of view. In fact the same physical ideas can be formulated for the case that the deformations and rotations are large (see also ref. 5). Singularities and phenomena at boundaries of systems are left out of consideration.

## 2. CONSERVATION LAWS AND BALANCE EQUATIONS

In the physics of continua the law of mass conservation, the equation of motion, the balance equation for energy and the balance equation for angular momentum play a fundamental role. Therefore, we shall start our discussion of plasticity and related phenomena with the formulation of these laws and with the introduction of the physical quantities which occur in them.

We shall consider a medium which may consist of several chemical components. However, diffusion phenomena will be left out of consideration. Chemical reactions are not excluded.

### 2.1. The law of conservation of mass

It will be supposed that mass is conserved. Hence, we shall assume that transformation of mass into other forms of energy does not occur.

By $\rho$ we denote the mass density (mass per unit of volume). In non-relativistic theories mass is a scalar field. Hence, we have

$$\rho = \rho(\vec{x},t) \,, \tag{2.1}$$

where $\vec{x}$ is the position vector and t is the time. We shall use an orthogonal Cartesian frame of axes which is fixed in space. Hence,

$$\vec{x} = \sum_{\alpha=1}^{3} x_\alpha \, \vec{e}_\alpha \,, \tag{2.2}$$

where $x_\alpha$ is the $\alpha$-coordinate of the position vector $\vec{x}$ and $\vec{e}_\alpha$ is the unit vector in the direction of the positive $x_\alpha$-axis. From the two preceding relations we have

$$\rho = \rho(x_1, x_2, x_3, t) \quad . \tag{2.3}$$

As is well known mass conservation is expressed by

$$\frac{\partial \rho}{\partial t} = - \operatorname{div} \rho \vec{v} \ , \tag{2.4}$$

where $\vec{v}$ is the velocity field of the medium.

The substantial (or material) derivative with respect to time d/dt is defined by

$$\frac{d}{dt} = \frac{\partial}{\partial t} + \sum_{\gamma=1}^{3} v_\gamma \frac{\partial}{\partial x_\gamma} \ , \tag{2.5}$$

where $v_\gamma$ is the $\gamma$-component of the velocity field $\vec{v}$.

From (2.4) and (2.5) one has the following alternative forms for the law of mass conservation

$$\frac{d\rho}{dt} = - \rho \operatorname{div} \vec{v} \tag{2.6}$$

and

$$\rho \frac{dv}{dt} = \operatorname{div} \vec{v} \ , \tag{2.7}$$

where

$$v = \rho^{-1} \tag{2.8}$$

is the specific volume (volume per unit of mass).

Finally, for an arbitrary quantity A (for instance, a component of a tensor field) one has

$$\rho \frac{dA}{dt} = \frac{\partial(\rho A)}{\partial t} + \mathrm{div}(\rho A\vec{v}) \ . \tag{2.9}$$

This relation follows from (2.5) and (2.6).

## 2.2. The balance equation for momentum (equation of motion)

The usual form for the equation of motion reads

$$\rho \frac{dv_\alpha}{dt} = - \sum_{\beta=1}^{3} \frac{\partial P_{\alpha\beta}}{\partial x_\beta} + \rho F_\alpha \ , \tag{2.10}$$

where $P_{\alpha\beta}$ is the mechanical pressure tensor and $\vec{F}$ is the force per unit of mass. Since $\rho\vec{v}$ is the density of momentum, $\vec{v}$, the velocity field, may be considered as the specific momentum (momentum per unit of mass). Hence, $P_{\alpha\beta}$ may be considered as the tensor of the density of the conductive flow of momentum and $\rho\vec{F}$ as the source of momentum (per unit of volume and per unit of time).

Using (2.9) one has

$$\rho \frac{dv_\alpha}{dt} = \frac{\partial}{\partial t}(\rho v_\alpha) + \sum_{\beta=1}^{3} \frac{\partial}{\partial x_\beta}(\rho\, v_\alpha v_\beta) \ . \tag{2.11}$$

Thus, one may write for (2.10)

$$\frac{\partial}{\partial t}(\rho v_\alpha) = - \sum_{\beta=1}^{3} \frac{\partial}{\partial x_\beta}(\rho\, v_\alpha v_\beta + P_{\alpha\beta}) + \rho F_\alpha \ . \tag{2.12}$$

It is seen from this equation that $\rho\, v_\alpha v_\beta$ is the tensor of the density of the convective flow of momentum and that $\rho\, v_\alpha v_\beta + P_{\alpha\beta}$ is the tensor of the total flow of momentum.

The mechanical stress tensor is defined by

$$\tau_{\alpha\beta} = - P_{\alpha\beta} \ . \tag{2.13}$$

Hence, one may also write for the equation of motion

$$\rho \frac{dv_\alpha}{dt} = \sum_{\beta=1}^{3} \frac{\partial \tau_{\alpha\beta}}{\partial x_\beta} + \rho F_\alpha \ . \tag{2.14}$$

See (2.10) and (2.13).

## 2.3. The balance equation for angular momentum

The substantial (or material) derivative with respect to time is defined by (2.5). The local derivative with respect to time $\partial/\partial t$ which occurs on the right-hand side of (2.5) is the derivative with respect to time at constant position vector $\vec{x}$. Hence, $\partial x_\alpha/\partial t = 0$. Moreover, $\partial x_\alpha/\partial x_\gamma = \delta_{\alpha\gamma}$, where the unit matrix $\delta_{\alpha\beta}$ is defined by

$$\delta_{\alpha\beta} = \begin{cases} 1 & \text{for } \alpha = \beta \ , \\ 0 & \text{for } \alpha \neq \beta \ . \end{cases} \tag{2.15}$$

Thus, it follows from (2.5) that

$$\frac{dx_\alpha}{dt} = v_\alpha \ . \tag{2.16}$$

Next, we consider the tensor field $x_\alpha v_\beta - x_\beta v_\alpha$. This antisymmetric tensor field is the specific external angular momentum with respect to the origin of the frame of axes. Using (2.16) we get with the aid of the equations of motion in the form (2.10) and the rules for partial differentiation

$$\rho \frac{d}{dt} (x_\alpha v_\beta - x_\beta v_\alpha) = \rho\left(x_\alpha \frac{dv_\beta}{dt} - x_\beta \frac{dv_\alpha}{dt}\right)$$

$$= x_\alpha\left(- \sum_{\gamma=1}^{3} \frac{\partial P_{\beta\gamma}}{\partial x_\gamma} + \rho F_\beta\right) - x_\beta\left(- \sum_{\gamma=1}^{3} \frac{\partial P_{\alpha\gamma}}{\partial x_\gamma} + \rho F_\alpha\right) =$$

$$= - \sum_{\gamma=1}^{3} \frac{\partial}{\partial x_\gamma} (x_\alpha P_{\beta\gamma} - x_\beta P_{\alpha\gamma}) +$$

$$+ \sum_{\gamma=1}^{3} \left( P_{\beta\gamma} \frac{\partial x_\alpha}{\partial x_\gamma} - P_{\alpha\gamma} \frac{\partial x_\beta}{\partial x_\gamma} \right) + \rho (x_\alpha F_\beta - x_\beta F_\alpha) ,$$

or, since $\partial x_\alpha / \partial x_\beta = \delta_{\alpha\beta}$,

$$\rho \frac{d}{dt} (x_\alpha v_\beta - x_\beta v_\alpha) = - \sum_{\gamma=1}^{3} \frac{\partial}{\partial x_\gamma} (x_\alpha P_{\beta\gamma} - x_\beta P_{\alpha\gamma}) + \\ + P_{\beta\alpha} - P_{\alpha\beta} + \rho (x_\alpha F_\beta - x_\beta F_\alpha) . \tag{2.17}$$

The tensor $x_\alpha P_{\beta\gamma} - x_\beta P_{\alpha\gamma}$ in (2.17) is the (third order) tensor of the density of the conductive flow of external angular momentum. We shall assume that there are no internal (or intrinsic) angular momenta, couple stresses etc. and that the only source of angular momentum is the torque due to the force per unit of mass $\vec{F}$. Hence, we must have

$$P_{\alpha\beta} = P_{\beta\alpha} , \tag{2.18}$$

as may be seen from (2.17). This relation expresses the symmetry of the mechanical pressure tensor. It is obvious that the mechanical stress tensor, defined by (2.13), is also symmetric. Thus,

$$\tau_{\alpha\beta} = \tau_{\beta\alpha} . \tag{2.19}$$

Hence, the balance equation for the (external) angular momentum reads

$$\rho \frac{d}{dt} (x_\alpha v_\beta - x_\beta v_\alpha) = - \sum_{\gamma=1}^{3} \frac{\partial}{\partial x_\gamma} (x_\alpha P_{\beta\gamma} - x_\beta P_{\alpha\gamma}) + \\ + \rho (x_\alpha F_\beta - x_\beta F_\alpha) . \tag{2.20}$$

It should be stressed that this balance equation is a consequence of the equations of motion and of the physical assumption (2.18) (or (2.19)).

## 2.4. The balance equation for the energy

The energy is the sum of the internal and kinetic energies. Hence, we have

$$e = u + \tfrac{1}{2}\vec{v}^2 , \tag{2.21}$$

where e is the specific energy (energy per unit of mass), u is the specific internal energy and $\frac{1}{2}\vec{v}^2$ is the specific kinetic energy.

It will be assumed that the only source of energy is the Newtonian work done by the force $\vec{F}$ (per unit of mass) which occurs in the equations of motion. This means that polarization and magnetization phenomena due to an electromagnetic field (if such a field occurs) are left out of consideration. Hence, one has

$$\rho \frac{de}{dt} = - \operatorname{div} \vec{J}^{(e)} + \rho\vec{v}\cdot\vec{F} , \tag{2.22}$$

where $\vec{J}^{(e)}$ is the density of the conductive flow of energy. Using (2.9), one may also write for the preceding equation

$$\frac{\partial}{\partial t} (\rho e) = - \operatorname{div}(\rho e\vec{v} + \vec{J}^{(e)}) + \rho\vec{v}\cdot\vec{F} , \tag{2.23}$$

where $\rho e$ is the energy density (energy per unit of volume) and $\rho e\vec{v}$ is the density of the convective flow of energy.

Let us consider a volume V bounded by a closed surface $\Omega$ which is fixed in space. We integrate both sides of (2.23) over the volume V. We then get with the aid of the divergence theorem

$$\frac{d}{dt} \int_V \rho e dV = - \int_\Omega \rho e(\vec{v}\cdot\vec{n}) d\Omega - \int_\Omega \vec{J}^{(e)}\cdot\vec{n} d\Omega + \int_V \rho\vec{v}\cdot\vec{F} dV ,$$

where $\vec{n}$ is the outward unit normal to the surface element $d\Omega$. This relation expresses that the rate of change of energy in V is caused by convective and conductive flows of energy through $\Omega$ (the first and second integrals on the right-hand side) and by the work done in V on the medium by the force $\vec{F}$.

## 2.5. The first law of thermodynamics (balance equation for internal energy)

Let us multiply both sides of the equation of motion (2.14) by $v_\alpha$ and sum over $\alpha$. One then gets by virtue of (2.5)

$$\rho \frac{d}{dt} (\tfrac{1}{2}\vec{v}^2) = \sum_{\alpha,\beta=1}^{3} v_\alpha \frac{\partial \tau_{\alpha\beta}}{\partial x_\beta} + \rho \vec{v}\cdot\vec{F} . \tag{2.24}$$

Hence, using (2.19) we have

$$\rho \frac{d}{dt} (\tfrac{1}{2}\vec{v}^2) = \sum_{\beta=1}^{3} \frac{\partial}{\partial x_\beta}\left(\sum_{\alpha=1}^{3} v_\alpha \tau_{\alpha\beta}\right) -$$

$$- \tfrac{1}{2} \sum_{\alpha,\beta=1}^{3} \tau_{\alpha\beta}\left(\frac{\partial v_\alpha}{\partial x_\beta} + \frac{\partial v_\beta}{\partial x_\alpha}\right) + \rho \vec{v}\cdot\vec{F} . \tag{2.25}$$

We shall define the heat flow $\vec{J}^{(q)}$ by (see also (2.13))

$$J_\beta^{(q)} = J_\beta^{(e)} - \sum_{\alpha=1}^{3} v_\alpha P_{\alpha\beta} = J_\beta^{(e)} + \sum_{\alpha=1}^{3} v_\alpha \tau_{\alpha\beta} . \tag{2.26}$$

Next, we subtract (2.25) from (2.22). We then obtain with the aid of the relations (2.21) and (2.26)

$$\rho \frac{du}{dt} = - \operatorname{div} \vec{J}^{(q)} + \tfrac{1}{2} \sum_{\alpha,\beta=1}^{3} \tau_{\alpha\beta}\left(\frac{\partial v_\alpha}{\partial x_\beta} + \frac{\partial v_\beta}{\partial x_\alpha}\right) . \tag{2.27}$$

This equation is a balance equation for the internal energy. The heat flow $\vec{J}^{(q)}$ is the conductive flow of internal energy and the second term on the right-hand side of (2.27) (containing the mechanical stress tensor $\tau_{\alpha\beta}$) is the source of internal energy (per unit of volume and per unit of time).

One may consider (2.25) as a balance equation for the kinetic energy. It is then seen that $-\Sigma_{\alpha=1}^{3} v_\alpha \tau_{\alpha\beta}$ is the conductive flow of the kinetic energy and that $\frac{1}{2}\Sigma_{\alpha,\beta=1}^{3} \tau_{\alpha\beta}(\partial v_\alpha/\partial x_\beta + \partial v_\beta/\partial x_\alpha)$ is the amount of energy which transforms per unit of volume and per unit of time from kinetic into internal energy.

Finally, we note that one may also write for (2.27)

$$\frac{\partial}{\partial t}(\rho u) = -\operatorname{div}(\rho u\vec{v} + \vec{J}^{(q)}) + \frac{1}{2}\sum_{\alpha,\beta=1}^{3} \tau_{\alpha\beta}\left(\frac{\partial v_\alpha}{\partial x_\beta} + \frac{\partial v_\beta}{\partial x_\alpha}\right), \tag{2.28}$$

where we used (2.9). Hence, $\rho u\vec{v}$ is the convective flow of internal energy.

It should be emphasized that the explicit forms for the first law discussed above are not influenced by the internal processes in a medium which will be discussed in the following. The internal processes manifest themselves, for instance, by the fact that the internal energy may be considered not only as a function of the temperature, the strain tensor or the specific volume, but also of certain parameters which describe the "internal state" of the medium.

## 3. SMALL DEFORMATIONS AND ROTATIONS AND THE STRAIN TENSOR

Let us choose an arbitrary time $t_0$. The state of the medium at this time will be referred to as the reference state. The position vector of a generic point P of the medium in the reference state will be denoted by $\vec{\xi}$. The displacement vector of P is defined by

$$\vec{u} = \vec{x} - \vec{\xi} \ . \tag{3.1}$$

We note that

$$\vec{x}(\vec{\xi}, t_0) = \vec{\xi} \ , \quad \vec{u}(\vec{\xi}, t_0) = \vec{0} \ . \tag{3.2}$$

We shall assume that the gradient of the displacement field is small. This implies that the deformations and the rotations are supposed to be small. However, the translations and the velocity of the medium may be large. Therefore, in general the substantial derivative with respect to time cannot be replaced by the local derivative with respect to time.

If the strains and the rotations are small the strain tensor $\varepsilon_{\alpha\beta}$ can be defined by

$$\varepsilon_{\alpha\beta} = \tfrac{1}{2}\left(\frac{\partial u_\alpha}{\partial x_\beta} + \frac{\partial u_\beta}{\partial x_\alpha}\right) . \tag{3.3}$$

We shall also call $\varepsilon_{\alpha\beta}$ the tensor of total strain. One has

$$\frac{d\varepsilon_{\alpha\beta}}{dt} = \tfrac{1}{2}\left(\frac{\partial v_\alpha}{\partial x_\beta} + \frac{\partial v_\beta}{\partial x_\alpha}\right) . \tag{3.4}$$

We note that

$$\varepsilon_{\alpha\beta} = \varepsilon_{\beta\alpha} \tag{3.5}$$

Throughout these lectures we shall limit ourselves for simplicity to small deformations; more precisely, deformations which are small from a kinematical (or geometrical) point of view. It should be emphasized, however, that the physical ideas which are developed here do not depend on the assumption that the deformations are small. The same physical ideas can also be formulated for the case the deformations are large from a kinematical point of view.

We shall also introduce the notion of strains which are small from a physical point of view. This notion will be discussed in section 23 on linear equations of state etc.

Using the expression (3.3) for the substantial time derivative of the strain tensor, one has from (2.27) the form

$$\rho \frac{du}{dt} = - \operatorname{div} \vec{J}^{(q)} + \sum_{\alpha,\beta=1}^{3} \tau_{\alpha\beta} \frac{d\varepsilon_{\alpha\beta}}{dt} \tag{3.6}$$

for the first law of thermodynamics.

## 4. DEVIATOR AND SCALAR PART OF A TENSOR

Suppose that $A_{\alpha\beta}$ is a tensor field of order 2. We denote by A the scalar part of $A_{\alpha\beta}$ which is defined by

$$A = \frac{1}{3} \sum_{\gamma=1}^{3} A_{\gamma\gamma} \tag{4.1}$$

and by $\tilde{A}_{\alpha\beta}$ the deviator of $A_{\alpha\beta}$ which is defined by

$$\tilde{A}_{\alpha\beta} = A_{\alpha\beta} - \frac{1}{3} \delta_{\alpha\beta} \sum_{\gamma=1}^{3} A_{\gamma\gamma} . \tag{4.2}$$

We note that 3A is called the trace of $A_{\alpha\beta}$.

From these definitions it follows that

$$A_{\alpha\beta} = \tilde{A}_{\alpha\beta} + A\delta_{\alpha\beta} , \tag{4.3}$$

$$\sum_{\gamma=1}^{3} \tilde{A}_{\gamma\gamma} = 0 \tag{4.4}$$

and that

$$\tilde{A}_{\alpha\beta} = \tilde{A}_{\beta\alpha} \quad \text{if and only if} \quad A_{\alpha\beta} = A_{\beta\alpha} . \tag{4.5}$$

From the definition (2.15) of $\delta_{\alpha\beta}$ it is seen that

$$\sum_{\alpha,\beta=1}^{3} \delta_{\alpha\beta}\delta_{\alpha\beta} = 3 \tag{4.6}$$

and with the aid of (4.4) we get

$$\sum_{\alpha,\beta=1}^{3} \delta_{\alpha\beta}\tilde{A}_{\alpha\beta} = 0 . \tag{4.7}$$

Hence, if we have two tensor fields $A_{\alpha\beta}$ and $B_{\alpha\beta}$ one has from (4.3), (4.6) and (4.7) the relation

$$\sum_{\alpha,\beta=1}^{3} A_{\alpha\beta}B_{\alpha\beta} = \sum_{\alpha,\beta=1}^{3} (\tilde{A}_{\alpha\beta} + A\delta_{\alpha\beta})(\tilde{B}_{\alpha\beta} + B\delta_{\alpha\beta}) =$$

$$= \sum_{\alpha,\beta=1}^{3} \tilde{A}_{\alpha\beta}\tilde{B}_{\alpha\beta} + 3AB . \tag{4.8}$$

In particular, we obtain from (4.1), (3.3) and (3.4) the relations

$$\varepsilon = \frac{1}{3} \operatorname{div} \vec{u} \tag{4.9}$$

and

$$\frac{d\varepsilon}{dt} = \frac{1}{3} \operatorname{div} \vec{v} . \tag{4.10}$$

For the deviator $\tilde{\varepsilon}_{\alpha\beta}$ of the strain tensor we get from (3.3), (4.3) and (4.9) the expression

$$\tilde{\varepsilon}_{\alpha\beta} = \varepsilon_{\alpha\beta} - \varepsilon\delta_{\alpha\beta} = \frac{1}{2}\left(\frac{\partial u_\alpha}{\partial x_\beta} + \frac{\partial u_\beta}{\partial x_\alpha} - \frac{2}{3}\delta_{\alpha\beta} \operatorname{div} \vec{u}\right) . \tag{4.11}$$

In an analogous manner one finds from (3.4), (4.3) and (4.10)

$$\frac{d\tilde{\varepsilon}_{\alpha\beta}}{dt} = \frac{1}{2}\left(\frac{\partial v_\alpha}{\partial x_\beta} + \frac{\partial v_\beta}{\partial x_\alpha} - \frac{2}{3}\delta_{\alpha\beta} \operatorname{div} \vec{v}\right) . \tag{4.12}$$

From the form (2.7) for the law of mass conservation and (4.10) we obtain

$$\rho \frac{dv}{dt} = 3 \frac{d\varepsilon}{dt} . \tag{4.13}$$

In a first approximation, in (4.13) $\rho$ may be replaced by $1/v_0$, where $v_0$ is the specific volume in the reference state. This leads to

$$v = v_0(1 + 3\varepsilon) . \tag{4.14}$$

Using the identity (4.8) one may write for the first law (3.6)

$$\rho \frac{du}{dt} = - \operatorname{div} \vec{J}^{(q)} + \sum_{\alpha,\beta=1}^{3} \tilde{\tau}_{\alpha\beta} \frac{d\tilde{\varepsilon}_{\alpha\beta}}{dt} + 3\tau \frac{d\varepsilon}{dt} , \tag{4.15}$$

or, using (4.13),

$$\rho \left(\frac{du}{dt} + P \frac{dv}{dt}\right) = - \operatorname{div} \vec{J}^{(q)} + \sum_{\alpha,\beta=1}^{3} \tilde{\tau}_{\alpha\beta} \frac{d\tilde{\varepsilon}_{\alpha\beta}}{dt} , \tag{4.16}$$

where

$$P = - \tau \tag{4.17}$$

is the scalar part of the pressure tensor.

## 5. ENTROPY, TEMPERATURE AND THERMODYNAMIC VARIABLES

In section 2 we introduced several general notions and laws which play an important role in the physics of continua. However, we introduced more variables than equations. For instance, stress-strain relations were left out of consideration. Nevertheless, the variables discussed in section 2 do not form a complete set to describe the (local) state of a system. For instance, we defined the heat flow (as the conductive flow of internal energy (see (2.26) and (2.27))) but the notion of temperature was not introduced.

A very important quantity is the entropy. By $s$ we denote the entropy per unit of mass. It is assumed that $s$ depends on the internal energy $u$ and on a set of scalar, vector and tensor fields denoted by $\Xi$. Hence,

$$s = s(u,\Xi) \ . \tag{5.1}$$

Next, we define the (absolute) temperature $T$ by

$$\frac{1}{T} = \frac{\partial}{\partial u} s(u,\Xi) \ . \tag{5.2}$$

It should be emphasized that it is supposed that the notion of entropy has a meaning not only in thermodynamic equilibrium but also outside of thermodynamic equilibrium (see also sections 19 and 22). Hence, by (5.2) also the temperature is defined both inside and outside of thermodynamic equilibrium.

Which quantities belong to the set $\Xi$ ? The most simple case is that $\Xi$ only consists of the specific volume $v$. This assumption characterizes a simple (ideal or viscous) fluid. Thus,

$$s = s(u,v) \quad \text{(fluid)}. \tag{5.3}$$

A thermoelastic medium can be characterized by the two relations

$$\begin{cases} s = s(u,\varepsilon_{\alpha\beta}) \,, & (5.4) \\ & \text{(thermoelastic medium)} \\ \tau_{\alpha\beta} = \rho T \dfrac{\partial}{\partial \varepsilon_{\alpha\beta}} s(u,\varepsilon_{\alpha\beta}) \,. & (5.5) \end{cases}$$

Hence, for a thermoelastic medium $\Xi$ only contains the strain tensor $\varepsilon_{\alpha\beta}$ (see (5.1) and (5.4)). Moreover, the mechanical stress tensor $\tau_{\alpha\beta}$ (which occurs in the equations of motion (2.14) and in the first law (2.27) (or (3.6))) is related by (5.5) to the partial derivatives of the entropy with respect to the components of the strain tensor.

More generally, a subset of all quantities on which the entropy depends is formed by quantities of which the substantial (material) derivatives with respect to time occur in the first law. See (3.6) and (4.16). The entropy will certainly depend on the specific internal energy u and on the specific volume v (or on $\varepsilon$, the scalar part of the strain tensor $\varepsilon_{\alpha\beta}$ (see (4.14))). There are substances of which the entropy depends on the deviator $\tilde{\varepsilon}_{\alpha\beta}$ of the strain tensor (for instance thermoelastic media) and there are substances of which s does not depend on $\tilde{\varepsilon}_{\alpha\beta}$ (for instance fluids). Hence, $\Xi$ will certainly contain v and possibly $\tilde{\varepsilon}_{\alpha\beta}$.

If a medium is influenced by an electric field and if it is polarizable, s depends on the polarization and the substantial time derivative of the polarization also occurs in the first law. Hence, in this case $\Xi$ also contains the polarization. An analogous remark holds for the magnetization. See, for instance, p. 386 of De Groot and Mazur.[15]

The entropy, variables on which the entropy depends, quantities which are obtained by partial differentiation of the entropy (as the temperature, see (5.2)) and functions of these quantities (as the free energy) will be called thermodynamic variables. Thermodynamic variables on which the entropy depends and of which the substantial time derivative occurs in the first law may be called external thermodynamic variables, because the values of these parameters can be prescribed by external influences.

## 6. INTERNAL THERMODYNAMIC VARIABLES

In the preceding section we discussed two new notions: entropy and temperature. The set of variables which we have now introduced (strain tensor, internal energy, specific volume, entropy and temperature) is sufficient to characterize the state of a thermoelastic medium or of a fluid. However, for the description of the processes which may occur in these media we still need laws as the Duhamel-Neumann law, Hooke's law or Newton's law for viscous fluid flow.

The set of variables is not complete if more complicated phenomena occur as chemical reactions, anelastic or plastic strains and dielectric and magnetic relaxation.

Let us consider a fluid mixture which consists of n chemical components. One then has for the mass density $\rho$ of the mixture $\rho = \Sigma_{k=1}^{n} \rho^{(k)}$, where $\rho^{(k)}$ is the mass density of chemical component k. The concentration $c^{(k)}$ of the chemical component k is defined by $c^{(k)} = \rho^{(k)}/\rho$. We note that it follows from these definitions that $\Sigma_{k=1}^{n} c^{(k)} = 1$. For such mixtures it is assumed that

$$s = s(u,v,c^{(1)},c^{(2)},\ldots,c^{(n-1)}) . \tag{6.1}$$

See, for instance, p. 23 of De Groot and Mazur.[15] Hence, the (local) thermodynamic state of a fluid mixture is completely specified by u, v and the additional scalar (macroscopic) thermodynamic variables $c^{(1)}$, $c^{(2)},c^{(3)},\ldots,c^{(n-1)}$.

In the Debye theory[15,43] for dielectric relaxation in polar liquids it is assumed that the polarization $\vec{P}$ is additively composed of a part $\vec{P}^{(0)}$ which is due to a deformation of the molecules and a part $\vec{P}^{(1)}$ which is due to the orientation of the polar molecules in an electric field. Hence, $\vec{P} = \vec{P}^{(0)} + \vec{P}^{(1)}$. Since $\vec{P}^{(0)}$ and $\vec{P}^{(1)}$ are due to different (microscopic) physical phenomena it is natural to suppose that both $\vec{P}^{(0)}$ and $\vec{P}^{(1)}$ are thermodynamic variables. Hence,

$$s = s(u,v,\vec{P}^{(0)},\vec{P}^{(1)}) , \tag{6.2}$$

or, for instance,

$$s = s(u,v,\vec{P},\vec{P}^{(1)}) , \tag{6.3}$$

since $\vec{P}^{(0)} = \vec{P} - \vec{P}^{(1)}$. It may be shown that (6.2) (or (6.3)) leads to the relaxation equation of Debye if methods of nonequilibrium thermodynamics are applied. Snoek's equation for magnetic relaxation may be derived in a similar way (see the references 40-43). As noted in the preceding section the polarization is a thermodynamic state variable in the case of media which are polarizable. In order to discuss dielectric relaxation, we introduce an additional vectorial thermodynamic variable $\vec{P}^{(1)}$ as may be seen from (6.3).

If both thermoelastic and inelastic strains occur (described by the strain tensors $\varepsilon_{\alpha\beta}^{(0)}$ and $\varepsilon_{\alpha\beta}^{(1)}$, respectively) one usually assumes that

$$\varepsilon_{\alpha\beta} = \varepsilon_{\alpha\beta}^{(0)} + \varepsilon_{\alpha\beta}^{(1)} , \tag{6.4}$$

where $\varepsilon_{\alpha\beta}$ is the tensor of the total strain (see also sections 7 and 8). Since thermoelastic and inelastic strains are due to different (microscopic) phenomena, we shall assume that

$$s = s(u,\varepsilon_{\alpha\beta}^{(0)},\varepsilon_{\alpha\beta}^{(1)}) , \tag{6.5}$$

or, by virtue of (6.4),

$$s = s(u,\varepsilon_{\alpha\beta},\varepsilon_{\alpha\beta}^{(1)}) . \tag{6.6}$$

Again, in (6.6) $\varepsilon^{(1)}_{\alpha\beta}$ is an additional tensorial thermodynamic variable which is used to describe the state of the medium.

Such additional variables on which the entropy depends and of which the substantial time derivatives do not occur in the first law may be called internal thermodynamic variables.

If extra variables are introduced Lebon and Rubi[44] speak of extended thermodynamics. Müller,[45] Lebon,[46] Lebon, Jou and Casas-Vazquez,[47] Jou, Casas-Vazquez and Lebon[48] and Gyarmati,[49] for instance, take as additional quantities the heat flux and the viscous stress. In these notes, however, we introduce extra variables of the nature discussed above (inelastic strains etc.).

## 7. HIDDEN TENSORIAL VARIABLES AS SPLITTING OPERATORS FOR THE STRAIN TENSOR

Let us now consider a medium for which the entropy depends on the internal energy u, the tensor of total strain $\varepsilon_{\alpha\beta}$ and some tensorial variable $Z_{\alpha\beta}$. Hence, we assume that

$$s = s(u, \varepsilon_{\alpha\beta}, Z_{\alpha\beta}) \ . \tag{7.1}$$

There are many substances for which the assumption that $\tilde{\varepsilon}_{\alpha\beta}$ is a thermodynamic state variable is not correct. Examples are fluids and Maxwell media. However, it will be shown in section 15 that such substances, which we shall term media without shape memory, can be considered as limiting cases of media for which relations of the type (7.1) hold. We shall assume that the tensorial variable $Z_{\alpha\beta}$ is a macroscopic quantity which we need to give a complete description of the state of the medium. We shall not specify the physical nature of $Z_{\alpha\beta}$. Therefore, we shall call $Z_{\alpha\beta}$ a hidden tensorial variable. We shall only assume that $Z_{\alpha\beta}$ influences the mechanical properties of the medium (this terminology will be explained in the following) and that it is a symmetric tensor field, i.e.

$$Z_{\alpha\beta} = Z_{\beta\alpha} \ . \tag{7.2}$$

Let us define the tensor fields $\tau^{(eq)}_{\alpha\beta}$ and $G_{\alpha\beta}$ by

$$\tau^{(eq)}_{\alpha\beta} = - \rho T \frac{\partial}{\partial \varepsilon_{\alpha\beta}} s(u, \varepsilon_{\alpha\beta}, Z_{\alpha\beta}) \tag{7.3}$$

and

$$G_{\alpha\beta} = \rho T \frac{\partial}{\partial Z_{\alpha\beta}} s(u,\varepsilon_{\alpha\beta},Z_{\alpha\beta}) \ . \tag{7.4}$$

With the aid of the four preceding relations and the definition (5.2) of the temperature, we obtain for the differential ds of s

$$Tds = du - v \sum_{\alpha,\beta=1}^{3} \tau_{\alpha\beta}^{(eq)} d\varepsilon_{\alpha\beta} + v \sum_{\alpha,\beta=1}^{3} G_{\alpha\beta} dZ_{\alpha\beta} \ , \tag{7.5}$$

where we used the relation $\rho = v^{-1}$ (see (2.8)). An equation of the type (7.5) is called Gibbs relation. We have

$$\tau_{\alpha\beta}^{(eq)} = \tau_{\beta\alpha}^{(eq)} \tag{7.6}$$

and

$$G_{\alpha\beta} = G_{\beta\alpha} \ . \tag{7.7}$$

It is seen from (7.5) that $\tau_{\alpha\beta}^{(eq)}$ has the dimension of stress. We shall call $\tau_{\alpha\beta}^{(eq)}$ the equilibrium-stress tensor.

Let us introduce the specific thermodynamic potential g which is defined by

$$g = u - Ts - v \sum_{\alpha,\beta=1}^{3} \tau_{\alpha\beta}^{(eq)} \varepsilon_{\alpha\beta} \ . \tag{7.8}$$

Using the Gibbs relation (7.5) one obtains

$$dg = - sdT - \sum_{\alpha,\beta=1}^{3} \varepsilon_{\alpha\beta} d(v\tau_{\alpha\beta}^{(eq)}) - v \sum_{\alpha,\beta=1}^{3} G_{\alpha\beta} dZ_{\alpha\beta} \ . \tag{7.9}$$

Hence,

$$g = g(T,v\tau_{\alpha\beta}^{(eq)},Z_{\alpha\beta}) \tag{7.10}$$

and

$$\varepsilon_{\alpha\beta} = - \frac{\partial}{\partial(v\tau^{(eq)}_{\alpha\beta})} g(T, v\tau^{(eq)}_{\alpha\beta}, Z_{\alpha\beta}) \ . \tag{7.11}$$

We shall say that the hidden tensorial variable $Z_{\alpha\beta}$ influences the mechanical properties of the medium if g is not the sum of two functions $g^{(1)}$ and $g^{(2)}$ where $g^{(1)}$ depends on T and $v\tau^{(eq)}_{\alpha\beta}$ and $g^{(2)}$ depends on T and $Z_{\alpha\beta}$. Hence, this means that the right-hand side of (7.11) depends not only on T and $v\tau^{(eq)}_{\alpha\beta}$ but also on $Z_{\alpha\beta}$. Thus, we have from (7.11) that there is a tensor-valued function $\varphi_{\alpha\beta}$ such that

$$\varepsilon_{\alpha\beta} = \varphi_{\alpha\beta}(T, v\tau^{(eq)}_{\alpha\beta}, Z_{\alpha\beta}) \ . \tag{7.12}$$

Next, we introduce the notation

$$\tau^{(eq)}_{(0)\alpha\beta} = 0 \ . \tag{7.13}$$

Furthermore, we introduce an arbitrary (but fixed) value $T_0$ for the temperature. We define the function $\varphi^{(1)}_{\alpha\beta}$ of $Z_{\alpha\beta}$ by

$$\varphi^{(1)}_{\alpha\beta}(Z_{\alpha\beta}) = \varphi_{\alpha\beta}(T_0, v_0\tau^{(eq)}_{(0)\alpha\beta}, Z_{\alpha\beta}) \ , \tag{7.14}$$

where $v_0$ is an arbitrary but fixed value for the specific volume, for instance, the specific volume in the reference state. We may define the strains $\varepsilon^{(0)}_{\alpha\beta}$ and $\varepsilon^{(1)}_{\alpha\beta}$ by

$$\varepsilon^{(1)}_{\alpha\beta} = \varphi^{(1)}_{\alpha\beta}(Z_{\alpha\beta}) \ , \tag{7.15}$$

$$\varepsilon^{(0)}_{\alpha\beta} = \varepsilon_{\alpha\beta} - \varphi^{(1)}_{\alpha\beta}(Z_{\alpha\beta}) \ , \tag{7.16}$$

or, using (7.12) and (7.14),

$$\varepsilon^{(1)}_{\alpha\beta} = \varphi_{\alpha\beta}(T_0, v_0\tau^{(eq)}_{(0)\alpha\beta}, Z_{\alpha\beta}) \ , \tag{7.17}$$

$$\varepsilon_{\alpha\beta}^{(0)} = \varphi_{\alpha\beta}(T, v\tau_{\alpha\beta}^{(eq)}, Z_{\alpha\beta}) - \varphi_{\alpha\beta}(T_0, v_0\tau_{(0)\alpha\beta}^{(eq)}, Z_{\alpha\beta}) . \tag{7.18}$$

Finally, we note that it follows from the two preceding equations that

$$\varepsilon_{\alpha\beta} = \varepsilon_{\alpha\beta}^{(0)} + \varepsilon_{\alpha\beta}^{(1)} , \tag{7.19}$$

that $\varepsilon_{\alpha\beta}^{(0)}$ and $\varepsilon_{\alpha\beta}^{(1)}$ are symmetric tensor fields and that

$$\varepsilon_{\alpha\beta}^{(0)} = 0 \text{ for all values of } Z_{\alpha\beta} \text{ if } T = T_0 \text{ and } \tau_{\alpha\beta}^{(eq)} = 0 \tag{7.20}$$

(see also (7.13)).

Thus, it is seen that with the aid of the hidden tensorial variable $Z_{\alpha\beta}$ the strain may be split into two parts $\varepsilon_{\alpha\beta}^{(0)}$ and $\varepsilon_{\alpha\beta}^{(1)}$. The strain $\varepsilon_{\alpha\beta}^{(0)}$ vanishes for all values of $Z_{\alpha\beta}$ if $T = T_0$ and $\tau_{\alpha\beta}^{(eq)} = 0$ (see (7.20)), while the strain $\varepsilon_{\alpha\beta}^{(1)}$ only depends on $Z_{\alpha\beta}$ (see (7.15)). Therefore, one may say that a hidden tensorial variable is a kind of operator which splits the strain into two parts (splitting operator for strains). However, the procedure is not unique. For instance, the splitting as described above depends on the choice for $T_0$ (see (7.14)). Later on we shall see that if the equilibrium-stress tensor $\tau_{\alpha\beta}^{(eq)}$ equals the mechanical stress tensor $\tau_{\alpha\beta}$, which occurs in the equations of motion and in the first law, $\varepsilon_{\alpha\beta}^{(0)}$, as defined above, is the thermoelastic part of the strain and $\varepsilon_{\alpha\beta}^{(1)}$ is the plastic part of the strain. If $\tau_{\alpha\beta}$ and $\tau_{\alpha\beta}^{(eq)}$ are different, no part of the strain is purely thermoelastic. Finally, we note that the procedure discussed in this section to introduce $\varepsilon_{\alpha\beta}^{(0)}$ and $\varepsilon_{\alpha\beta}^{(1)}$ is analogous to the procedure which was used in the references 40 and 41 to split the polarization (or the magnetization) into two parts in case of vectorial internal variables which are connected with dielectric (or magnetic) relaxation phenomena (see also reference 42).

## 8. INELASTIC STRAINS AS INTERNAL VARIABLES, MECHANICAL STRESS, EQUILIBRIUM STRESS, VISCOUS STRESS, AFFINITY STRESSES AND MEMORY STRESSES

We shall now assume that the tensor-valued function $\varphi^{(1)}_{\alpha\beta}$ which occurs in (7.15) has an inverse $\psi^{(1)}_{\alpha\beta}$. Hence,

$$Z_{\alpha\beta} = \psi^{(1)}_{\alpha\beta}(\varepsilon^{(1)}_{\alpha\beta}) . \tag{8.1}$$

We recall the assumption (7.1)

$$s = s(u,\varepsilon_{\alpha\beta},Z_{\alpha\beta}) . \tag{8.2}$$

From the two preceding relations it follows that one may also write

$$s = s(u,\varepsilon_{\alpha\beta},\varepsilon^{(1)}_{\alpha\beta}) \tag{8.3}$$

and by virtue of (7.19) one then has

$$s = s(u,\varepsilon^{(0)}_{\alpha\beta},\varepsilon^{(1)}_{\alpha\beta}) . \tag{8.4}$$

If the entropy is written in the form (8.3) it is seen that $\varepsilon^{(1)}_{\alpha\beta}$ plays the role of an internal variable.

Next, we shall generalize the ideas developed in the preceding section. We shall assume that there are several microscopic phenomena which influence the mechanical properties of the medium under consideration and that the (thermodynamic) state of the medium may be described by the internal energy, the total strain and n macroscopic tensorial internal variables. This includes the possibility of scalar internal variables. Indeed, a tensorial variable of the form $Z\delta_{\alpha\beta}$, where Z is a scalar field, is in fact a scalar variable. Hence, in this way it is possible to include, for instance, chemical reactions. In principle it is also possible to introduce vectorial internal variables. Such vectorial internal parameters play a role in the thermodynamic theory of dielectric and magnetic relaxation. See the references 40-43. However, as far as the author knows, there is no experimental evidence that vector fields play a role in plasticity phenomena as internal variables.

Hence, we assume that

$$s = s(u, \varepsilon_{\alpha\beta}, Z^{(1)}_{\alpha\beta}, \ldots, Z^{(n)}_{\alpha\beta}) \; , \tag{8.5}$$

where $Z^{(1)}_{\alpha\beta}, \ldots, Z^{(n)}_{\alpha\beta}$ are symmetric tensor fields, i.e.

$$Z^{(k)}_{\alpha\beta} = Z^{(k)}_{\beta\alpha} \qquad (k = 1, \ldots, n) \; . \tag{8.6}$$

We define the equilibrium-stress tensor $\tau^{(eq)}_{\alpha\beta}$ by

$$\tau^{(eq)}_{\alpha\beta} = - \rho T \frac{\partial}{\partial \varepsilon_{\alpha\beta}} s(u, \varepsilon_{\alpha\beta}, Z^{(1)}_{\alpha\beta}, Z^{(2)}_{\alpha\beta}, \ldots, Z^{(n)}_{\alpha\beta}) \tag{8.7}$$

and the definition (5.2) for the temperature becomes

$$\frac{1}{T} = \frac{\partial}{\partial u} s(u, \varepsilon_{\alpha\beta}, Z^{(1)}_{\alpha\beta}, \ldots, Z^{(n)}_{\alpha\beta}) \; . \tag{8.8}$$

Furthermore, we define the tensors $G^{(k)}_{\alpha\beta}$ by

$$G^{(k)}_{\alpha\beta} = \rho T \frac{\partial}{\partial Z^{(k)}_{\alpha\beta}} s(u, \varepsilon_{\alpha\beta}, Z^{(1)}_{\alpha\beta}, \ldots, Z^{(n)}_{\alpha\beta}) \; . \tag{8.9}$$

One then has from (8.5), (8.7), (8.8) and (8.9)

$$Tds = du - v \sum_{\alpha,\beta=1}^{3} \tau^{(eq)}_{\alpha\beta} d\varepsilon_{\alpha\beta} + v \sum_{k=1}^{n} \sum_{\alpha,\beta=1}^{3} G^{(k)}_{\alpha\beta} dZ^{(k)}_{\alpha\beta} \; . \tag{8.10}$$

This expression for ds is analogous to (7.5). It should be noted that

$$\tau^{(eq)}_{\alpha\beta} = \tau^{(eq)}_{\beta\alpha} \ , \qquad \tilde{\tau}^{(eq)}_{\alpha\beta} = \tilde{\tau}^{(eq)}_{\beta\alpha} \tag{8.11}$$

and that

$$G^{(k)}_{\alpha\beta} = G^{(k)}_{\beta\alpha} \ , \qquad \tilde{G}^{(k)}_{\alpha\beta} = \tilde{G}^{(k)}_{\beta\alpha} \qquad (k = 1,\dots,n) \ . \tag{8.12}$$

In the same way as in the preceding section we may introduce the thermodynamic potential g and we can obtain a function $\varphi_{\alpha\beta}$ such that

$$\varepsilon_{\alpha\beta} = \varphi_{\alpha\beta}(T, v\tau^{(eq)}_{\alpha\beta}, Z^{(1)}_{\alpha\beta}, Z^{(2)}_{\alpha\beta},\dots,Z^{(n)}_{\alpha\beta}) \tag{8.13}$$

(cf. (7.12)). By $Z^{(k)}_{(0)\alpha\beta}$ we denote arbitrary but fixed values for the tensors $Z^{(k)}_{\alpha\beta}$ $(k = 1,\dots,n-1)$. As in section 7 (see (7.13)) $\tau^{(eq)}_{(0)\alpha\beta} = 0$, $T_0$ is an arbitrary but fixed temperature and $v_0$ is an arbitrary but fixed value for the specific volume (for instance, the specific volume in the reference state).

Next, we define $\varepsilon^{(0)}_{\alpha\beta}, \varepsilon^{(1)}_{\alpha\beta},\dots,\varepsilon^{(n)}_{\alpha\beta}$ by

$$\varepsilon^{(0)}_{\alpha\beta} = \varphi_{\alpha\beta}(T, v\tau^{(eq)}_{\alpha\beta}, Z^{(1)}_{\alpha\beta},\dots,Z^{(n)}_{\alpha\beta}) -$$

$$- \varphi_{\alpha\beta}(T_0, v_0\tau^{(eq)}_{(0)\alpha\beta}, Z^{(1)}_{\alpha\beta},\dots,Z^{(n)}_{\alpha\beta}) \ , \tag{8.14}$$

$$\varepsilon^{(1)}_{\alpha\beta} = \varphi_{\alpha\beta}(T_0, v_0\tau^{(eq)}_{(0)\alpha\beta}, Z^{(1)}_{\alpha\beta}, Z^{(2)}_{\alpha\beta},\dots,Z^{(n)}_{\alpha\beta}) -$$

$$- \varphi_{\alpha\beta}(T_0, v_0\tau^{(eq)}_{(0)\alpha\beta}, Z^{(1)}_{(0)\alpha\beta}, Z^{(2)}_{\alpha\beta},\dots,Z^{(n)}_{\alpha\beta}) \ , \tag{8.15}$$

$$\varepsilon^{(k)}_{\alpha\beta} = \varphi_{\alpha\beta}(T_0, v_0\tau^{(eq)}_{(0)\alpha\beta}, Z^{(1)}_{(0)\alpha\beta},\dots,Z^{(k-1)}_{(0)\alpha\beta}, Z^{(k)}_{\alpha\beta}, Z^{(k+1)}_{\alpha\beta},\dots,Z^{(n)}_{\alpha\beta}) -$$

$$- \varphi_{\alpha\beta}(T_0, v_0\tau^{(eq)}_{(0)\alpha\beta}, Z^{(1)}_{(0)\alpha\beta},\dots,Z^{(k-1)}_{(0)\alpha\beta}, Z^{(k)}_{(0)\alpha\beta}, Z^{(k+1)}_{\alpha\beta},\dots,Z^{(n)}_{\alpha\beta})$$

$$(k = 2,\dots,n-1) \ , \tag{8.16}$$

$$\varepsilon_{\alpha\beta}^{(n)} = \varphi_{\alpha\beta}(T_0, v_0\tau_{(0)\alpha\beta}^{(eq)}, Z_{(0)\alpha\beta}^{(1)}, \dots, Z_{(0)\alpha\beta}^{(n-1)}, Z_{\alpha\beta}^{(n)}) \ . \tag{8.17}$$

From these definitions it is seen that

$$\varepsilon_{\alpha\beta} = \varepsilon_{\alpha\beta}^{(0)} + \varepsilon_{\alpha\beta}^{(1)} + \dots + \varepsilon_{\alpha\beta}^{(n)} \ . \tag{8.18}$$

Because of the latter relation we call the $\varepsilon_{\alpha\beta}^{(k)}$ $(k = 0,1,\dots,n)$ partial strain tensors. Furthermore, $\varepsilon_{\alpha\beta}^{(0)},\dots,\varepsilon_{\alpha\beta}^{(n)}$ have the properties

$$\varepsilon_{\alpha\beta}^{(0)} = 0 \text{ for all values of } Z_{\alpha\beta}^{(1)},\dots,Z_{\alpha\beta}^{(n)} \text{ if } T = T_0 \text{ and } \tau_{\alpha\beta}^{(eq)} = 0 \ , \tag{8.19}$$

$$\varepsilon_{\alpha\beta}^{(k)} = 0 \text{ for all values of } Z_{\alpha\beta}^{(k+1)},\dots,Z_{\alpha\beta}^{(n)} \text{ if } Z_{\alpha\beta}^{(k)} = Z_{(0)\alpha\beta}^{(k)} \quad (k = 1,\dots,n-1) \ , \tag{8.20}$$

$$\varepsilon_{\alpha\beta}^{(n)} \text{ is a function only of } Z_{\alpha\beta}^{(n)} \ . \tag{8.21}$$

It is also important to note that it follows from (8.15), (8.16) and (8.17) that there exist functions $\varphi_{\alpha\beta}^{(k)}$ $(k = 1,\dots,n)$ such that

$$\varepsilon_{\alpha\beta}^{(k)} = \varphi_{\alpha\beta}^{(k)}(Z_{\alpha\beta}^{(k)},\dots,Z_{\alpha\beta}^{(n)}) \qquad (k = 1,\dots,n) \ . \tag{8.22}$$

These functions are such that the $\varepsilon_{\alpha\beta}^{(k)}$ have the properties (8.20) and (8.21).

Next, we assume that the functions $\varphi_{\alpha\beta}^{(k)}$ have inverse functions $\psi_{\alpha\beta}^{(k)}$ such that

$$Z_{\alpha\beta}^{(k)} = \psi_{\alpha\beta}^{(k)}(\varepsilon_{\alpha\beta}^{(k)},\dots,\varepsilon_{\alpha\beta}^{(n)}) \qquad (k = 1,\dots,n) \ . \tag{8.23}$$

One may then write for the entropy (see (8.5))

$$s = s(u, \varepsilon_{\alpha\beta}, \varepsilon_{\alpha\beta}^{(1)},\dots,\varepsilon_{\alpha\beta}^{(n)}) \tag{8.24}$$

and (8.19) may be written in the form

$$\varepsilon_{\alpha\beta}^{(0)} = 0 \text{ for all values of } \varepsilon_{\alpha\beta}^{(1)},\ldots,\varepsilon_{\alpha\beta}^{(n)} \text{ if } T = T_0 \text{ and } \tau_{\alpha\beta}^{(eq)} = 0 \ . \tag{8.25}$$

By virtue of (8.18) one also has

$$s = s(u,\varepsilon_{\alpha\beta}^{(0)},\varepsilon_{\alpha\beta}^{(1)},\ldots,\varepsilon_{\alpha\beta}^{(n)}) \ . \tag{8.26}$$

From (8.7) and (8.23) it follows that $\tau_{\alpha\beta}^{(eq)}$ is also given by

$$\tau_{\alpha\beta}^{(eq)} = -\rho T \frac{\partial}{\partial \varepsilon_{\alpha\beta}} s(u,\varepsilon_{\alpha\beta},\varepsilon_{\alpha\beta}^{(1)},\ldots,\varepsilon_{\alpha\beta}^{(n)}) \ . \tag{8.27}$$

In an analogous way one has for the temperature (see (8.8))

$$\frac{1}{T} = \frac{\partial}{\partial u} s(u,\varepsilon_{\alpha\beta},\varepsilon_{\alpha\beta}^{(1)},\ldots,\varepsilon_{\alpha\beta}^{(n)}) \ . \tag{8.28}$$

We define the tensor fields $\tau_{\alpha\beta}^{(1)},\ldots,\tau_{\alpha\beta}^{(n)}$ by

$$\tau_{\alpha\beta}^{(k)} = \rho T \frac{\partial}{\partial \varepsilon_{\alpha\beta}^{(k)}} s(u,\varepsilon_{\alpha\beta},\varepsilon_{\alpha\beta}^{(1)},\ldots,\varepsilon_{\alpha\beta}^{(n)}) \qquad (k = 1,\ldots,n) \ . \tag{8.29}$$

From (8.24) and the preceding relations one gets

$$T ds = du - v \sum_{\alpha,\beta=1}^{3} \tau_{\alpha\beta}^{(eq)} d\varepsilon_{\alpha\beta} + v \sum_{k=1}^{n} \sum_{\alpha,\beta=1}^{3} \tau_{\alpha\beta}^{(k)} d\varepsilon_{\alpha\beta}^{(k)} \ . \tag{8.30}$$

Using (8.18) one may also write for (8.30)

$$T ds = du - v \sum_{\alpha,\beta=1}^{3} \tau_{\alpha\beta}^{(eq)} d\varepsilon_{\alpha\beta}^{(0)} + v \sum_{k=1}^{n} \sum_{\alpha,\beta=1}^{3} (\tau_{\alpha\beta}^{(k)} - \tau_{\alpha\beta}^{(eq)}) d\varepsilon_{\alpha\beta}^{(k)} \ . \tag{8.31}$$

We define the tensor fields $\tau_{(m)\alpha\beta}^{(k)}$ $(k = 1,\ldots,n)$ by

$$\tau^{(k)}_{(m)\alpha\beta} = \tau^{(k)}_{\alpha\beta} - \tau^{(eq)}_{\alpha\beta} \qquad (k = 1,\ldots,n) \ . \tag{8.32}$$

Thus,

$$\tau^{(k)}_{\alpha\beta} = \tau^{(eq)}_{\alpha\beta} + \tau^{(k)}_{(m)\alpha\beta} \qquad (k = 1,\ldots,n) \ . \tag{8.33}$$

Hence, (8.31) becomes

$$Tds = du - v \sum_{\alpha,\beta=1}^{3} \tau^{(eq)}_{\alpha\beta} d\varepsilon^{(0)}_{\alpha\beta} + v \sum_{k=1}^{n} \sum_{\alpha,\beta=1}^{3} \tau^{(k)}_{(m)\alpha\beta} d\varepsilon^{(k)}_{\alpha\beta} \ . \tag{8.34}$$

We now have

$$\frac{1}{T} = \frac{\partial}{\partial u} s(u, \varepsilon^{(0)}_{\alpha\beta}, \varepsilon^{(1)}_{\alpha\beta}, \ldots, \varepsilon^{(n)}_{\alpha\beta}) \ , \tag{8.35}$$

$$\tau^{(eq)}_{\alpha\beta} = - \rho T \frac{\partial}{\partial \varepsilon^{(0)}_{\alpha\beta}} s(u, \varepsilon^{(0)}_{\alpha\beta}, \varepsilon^{(1)}_{\alpha\beta}, \ldots, \varepsilon^{(n)}_{\alpha\beta}) \ , \tag{8.36}$$

$$\tau^{(k)}_{(m)\alpha\beta} = \rho T \frac{\partial}{\partial \varepsilon^{(k)}_{\alpha\beta}} s(u, \varepsilon^{(0)}_{\alpha\beta}, \varepsilon^{(1)}_{\alpha\beta}, \ldots, \varepsilon^{(n)}_{\alpha\beta}) \qquad (k = 1,\ldots,n) \ . \tag{8.37}$$

Furthermore, we define the tensor field $\tau^{(vi)}_{\alpha\beta}$ by

$$\tau^{(vi)}_{\alpha\beta} = \tau_{\alpha\beta} - \tau^{(eq)}_{\alpha\beta} \ , \tag{8.38}$$

or

$$\tau_{\alpha\beta} = \tau^{(eq)}_{\alpha\beta} + \tau^{(vi)}_{\alpha\beta} \ . \tag{8.39}$$

We shall call the tensors $\tau^{(k)}_{\alpha\beta}$ $(k = 1,\ldots,n)$ affinity-stress tensors, the tensors $\tau^{(k)}_{(m)\alpha\beta}$ $(k = 1,\ldots,n)$ memory-stress tensors and $\tau^{(vi)}_{\alpha\beta}$ the viscous stress tensor. We recall that $\tau_{\alpha\beta}$ is the mechanical stress tensor, which occurs in the equations of motion and in the first law and that $\tau^{(eq)}_{\alpha\beta}$ is the equilibrium-stress tensor.

Finally, we note that all strains and stresses which we have introduced are represented by symmetric tensors, i.e.

$$\varepsilon^{(k)}_{\alpha\beta} = \varepsilon^{(k)}_{\beta\alpha} \quad , \quad \tilde{\varepsilon}^{(k)}_{\alpha\beta} = \tilde{\varepsilon}^{(k)}_{\beta\alpha} \qquad (k = 0,1,\ldots,n) \; , \tag{8.40}$$

$$\tau^{(k)}_{\alpha\beta} = \tau^{(k)}_{\beta\alpha} \qquad \tilde{\tau}^{(k)}_{\alpha\beta} = \tilde{\tau}^{(k)}_{\beta\alpha} \qquad (k = 1,\ldots,n) \; , \tag{8.41}$$

$$\tau^{(k)}_{(m)\alpha\beta} = \tau^{(k)}_{(m)\beta\alpha} \; , \quad \tilde{\tau}^{(k)}_{(m)\alpha\beta} = \tilde{\tau}^{(k)}_{(m)\beta\alpha} \qquad (k = 1,\ldots,n) \; , \tag{8.42}$$

$$\tau^{(eq)}_{\alpha\beta} = \tau^{(eq)}_{\beta\alpha} \quad , \quad \tilde{\tau}^{(eq)}_{\alpha\beta} = \tilde{\tau}^{(eq)}_{\beta\alpha} \; , \tag{8.43}$$

$$\tau^{(vi)}_{\alpha\beta} = \tau^{(vi)}_{\beta\alpha} \quad , \quad \tilde{\tau}^{(vi)}_{\alpha\beta} = \tilde{\tau}^{(vi)}_{\beta\alpha} \; , \tag{8.44}$$

$$\tau_{\alpha\beta} = \tau_{\beta\alpha} \quad , \quad \tilde{\tau}_{\alpha\beta} = \tilde{\tau}_{\beta\alpha} \; , \tag{8.45}$$

$$\varepsilon_{\alpha\beta} = \varepsilon_{\beta\alpha} \quad , \quad \tilde{\varepsilon}_{\alpha\beta} = \tilde{\varepsilon}_{\beta\alpha} \; . \tag{8.46}$$

A more detailed discussion of the physical meaning of these quantities will be given in section 27.

## 9. ENTROPY BALANCE AND ENTROPY PRODUCTION

We recall the form (3.6) for the first law:

$$\rho \frac{du}{dt} = - \operatorname{div} \vec{J}^{(q)} + \sum_{\alpha,\beta=1}^{3} \tau_{\alpha\beta} \frac{d\varepsilon_{\alpha\beta}}{dt} . \tag{9.1}$$

From the form (8.10) for the Gibbs relation one obtains

$$T \frac{ds}{dt} = \frac{du}{dt} - v \sum_{\alpha,\beta=1}^{3} \tau_{\alpha\beta}^{(eq)} \frac{d\varepsilon_{\alpha\beta}}{dt} + v \sum_{k=1}^{n} \sum_{\alpha,\beta=1}^{3} G_{\alpha\beta}^{(k)} \frac{dZ_{\alpha\beta}^{(k)}}{dt} . \tag{9.2}$$

Let us eliminate du/dt from (9.1) and (9.2). To this end we multiply (9.2) by $\rho/T$. One then gets

$$\rho \frac{ds}{dt} = \frac{\rho}{T} \frac{du}{dt} - \frac{1}{T} \sum_{\alpha,\beta=1}^{3} \tau_{\alpha\beta}^{(eq)} \frac{d\varepsilon_{\alpha\beta}}{dt} + \frac{1}{T} \sum_{k=1}^{n} \sum_{\alpha,\beta=1}^{3} G_{\alpha\beta}^{(k)} \frac{dZ_{\alpha\beta}^{(k)}}{dt} . \tag{9.3}$$

Using (9.1) one may write for (9.3)

$$\rho \frac{ds}{dt} = - \frac{1}{T} \operatorname{div} \vec{J}^{(q)} + \frac{1}{T} \sum_{\alpha,\beta=1}^{3} (\tau_{\alpha\beta} - \tau_{\alpha\beta}^{(eq)}) \frac{d\varepsilon_{\alpha\beta}}{dt} +$$

$$+ \frac{1}{T} \sum_{k=1}^{n} \sum_{\alpha,\beta=1}^{3} G_{\alpha\beta}^{(k)} \frac{dZ_{\alpha\beta}^{(k)}}{dt} . \tag{9.4}$$

We have

$$- \frac{1}{T} \operatorname{div} \vec{J}^{(q)} = - \operatorname{div} \frac{\vec{J}^{(q)}}{T} + \vec{J}^{(q)} \cdot \operatorname{grad} \frac{1}{T} =$$

$$= - \operatorname{div} \frac{\vec{J}^{(q)}}{T} - \frac{1}{T^2} \vec{J}^{(q)} \cdot \operatorname{grad} T \; . \tag{9.5}$$

Moreover, we have $\tau^{(vi)}_{\alpha\beta} = \tau_{\alpha\beta} - \tau^{(eq)}_{\alpha\beta}$ (see (8.38)). Using the latter relation and (9.4) one obtains

$$\rho \cdot \frac{ds}{dt} = - \operatorname{div} \frac{\vec{J}^{(q)}}{T} + \sigma^{(s)} \; , \tag{9.6}$$

where

$$\sigma^{(s)} = \frac{1}{T} \left( -\frac{1}{T} \vec{J}^{(q)} \cdot \operatorname{grad} T + \sum_{\alpha,\beta=1}^{3} \tau^{(vi)}_{\alpha\beta} \frac{d\varepsilon_{\alpha\beta}}{dt} + \sum_{k=1}^{n} \sum_{\alpha,\beta=1}^{3} G^{(k)}_{\alpha\beta} \frac{dZ^{(k)}_{\alpha\beta}}{dt} \right) . \tag{9.7}$$

It is seen that (9.6) has the form of a balance equation. The quantity $\vec{J}^{(s)}$, defined by

$$\vec{J}^{(s)} = \frac{\vec{J}^{(q)}}{T} \; , \tag{9.8}$$

is the conductive flow of entropy and $\sigma^{(s)}$ is the entropy source strength, i.e. the entropy production per unit of volume and per unit of time.

In the next section we discuss the relation between (9.8) and the Carnot-Clausius theorem. If we use the form (8.30) for the Gibbs relation we also obtain (9.6), but for the entropy production we get the expression

$$\sigma^{(s)} = \frac{1}{T} \left( -\frac{1}{T} \vec{J}^{(q)} \cdot \operatorname{grad} T + \sum_{\alpha,\beta=1}^{3} \tau^{(vi)}_{\alpha\beta} \frac{d\varepsilon_{\alpha\beta}}{dt} + \sum_{k=1}^{n} \sum_{\alpha,\beta=1}^{3} \tau^{(k)}_{\alpha\beta} \frac{d\varepsilon^{(k)}_{\alpha\beta}}{dt} \right) . \tag{9.9}$$

Furthermore, using (8.18) one may also write for (9.9)

$$\sigma^{(s)} = \frac{1}{T} \left\{ - \frac{1}{T} \vec{J}^{(q)} \cdot \operatorname{grad} T + \sum_{\alpha,\beta=1}^{3} \tau^{(vi)}_{\alpha\beta} \frac{d\varepsilon^{(0)}_{\alpha\beta}}{dt} + \right.$$

$$+ \sum_{k=1}^{n} \sum_{\alpha,\beta=1}^{3} (\tau_{\alpha\beta}^{(k)} + \tau_{\alpha\beta}^{(vi)}) \frac{d\varepsilon_{\alpha\beta}^{(k)}}{dt} \Big\} . \tag{9.10}$$

From (8.33) and (8.39) we have

$$\tau_{\alpha\beta}^{(k)} + \tau_{\alpha\beta}^{(vi)} = \tau_{\alpha\beta}^{(eq)} + \tau_{(m)\alpha\beta}^{(k)} + \tau_{\alpha\beta}^{(vi)} = \tau_{\alpha\beta} + \tau_{(m)\alpha\beta}^{(k)} . \tag{9.11}$$

Hence, with the aid of (9.10) we obtain

$$\sigma^{(s)} = \frac{1}{T} \Big\{ - \frac{1}{T} \vec{J}^{(q)} \cdot \text{grad } T + \sum_{\alpha,\beta=1}^{3} \tau_{\alpha\beta}^{(vi)} \frac{d\varepsilon_{\alpha\beta}^{(0)}}{dt} +$$

$$+ \sum_{k=1}^{n} \sum_{\alpha,\beta=1}^{3} (\tau_{\alpha\beta} + \tau_{(m)\alpha\beta}^{(k)}) \frac{d\varepsilon_{\alpha\beta}^{(k)}}{dt} \Big\} . \tag{9.12}$$

The first terms on the right-hand sides of (9.7), (9.9), (9.10) and (9.12), containing the heat flow $\vec{J}^{(q)}$, represent the entropy production due to heat conduction. The remaining sums on the right-hand sides of these four expressions for $\sigma^{(s)}$ give the entropy production due to mechanical phenomena. It is seen from (9.10) and (9.12) that all changes in the partial strain tensors $\varepsilon_{\alpha\beta}^{(k)}$ $(k = 0,1,\ldots,n)$ are irreversible phenomena, i.e. they contribute to the entropy production. Therefore, we shall call these partial strains inelastic. Later on (see the sections 15 and 16) we shall discuss some special cases. For instance, we have the limiting case that one of the partial inelastic strains is reversible, i.e. this partial strain can be considered as the thermoelastic part of the total strain (see section 16). Secondly, it may occur that the deviator of one of the memory-stress tensors vanishes. In this case (see section 15) $\tilde{\varepsilon}_{\alpha\beta}$ is not a thermodynamic state variable. Finally, we remark that the inelastic partial strains must be divided into anelastic strains and plastic strains (see section 25).

## 10. THE NONNEGATIVE CHARACTER OF THE ENTROPY PRODUCTION, THE CARNOT-CLAUSIUS THEOREM AND THE SECOND LAW

As usually we postulate that the entropy source strength is a non-negative quantity. Hence,

$$\sigma^{(s)} \geq 0 \; . \tag{10.1}$$

Furthermore, we recall the relation (9.8):

$$\vec{J}^{(s)} = \frac{\vec{J}^{(q)}}{T} \; . \tag{10.2}$$

We shall now show that (10.1) is the local formulation for the second law and that (10.2) can be considered as the local formulation of the Carnot-Clausius theorem.

Using (2.9) the balance equation (9.6) for the entropy becomes

$$\frac{\partial(\rho s)}{\partial t} = - \operatorname{div}\left(\frac{\vec{J}^{(q)}}{T} + \rho s\vec{v}\right) + \sigma^{(s)} \; . \tag{10.3}$$

Let us consider a volume V bounded by a closed surface $\Omega$ which is fixed in space. Using Gauss' theorem one obtains

$$\frac{d}{dt}\int_V \rho s \, dV = - \int_\Omega \left\{\left(\frac{\vec{J}^{(q)}}{T} + \rho s\vec{v}\right)\cdot\vec{n}\right\} d\Omega + \int_V \sigma^{(s)} dV \; , \tag{10.4}$$

where $\vec{n}$ is the outward unit normal to the surface element $d\Omega$. Since $\rho$ is the mass density and s is the entropy per unit of mass, the quantity S, defined by

$$S = \int_V \rho s \, dV \, , \tag{10.5}$$

is the entropy of the system in V. Let us multiply (10.4) by a (infinitesimal) time dt. One then obtains

$$dS = d_e S + d_i S \, , \tag{10.6}$$

where $d_e S$, defined by

$$d_e S = -\left[ \int_\Omega \left\{ \left( \frac{\vec{J}^{(q)}}{T} + \rho s \vec{v} \right) \cdot \vec{n} \right\} d\Omega \right] dt \, , \tag{10.7}$$

is the change in the entropy due to interaction of the system with its surroundings, and $d_i S$, defined by

$$d_i S = \left( \int_V \sigma^{(s)} dV \right) dt \, , \tag{10.8}$$

is the change due to internal processes. Furthermore,

$$dQ = -\left\{ \int_\Omega (\vec{J}^{(q)} \cdot \vec{n}) \, d\Omega \right\} dt \tag{10.9}$$

is the heat added to the system by its surroundings during the time dt.

By a closed system we mean a system which can only exchange heat with its surroundings (and no matter). Hence, at $\Omega$ we have $\vec{v} \cdot \vec{n} = 0$ for such a system and (10.7) becomes

$$d_e S = -\left\{ \int_\Omega \left( \frac{\vec{J}^{(q)}}{T} \cdot \vec{n} \right) d\Omega \right\} dt \quad \text{for a closed system.} \tag{10.10}$$

Suppose that T at the boundary $\Omega$ of the system is uniform. One then obtains from (10.10) with the aid of (10.9)

$$d_e S = \frac{dQ}{T} \quad \text{for a closed system with uniform boundary temperature T.} \qquad (10.11)$$

For instance, one may think of a system in a heat bath with uniform temperature.

The formula $d_e S = dQ/T$ expresses the Carnot-Clausius theorem. In fact, this theorem justifies the interpretation of $\vec{J}^{(q)}/T$ as the conductive flow $\vec{J}^{(s)}$ of the entropy, i.e. it justifies (10.2).

Since $\sigma^{(s)} \geq 0$ (see (10.1)), we have $\int_V \sigma^{(s)} dV \geq 0$ and, thus, from (10.8) one obtains

$$d_i S \geq 0 . \qquad (10.12)$$

This inequality is known as the second law of thermodynamics. From (10.6) it is seen that the second law may also be written in the form

$$dS \geq d_e S . \qquad (10.13)$$

From the latter inequality we have by virtue of (10.10) and (10.11)

$$dS \geq - \left\{ \int_\Omega \left( \frac{\vec{J}^{(q)}}{T} \cdot \vec{n} \right) d\Omega \right\} dt \quad \text{for a closed system} \qquad (10.14)$$

and

$$dS \geq \frac{dQ}{T} \quad \text{for a closed system with uniform boundary temperature T.} \qquad (10.15)$$

It is said that a system is adiabatically insulated if the system can exchange neither heat nor matter with its surroundings. Hence, at the boundary both $\vec{J}^{(q)} \cdot \vec{n}$ and $\vec{v} \cdot \vec{n}$ vanish. From (10.7) it then follows that $d_e S = 0$. Hence, the inequality (10.13) becomes

$$dS \geq 0 \quad \text{for an adiabatically insulated system.} \qquad (10.16)$$

This formulation of the second law is often used.

## 11. PHENOMENOLOGICAL EQUATIONS (FOURIER's LAW AND MECHANICAL FLOW LAWS)

We recall the form (9.9) for the entropy source strength:

$$T\sigma^{(s)} = -\frac{1}{T}\vec{J}^{(q)}\cdot\text{grad } T + \sum_{\alpha,\beta=1}^{3} \tau_{\alpha\beta}^{(vi)} \frac{d\varepsilon_{\alpha\beta}}{dt} + \sum_{k=1}^{n} \sum_{\alpha,\beta=1}^{3} \tau_{\alpha\beta}^{(k)} \frac{d\varepsilon_{\alpha\beta}^{(k)}}{dt} . \quad (11.1)$$

It is seen from (11.1) that the entropy production is a sum of terms, where each term is the inner product of two vectors or two tensors. The first term on the right-hand side of (11.1) is the scalar product of the heat flow $\vec{J}^{(q)}$ and $-1/T$ grad T. The second terms is the inner product of the viscous stress tensor and the substantial derivative with respect to time of the tensor of total strain. Finally, the last term is a sum of inner products of affinity-stress tensors and substantial time derivatives of partial strain tensors.

In section 8 we have seen that the mechanical stress tensor $\tau_{\alpha\beta}$ is the sum of the equilibrium-stress tensor $\tau_{\alpha\beta}^{(eq)}$ and the viscous stress tensor $\tau_{\alpha\beta}^{(vi)}$ (see (8.39)). As we noted in section 2.2 we have $P_{\alpha\beta} = -\tau_{\alpha\beta} = -\tau_{\alpha\beta}^{(eq)} - \tau_{\alpha\beta}^{(vi)}$ (see (2.13)) and $P_{\alpha\beta}$, the mechanical pressure tensor, is the conductive flow of momentum. Hence, $-\tau_{\alpha\beta}^{(vi)}$ is that part of the conductive flow of momentum which contributes to the entropy production, whereas $-\tau_{\alpha\beta}^{(eq)}$ is a thermodynamic state variable determined by (8.27). (See for the definition of thermodynamic state variable the last paragraph of section 5.) Therefore, one may call $\tau_{\alpha\beta}^{(eq)}$ the reversible part of the stress and $\tau_{\alpha\beta}^{(vi)}$ the irreversible part. We shall call such an irreversible conductive flow as $\tau_{\alpha\beta}^{(vi)}$ a flux (or flow). As we have seen in section 2.5 the heat flow $\vec{J}^{(q)}$ is the conductive flow of internal energy. Since $\vec{J}^{(q)}$ occurs in the entropy source strength it is like $\tau_{\alpha\beta}^{(vi)}$

an irreversible conductive flow. Hence, we also call it a flux.

Another class of quantities which we shall call fluxes is formed by time derivatives of internal thermodynamic variables. Thus, in (11.1) the time derivatives $d\varepsilon_{\alpha\beta}^{(k)}/dt$ of the partial strain tensors $\varepsilon_{\alpha\beta}^{(k)}$ $(k = 1,\ldots,n)$ are also fluxes.

Hence, in each product on the right-hand side of (11.1) one of the factors is a flux. By definition the other factor is the affinity conjugate to the flux. Affinities are also called thermodynamic forces. The time derivative $d\varepsilon_{\alpha\beta}/dt$ of the tensor of total strain is the affinity conjugate to the viscous stress $\tau_{\alpha\beta}^{(vi)}$, $-1/T$ grad T is the affinity conjugate to the heat flow $\vec{J}^{(q)}$ and the affinity-stress tensor $\tau_{\alpha\beta}^{(k)}$ is the affinity conjugate to the flux $d\varepsilon_{\alpha\beta}^{(k)}/dt$.

Next, we introduce the physical assumption that the fluxes depend on the affinities by the following equations:

$$\frac{d\varepsilon_{\alpha\beta}^{(j)}}{dt} = \sum_{k=1}^{n} \sum_{\gamma,\zeta=1}^{3} L_{\alpha\beta\gamma\zeta}^{(j)(k)} \tau_{\gamma\zeta}^{(k)} + \sum_{\gamma,\zeta=1}^{3} L_{\alpha\beta\gamma\zeta}^{(j)(0)} \frac{d\varepsilon_{\gamma\zeta}}{dt} - T^{-1} \sum_{\gamma=1}^{3} L_{\alpha\beta\gamma}^{(j)(q)} \frac{\partial T}{\partial x_\gamma} \qquad (j = 1,\ldots,n) , \tag{11.2}$$

$$\tau_{\alpha\beta}^{(vi)} = \sum_{k=1}^{n} \sum_{\gamma,\zeta=1}^{3} L_{\alpha\beta\gamma\zeta}^{(0)(k)} \tau_{\gamma\zeta}^{(k)} + \sum_{\gamma,\zeta=1}^{3} L_{\alpha\beta\gamma\zeta}^{(0)(0)} \frac{d\varepsilon_{\gamma\zeta}}{dt} - T^{-1} \sum_{\gamma=1}^{3} L_{\alpha\beta\gamma}^{(0)(q)} \frac{\partial T}{\partial x_\gamma} , \tag{11.3}$$

$$J_\alpha^{(q)} = \sum_{k=1}^{n} \sum_{\beta,\gamma=1}^{3} L_{\alpha\beta\gamma}^{(q)(k)} \tau_{\beta\gamma}^{(k)} + \sum_{\beta,\gamma=1}^{3} L_{\alpha\beta\gamma}^{(q)(0)} \frac{d\varepsilon_{\beta\gamma}}{dt} - T^{-1} \sum_{\beta=1}^{3} L_{\alpha\beta}^{(q)(q)} \frac{\partial T}{\partial x_\beta} . \tag{11.4}$$

The tensors $L^{(j)(k)}_{\alpha\beta\gamma\zeta}$ $(j,k = 0,1,\ldots,n)$, $L^{(j)(q)}_{\alpha\beta\gamma}$ $(j = 0,1,\ldots,n)$, $L^{(q)(j)}_{\alpha\beta\gamma}$ $(j = 0,1,\ldots,n)$ and $L^{(q)(q)}_{\alpha\beta}$ are called phenomenological tensors and the equations (11.2), (11.3) and (11.4) are called phenomenological equations. The first of these equations is the equation for the irreversible process of inelastic flow. This equation may be regarded as a generalization of Lévy's law. Furthermore, (11.3) describes a viscous flow phenomenon and this relation may be considered to be a generalization of Newton's law for ordinary viscous fluids. Finally, (11.4) is a generalization of Fourier's law for heat conduction. The second and third sums on the right-hand side of (11.2) represent a possible influence of the affinities conjugate to viscous flow and heat conduction on inelastic flow. Such mutual influences among irreversible phenomena are called cross effects. The first and third sums on the right-hand side of (11.3) and the first and second sums on the right-hand side of (11.4) also represent possible cross effects. Moreover, there are n different inelastic flows and among these flows cross effects may also occur. The cross effects are represented by the terms on the right-hand side of (11.2) which contain the phenomenological tensors $L^{(j)(k)}_{\alpha\beta\gamma\zeta}$ $(j,k = 1,\ldots,n;\ j \neq k)$.

For the formulation of phenomenological equations one may also use one of the other expressions which are given for the entropy production in section 9. These expressions for $\sigma^{(s)}$ are also sums of products of fluxes and affinities, but, which factor must be considered as a flux and which as an affinity is not always uniquely decidable. This, however, is physically irrelevant, since the physical statement is that fluxes and affinities determine each other uniquely (independent of the answer to the question which factors must be chosen as fluxes in the products of which the entropy production is additively composed).

Since all second order tensors which occur in (11.2), (11.3) and (11.4) as fluxes or affinities are symmetric, the phenomenological tensors may be chosen so that they satisfy the symmetry relations

$$L^{(j)(k)}_{\alpha\beta\gamma\zeta} = L^{(j)(k)}_{\beta\alpha\gamma\zeta} = L^{(j)(k)}_{\alpha\beta\zeta\gamma} = L^{(j)(k)}_{\beta\alpha\zeta\gamma} \quad (j,k = 0,1,\ldots,n)\ , \tag{11.5}$$

$$L^{(j)(q)}_{\alpha\beta\gamma} = L^{(j)(q)}_{\beta\alpha\gamma} \qquad (j = 0,1,\ldots,n) \;, \tag{11.6}$$

$$L^{(q)(j)}_{\alpha\beta\gamma} = L^{(q)(j)}_{\alpha\gamma\beta} \qquad (j = 0,1,\ldots,n) \;. \tag{11.7}$$

Another type of symmetry relations for the phenomenological tensors (Onsager relations) will be discussed in the next section.

As noted above one may also use one of the other expressions which are given in section 9 (with different fluxes and affinities) for the formulation of the phenomenological equations. Such equations, however, are equivalent with (11.2)-(11.4) if one uses the fluxes and affinities which occur in (9.10) or (9.12). From a physical point of view they are less satisfactory if one uses the fluxes and affinities which occur in (9.7) unless it turns out to be possible to give physical interpretations for the hidden internal variables.

## 12. ONSAGER RELATIONS AND THE CASIMIR MINUS SIGN

All microscopic quantities which occur in a phenomenological theory are (ensemble) averages of functions of the (generalized) coordinates and momenta of the constituting microscopic particles of the system. In some of such averages the momenta do not play any role (for instance, the displacement field, the gradient of the displacement field, the strain tensor, the mass density). Some macroscopic quantities (for instance, energy, temperature) depend on the momenta but are even functions of the velocities of the particles. Finally, there are macroscopic quantities which are odd functions of the particle velocities (heat flow, matter flow, time derivatives of quantities which are even functions of the particle velocities as strain rates and (macroscopic) local velocities).

We shall call a macroscopic quantity an even (odd) quantity if it is an even (odd) function of the particle velocities. Of course, macroscopic quantities which do not depend on the microscopic velocities are considered as even functions of these velocities.

Some of the fluxes which we have introduced in the preceding section are odd quantities (the heat flow and the time derivatives of the partial strains) and one of them is even (the viscous stress tensor). All stress tensors which we have introduced in our discussion (mechanical, viscous, equilibrium, affinity and memory stresses) are even functions of the particle velocities. (This is closely connected with the fact that $\rho v_\alpha v_\beta$, the convective flow of momentum, which occurs in the form (2.12) for the equations of motion, is obviously an even quantity.)

Furthermore, the affinity conjugate to an even flow is an odd quantity and the affinity conjugate to an odd flow is an even quantity.

Indeed, the affinity conjugate to the viscous stress tensor (an even quantity) is the time derivative of the tensor of total strain which is an odd quantity. The odd fluxes are the heat flow and the time derivatives of the partial strain tensors and the affinities conjugate to these fluxes are $-1/T$ grad $T$ and the affinity-stress tensors $\tau_{\alpha\beta}^{(j)}$ (cf. (11.1)) which are even quantities.

Hence, one may write for the entropy production

$$T\sigma^{(s)} = \sum_{i=1}^{n} J_i^{(o)} X_i^{(e)} + \sum_{k=1}^{m} J_k^{(e)} X_k^{(o)} , \qquad (12.1)$$

where the $J_i^{(0)}$ and $J_i^{(e)}$ are components of vectors and tensors which represent fluxes and $X_i^{(o)}$ and $X_i^{(e)}$ are components of vectors and tensors which represent affinities. The $J_i^{(o)}$ and $X_i^{(o)}$ are odd quantities and the $J_i^{(e)}$ and $X_i^{(e)}$ are even quantities. Since $T$ is even, it is seen from (12.1) that the entropy production $\sigma^{(s)}$ is an odd quantity. Hence, if one could reverse the velocities of all microscopic particles one would have $\sigma^{(s)} \leq 0$. In principle such a "reversed process" is possible (microscopic reversibility) if one also reverses the external magnetic fields. However, for natural processes $\sigma^{(s)} \geq 0$ (second law).

One may now write for the phenomenological equations

$$J_i^{(o)} = \sum_{j=1}^{n} L_{ij}^{(oe)} X_j^{(e)} + \sum_{k=1}^{m} L_{ik}^{(oo)} X_k^{(o)} \qquad (i = 1,\dots,n) , \qquad (12.2)$$

$$J_k^{(e)} = \sum_{i=1}^{n} L_{ki}^{(ee)} X_i^{(e)} + \sum_{\ell=1}^{m} L_{k\ell}^{(eo)} X_\ell^{(o)} \qquad (k = 1,\dots,m) . \qquad (12.3)$$

The Onsager relations are the following symmetry relations for the coefficients which occur in the two preceding equations:

$$L_{ij}^{(oe)} = L_{ji}^{(oe)} \qquad (i,j = 1,\dots,n) , \qquad (12.4)$$

$$L^{(eo)}_{k\ell} = L^{(eo)}_{\ell k} \qquad (k,\ell = 1,\dots,m) \ , \tag{12.5}$$

$$L^{(oo)}_{ik} = - L^{(ee)}_{ki} \qquad (i = 1,\dots,n;\ k = 1,\dots,m) \ . \tag{12.6}$$

The minus sign which occurs in (12.6) is called the Casimir minus sign.

A detailed discussion of the Onsager relations is given by De Groot and Mazur.[15] The Onsager relations can be derived from microscopic considerations. They cannot be deduced from phenomenological principles. In a macroscopic theory they are postulated and, if possible, they must be verified with the aid of suitable experiments. On the other hand, it is seen that for the formulation of the Onsager relations we had to introduce even and odd variables. Hence, it is hardly possible to discuss the Onsager relations without referring to microscopic notions.

The only flux on the left-hand sides of the phenomenological equations (11.2)-(11.4) which is even is the viscous stress tensor $\tau^{(vi)}_{\alpha\beta}$. The remaining fluxes are odd quantities. Hence, for the type of system which we study the Onsager relations read

$$L^{(j)(k)}_{\alpha\beta\gamma\zeta} = L^{(k)(j)}_{\gamma\zeta\alpha\beta} \qquad (j,k = 1,\dots,n) \ , \tag{12.7}$$

$$L^{(j)(o)}_{\alpha\beta\gamma\zeta} = - L^{(o)(j)}_{\gamma\zeta\alpha\beta} \qquad (j = 1.\dots.n) \ , \tag{12.8}$$

$$L^{(o)(o)}_{\alpha\beta\gamma\zeta} = L^{(o)(o)}_{\gamma\zeta\alpha\beta} \ , \tag{12.9}$$

$$L^{(j)(q)}_{\alpha\beta\gamma} = L^{(q)(j)}_{\gamma\alpha\beta} \qquad (j = 1,\dots,n) \ , \tag{12.10}$$

$$L^{(o)(q)}_{\alpha\beta\gamma} = - L^{(q)(o)}_{\gamma\alpha\beta} \ , \tag{12.11}$$

$$L^{(q)(q)}_{\alpha\beta} = L^{(q)(q)}_{\beta\alpha} \ . \tag{12.12}$$

Thus, the Onsager relations entail the symmetry of the heat conductivity tensor $L^{(q)(q)}_{\alpha\beta}$ (see (12.12)). An analogous remark holds for the viscosity tensor $L^{(o)(o)}_{\alpha\beta\gamma\zeta}$ and for the tensors $L^{(k)(k)}_{\alpha\beta\gamma\zeta}$ which represent fluidities (see (12.7) and (12.9)). The tensors $L^{(o)(q)}_{\alpha\beta\gamma}$ and $L^{(q)(o)}_{\gamma\alpha\beta}$ are connected with cross effects between heat conduction and viscous flow and these cross effects are related by (12.11). The tensors $L^{(j)(q)}_{\alpha\beta\gamma}$ and $L^{(q)(j)}_{\gamma\alpha\beta}$ are related with cross effects among heat conduction and inelastic flows. The tensors $L^{(j)(k)}_{\alpha\beta\gamma\zeta}$ $(j = 0,1,\ldots,n)$ describe for $j \neq k$ several possible cross effects among mechanical flow phenomena and the relation among these effects is given by (12.7) and (12.8).

The number of independent components of the phenomenological tensors is reduced by (11.5)-(11.7) and (12.7)-(12.12). Further reductions may occur as a consequence of symmetries in the medium (as in crystal lattices).

The Onsager relations do not play an important role in our discussion of plasticity since, for simplicity, in general we neglect cross effects.

Using the same methods as in the references 40 and 42 it may be shown that the Onsager relations are invariant under the type of transformations discussed in sections 7 and 8.

## 13. PHENOMENOLOGICAL EQUATIONS AND ONSAGER RELATIONS FOR ISOTROPIC MEDIA

The phenomenological tensors are determined by the physical properties of the medium. If the medium is isotropic these tensors must be invariant with respect to all rotations and to inversion of the axis frame. It can be shown that isotropic phenomenological tensors of order four, which also satisfy the symmetry relations (11.5), must have the form

$$L^{(j)(k)}_{\alpha\beta\gamma\zeta} = \frac{1}{2}\,\eta_s^{(j,k)}(\delta_{\alpha\gamma}\delta_{\beta\zeta} + \delta_{\beta\gamma}\delta_{\alpha\zeta}) + \frac{1}{3}\,(\eta_v^{(j,k)} - \eta_s^{(j,k)})\delta_{\alpha\beta}\delta_{\gamma\zeta}$$

$$(j,k = 0,1,\dots,n)\;, \qquad (13.1)$$

where $\eta_s^{(j,k)}$ and $\eta_v^{(j,k)}$ are scalars. Furthermore, one has in case of isotropy

$$L^{(j)(q)}_{\alpha\beta\gamma} = L^{(q)(j)}_{\alpha\beta\gamma} = 0 \qquad (j = 0,1,\dots,n)\;, \qquad (13.2)$$

$$L^{(q)(q)}_{\alpha\beta} = \lambda T\delta_{\alpha\beta}\;, \qquad (13.3)$$

where $\lambda$ is the (scalar) heat conductivity.

The Onsager relations (12.7)-(12.12) are satisfied if

$$\eta_s^{(j,k)} = \eta_s^{(k,j)} \qquad (j,k = 1,\dots,n)\;, \qquad (13.4)$$

$$\eta_s^{(j,0)} = -\,\eta_s^{(0,j)} \qquad (j = 1,\dots,n)\;, \qquad (13.5)$$

$$\eta_v^{(j,k)} = \eta_v^{(k,j)} \qquad (j,k = 1,\dots,n)\;, \qquad (13.6)$$

$$\eta_v^{(j,0)} = - \eta_v^{(0,j)} \qquad (j = 1,\ldots,n) \ . \qquad (13.7)$$

Using (13.1), (13.2) and (13.3), the phenomenological equations (11.2), (11.3) and (11.4) become

$$\frac{d\tilde{\varepsilon}_{\alpha\beta}^{(j)}}{dt} = \sum_{k=1}^{n} \eta_s^{(j,k)}\tilde{\tau}_{\alpha\beta}^{(k)} + \eta_s^{(j,0)} \frac{d\tilde{\varepsilon}_{\alpha\beta}}{dt} \qquad (j = 1,\ldots,n) \ , \qquad (13.8)$$

$$\tilde{\tau}_{\alpha\beta}^{(vi)} = \sum_{k=1}^{n} \eta_s^{(0,k)}\tilde{\tau}_{\alpha\beta}^{(k)} + \eta_s^{(0,0)} \frac{d\tilde{\varepsilon}_{\alpha\beta}}{dt} \ , \qquad (13.9)$$

$$\frac{d\varepsilon^{(j)}}{dt} = \sum_{k=1}^{n} \eta_v^{(j,k)}\tau^{(k)} + \eta_v^{(j,0)} \frac{d\varepsilon}{dt} \qquad (j = 1,\ldots,n) \ , \qquad (13.10)$$

$$\tau^{(vi)} = \sum_{k=1}^{n} \eta_v^{(0,k)}\tau^{(k)} + \eta_v^{(0,0)} \frac{d\varepsilon}{dt} \ , \qquad (13.11)$$

$$\vec{J}^{(q)} = - \lambda \ \text{grad} \ T \ , \qquad (13.12)$$

where we also used the symmetry of $\varepsilon_{\alpha\beta}$ and $\tau_{\alpha\beta}^{(k)}$ and the definitions of scalar part and deviator of a tensor (see section 4). The phenomenological coefficients with subscript s occur in the equations (13.8) and (13.9) which describe irreversible shear phenomena and the coefficients with subscript v occur in the volumetric equations (13.10) and (13.11). In particular $\eta_s^{(0,0)}$ and $\eta_v^{(0,0)}$ may be called the shear viscosity and the volume viscosity, respectively. Indeed, using (4.12), (13.9) becomes

$$\tilde{\tau}_{\alpha\beta}^{(vi)} = \frac{1}{2} \eta_s^{(0,0)}\left(\frac{\partial v_\alpha}{\partial x_\beta} + \frac{\partial v_\beta}{\partial x_\alpha} - \frac{2}{3} \delta_{\alpha\beta} \ \text{div} \ \vec{v}\right) + \sum_{k=1}^{n} \eta_s^{(0,k)}\tilde{\tau}_{\alpha\beta}^{(k)} \ . \qquad (13.13)$$

Apart from the second term on the right-hand side of this equation, (13.13) is Newton's flow law for viscous fluids. Using (4.10) or (4.13) an analogous reformulation may be applied to (13.11). The coefficients

$\eta_s^{(j,k)}$ and $\eta_v^{(j,k)}$ are fluidities and $\eta_s^{(j,0)}$, $\eta_s^{(0,j)}$, $\eta_v^{(j,0)}$ and $\eta_v^{(0,j)}$ $(j = 1,\ldots,n)$ are dimensionless coefficients connected with possible cross effects among viscous flow and inelastic flows. The coefficients $\eta_s^{(j,k)}$ and $\eta_v^{(j,k)}$ $(j,k = 1,\ldots,n;\ j \neq k)$ represent possible cross effects among the n inelastic flows. Finally, (13.12) is a well-known form for Fourier's law in isotropic substances and $\lambda$ is the coefficient of heat conductivity.

## 14. THE FREE ENERGY, THE THERMODYNAMIC POTENTIAL, THE ENTHALPY AND SOME OTHER THERMODYNAMIC FUNCTIONS

In the usual way we define the specific free energy f, the specific thermodynamic potential g and the specific enthalpy h by

$$f = u - Ts \,, \tag{14.1}$$

$$g = u - Ts - v \sum_{\alpha,\beta=1}^{3} \tau_{\alpha\beta}^{(eq)} \varepsilon_{\alpha\beta} \,, \tag{14.2}$$

$$h = u - v \sum_{\alpha,\beta=1}^{3} \tau_{\alpha\beta}^{(eq)} \varepsilon_{\alpha\beta} \,. \tag{14.3}$$

Moreover, we define the thermodynamic functions $f^*$, $g^*$ and $h^*$ by

$$f^* = f + v \sum_{k=1}^{n} \sum_{\alpha,\beta=1}^{3} \tau_{\alpha\beta}^{(k)} \varepsilon_{\alpha\beta}^{(k)} \,, \tag{14.4}$$

$$g^* = g + v \sum_{k=1}^{n} \sum_{\alpha,\beta=1}^{3} \tau_{\alpha\beta}^{(k)} \varepsilon_{\alpha\beta}^{(k)} \,, \tag{14.5}$$

$$h^* = h + v \sum_{k=1}^{n} \sum_{\alpha,\beta=1}^{3} \tau_{\alpha\beta}^{(k)} \varepsilon_{\alpha\beta}^{(k)} \,. \tag{14.6}$$

From these definitions we obtain with the aid of the Gibbs relation (8.30) for the differentials of f, g, h, $f^*$, $g^*$ and $h^*$

$$df = - sdT + v \sum_{\alpha,\beta=1}^{3} \tau_{\alpha\beta}^{(eq)} d\varepsilon_{\alpha\beta} - v \sum_{k=1}^{n} \sum_{\alpha,\beta=1}^{3} \tau_{\alpha\beta}^{(k)} d\varepsilon_{\alpha\beta}^{(k)} \,, \tag{14.7}$$

$$dg = - sdT - \sum_{\alpha,\beta=1}^{3} \varepsilon_{\alpha\beta} d(v\tau^{(eq)}_{\alpha\beta}) - v \sum_{k=1}^{n} \sum_{\alpha,\beta=1}^{3} \tau^{(k)}_{\alpha\beta} d\varepsilon^{(k)}_{\alpha\beta} , \qquad (14.8)$$

$$dh = Tds - \sum_{\alpha,\beta=1}^{3} \varepsilon_{\alpha\beta} d(v\tau^{(eq)}_{\alpha\beta}) - v \sum_{k=1}^{n} \sum_{\alpha,\beta=1}^{3} \tau^{(k)}_{\alpha\beta} d\varepsilon^{(k)}_{\alpha\beta} , \qquad (14.9)$$

$$df^{*} = - sdT + v \sum_{\alpha,\beta=1}^{3} \tau^{(eq)}_{\alpha\beta} d\varepsilon_{\alpha\beta} + \sum_{k=1}^{n} \sum_{\alpha,\beta=1}^{3} \varepsilon^{(k)}_{\alpha\beta} d(v\tau^{(k)}_{\alpha\beta}) , \qquad (14.10)$$

$$dg^{*} = - sdT - \sum_{\alpha,\beta=1}^{3} \varepsilon_{\alpha\beta} d(v\tau^{(eq)}_{\alpha\beta}) + \sum_{k=1}^{n} \sum_{\alpha,\beta=1}^{3} \varepsilon^{(k)}_{\alpha\beta} d(v\tau^{(k)}_{\alpha\beta}) , \qquad (14.11)$$

$$dh^{*} = Tds - \sum_{\alpha,\beta=1}^{3} \varepsilon_{\alpha\beta} d(v\tau^{(eq)}_{\alpha\beta}) + \sum_{k=1}^{n} \sum_{\alpha,\beta=1}^{3} \varepsilon^{(k)}_{\alpha\beta} d(v\tau^{(k)}_{\alpha\beta}) . \qquad (14.12)$$

Using the relations $\tau^{(k)}_{(m)\alpha\beta} = \tau^{(k)}_{\alpha\beta} - \tau^{(eq)}_{\alpha\beta}$ and $\varepsilon_{\alpha\beta} = \varepsilon^{(0)}_{\alpha\beta} + \varepsilon^{(1)}_{\alpha\beta} + \ldots + \varepsilon^{(n)}_{\alpha\beta}$ (see (8.32) and (8.18)) one obtains from (14.7), (14.11) and (14.12)

$$df = - sdT + v \sum_{\alpha,\beta=1}^{3} \tau^{(eq)}_{\alpha\beta} d\varepsilon^{(0)}_{\alpha\beta} - v \sum_{k=1}^{n} \sum_{\alpha,\beta=1}^{3} \tau^{(k)}_{(m)\alpha\beta} d\varepsilon^{(k)}_{\alpha\beta} , \qquad (14.13)$$

$$dg^{*} = - sdT - \sum_{\alpha,\beta=1}^{3} \varepsilon^{(0)}_{\alpha\beta} d(v\tau^{(eq)}_{\alpha\beta}) + \sum_{k=1}^{n} \sum_{\alpha,\beta=1}^{3} \varepsilon^{(k)}_{\alpha\beta} d(v\tau^{(k)}_{(m)\alpha\beta}) , \qquad (14.14)$$

$$dh^{*} = Tds - \sum_{\alpha,\beta=1}^{3} \varepsilon^{(0)}_{\alpha\beta} d(v\tau^{(eq)}_{\alpha\beta}) + \sum_{k=1}^{n} \sum_{\alpha,\beta=1}^{3} \varepsilon^{(k)}_{\alpha\beta} d(v\tau^{(k)}_{(m)\alpha\beta}) . \qquad (14.15)$$

From (14.7) it follows that

$$s = - \frac{\partial}{\partial T} f(T, \varepsilon_{\alpha\beta}, \varepsilon^{(1)}_{\alpha\beta}, \varepsilon^{(2)}_{\alpha\beta}, \ldots, \varepsilon^{(n)}_{\alpha\beta}) , \qquad (14.16)$$

$$\tau^{(eq)}_{\alpha\beta} = \rho \frac{\partial}{\partial \varepsilon_{\alpha\beta}} f(T, \varepsilon_{\alpha\beta}, \varepsilon^{(1)}_{\alpha\beta}, \varepsilon^{(2)}_{\alpha\beta}, \ldots, \varepsilon^{(n)}_{\alpha\beta}) , \qquad (14.17)$$

$$\tau_{\alpha\beta}^{(k)} = - \rho \frac{\partial}{\partial \varepsilon_{\alpha\beta}^{(k)}} f(T, \varepsilon_{\alpha\beta}, \varepsilon_{\alpha\beta}^{(1)}, \varepsilon_{\alpha\beta}^{(2)}, \dots, \varepsilon_{\alpha\beta}^{(n)}) \quad (k = 1, \dots, n) \ . \tag{14.18}$$

Analogous results may be derived from (14.8)-(14.15).

If the free energy is known as function of $T, \varepsilon_{\alpha\beta}, \varepsilon_{\alpha\beta}^{(1)}, \dots, \varepsilon_{\alpha\beta}^{(n)}$, the entropy, the equilibrium-stress tensor and the affinity-stress tensor are also known as functions of the temperature and the strains. Such relations among stresses, strains and temperature are called equations of state.

In sections 16 and 20 we shall give some further results concerning equations of state and thermodynamic functions. Linear equations of state are considered in section 23.

## 15. MEDIA WITHOUT SHAPE MEMORY

We shall call a substance a medium without shape memory if the deviator of one of the memory-stress tensors, say $\tilde{\tau}^{(n)}_{(m)\alpha\beta}$, vanishes identically. If the deviator of none of the memory-stress tensors vanishes we speak of a medium with shape memory.

Let us give some equivalent formulations for the statement that a medium has no shape memory. In the first place we have, as noted above,

$$\text{no shape memory: } \tilde{\tau}^{(n)}_{(m)\alpha\beta} = 0 \, . \qquad (15.1)$$

This is equivalent with

$$\text{no shape memory: } \tau^{(n)}_{(m)\alpha\beta} = \tau^{(n)}_{(m)}\delta_{\alpha\beta} \, , \qquad (15.2)$$

where $\tau^{(n)}_{(m)}$ is the scalar part of $\tau^{(n)}_{(m)\alpha\beta}$ (see section 4). Furthermore, we have from (15.1) and (8.33)

$$\text{no shape memory: } \tilde{\tau}^{(n)}_{\alpha\beta} = \tilde{\tau}^{(eq)}_{\alpha\beta} \, , \qquad (15.3)$$

or, using (15.2) and (8.33),

$$\text{no shape memory: } \tau^{(n)}_{\alpha\beta} = \tau^{(eq)}_{\alpha\beta} + \tau^{(n)}_{(m)}\delta_{\alpha\beta} \, . \qquad (15.4)$$

Next, we consider the Gibbs relation for media without shape memory. From (8.34) and (15.2) we obtain with the aid of the definition (4.1) of the scalar part of a tensor

$$Tds = du - v \sum_{\alpha,\beta=1}^{3} \tau_{\alpha\beta}^{(eq)} d\varepsilon_{\alpha\beta}^{(0)} + v \sum_{k=1}^{n-1} \sum_{\alpha,\beta=1}^{3} \tau_{(m)\alpha\beta}^{(k)} d\varepsilon_{\alpha\beta}^{(k)} + 3v\tau_{(m)}^{(n)} d\varepsilon^{(n)} . \quad (15.5)$$

From (8.30) and (15.4) we obtain

$$Tds = du - v \sum_{\alpha,\beta=1}^{3} \tau_{\alpha\beta}^{(eq)} d\varepsilon_{\alpha\beta}^{*} + v \sum_{k=1}^{n-1} \sum_{\alpha,\beta=1}^{3} \tau_{\alpha\beta}^{(k)} d\varepsilon_{\alpha\beta}^{(k)} + 3v\tau_{(m)}^{(n)} d\varepsilon^{(n)} , \quad (15.6)$$

where

$$\varepsilon_{\alpha\beta}^{*} = \varepsilon_{\alpha\beta} - \varepsilon_{\alpha\beta}^{(n)} . \quad (15.7)$$

It follows from (15.5) and (15.6) that for media without shape memory one has for the entropy

$$s = s(u, \varepsilon_{\alpha\beta}^{(0)}, \varepsilon_{\alpha\beta}^{(1)}, \ldots, \varepsilon_{\alpha\beta}^{(n-1)}, \varepsilon^{(n)}) , \quad (15.8)$$

$$s = s(u, \varepsilon_{\alpha\beta}^{*}, \varepsilon_{\alpha\beta}^{(1)}, \ldots, \varepsilon_{\alpha\beta}^{(n-1)}, \varepsilon^{(n)}) . \quad (15.9)$$

The relations (15.7) and (15.9) show that the entropy does not depend on the components of $\tilde{\varepsilon}_{\alpha\beta}$. Therefore, for media without shape memory, the tensor of total strain is not a thermodynamic state variable. The specific volume, however, is a thermodynamic state variable. It is seen from (15.8) that s depends on $\varepsilon^{(0)}, \varepsilon^{(1)}, \ldots, \varepsilon^{(n-1)}, \varepsilon^{(n)}$. Thus, $\varepsilon = \varepsilon^{(0)} + \varepsilon^{(1)} + \ldots + \varepsilon^{(n)}$ is a state variable and $v = v_0(1 + 3\varepsilon)$ (see (4.14)) has the same property. Hence, one may also write for (15.8)

$$s = s(u, v, \varepsilon_{\alpha\beta}^{(0)}, \varepsilon_{\alpha\beta}^{(1)}, \ldots, \varepsilon_{\alpha\beta}^{(n-1)}) . \quad (15.10)$$

If a medium has no shape memory this has also consequences for the phenomenological equations and the form of the entropy production. In the phenomenological equations (11.2)-(11.4) $\tau_{\alpha\beta}^{(n)}$ may be replaced by the right-hand side of (15.4) and in (13.8) and (13.9) $\tilde{\tau}_{\alpha\beta}^{(n)}$ may be replaced by $\tilde{\tau}_{\alpha\beta}^{(eq)}$ (see (15.3)). Analogous remarks hold for the expressions (9.9), (9.10) and (9.12) for the entropy production.

## 16. NONVISCOUS MEDIA, ELASTIC STRAINS, HOOKE's LAW, THE DUHAMEL-NEUMANN LAW AND ECKART's PRINCIPLE OF RELAXABILITY-IN-THE-SMALL

A substance may be called a medium without shear viscosity if $\tilde{\tau}_{\alpha\beta}^{(vi)}$ vanishes, a medium without volume viscosity if $\tau^{(vi)}$ vanishes and a nonviscous medium if $\tau_{\alpha\beta}^{(vi)}$ vanishes. Hence,

$$\text{no shear viscosity: } \tilde{\tau}_{\alpha\beta}^{(vi)} = 0 , \tag{16.1}$$

$$\text{no volume viscosity: } \tau^{(vi)} = 0 , \tag{16.2}$$

$$\text{nonviscous: } \tau_{\alpha\beta}^{(vi)} = 0 . \tag{16.3}$$

Assumptions of this type have some consequences for the phenomenological tensors. For instance, one has from (11.3)

$$\text{nonviscous: } L_{\alpha\beta\gamma\zeta}^{(o)(k)} = 0 \quad (k = 0,1,\ldots,n) , \quad L_{\alpha\beta\gamma}^{(o)(q)} = 0 . \tag{16.4}$$

By virtue of the Onsager relations (12.8) and (12.11) one also has

$$\text{nonviscous: } L_{\alpha\beta\gamma\zeta}^{(k)(o)} = 0 \quad (k = 1,\ldots,n) , \quad L_{\alpha\beta\gamma}^{(q)(o)} = 0 . \tag{16.5}$$

For isotropic media we have from (13.9), (13.11) and the Onsager relations (13.5) and (13.7)

$$\text{no shear viscosity: } \eta_s^{(o,k)} = \eta_s^{(k,o)} = 0 \quad (k = 0,1,\ldots,n) , \tag{16.6}$$

$$\text{no volume viscosity: } \eta_v^{(o,k)} = \eta_v^{(k,o)} = 0 \quad (k = 0,1,\ldots,n) . \tag{16.7}$$

Hence, for isotropic media without shear viscosity the flow law (13.8) becomes

$$\text{no shear viscosity:}\quad \frac{d\tilde{\varepsilon}_{\alpha\beta}^{(j)}}{dt} = \sum_{k=1}^{n} \eta_s^{(j,k)} \tilde{\tau}_{\alpha\beta}^{(k)} \qquad (j = 1,\ldots,n) \tag{16.8}$$

and for media without volume viscosity (13.10) reduces to

$$\text{no volume viscosity:}\quad \frac{d\varepsilon^{(j)}}{dt} = \sum_{k=1}^{n} \eta_v^{(j,k)} \tau^{(j)} \qquad (j = 1,\ldots,n)\ . \tag{16.9}$$

If an anisotropic medium has no volume viscosity we must have $\Sigma_{\alpha=1}^{3} L_{\alpha\alpha\gamma\zeta}^{(o)(k)} = 0$ $(k = 0,1,\ldots,n)$ and $\Sigma_{\alpha=1}^{3} L_{\alpha\alpha\gamma}^{(o)(q)} = 0$ (see (11.3)).

From (8.39) it is seen that

$$\text{nonviscous:}\qquad \tau_{\alpha\beta} = \tau_{\alpha\beta}^{(eq)}\ , \tag{16.10}$$

$$\text{no shear viscosity:}\quad \tilde{\tau}_{\alpha\beta} = \tilde{\tau}_{\alpha\beta}^{(eq)}\ , \tag{16.11}$$

$$\text{no volume viscosity:}\quad \tau = \tau^{(eq)}\ . \tag{16.12}$$

If a medium is nonviscous both the expressions (9.9) and (9.10) for the entropy production reduce to

$$T\sigma^{(s)} = -\frac{1}{T}\vec{J}^{(q)}\cdot \text{grad}\, T + \sum_{k=1}^{n}\sum_{\alpha,\beta=1}^{3} \tau_{\alpha\beta}^{(k)} \frac{d\varepsilon_{\alpha\beta}^{(k)}}{dt}\ . \tag{16.13}$$

Furthermore, one may write for (8.27), (8.30), (14.7) and (14.17)

$$\tau_{\alpha\beta} = -\rho T \frac{\partial}{\partial\varepsilon_{\alpha\beta}} s(u,\varepsilon_{\alpha\beta},\varepsilon_{\alpha\beta}^{(1)},\ldots,\varepsilon_{\alpha\beta}^{(n)})\ , \tag{16.14}$$

$$T ds = du - v\sum_{\alpha,\beta=1}^{3} \tau_{\alpha\beta}\, d\varepsilon_{\alpha\beta} + v\sum_{k=1}^{n}\sum_{\alpha,\beta=1}^{3} \tau_{\alpha\beta}^{(k)}\, d\varepsilon_{\alpha\beta}^{(k)}\ , \tag{16.15}$$

$$df = -\,s\,dT + v \sum_{\alpha,\beta=1}^{3} \tau_{\alpha\beta}\, d\varepsilon_{\alpha\beta} - v \sum_{k=1}^{n} \sum_{\alpha,\beta=1}^{3} \tau_{\alpha\beta}^{(k)}\, d\varepsilon_{\alpha\beta}^{(k)} \,, \tag{16.16}$$

$$\tau_{\alpha\beta} = \rho \frac{\partial}{\partial \varepsilon_{\alpha\beta}} f(T, \varepsilon_{\alpha\beta}, \varepsilon_{\alpha\beta}^{(1)}, \ldots, \varepsilon_{\alpha\beta}^{(n)}) \,. \tag{16.17}$$

Hence, $\tau_{\alpha\beta}$ is a thermodynamic state variable. In an analogous way one has from (14.13)

$$\tau_{\alpha\beta} = \rho \frac{\partial}{\partial \varepsilon_{\alpha\beta}^{(0)}} f(T, \varepsilon_{\alpha\beta}^{(0)}, \varepsilon_{\alpha\beta}^{(1)}, \ldots, \varepsilon_{\alpha\beta}^{(n)}) \,. \tag{16.18}$$

If, in a first approximation, it is permissible to linearize the right-hand side of (16.18) (linear equation of state) one obtains

$$\tau_{\alpha\beta} = \sum_{\gamma,\zeta=1}^{3} a_{\alpha\beta\gamma\zeta}^{(0)(0)} \varepsilon_{\gamma\zeta}^{(0)} + a_{\alpha\beta}^{(0)} (T - T_0) \,, \tag{16.19}$$

where $T_0$ is the temperature introduced in the sections 7 and 8 and $a_{\alpha\beta\gamma\zeta}^{(0)(0)}$ and $a_{\alpha\beta}^{(0)}$ are two tensors of which the components are constants. On the right-hand side of (16.19) terms with the components of $\varepsilon_{\alpha\beta}^{(1)}, \ldots, \varepsilon_{\alpha\beta}^{(n)}$ do not occur because $\varepsilon_{\alpha\beta}^{(0)}$ is defined in sections 7 and 8 so that it must vanish for all values of $\varepsilon_{\alpha\beta}^{(1)}, \ldots, \varepsilon_{\alpha\beta}^{(n)}$ if $T = T_0$ and $\tau_{\alpha\beta} = 0$ (see (8.25)). Hence, for a nonviscous medium $\varepsilon_{\alpha\beta}^{(0)}$ is the thermoelastic part of the strain and (16.19) is the Duhamel-Neumann law. If one neglects thermal effects it reduces to Hooke's law. The conclusion that for a nonviscous medium $\varepsilon_{\alpha\beta}^{(0)}$ may be considered as the thermoelastic part of the strain may also be drawn from the (not necessarily linear) equation (16.18). If one considers a sufficiently small bit of matter one may quickly change the temperature. By external influences one may also quickly change the mechanical stresses. Instantaneous changes in $\varepsilon_{\alpha\beta}^{(1)}, \ldots, \varepsilon_{\alpha\beta}^{(n)}$ are impossible because of the flow law (16.8). The only part of the strain which can quickly change is $\varepsilon_{\alpha\beta}^{(0)}$. Therefore, one may consider $\varepsilon_{\alpha\beta}^{(0)}$ as the (thermo) elastic strain tensor. It should be noted that $d\varepsilon_{\alpha\beta}^{(0)}/dt$ does not occur in the expression (16.13) for the entropy production. Hence, one may say that for a nonviscous medium $\varepsilon_{\alpha\beta}^{(0)}$ is the reversible part of the strain.

The remarks above are closely connected with Eckart's principle of relaxability-in-the-small.[27] In fact, this principle is a hypothetical experiment. We consider a small bit of matter (surrounding some point of the medium), which is large from the microscopic point of view (i.e. it contains still a great many molecules (or atoms)) and small from the macroscopic point of view (i.e. the variations within the bit of such quantities as the temperature and the strains may be neglected). If, when the medium is in an arbitrary state, the bit is cut out, all strains will be relaxed, since on the surface of the bit no forces work. (As we have seen before such a sudden strain relaxation can only occur in nonviscous materials.) The part of the strain which is suddenly relaxed is the elastic strain and the remaining part is the total inelastic strain.

It is seen that (8.14) is the generalization of this idea to the case of thermoelastic strains and that the total inelastic strain is given by $\varepsilon_{\alpha\beta}^{(1)} + \ldots + \varepsilon_{\alpha\beta}^{(n)}$ (see (8.15)-(8.18)).

## 17. NONVISCOUS MEDIA WITHOUT SHAPE MEMORY, LÉVY's LAW

If a medium is nonviscous and if, moreover, it has no shape memory, it follows from (15.3) and (16.11) that

$$\tilde{\tau}_{\alpha\beta} = \tilde{\tau}^{(eq)}_{\alpha\beta} = \tilde{\tau}^{(n)}_{\alpha\beta} \tag{17.1}$$

and from (15.4) and (16.12) one obtains

$$\tau^{(n)} = \tau^{(eq)} + \tau^{(n)}_{(m)} = \tau + \tau^{(n)}_{(m)} \ . \tag{17.2}$$

In particular, we shall investigate the case n = 1. We then have

$$\varepsilon_{\alpha\beta} = \varepsilon^{(0)}_{\alpha\beta} + \varepsilon^{(1)}_{\alpha\beta} \tag{17.3}$$

(see section 7, where the case n = 1 is discussed, and in particular (7.19)). From (15.8) we obtain

$$s = s(u, \varepsilon^{(0)}_{\alpha\beta}, \varepsilon^{(1)}) \ , \tag{17.4}$$

or (see (15.10))

$$s = s(u, v, \varepsilon^{(0)}_{\alpha\beta}) \ . \tag{17.5}$$

Since we assume that the medium is nonviscous we have $\eta_s^{(1,0)} = 0$ (see (16.6)) and, hence, the flow law (13.8) becomes

$$\frac{d\tilde{\varepsilon}^{(1)}_{\alpha\beta}}{dt} = \eta_s^{(1,1)}\tilde{\tau}^{(1)}_{\alpha\beta} , \tag{17.6}$$

or, using (17.1) with n = 1,

$$\frac{d\tilde{\varepsilon}^{(1)}_{\alpha\beta}}{dt} = \eta_s^{(1,1)}\tilde{\tau}_{\alpha\beta} . \tag{17.7}$$

The latter formula has been proposed by Lévy[50] in 1870 and is called Lévy's law. It states that the material derivative with respect to time of the deviatoric inelastic strain is proportional to the deviatoric mechanical stress.

Finally, we note that it follows from (17.1) and (17.2) that the entropy production (16.13) for the case n = 1 is given by

$$T\sigma^{(s)} = -\frac{1}{T}\vec{J}^{(q)}\cdot \text{grad } T + \sum_{\alpha,\beta=1}^{3} \tilde{\tau}_{\alpha\beta}\frac{d\tilde{\varepsilon}^{(1)}_{\alpha\beta}}{dt} + 3\tau^{(1)}\frac{d\varepsilon^{(1)}}{dt} , \tag{17.8}$$

where we also used (4.8).

# 18. SOURCES OF NONLINEARITY

The phenomenological equations which we have discussed in sections 11 and 12 are often denoted as the linear laws of nonequilibrium thermodynamics. Indeed these laws are linear relations among fluxes and affinities. However, the phenomenological tensors cannot be considered as constants.

As an example we take the viscosity of a fluid mixture which consists of an arbitrary number of chemical components. It is known from experimental data that in general the viscosity of such a mixture depends on the temperature, the pressure and the concentrations of the chemical components (see section 6), i.e. the viscosity depends on all thermodynamic state variables. Related phenomena are structural viscosity, thixotropy and antithixotropy. An analogous remark (as for the viscosity) holds for the heat conductivity of a mixture. The phenomenological tensors may even show jumps (superconductivity). We also note that the possible dependence of the phenomenological tensors on the magnetic field (if such a field is present) may not be neglected if one formulates the Onsager relations. If $\vec{B}$ is the magnetic field (12.4)-(12.6) become

$$L^{(oe)}_{ij}(\vec{B}) = L^{(oe)}_{ji}(-\vec{B}) \qquad (i,j = 1,\ldots,n) \, , \tag{18.1}$$

$$L^{(eo)}_{k\ell}(\vec{B}) = L^{(eo)}_{\ell k}(-\vec{B}) \qquad (k,\ell = 1,\ldots,m) \, , \tag{18.2}$$

$$L^{(oo)}_{ik}(\vec{B}) = -L^{(ee)}_{ki}(-\vec{B}) \quad (i = 1,\ldots,n; \; k = 1,\ldots,m) \, . \tag{18.3}$$

In section 14 we gave some examples of equations of state (see (14.17) and (14.18)). Such equations of state (as Hooke's law or the Duhamel-Neumann law) are often assumed to be linear in a first approximation. According to Reiner[26] Hooke's law for elasticity is not general enough, even not for strains which may be considered as small from a geometrical point of view. Reiner[26] refers to observations by Poynting[51,52] (Poynting effect). It may be that, also for the more complicated substances considered in these lectures simple linearizations of the equations of state as (16.19) are not always permissible.

In rheology one has also introduced nonlinear relations among fluxes and affinities. For instance, Reiner-Rivlin fluids are defined by the constitutive equation

$$\tau_{\alpha\beta}^{(vi)} = f_{(1)}\delta_{\alpha\beta} + f_{(2)} \frac{d\varepsilon_{\alpha\beta}}{dt} + f_{(3)} \sum_{\gamma=1}^{3} \frac{d\varepsilon_{\alpha\gamma}}{dt} \frac{d\varepsilon_{\gamma\beta}}{dt} , \qquad (18.4)$$

where the scalars $f_{(k)}$ $(k = 1,2,3)$ may still depend on the invariants of the tensor $d\varepsilon_{\alpha\beta}/dt$ (and on $T$). There is, however, no definite experimental evidence that such fluids exist.[26]

Finally, it should be noted that the fact that in many equations the substantial derivative with respect to time occurs (and not the local derivative) is a fourth source of nonlinear behaviour. Moreover, if the deformations and rotations are large from a kinematical point of view, this may also give rise to nonlinearities.

## 19. REMARKS ON THE DOMAIN OF VALIDITY OF THERMODYNAMIC THEORIES

If a medium is too far outside of thermodynamic equilibrium, the methods of nonequilibrium thermodynamics cannot be applied. Let us consider a gas. If the variations in temperature over a distance comparable with the mean free path of the molecules is of the order of magnitude of the temperature itself the gas is outside of the domain of applicability of nonequilibrium thermodynamics. Furthermore, many chemical reactions cannot be described by the type of "linear laws" discussed in section 11. The domain of validity does not follow from macroscopic considerations. Either it must be determined from experimental data or it must be derived with the aid of microscopic theories.

If a system moves with high velocities with respect to the observer also relativistic effects must be taken into account.

On the other hand the developed equations do not hold if the medium is not sufficiently far outside of thermodynamic equilibrium. Suppose that there is a temperature difference $\Delta T$ over a distance of, say, one meter. If $\Delta T$ is so small that it is comparable with the mean value of the temperature fluctuations in the medium one cannot expect that Fourier's law holds in its usual form (see, for instance, (13.12)). In the presence of fluctuations Landau and Lifschitz write for Fourier's law (see p. 524 of reference 19)

$$\vec{J}^{(q)} = - \lambda \text{ grad } T + \vec{g} , \tag{19.1}$$

where $\vec{g}$ is a random vectorial quantity which is not related to velocity and temperature gradients.

If one considers the solution of the equation for the heat conduction, for instance for a point source, the speed of propagation of thermal disturbances is infinite. However, at large distances of the source grad T is very small for such disturbances and the term $\vec{g}$ on the right-hand side of (19.1) becomes the dominant term. Hence, the parabolic equation for the heat conduction fails to give a correct description. Analogous remarks also hold for the other phenomenological equations if the medium is not sufficiently far outside of thermodynamic equilibrium.

The idea to investigate the consequences of the assumption that the entropy also depends on fluxes (as the heat flow) is probably of great importance for the study of systems which are far outside of thermodynamic equilibrium. However, in order to explain the paradox of the parabolic equations it may be of importance to take into account the influence of fluctuations, as indicated above. Under normal laboratory conditions ("moderate non-equilibrium states") parabolic equations as the equation for the heat conduction give a good description of physical reality.

## 20. ELASTIC-PLASTIC MEDIA, SINGULAR YIELD CONDITIONS

In section 18 we have discussed four possible sources of nonlinear behaviour of materials: nonconstant phenomenological tensors, nonlinear equations of state, nonlinear phenomenological equations and the substantial derivative with respect to time. We shall now show that a plasticity theory can be developed which contains well-known formulae for plastic flow (as the Von Mises flow law) as special cases if we introduce the inelastic strains as internal thermodynamic variables and if we make an assumptiom concerning the dependence of the phenomenological tensors on the thermodynamic state variables.

Let us consider an isotropic nonviscous medium. For such media the definition (14.2) of the thermodynamic potential becomes

$$g = u - Ts - v \sum_{\alpha,\beta=1}^{3} \tau_{\alpha\beta}\, \varepsilon_{\alpha\beta} \,. \tag{20.1}$$

For the differential of g we get (see (14.8))

$$dg = - sdT - \sum_{\alpha,\beta=1}^{3} \varepsilon_{\alpha\beta}\, d(v\tau_{\alpha\beta}) - v \sum_{k=1}^{n} \sum_{\alpha,\beta=1}^{3} \tau_{\alpha\beta}^{(k)}\, d\varepsilon_{\alpha\beta}^{(k)} \,. \tag{20.2}$$

We recall that the free energy is defined by f = u - Ts. For nonviscous media its differential is given by (see (14.7))

$$df = - sdT + v \sum_{\alpha,\beta=1}^{3} \tau_{\alpha\beta}\, d\varepsilon_{\alpha\beta} - v \sum_{k=1}^{n} \sum_{\alpha,\beta=1}^{3} \tau_{\alpha\beta}^{(k)}\, d\varepsilon_{\alpha\beta}^{(k)} \,. \tag{20.3}$$

Since $\varepsilon_{\alpha\beta} = \varepsilon^{(0)}_{\alpha\beta} + \varepsilon^{(1)}_{\alpha\beta} + \ldots + \varepsilon^{(n)}_{\alpha\beta}$ (see (8.18)) and since for nonviscous media we have from (9.11)

$$\tau^{(k)}_{\alpha\beta} = \tau_{\alpha\beta} + \tau^{(k)}_{(m)\alpha\beta} , \tag{20.4}$$

one may also write for (20.3)

$$df = - sdT + v \sum_{\alpha,\beta=1}^{3} \tau_{\alpha\beta} d\varepsilon^{(0)}_{\alpha\beta} - v \sum_{k=1}^{n} \sum_{\alpha,\beta=1}^{3} \tau^{(k)}_{(m)\alpha\beta} d\varepsilon^{(k)}_{\alpha\beta} . \tag{20.5}$$

Let us define $g^{**}$ by

$$g^{**} = u - Ts - v \sum_{\alpha,\beta=1}^{3} \tau_{\alpha\beta} \varepsilon^{(0)}_{\alpha\beta} . \tag{20.6}$$

From $f = u - Ts$, (20.5) and (20.6) we then obtain

$$dg^{**} = - sdT - \sum_{\alpha,\beta=1}^{3} \varepsilon^{(0)}_{\alpha\beta} d(v\tau_{\alpha\beta}) - v \sum_{k=1}^{n} \sum_{\alpha,\beta=1}^{3} \tau^{(k)}_{(m)\alpha\beta} d\varepsilon^{(k)}_{\alpha\beta} . \tag{20.7}$$

Next, we list some consequences of the preceding definitions and relations. From (20.5) we have

$$\tau_{\alpha\beta} = \rho \frac{\partial}{\partial \varepsilon^{(0)}_{\alpha\beta}} f(T,\varepsilon^{(0)}_{\alpha\beta},\varepsilon^{(1)}_{\alpha\beta},\ldots,\varepsilon^{(n)}_{\alpha\beta}) \tag{20.8}$$

and

$$\tau^{(k)}_{(m)\alpha\beta} = - \rho \frac{\partial}{\partial \varepsilon^{(k)}_{\alpha\beta}} f(T,\varepsilon^{(0)}_{\alpha\beta},\varepsilon^{(1)}_{\alpha\beta},\ldots,\varepsilon^{(n)}_{\alpha\beta}) , \tag{20.9}$$

where we also used $\rho = v^{-1}$. Furthermore, we obtain from (20.2)

$$\tau^{(k)}_{\alpha\beta} = - \rho \frac{\partial}{\partial \varepsilon^{(k)}_{\alpha\beta}} g(T,v\tau_{\alpha\beta},\varepsilon^{(1)}_{\alpha\beta},\ldots,\varepsilon^{(n)}_{\alpha\beta}) \tag{20.10}$$

and from (20.7)

$$\tau^{(k)}_{(m)\alpha\beta} = - \rho \frac{\partial}{\partial\varepsilon^{(k)}_{\alpha\beta}} g^{**}(T, v\tau_{\alpha\beta}, \varepsilon^{(1)}_{\alpha\beta}, \ldots, \varepsilon^{(n)}_{\alpha\beta}) \ . \tag{20.11}$$

Finally, it is seen from (20.2) that

$$g = g(T, v\tau_{\alpha\beta}, \varepsilon^{(1)}_{\alpha\beta}, \ldots, \varepsilon^{(n)}_{\alpha\beta}) \ , \tag{20.12}$$

i.e. the quantities $T, v\tau_{\alpha\beta}, \varepsilon^{(1)}_{\alpha\beta}, \ldots, \varepsilon^{(n)}_{\alpha\beta}$ form a complete set of thermodynamic state variables.

Next, we consider the flow law for isotropic nonviscous media. We shall assume that there are no cross effects among the n inelastic flows (i.e. $\eta_s^{(j,k)} = 0$ $(j,k = 1,\ldots,n;\ j \neq k)$). One then gets from (16.8)

$$\frac{d\tilde{\varepsilon}^{(j)}_{\alpha\beta}}{dt} = \eta_s^{(j,j)} \tilde{\tau}^{(j)}_{\alpha\beta} \qquad (j = 1,\ldots,n) \ . \tag{20.13}$$

In general one may assume that plasticity does not affect volumetric phenomena, i.e. the $\eta_v^{(j,k)}$ vanish and (16.9) becomes

$$\frac{d\varepsilon^{(j)}}{dt} = 0 \qquad (j = 1,\ldots,n) \ . \tag{20.14}$$

We also recall Fourier's law (see (13.12))

$$\vec{J}^{(q)} = - \lambda \ \mathrm{grad}\ T \ . \tag{20.15}$$

Next, we consider the entropy production. Using the three above flow laws (16.13) becomes

$$T\sigma^{(s)} = \frac{\lambda}{T} (\mathrm{grad}\ T)^2 + \sum_{j=1}^{n} \eta_s^{(j,j)} \sum_{\alpha,\beta=1}^{3} (\tilde{\tau}^{(j)}_{\alpha\beta})^2 \ . \tag{20.16}$$

We shall assume that

$$\lambda \geq 0 \ , \quad \eta_s^{(j,j)} \geq 0 \qquad (j = 1,\ldots,n) \ . \tag{20.17}$$

These inequalities ensure the nonnegative character of $\sigma^{(s)}$.

It is seen from (20.12) that $T, v\tau_{\alpha\beta}, \varepsilon_{\alpha\beta}^{(1)},\ldots,\varepsilon_{\alpha\beta}^{(n)}$ form a complete set of thermodynamic state variables. In general the phenomenological coefficients may depend on these state variables. Thus, one may assume that, if v is constant in a first approximation,

$$\eta_s^{(i,i)} = \eta_s^{(i,i)}(T,\tau_{\alpha\beta},\varepsilon_{\alpha\beta}^{(1)},\ldots,\varepsilon_{\alpha\beta}^{(n)}) \qquad (i = 1,\ldots,n) \ . \tag{20.18}$$

Phenomenological theories (as thermodynamics) do not provide any information on the explicit form of the functions on the right-hand sides of (20.18). In order to explain plasticity phenomena we now introduce a physical assumption concerning this dependence. We suppose that the functions on the right-hand sides of (20.18) are such that there exist functions $\Phi^{(1,1)},\Phi^{(2,2)},\ldots,\Phi^{(n,n)}$ of the thermodynamic state variables, i.e.

$$\Phi^{(i,i)} = \Phi^{(i,i)}(T,\tau_{\alpha\beta},\varepsilon_{\alpha\beta}^{(1)},\ldots,\varepsilon_{\alpha\beta}^{(n)}) \qquad (i = 1,\ldots,n) \tag{20.19}$$

such that

$$\eta_s^{(i,i)} = 0 \quad \text{if} \quad \Phi^{(i,i)} \leq 0 \ , \tag{20.20}$$

$$\eta_s^{(i,i)} \geq 0 \quad \text{if} \quad \Phi^{(i,i)} = 0 \ , \tag{20.21}$$

$$\eta_s^{(i,i)} > 0 \quad \text{if} \quad \Phi^{(i,i)} > 0 \ . \tag{20.22}$$

We shall call the functions $\Phi^{(i,i)}$ partial yield functions. We shall say that the medium is in the preplastic range if all the partial yield functions are negative. The surfaces $\Phi^{(i,i)} = 0$ in $\tau_{\alpha\beta}$-space may be called partial yield surfaces. The boundary of the preplastic range in $\tau_{\alpha\beta}$-space consists of parts of the partial yield surfaces and may be called the effective yield surface. At intersections of the partial yield surfaces the effective yield surface may have edges and corners. Thus, edges and corners in the effective yield surface may be a consequence of the fact that several microscopic processes may give rise to plastic deformations, where each of these phenomena has its own partial yield condition. Of course, this does not exclude the possibility that one or more of the partial yield surfaces may also show singularities.

Since it is assumed that the medium is nonviscous, it is purely (thermo) elastic in the preplastic range. The plastic strain $\varepsilon_{\alpha\beta}^{(pl)}$ may be defined by

$$\varepsilon_{\alpha\beta}^{(pl)} = \sum_{k=1}^{n} \varepsilon_{\alpha\beta}^{(k)} \tag{20.23}$$

and in this case (where the phenomenological coefficients have the property expressed by (20.20)-(20.22)) the $\varepsilon_{\alpha\beta}^{(k)}$ may be called the partial plastic strain tensors.

## 21. PARTIAL YIELD FUNCTIONS, KOITER's FORMULA, STRAIN HARDENING, THE BAUSCHINGER EFFECT AS CONSEQUENCE OF MEMORY STRESSES

As noted in the preceding section, the explicit forms of $\eta_s^{(i,i)}$ and, hence, also of $\Phi^{(i,i)}$ as functions of $T,\tau_{\alpha\beta},\varepsilon_{\alpha\beta}^{(1)},\ldots,\varepsilon_{\alpha\beta}^{(n)}$ cannot be deduced from arguments of a phenomenological nature. However, let us investigate the following form for $\Phi^{(i,i)}$

$$\Phi^{(i,i)} = \tfrac{1}{2}\sum_{\alpha,\beta=1}^{3}(\tilde{\tau}_{\alpha\beta}^{(i)})^2 - (K^{(i,i)})^2 \qquad (i = 1,\ldots,n) , \tag{21.1}$$

where the $K^{(i,i)}$ are functions of the temperature and of the partial plastic strain tensors, i.e.

$$K^{(i,i)} = K^{(i,i)}(T,\varepsilon_{\alpha\beta}^{(1)},\ldots,\varepsilon_{\alpha\beta}^{(n)}) \qquad (i = 1,\ldots,n) . \tag{21.2}$$

Using (21.2) and (20.10) it is seen from (21.1) that indeed the $\Phi^{(i,i)}$ have the form (20.19). Using (20.4) we obtain from (21.1)

$$\Phi^{(i,i)} = \tfrac{1}{2}\sum_{\alpha,\beta=1}^{3}(\tilde{\tau}_{\alpha\beta})^2 + \sum_{\alpha,\beta=1}^{3}\tilde{\tau}_{\alpha\beta}\tilde{\tau}_{(m)\alpha\beta}^{(i)} - (k^{(i,i)})^2 \quad (i = 1,\ldots,n) , \tag{21.3}$$

where

$$(k^{(i,i)})^2 = (K^{(i,i)})^2 - \tfrac{1}{2}\sum_{\alpha,\beta=1}^{3}(\tilde{\tau}_{(m)\alpha\beta}^{(i)})^2 \qquad (i = 1,\ldots,n) . \tag{21.4}$$

If one assumes that the specific free energy f has the form

$$f = f^{(1)} + f^{(2)} , \tag{21.5}$$

where

$$f^{(1)} = f^{(1)}(T,\varepsilon_{\alpha\beta}^{(0)}) \tag{21.6}$$

and

$$f^{(2)} = f^{(2)}(T,\varepsilon_{\alpha\beta}^{(1)},\ldots,\varepsilon_{\alpha\beta}^{(n)}) , \tag{21.7}$$

it follows from (20.8) and (20.9) that

$$\tau_{\alpha\beta} = \tau_{\alpha\beta}(T,\varepsilon_{\alpha\beta}^{(0)}) \tag{21.8}$$

and that

$$\tau_{(m)\alpha\beta}^{(i)} = \tau_{(m)\alpha\beta}^{(1)}(T,\varepsilon_{\alpha\beta}^{(1)},\ldots,\varepsilon_{\alpha\beta}^{(n)}) , \tag{21.9}$$

if $\rho = v^{-1}$ is considered as a constant in a first approximation. From (21.4) we have with the aid of (21.2) and (21.9)

$$k^{(i,i)} = k^{(i,i)}(T,\varepsilon_{\alpha\beta}^{(1)},\ldots,\varepsilon_{\alpha\beta}^{(n)}) \qquad (i = 1,\ldots,n) . \tag{21.10}$$

Hence, the assumption that (21.5), (21.6) and (21.7) hold leads to the result that the relation among mechanical stress, elastic strain and temperature is not influenced by the plastic strains (see (21.8)). In a first approximation this is in agreement with experimental findings (see the references 53, 54 and 55). As we have shown in section 16, an equation of state of the type of (21.8) is a consequence of (8.25) and (16.18). Indeed, the Duhamel-Neumann law (16.19) has the structure of (21.8).

By virtue of (21.9), (21.10) and (20.4) we have from (21.3)

$$\frac{\partial}{\partial\tau_{\alpha\beta}} \Phi^{(i,i)}(T,\tau_{\alpha\beta},\varepsilon_{\alpha\beta}^{(1)},\ldots,\varepsilon_{\alpha\beta}^{(n)}) = \tilde{\tau}_{\alpha\beta} + \tilde{\tau}_{(m)\alpha\beta}^{(i)} = \tilde{\tau}_{\alpha\beta}^{(i)} . \tag{21.11}$$

Hence, one may write for the flow laws (20.13)

$$\frac{d\tilde{\varepsilon}^{(i)}_{\alpha\beta}}{dt} = \eta_s^{(i,i)} \frac{\partial \Phi^{(i,i)}}{\partial \tau_{\alpha\beta}} \qquad (i = 1,\dots,n) \ . \tag{21.12}$$

For the time derivative of the deviator of the plastic strain tensor $\varepsilon^{(pl)}_{\alpha\beta}$ defined by (20.23) one gets with the aid of the flow laws (21.12)

$$\frac{d\tilde{\varepsilon}^{(pl)}_{\alpha\beta}}{dt} = \sum_{j=1}^{n} \eta_s^{(j,j)} \frac{\partial \Phi^{(j,j)}}{\partial \tau_{\alpha\beta}} , \tag{21.13}$$

where the coefficients $\eta_s^{(j,j)}$ satisfy (20.18), (20.20), (20.21) and (20.22).

A formula of the type of (21.13) plays an important role in Koiter's theory for plasticity phenomena in media with singular yield surfaces. It should be emphasized, however, that there are several differences between Koiter's theory and the present theory. In Koiter's theory a formula of the type of (21.13) is introduced as a basic law for perfectly plastic materials. These materials are defined by the property that they cannot support stresses in excess of a certain fixed yield limit. For work-hardening materials with singular yield surfaces Koiter proposes a different formula. In the theory presented here, it is assumed that $\Phi^{(i,i)}$ and $\eta_s^{(i,i)}$ are functions of the thermodynamic state parameters (for instance, $T, \tau_{\alpha\beta}, \varepsilon^{(1)}_{\alpha\beta}, \dots, \varepsilon^{(n)}_{\alpha\beta}$). Furthermore, is is assumed that $\varepsilon^{(1)}_{\alpha\beta}, \varepsilon^{(2)}_{\alpha\beta}, \dots, \varepsilon^{(n)}_{\alpha\beta}$ are internal thermodynamic degrees of freedom, that methods of nonequilibrium thermodynamics may be applied, that the total strain is given by $\varepsilon_{\alpha\beta} = \varepsilon^{(0)}_{\alpha\beta} + \varepsilon^{(1)}_{\alpha\beta} + \dots + \varepsilon^{(n)}_{\alpha\beta}$ and that the coefficients $\eta_s^{(1,1)}, \eta_s^{(2,2)}, \dots, \eta_s^{(n,n)}$ satisfy the relations (20.20), (20.21) and (20.22). We also supposed that $\Phi^{(i,i)}$ has the form (21.1). If, moreover, it is assumed that the free energy f has the form determined by (21.5)-(21.7), the thermodynamic flow laws (20.13) may be written in the forms (21.12) and (21.13).

The form (21.1) for the partial yield functions has the following characteristics. As noted in the preceding section the effective yield surface may have singularities at intersections of the partial yield surfaces. Secondly, it is seen from (21.3) that

$$\phi^{(i,i)}(T,-\tilde{\tau}_{\alpha\beta},\varepsilon^{(1)}_{\alpha\beta},\ldots,\varepsilon^{(n)}_{\alpha\beta}) \neq \phi^{(i,i)}(T,\tilde{\tau}_{\alpha\beta},\varepsilon^{(1)}_{\alpha\beta},\ldots,\varepsilon^{(n)}_{\alpha\beta}) \; . \qquad (21.14)$$

This inequality is called the Bauschinger effect. It is seen from (21.3) that this effect is caused by the internal product of the deviator of the mechanical stress tensor $\tau_{\alpha\beta}$ and the deviator of the memory-stress tensor $\tau^{(i)}_{(m)\alpha\beta}$. Hence, each of the partial plastic strains has its own Bauschinger effect which is caused by the deviatoric memory stress. If in some state the deviatoric memory stress vanishes the Bauschinger effect does not occur in that state. Since $\tau^{(i)}_{(m)\alpha\beta}$ and $k^{(i,i)}$ may depend on the partial plastic strains (see (21.9) and (21.10)) strain hardening is included. Finally, we note that it is seen from (21.12) that each partial plastic flow is perpendicular to its partial yield surface.

## 22. THERMODYNAMIC EQUILIBRIUM AND STATES OF PERFECT THERMODYNAMIC EQUILIBRIUM OF ELASTIC-PLASTIC MEDIA, THE REFERENCE STATE

A medium is in a state of thermodynamic equilibrium if the entropy production vanishes. For elastic-plastic media satisfying the flow laws (20.13)-(20.15) the entropy production is given by (20.16). It is seen from this equation that $\sigma^{(s)}$ vanishes if grad T and $\eta_s^{(i,i)}$ $(i = 1,\dots,n)$ vanish. Hence, the medium is in thermodynamic equilibrium if it is at uniform temperature in the preplastic range (characterized by $\Phi^{(i,i)} < 0$ for $i = 1,\dots,n$; see (20.20)). We note that it is not necessary for thermodynamic equilibrium that $\tilde{\tau}_{\alpha\beta}$ and $\tilde{\varepsilon}^{(0)}_{\alpha\beta}$ are constant or that they vanish.

We shall say that a medium is in a state of perfect thermodynamic equilibrium at temperature T if T is uniform and if $\tau_{\alpha\beta}, \tau^{(1)}_{\alpha\beta}, \dots, \tau^{(n)}_{\alpha\beta}$ all vanish. Hence, in this case the entropy production (20.16) also vanishes. It is seen from (20.4) that in perfect equilibrium the memory-stresses $\tau^{(i)}_{(m)\alpha\beta}$ also vanish. Hence, one may determine for any temperature the strains $\varepsilon^{(0)}_{\alpha\beta}, \dots, \varepsilon^{(n)}_{\alpha\beta}$ in a state of perfect thermodynamic equilibrium from the relations (20.8) and (20.9) (with vanishing left-hand sides). If, however, a medium has no shape memory the deviator of one of the memory-stress tensors, say $\tilde{\tau}^{(n)}_{(m)\alpha\beta}$, vanishes identically and $\tilde{\varepsilon}^{(n)}_{\alpha\beta}$ is not a thermodynamic state variable. Hence, $\tilde{\varepsilon}^{(n)}_{\alpha\beta}$ is not determined by the conditions for perfect equilibrium and can be chosen arbitrarily.

In section 3 we introduced the notion of reference state. We shall take as reference state, a state of perfect thermodynamic equilibrium at a fixed temperature $T_0$. By means of transformations of variables one may choose the $\varepsilon^{(k)}_{\alpha\beta}$ so that they all vanish in the reference state.

# 23. LINEAR EQUATIONS OF STATE FOR ISOTROPIC MEDIA, FREUDENTHAL's YIELD FUNCTION, A STRESS-STRAIN RELATION

Let us suppose that the specific free energy f for isotropic media has the form

$$f = v_0\Big[\frac{1}{2}\,a^{(0,0)}\sum_{\alpha,\beta=1}^{3}(\tilde{\varepsilon}^{(0)}_{\alpha\beta})^2 + \frac{1}{2}\sum_{j,k=1}^{n}(a^{(j,k)} - a^{(0,0)})\sum_{\alpha,\beta=1}^{3}\tilde{\varepsilon}^{(j)}_{\alpha\beta}\tilde{\varepsilon}^{(k)}_{\alpha\beta} +$$

$$+ \frac{3}{2}\,b^{(0,0)}(\varepsilon^{(0)})^2 + \frac{3}{2}\sum_{j,k=1}^{n}(b^{(j,k)} - b^{(0,0)})\varepsilon^{(j)}\varepsilon^{(k)} +$$

$$+ 3(T - T_0)\Big\{c^{(0)}\varepsilon^{(0)} + \sum_{j=1}^{n}(c^{(0)} - c^{(j)})\varepsilon^{(j)}\Big\}\Big] - \psi\,, \tag{23.1}$$

where $\psi$ is some function of the temperature, $a^{(0,0)}$, $b^{(0,0)}$, $a^{(j,k)}$, $b^{(j,k)}$ $(j,k = 1,\dots,n)$ and $c^{(j)}$ $(j = 0,1,\dots,n)$ are constants and $T_0$ and $v_0$ are the temperature and the specific volume introduced in sections 7 and 8 for which we take the temperature and the specific volume of the reference state. In a first approximation one may replace $\rho$ by $\rho_0 = 1/v_0$ in (20.8) and (20.9). Using the rules for deviators and scalar parts of tensors discussed in section 4, we obtain from (23.1), (20.8) and (20.9)

$$\tilde{\tau}_{\alpha\beta} = a^{(0,0)}\tilde{\varepsilon}^{(0)}_{\alpha\beta}\,, \tag{23.2}$$

$$\tilde{\tau}^{(j)}_{(m)\alpha\beta} = \sum_{k=1}^{n}(a^{(0,0)} - a^{(j,k)})\tilde{\varepsilon}^{(k)}_{\alpha\beta}\,, \tag{23.3}$$

$$\tau = b^{(0,0)}\varepsilon^{(0)} + c^{(0)}(T - T_0) \tag{23.4}$$

and

$$\tau_{(m)}^{(j)} = \sum_{k=1}^{n} (b^{(0,0)} - b^{(j,k)})\varepsilon^{(k)} + (c^{(j)} - c^{(0)})(T - T_0) \; . \tag{23.5}$$

We note that (23.2) and (23.4) express Hooke's law for thermoelastic media (Duhamel-Neumann law).

In the reference state all the partial strain tensors vanish (see the last paragraph of the preceding section). Hence, it is seen that (at temperature $T_0$) all stresses in the reference state vanish, i.e. indeed it is a state of perfect equilibrium. It is also seen from the five preceding relations that (8.25) and (21.5)-(21.9) are satisfied. Using the relation $\varepsilon_{\alpha\beta} = \varepsilon_{\alpha\beta}^{(0)} + \varepsilon_{\alpha\beta}^{(1)} + \ldots + \varepsilon_{\alpha\beta}^{(n)}$ one obtains from (20.4), (23.2) and (23.3) for the deviators of the affinity-stress tensors

$$\tilde{\tau}_{\alpha\beta}^{(j)} = \tilde{\tau}_{\alpha\beta} + \tilde{\tau}_{(m)\alpha\beta}^{(j)} = a^{(0,0)}\tilde{\varepsilon}_{\alpha\beta} - \sum_{k=1}^{n} a^{(j,k)}\tilde{\varepsilon}_{\alpha\beta}^{(k)} \; . \tag{23.6}$$

In an analogous way one has for the scalar parts of these tensors

$$\tau^{(j)} = \tau + \tau_{(m)}^{(j)} = b^{(0,0)}\varepsilon - \sum_{k=1}^{n} b^{(j,k)}\varepsilon^{(k)} + c^{(j)}(T - T_0) \; . \tag{23.7}$$

If linear equations of state hold (see (23.2)-(23.7)) the strains are said to be small from a physical point of view.

It should be noted that if the viscous stress does not vanish one must replace $\tau_{\alpha\beta}$ by $\tau_{\alpha\beta}^{(eq)}$ in (23.2), (23.4), (23.6) and (23.7).

In case of linear equations of state the forms (21.3) of the partial yield functions $\Phi^{(i,i)}$ become with the aid of (23.3)

$$\Phi^{(i,i)} = \tfrac{1}{2} \sum_{\alpha,\beta=1}^{3} (\tilde{\tau}_{\alpha\beta})^2 + \sum_{k=1}^{n} (a^{(0,0)} - a^{(i,k)}) \sum_{\alpha,\beta=1}^{3} \tilde{\varepsilon}_{\alpha\beta}^{(k)}\tilde{\tau}_{\alpha\beta} - (k^{(i,i)})^2 \; . \tag{23.8}$$

In the rest of this section we shall consider the case $n = 1$. Hence, in this case we have from (23.8)

$$\Phi = \Phi^{(1,1)} =$$

$$= \tfrac{1}{2} \sum_{\alpha,\beta=1}^{3} (\tilde{\tau}_{\alpha\beta})^2 + (a^{(0,0)} - a^{(1,1)}) \sum_{\alpha,\beta=1}^{3} \tilde{\varepsilon}^{(1)}_{\alpha\beta} \tilde{\tau}_{\alpha\beta} - (k^{(1,1)})^2 . \quad (23.9)$$

A yield function of this type has been proposed by Freudenthal (see p. 282 of reference 25). In this form for the yield function the Bauschinger effect is proportional to the plastic strain. An analogous remark holds for (23.8).

For $n = 1$ one has for the flow law determined by (21.11), (21.12), (23.9) and (23.3)

$$\frac{d\tilde{\varepsilon}^{(1)}_{\alpha\beta}}{dt} = \eta_s^{(1,1)} \frac{\partial \Phi}{\partial \tau_{\alpha\beta}} = \eta_s^{(1,1)} \tilde{\tau}^{(1)}_{\alpha\beta} = \eta_s^{(1,1)} (\tilde{\tau}_{\alpha\beta} + \tilde{\tau}^{(1)}_{(m)\alpha\beta}) =$$

$$= \eta_s^{(1,1)} \{\tilde{\tau}_{\alpha\beta} - (a^{(1,1)} - a^{(0,0)}) \tilde{\varepsilon}^{(1)}_{\alpha\beta}\} \quad (23.10)$$

and from (21.10) we have

$$k^{(1,1)} = k^{(1,1)}(T, \varepsilon^{(1)}_{\alpha\beta}) . \quad (23.11)$$

Next, we shall show that a stress-strain relation may be obtained by eliminating the plastic strain tensor $\tilde{\varepsilon}^{(1)}_{\alpha\beta}$ from the equations of state (23.2), (23.3) and (23.6) and the flow law (20.13). Since $\tilde{\varepsilon}_{\alpha\beta} = \tilde{\varepsilon}^{(0)}_{\alpha\beta} + \tilde{\varepsilon}^{(1)}_{\alpha\beta}$ we have from the equation of state (23.2)

$$a^{(0,0)} \tilde{\varepsilon}^{(1)}_{\alpha\beta} = a^{(0,0)} \tilde{\varepsilon}_{\alpha\beta} - a^{(0,0)} \tilde{\varepsilon}^{(0)}_{\alpha\beta} = a^{(0,0)} \tilde{\varepsilon}_{\alpha\beta} - \tilde{\tau}_{\alpha\beta} . \quad (23.12)$$

Since we suppose that $a^{(0,0)}$ is constant it follows from (23.12) that

$$a^{(0,0)} \frac{d\tilde{\varepsilon}^{(1)}_{\alpha\beta}}{dt} = a^{(0,0)} \frac{d\tilde{\varepsilon}_{\alpha\beta}}{dt} - \frac{d\tilde{\tau}_{\alpha\beta}}{dt} \,. \tag{23.13}$$

On the other hand one has from the flow law (20.13) with the aid of (23.6) and (23.12)

$$a^{(0,0)} \frac{d\tilde{\varepsilon}^{(1)}_{\alpha\beta}}{dt} = a^{(0,0)} \eta_s^{(1,1)} \tilde{\tau}^{(1)}_{\alpha\beta}$$

$$= a^{(0,0)} \eta_s^{(1,1)} (a^{(0,0)} \tilde{\varepsilon}_{\alpha\beta} - a^{(1,1)} \tilde{\varepsilon}^{(1)}_{\alpha\beta})$$

$$= (a^{(0,0)})^2 \eta_s^{(1,1)} \tilde{\varepsilon}_{\alpha\beta} - a^{(1,1)} \eta_s^{(1,1)} (a^{(0,0)} \tilde{\varepsilon}_{\alpha\beta} - \tilde{\tau}_{\alpha\beta})$$

$$= a^{(0,0)} (a^{(0,0)} - a^{(1,1)}) \eta_s^{(1,1)} \tilde{\varepsilon}_{\alpha\beta} + a^{(1,1)} \eta_s^{(1,1)} \tilde{\tau}_{\alpha\beta} \,. \tag{23.14}$$

Finally, one may combine (23.13) and (23.14) and this leads to the stress-strain relation

$$a^{(1,1)} \eta_s^{(1,1)} \tilde{\tau}_{\alpha\beta} + \frac{d\tilde{\tau}_{\alpha\beta}}{dt} = a^{(0,0)} \left\{ (a^{(1,1)} - a^{(0,0)}) \eta_s^{(1,1)} \tilde{\varepsilon}_{\alpha\beta} + \frac{d\tilde{\varepsilon}_{\alpha\beta}}{dt} \right\} , \tag{23.15}$$

where $\eta_s^{(1,1)}$ vanishes for $\Phi < 0$. Hence, (23.15) implies

$$\frac{d\tilde{\tau}_{\alpha\beta}}{dt} = a^{(0,0)} \frac{d\tilde{\varepsilon}_{\alpha\beta}}{dt} \quad \text{if } \Phi < 0 \,, \tag{23.16}$$

i.e. in the preplastic range the medium is elastic. It should be noted that in the derivation of (23.14) we did not use any assumption concerning the explicit form of the yield function $\Phi$.

We note that

$$a^{(1,1)} \geq a^{(0,0)} \geq 0 \,. \tag{23.17}$$

These inequalities ensure that f increases if, starting from the reference state, the deviatoric strains $\tilde{\varepsilon}^{(0)}_{\alpha\beta}$ and $\tilde{\varepsilon}^{(1)}_{\alpha\beta}$ get values different from zero by loading the medium at constant temperature (see (23.1) for n = 1).

## 24. ELASTIC-PLASTIC MEDIA WITHOUT SHAPE MEMORY, THE LÉVY-VON MISES THEORY, THE PRANDTL-REUSS EQUATION

If a medium has no shape memory the deviator of one of the memory-stress tensors, say $\tilde{\tau}^{(n)}_{(m)\alpha\beta}$, vanishes. We recall (17.1):

$$\tilde{\tau}^{(n)}_{\alpha\beta} = \tilde{\tau}^{(eq)}_{\alpha\beta} = \tilde{\tau}_{\alpha\beta} \ . \tag{24.1}$$

In particular we shall apply this to media for which n = 1 and for which the equations of state may be linearized. Thus, we have

$$\tilde{\tau}^{(1)}_{(m)\alpha\beta} = 0 \ , \quad \tilde{\tau}^{(1)}_{\alpha\beta} = \tilde{\tau}^{(eq)}_{\alpha\beta} = \tilde{\tau}_{\alpha\beta} \ . \tag{24.2}$$

It is seen from (23.3) that it follows from $\tilde{\tau}^{(1)}_{(m)\alpha\beta} = 0$ that

$$a^{(1,1)} = a^{(0,0)} \ . \tag{24.3}$$

Hence, the yield function (23.9) becomes

$$\Phi = \tfrac{1}{2} \sum_{\alpha,\beta=1}^{3} (\tilde{\tau}_{\alpha\beta})^2 - (k^{(1,1)})^2 \ . \tag{24.4}$$

Since $\tilde{\varepsilon}^{(1)}_{\alpha\beta}$ is not a thermodynamic state variable (see section 15) one has from (23.11) if $\varepsilon^{(1)} = 0$ (no inelastic (plastic) volumetric phenomena)

$$k^{(1,1)} = k^{(1,1)}(T) \ . \tag{24.5}$$

The function given by (24.4) is the Von Mises yield function. The flow law (23.10) reduces to

$$\frac{d\tilde{\varepsilon}_{\alpha\beta}^{(1)}}{dt} = \eta_s^{(1,1)} \tilde{\tau}_{\alpha\beta} = \eta_s^{(1,1)} \frac{\partial\Phi}{\partial\tau_{\alpha\beta}} . \tag{24.6}$$

This is the Lévy-Von Mises flow law. Finally, the stress-strain relation (23.15) becomes

$$a^{(0,0)}\eta_s^{(1,1)}\tau_{\alpha\beta} + \frac{d\tilde{\tau}_{\alpha\beta}}{dt} = a^{(0,0)} \frac{d\tilde{\varepsilon}_{\alpha\beta}}{dt} . \tag{24.7}$$

The latter relation is known as the Prandtl-Reuss equation.

## 25. VISCOPLASTICITY, SCHOFIELD-SCOTT BLAIR MEDIA, BINGHAM MEDIA

So far we considered media which were purely elastic in the preplastic range. Let us recall the flow laws (13.8) and (13.9) for distortional phenomena in isotropic media

$$\frac{d\tilde{\varepsilon}^{(j)}_{\alpha\beta}}{dt} = \sum_{k=1}^{n} \eta_s^{(j,k)} \tilde{\tau}^{(k)}_{\alpha\beta} + \eta_s^{(j,0)} \frac{d\tilde{\varepsilon}_{\alpha\beta}}{dt} \quad (j = 1,\dots,n) \; , \tag{25.1}$$

$$\tilde{\tau}^{(vi)}_{\alpha\beta} = \sum_{k=1}^{n} \eta_s^{(0,k)} \tilde{\tau}^{(k)}_{\alpha\beta} + \eta_s^{(0,0)} \frac{d\tilde{\varepsilon}_{\alpha\beta}}{dt} \; . \tag{25.2}$$

We now generalize the ideas developed in section 20 and we assume that there are partial yield functions $\Phi^{(j,k)}$ ($j,k = m+1,\dots,n$, where $1 \le m < n$) such that $\eta_s^{(j,k)}$ ($j,k = m+1,\dots,n$, where $1 \le m < n$) vanishes if $\Phi^{(j,k)} < 0$, while for the remaining $\eta_s^{(j,k)}$, connected with elastic relaxation, such yield conditions do not exist. We shall call the partial strains $\varepsilon^{(0)}_{\alpha\beta}, \varepsilon^{(1)}_{\alpha\beta}, \dots, \varepsilon^{(m)}_{\alpha\beta}$ anelastic strains and the strains $\varepsilon^{(m+1)}_{\alpha\beta}, \dots$ $\dots, \varepsilon^{(n)}_{\alpha\beta}$ plastic strains. The tensor $\varepsilon^{(0)}_{\alpha\beta}$ describes the elastic strain if the medium is nonviscous. For simplicity we shall neglect all cross effects among the anelastic flows and plastic flows, i.e. we shall assume that $\eta_s^{(i,j)}$ vanishes for $i = 1,\dots,m$ and $j = m+1,\dots,n$ and for $i = m+1,\dots,n$ and $j = 1,\dots,m$. In the preplastic range, where all $\Phi^{(j,k)} < 0$ and, hence, all $\eta_s^{(j,k)}$ vanish for $j,k = m+1,\dots,n$, (25.1) reduces to

$$\frac{d\tilde{\varepsilon}_{\alpha\beta}^{(j)}}{dt} = \sum_{k=1}^{m} \eta_s^{(j,k)} \tilde{\tau}_{\alpha\beta}^{(k)} + \eta_s^{(j,0)} \frac{d\tilde{\varepsilon}_{\alpha\beta}}{dt} \qquad (j = 1,\ldots,m) . \tag{25.3}$$

This shows that in the preplastic range elastic relaxation may occur. The case of a nonviscous medium with m = 1 and n = 2 may be considered as a Schofield-Scott Blair medium (see the references 9 and 26).

The partial yield functions $\Phi^{(j,k)}$ cannot be independent of each other. In the first place $\Phi^{(j,k)} = \Phi^{(k,j)}$ since by virtue of the Onsager relations $\eta_s^{(j,k)} = \eta_s^{(k,j)}$. Moreover, it is required that the entropy source strength is nonnegative.

For anisotropic media, there may exist a separate yield condition for each independent component of the phenomenological tensors for plastic flow.

A Bingham medium is a typical example of a viscoplastic substance. These media are rigid in the preplastic range (i.e. $\tilde{\varepsilon}_{\alpha\beta}^{(0)} = 0$ and there are no elastic relaxation phenomena) and at stresses exceeding the yield limit they behave as viscous fluids. For a discussion from the point of view of the developed formalism see section 8 of reference 4.

## 26. SHAPE-MEMORY EFFECTS

Again, we consider an isotropic elastic-plastic medium with $n = 1$, i.e. $\varepsilon_{\alpha\beta} = \varepsilon^{(0)}_{\alpha\beta} + \varepsilon^{(1)}_{\alpha\beta}$ and we assume that the medium has shape memory. Since the medium is assumed to be an elastic-plastic medium it is non-viscous and $\varepsilon^{(0)}_{\alpha\beta}$ and $\varepsilon^{(1)}_{\alpha\beta}$ are the elastic strain tensor and the plastic strain tensor, respectively.

If $n = 1$ we have from (21.11) and (21.12)

$$\frac{d\tilde{\varepsilon}^{(1)}_{\alpha\beta}}{dt} = \eta_s^{(1,1)} \frac{\partial\Phi}{\partial\tau_{\alpha\beta}} = \eta_s^{(1,1)}(\tilde{\tau}_{\alpha\beta} + \tilde{\tau}^{(1)}_{(m)\alpha\beta}) = \eta_s^{(1,1)}\tilde{\tau}^{(1)}_{\alpha\beta} , \tag{26.1}$$

if we assume that the yield function $\Phi^{(1,1)}$ has the form (21.1):

$$\Phi^{(1,1)} = \tfrac{1}{2} \sum_{\alpha,\beta=1}^{3} (\tilde{\tau}^{(1)}_{\alpha\beta})^2 - (K^{(1,1)})^2 . \tag{26.2}$$

Let us consider these relations for the case that the mechanical stress tensor $\tau_{\alpha\beta}$ vanishes. One then has

$$\frac{d\tilde{\varepsilon}^{(1)}_{\alpha\beta}}{dt} = \eta_s^{(1,1)}\tilde{\tau}^{(1)}_{(m)\alpha\beta} \tag{26.3}$$

and

$$\Phi^{(1,1)} = \tfrac{1}{2} \sum_{\alpha,\beta=1}^{3} (\tilde{\tau}^{(1)}_{(m)\alpha\beta})^2 - (K^{(1,1)})^2 , \tag{26.4}$$

since in this case $\tilde{\tau}^{(1)}_{\alpha\beta} = \tilde{\tau}^{(1)}_{(m)\alpha\beta}$ (see (20.4)).

Suppose that the medium is in the preplastic range (i.e. $\Phi^{(1,1)} < 0$) at the temperature T' with vanishing mechanical stresses. Next, we heat the medium (without applying external forces) to a temperature T". From (21.1) we have

$$K^{(1,1)} = K^{(1,1)}(T,\varepsilon_{\alpha\beta}^{(1)}) \ . \tag{26.5}$$

As noted in section 21 the explicit form of the yield functions and, hence, the explicit form of the function on the right-hand side of (26.5) cannot be deduced from thermodynamic or other phenomenological arguments. However, we may assume that

$$K^{(1,1)}(T'',\varepsilon_{\alpha\beta}^{(1)}) < K^{(1,1)}(T',\varepsilon_{\alpha\beta}^{(1)}) \quad \text{if} \quad T'' > T' \ . \tag{26.6}$$

This means that the yield stress in shear decreases if the temperature increases (at constant plastic strain). Hence, it may occur that (neglecting in a first approximation the temperature dependence of $\tilde{\tau}^{(1)}_{(m)\alpha\beta}$ (see, for instance, (23.3)), $\Phi$ becomes positive if T is sufficiently high. Thus, the medium starts to flow. The same phenomenon occurs if, at sufficiently high temperature, $\eta_s^{(1,1)}$ is not exactly equal to zero in the preplastic range. These types of flow processes do not occur if the medium has returned to a state of perfect thermodynamic equilibrium because in this state $\tilde{\tau}^{(1)}_{(m)\alpha\beta}$ vanishes (see section 22). Hence, one may say that at sufficiently high temperatures there is a tendency to return to a state of perfect equilibrium. Such shape-memory effects have been observed in plastics and alloys.[56,57] We note that it may be seen from (26.3) that if a medium has no shape memory and if n = 1 such effects of shape recovery cannot occur. If n ≥ 2 and if the medium has no shape memory a partial recovery of shape is possible. We note that in a medium with shape memory a complete recovery of shape is also possible if n ≥ 2. Finally, we remark that if (at sufficiently high temperatures) a medium shows shape-memory effects, this is associated with a kind of pseudo-elastic mechanical behaviour (in particular if complete recovery of shape occurs).

## 27. A REVIEW OF STRAINS AND STRESSES

In section 8 we obtained the relations

$$\varepsilon_{\alpha\beta} = \varepsilon^{(0)}_{\alpha\beta} + \varepsilon^{(1)}_{\alpha\beta} + \dots + \varepsilon^{(n)}_{\alpha\beta} , \tag{27.1}$$

$$\tau_{\alpha\beta} = \tau^{(eq)}_{\alpha\beta} + \tau^{(vi)}_{\alpha\beta} \tag{27.2}$$

and

$$\tau^{(k)}_{\alpha\beta} = \tau^{(eq)}_{\alpha\beta} + \tau^{(k)}_{(m)\alpha\beta} \qquad (k = 1,\dots,n) \tag{27.3}$$

(see (8.18), (8.39) and (8.33)). We shall now discuss the physical meaning of the strains and stresses which occur in the three above relations.

The mechanical stress tensor $\tau_{\alpha\beta}$ occurs in the equation of motion (see section 2.2). As may be seen from (27.2), the mechanical stress tensor is the sum of the equilibrium-stress tensor $\tau^{(eq)}_{\alpha\beta}$ and the viscous stress tensor $\tau^{(vi)}_{\alpha\beta}$. In the second paragraph of section 11 we noted that $\tau^{(eq)}_{\alpha\beta}$ may be considered as the reversible part of the mechanical stress and $\tau^{(vi)}_{\alpha\beta}$ as the irreversible part. In section 8, $\varepsilon^{(0)}_{\alpha\beta}$ has been introduced in such a way that it vanishes for some (arbitrarily chosen) temperature $T_0$ if the equilibrium stress $\tau^{(eq)}_{\alpha\beta}$ vanishes (see (8.25)). As we have shown in section 16, $\varepsilon^{(0)}_{\alpha\beta}$ is the thermoelastic strain if the medium is nonviscous, i.e. if $\tau^{(vi)}_{\alpha\beta}$ is always equal to zero. In media such that during flow processes $\tau^{(vi)}_{\alpha\beta} \neq 0$ there is no elastic part of the strain, i.e. there is no instantaneous elastic response if the medium is suddenly loaded. Examples of such media are viscous fluids and Kelvin-Voigt media.[2]

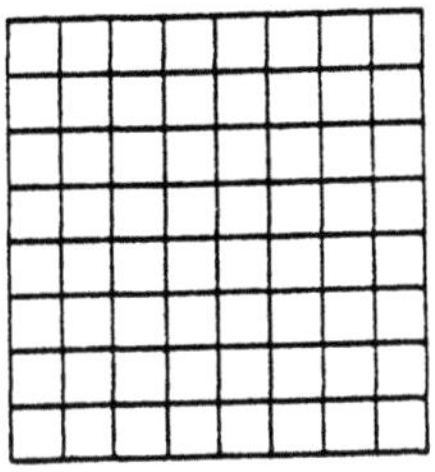

Fig. 1

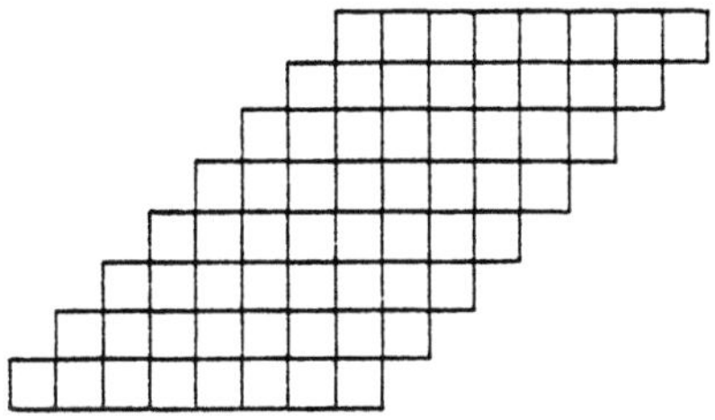

Fig. 2

The strains $\varepsilon^{(1)}_{\alpha\beta},\ldots,\varepsilon^{(n)}_{\alpha\beta}$ are connected with irreversible phenomena and are governed by the flow laws discussed in sections 11 and 13. In non-viscous media $\varepsilon^{(0)}_{\alpha\beta}$ is the reversible part of the strain. In viscous media all strains on the right-hand side of (27.1) are irreversible.

Let us give an example of a medium without shape memory. In figure 1 a crystal lattice is represented schematically. Let us suppose that there occurs only one microscopic phenomenon that gives rise to inelastic deformations (i.e. $n = 1$) and that this phenomenon is slip along the horizontal and vertical lines in figure 1. In figure 2 a possible result of such a slip phenomenon is shown. In this case the inelastic deformations are not associated with changes in the microscopic structure of the medium and if the medium in the two states (shown in figure 1 and figure 2) undergoes the same elastic deformations and changes in temperature, the thermodynamic state variables will have the same values in both cases (i.e. all thermodynamic state variables may be considered to be functions of $T$ and $\varepsilon^{(0)}_{\alpha\beta}$). This means that the medium has no shape memory and, hence, $\tilde{\tau}^{(1)}_{(m)\alpha\beta} = 0$. We note that in this case no lattice defects are associated with the inelastic deformations (see figure 2) and, hence, there are no microscopic stress fields surrounding imperfections in the medium. Moreover, neither the components of $\tilde{\varepsilon}_{\alpha\beta}$ nor the components of $\tilde{\varepsilon}^{(1)}_{\alpha\beta}$ are thermodynamic state variables.

If inelastic strains are caused by microscopic defects, these strains are associated with changes in the microscopic structure of the medium and must be considered as thermodynamic state variables. In this case the memory stresses do not vanish and, as we have seen for elastic-plastic media in the preceding section, under special conditions (concerning the phenomenological coefficients and/or the yield condition) they may give rise to shape-memory effects. See also the discussion of Poynting-Thomson media (standard linear solids) in section 4 of reference 4.

The remarks in the two preceding paragraphs suggest that the memory-stress tensors $\tau^{(k)}_{(m)\alpha\beta}$ give a macroscopic description of the influence of microscopic stress fields (surrounding imperfections in the medium) on the mechanical and thermodynamic behaviour of the medium.

In section 16 we have shown that for nonviscous media the mechanical stress tensor is connected with the elastic strains $\varepsilon^{(0)}_{\alpha\beta}$ and the temperature by (16.19) if the equations of state may be linearized. If the medium shows viscous effects it may be shown in the same way that in a first approximation

$$\tau^{(eq)}_{\alpha\beta} = \sum_{\gamma,\zeta=1}^{3} a^{(0)(0)}_{\alpha\beta\gamma\zeta}\,\varepsilon^{(0)}_{\gamma\zeta} + a^{(0)}_{\alpha\beta}(T - T_0) \; . \tag{27.4}$$

Hence, $\tau^{(eq)}_{\alpha\beta}$ may be considered as a thermoelastic stress.

In the flow laws introduced in sections 11 and 13 the thermodynamic forces for changes in the inelastic strains are the affinity-stress tensors $\tau^{(k)}_{\alpha\beta}$ given by (27.3). It is seen from (27.3) that $\tau^{(k)}_{\alpha\beta}$ is the sum of the equilibrium stress and the k-th memory stress. In other words, the thermodynamic forces which cause changes in the plastic strains are the sums of the thermoelastic stress and a stress which describes in the macroscopic theory the influence of microscopic stress fields surrounding imperfections in the medium.

## REFERENCES

1. Kluitenberg, G.A., Thermodynamical theory of elasticity and plasticity, *Physica*, 28, 217, 1962.

2. Kluitenberg, G.A., A note on the thermodynamics of Maxwell bodies, Kelvin bodies (Voigt bodies), and fluids, *Physica*, 28, 561, 1962.

3. Kluitenberg, G.A., On rheology and thermodynamics of irreversible processes, *Physica*, 28, 1173, 1962.

4. Kluitenberg, G.A., On the thermodynamics of viscosity and plasticity, *Physica*, 29, 633, 1963.

5. Kluitenberg, G.A., A unified thermodynamic theory for large deformations in elastic media and in Kelvin (Voigt) media, and for viscous fluid flow, *Physica*, 30, 1945, 1964.

6. Kluitenberg, G.A., Application of the thermodynamics of irreversible processes to continuum mechanics, in *Non-Equilibrium Thermodynamics, Variational Techniques, and Stability*, Donnelly, R.J., Herman, R. and Prigogine, I., Eds., University of Chicago Press, Chicago, Il., 1966, 91.

7. Kluitenberg, G.A., On heat dissipation due to irreversible mechanical phenomena in continuous media, *Physica*, 35, 177, 1967.

8. Kluitenberg, G.A., A thermodynamic derivation of the stress-strain relations for Burgers media and related substances, *Physica*, 38, 513, 1968.

9. Kluitenberg, G.A., On plasticity, elastic relaxation phenomena, and a stress-strain relation which characterizes Schofield-Scott Blair media, *Appl.Sci.Res.*, 25, 383, 1972.

10. Kluitenberg, G.A., A thermodynamic discussion of the possibility of singular yield conditions in plasticity theory, *Physica*, 88A, 122, 1977.

11. Kluitenberg, G.A. and Ciancio, V., On linear dynamical equations of state for isotropic media I, General formalism, *Physica*, 93A, 273, 1978.

12. Ciancio, V. and Kluitenberg, G.A., On linear dynamical equations of state for isotropic media II, Some cases of special interest, *Physica*, 99A, 592, 1979.

13. Kluitenberg, G.A., Turrisi, E. and Ciancio, V., On the propagation of linear transverse acoustic waves in isotropic media with mechanical relaxation phenomena due to viscosity and a tensorial internal variable, I. General formalism, *Physica*, 110A, 361, 1982.

14. De Groot, S.R., *Thermodynamics of Irreversible Processes*, North-Holland Publishing Company, Amsterdam, and Interscience Publishers Inc., New York, 1951.

15. De Groot, S.R. and Mazur, P., *Non-Equilibrium Thermodynamics*, North-Holland Publishing Company, Amsterdam, and Interscience Publishers Inc., New York, 1962.

16. Prigogine, I., *Etude Thermodynamique des Phénomènes irréversibles*, Dunod, Paris, et Editions Desoer, Liège, 1947.

17. Prigogine, I., *Introduction to Thermodynamics of Irreversible Processes*, Interscience Publishers - John Wiley & Sons, New York - London, 1961.

18. Meixner, J. and Reik, H.G., *Thermodynamik der irreversiblen Prozesse, Handbuch der Physik*, Band III/2, Springer, Berlin, 1959, 413.

19. Landau, L.D. and Lifshitz, E.M., *Fluid Mechanics*, Volume 6 of *Course of Theoretical Physics*, Pergamon Press, London, 1959.

20. Landau, L.D. and Lifshitz, E.M., *Theory of Elasticity*, Volume 7 of *Course of Theoretical Physics*, Pergamon Press, London, 1959.

21. Sedov, L.I., *Foundations of the Non-Linear Mechanics of Continua*, Pergamon Press, Oxford, 1966.

22. Germain, P., *Cours de Mécanique des Milieux Continus*, Tome I, *Théorie Générale*, Masson et Cie, Paris, 1973.

23. Carini, G., *Appunti di Istituzioni di Fisica Matematica*, Vincenzo Ferrara, Messina, 1970.

24. Koiter, W.T., General theorems for elastic-plastic solids, in *Progress in Solid Mechanics*, Volume I, Sneddon, I.N. and Hill, R., Eds., North-Holland Publishing Company, Amsterdam, 1960, Chapter IV, 165.

25. Freudenthal, A.M. and Geiringer, H., *The Mathematical Theories of the Inelastic Continuum, Handbuch der Physik*, Band VI, Springer Verlag, Berlin, 1958, 229.

26. Reiner, M., *Rheology, Handbuch der Physik*, Band VI, Springer Verlag, Berlin, 1958, 434.

27. Eckart, C., The thermodynamics of irreversible processes, IV. The theory of elasticity and anelasticity, *Phys. Rev.*, 73, 373, 1948.

28. Meixner, J., Thermodynamik und Relaxationserscheinungen, *Z. Naturforschung*, 4a, 594, 1949.

29. Meixner, J., Allgemeine Theorie der Schallabsorption in Gasen und Flüssigkeiten unter Berücksichtigung der Transporterscheinungen, *Acustica*, 2, 101, 1952

30. Meixner, J., Die thermodynamische Theorie der Relaxationserscheinungen und ihr Zusammenhang mit der Nachwirkungstheorie, *Kolloid-Z.*, 134, 3, 1953.

31. Meixner, J., Thermodynamische Theorie der elastischen Relaxation, *Z. Naturforschung*, 9A, 654, 1954.

32. Meixner, J., On the thermodynamic theory of the second viscosity, *Proc. Roy. Soc. London*, A226, 51, 1954.

33. Meixner, J., Thermodynamische Erweiterung der Nachwirkungstheorie, *Z. für Physik*, 139, 30, 1954.

34. Staverman, A.J. and Schwarzl, F., Thermodynamics of viscoelastic behaviour (model theory), *Proc. Roy. Soc. Amsterdam*, B55, 474, 1952.

35. Staverman, A.J. and Schwarzl, F., Non-equilibrium thermodynamics of visco-elastic behaviour, *Proc. Roy. Soc. Amsterdam*, B55, 486, 1952

36. Polder, D., Theory of the elastic aftereffect and the diffusion of carbon in alpha-iron, *Philips Res. Rep.*, 1, 5, 1945.

37. Falk, G. and Meixner, J., Klassifikation von Relaxationserscheinungen in Kristallen, *Z. Naturforschung*, 11A, 782, 1956.

38. König, H. and Meixner, J., Lineare Systeme und lineare Transformationen, *Math. Nachr.*, 19, 265, 1958.

39. Meixner, J. and König, H., Zur Theorie der linearen dissipativen Systeme, *Rheologica Acta*, 1, 190, 1958.

40. Kluitenberg, G.A., On dielectric and magnetic relaxation phenomena and vectorial internal degrees of freedom in thermodynamics, *Physica*, 87A, 302, 1977.

41. Kluitenberg, G.A., On vectorial internal variables and dielectric and magnetic relaxation phenomena, *Physica*, 109A, 91, 1981.

42. Veenendaal, W. van, and Kluitenberg, G.A., On transformations of internal vectorial variables in the thermodynamic theory of dielectric relaxation and the invariance of Onsager's reciprocal relations, *Physica*, 109A, 123, 1981.

43. Kluitenberg, G.A., On dielectric and magnetic relaxation phenomena and non-equilibrium thermodynamics, *Physica*, 68, 75, 1973.

44. Lebon, G. and Rubi, J.M., A generalized theory of thermoviscous fluids, *J. Non-Equilib. Thermodyn.*, 5, 285, 1980.

45. Müller, I., Zum Paradoxon der Wärmeleitungstheorie, *Z. für Physik*, 198, 329, 1967.

46. Lebon, G., Derivation of generalized Fourier and Stokes-Newton equations based on the thermodynamics of irreversible processes, *Bull. Acad. Roy. Sci. Belgique*, 64, 456, 1978.

47. Lebon, G., Jou, D. and Casas-Vazquez, J., An extension of the local equilibrium hypothesis, *J. of Physics*, 13A, 275, 1980.

48. Jou, D., Casas-Vazquez, J. and Lebon, G., A dynamical interpretation of second order constitutive equations of hydrodynamics, *J. Non-Equilib. Thermodyn.*, 4, 349, 1979.

49. Gyarmati, I., On the wave approach of thermodynamics and some problems of non-linear theories, *J. Non-Equilib. Thermodyn.*, 2, 233, 1977.

50. Lévy, M., Mémoire sur les équations générales des mouvements intérieurs des corps solides ductiles au delà des limites où l'élasticité pourrait les ramener à leur premier état, *C. Rend. Paris*, 70, 1323, 1870.

51. Poynting, J.H., On pressure perpendicular to the shear planes in finite pure shears, and on the lengthening of loaded wires when twisted, *Proc. Roy. Soc. London*, 82A, 546, 1909.

52. Poynting, J.H., On the changes in the dimensions of a steel wire when twisted, and on the pressure of distortional waves in steel, *Proc. Roy. Soc. London*, 86A, 534, 1912.

53. Dalby, W.E., Researches on the elastic properties and the plastic extension of metals, *Phil. Trans. Roy. Soc. London*, 221A, 117, 1921.

54. Howard, J.V. and Smith, S.L., Recent developments in tensile testing, *Proc. Roy. Soc. London*, 107A, 113, 1925.

55. Lee, E.H. and Liu, D.T., Finite strain elastic-plastic theory, in *Irreversible Aspects of Continuum Mechanics and Transfer of Physical Characteristics in Moving Fluids*, Parkus, H. and Sedov, L.I., Eds., Springer-Verlag, Wien - New York, 1968, 213.

56. Warlimont, H. and Delaey, L., *Martensitic Transformations in Copper-Silver- and Gold-Based Alloys*, Pergamon Press, Oxford, 1974.

57. *Shape Memory Effects in Alloys*, Perkins, J., Ed., Plenum Press, New York, 1975.

# THERMODYNAMIC ASPECTS OF CYCLIC AND MONOTONE PLASTICITY

B. RANIECKI
Institute of Fundamental
Technological Research
00-049 Warsaw, Swietokrzyska 21
Poland

## I. INTRODUCTION.

1. Consider a small element of plastically strained metal whose elastic-plastic properties are the same at every particle (at least in statistical sense) and at every instant of a process (homogeneous process). To develop the general frame for entirely phenomenological description of its behaviour we shall regard the element as a closed and chemically inert thermodynamical system, and use the modified concepts of classical thermostatics, despite of incompletely resolved controversy as to correctness of such procedure.

The application of classical concepts to the plastically deformed material element encounters the difficulties at a very beginning of the development of a theory. Let us mention two of them:

(i) According to the phenomenological thermodynamics [1] the notion of an internal energy is conceptually related to the notion of an equilibrium state. From phenomenological observations of the behaviour of plastically strained metals (and alloys) all what is evident, as regards the state are the following facts: (a) Empirical temperature $\Theta$ and total strain $\underset{\sim}{\varepsilon}$ by themself do not represent the thermodynamical state. (b) Similarly, $\Theta$, elastic strain tensor $\underset{\sim}{\varepsilon}^e$, and plastic strain tensor $\underset{\sim}{\varepsilon}^p$ by themself are not sufficient to describe the state. This follows from the experimental observation that heat exchanged with a surrounding during an isothermal closed cyclic process in $\underset{\sim}{\varepsilon}^e$ - $\underset{\sim}{\varepsilon}^p$ space is in general not equal to the inelastic work.

One is aware, therefore, of existence of internal state, that is usually characterized by a set of internal state parameters. Viewed macroscopically the internal parameters are to signify the most dominant aspects of the phenomena occuring at the microscale. Such aspects exibit at macro-level in the form of hardening, Bauschinger phenomenon, anisotropy caused by prior plastic deformations, temperature history effects, strain-rate history effects, cycle-dependent softening etc., and are experimentally well recognized. Internal parameters, however, are not well defined physical quantities insomuch they are not directly measured and can not be controlled in an actual experiment. Generally, there also still exist a lot of uncertainites concerning the number and tensorial character of internal parameters. In a very simple situation when only one or a few aspects of macroscopic behaviour of elastic-plastic materials is intended to account for (such as e.g. isotropic hardening defined in usual rate-independent plasticity theory) one can take one scalar parameter and indicate indirect way of its measurement. The example is Hart's "hardness parameter" [2,3] or a parameter defined in [4]. When more effects are to be accounted for, the number of internal parameters must be increased. Then, naturally arises the problem of their operational definition. While I am not able to provide appreciable solution to this problem, I propose here (Section 5) to use the usual "normality postulate" [5-10] not only with regards to the rates of external parameters (i.e. $\dot{\underset{\sim}{\varepsilon}}^{p}$) but also with reference to the rates of internal parameters and associated irreversible forces (see also [11-13]). This assumption couples the stored free energy of cold-work with the hardening (softening) effects. Although it restricts the class of possible rate-equations it is very helpful in an application of the discussed scheme*) to

*) The scheme presented here is an example of a thermodynamic approach to modelling of the elastic-plastic behaviour of solids rather than the rigorous thermodynamic theory of thermoplasticity.

practical situations, and it can be regarded as a guide in discovering the appreciable equations from the "sea of possibilities".

Once the general aspects (of the macroscopic behaviour of plastically strained metal) that are to be accounted for in theoretical description are known, one can undertake the decision as to what minimum number and what type of internal variables are adequate. When the general frame of the description is esthablished one can make an effort to identify parameters, at least in the class of proportional loadings, and then to quantify and to verify the description with the aid of experimental procedure.

Vakulenko [6], Kestin [14], Kestin and Rice [11] when applying the concepts of classical thermostatics to plastically strained metals treat the element as thermodynamic system being in "constrained equilibrium". It is believed that equilibrium is maintained not only by external constraints (various idealized walls or partitions) but also by internal one. The internal constraints may be developed by system itself in the course of a process, and they do not need to be of a concrete nature [14]. Here we adopt this type of reasoning.

(ii) One of the basic notion of the classical thermostatics is that of an idealized "reversible process" which is constituted of succesion of (constrained) equilibrium states, and in the course of which the work (reversible) is done precisely by the forces which hold the system in (constrained) equilibrium [1,15].

During an actual process of elastic-plastic deformations of metal, however, one can manipulate merely the external constraints, changing in this manner strains (elastic and plastic) and external forces. With the aid of this means it is impossible to contrive a real process that has a close relation to a reversible one, no matter how slow the real process is. To overcome this difficulty, Kestin [14] suggests to consider the notion of a reversible process as purely conceptual. Accordingly, to make use of this notion it is sufficient to imagine that the internal forces associated with internal parameters are balanced by fictitious external forces, so that the material element can be made to perform a reversible process in thought. The work done in such imaginary reversible process (i.e. reversible work) is naturally different from that done in an actual process. During the course of the latter, the internal forces are completely unbalanced.

Here we follow this way of thinking. It enables to make use of other concepts of classical thermostatics while developing the general frame of the description of a behaviour of plastically strained metals, in particular, the "principle of Carathéodory" and all conclusions following from this principle as regards the absolute temperature T, entropy function s and its additivity. In result, one obtains the Gibbs fundamental equation of state (cf [1,16]) which is strictly valid only during the hypothetical reversible processes. Following the common practice (cf e.g. [5-8, 9,17-19] it is then postulated that this equation is approximately valid in an actual homogeneous process of elastic--plastic deformations. The postulated form of the Gibbs equation is presented in Sec.1.

Finally, it is worthwhile to mention that the direct consequence of our inability to perform the reversible process is the following experimental problem: the heat that participates in the course of an actual isothermal process of e-p deformations consists of two parts (a) heat due to the dissipation of mechanical work (b) heat due to the change of the entropy*). Since during a process of plastic straining one of the state-parameter, say hardening parameter, never decreases, the difficulty arises in distinguishing (at macro-level) the latter heat from the former. While cold work certainly produces change in the entropy of metals it is believed that such change (when multiplied by T) is small in comparisson to energy stored. Therefore in

*) termed "reversible" heat of structural hardening in Sec. 4., an effect analogous to that of heat of elastic strains in usual thermoelasticity theory.

practical applications of the theory one frequently identifies the entropy of elastic-plastic solid with the entropy of corresponding elastic solid [20].

2. The constitutive equations derived with the help of the concepts of thermodynamics consist of two groups. The first group contains so called thermal equations of state. They follow directly from the adopted Gibbs equation. We present them in Part II where the bulk thermostatic properties of e-p solids are briefly discussed including the equation for temperature.

The other group contains rate equations (or phenomenological equations) which link the dissipative forces with the rates of internal and external flows occuring in the expression for the entropy production. We express them in terms of the dissipative potential in Part III. Since we take into account only rate-independent changes in properties of plastically deformed solid, the dissipative potential must be homogeneous function of order one of the rates of flows. Accordingly, the inverse form of the rate equations is expressed in terms of "yield functions" and constitutes the generalized plastic flow law and hardening "law". In fact our main purpose is to generalize the familar plastic flow law to encompass some typical effects observed in solids under monotonic and cyclic loading conditions including storage-release-storage sequence of energy exchange [21], and to link the hardening (softening) with the stored (release) free energy. In parts III and IV the preliminary results of my recent joint work with Z.Mróz [22] are presented. They concern the qualitative analysis of the bi-yield surface model with asymptotically linear kinematic hardening. The concept of multi-yield surface has recently proved to be convenient in the description of the behaviour of metals under cyclic loadings [23-26]. It dates back to an idea of Besseling [27].

3. The behaviour of plastically strained solids is very complex even when all rate effects are neglected. We present hereafter the list of some phenomenological aspects of material behaviour that are to be considered in theoretical description.

(i) Some observed effects, associated with a prior monotonic straining:

(a) The hardening (softening).

(b) Changes in the rate of hardening (softening).

(c) Anisotropy induced by a prior plastic deformation.

(d) Bauschinger effect.

(e) Temperature history effect [28-32] - yield stress and rate of hardening depend on temperature of prior straining; more strain hardening is caused by certain strain introduced at a low temperature than is caused by the same strain introduced at a higher temperature.

(f) Asymmetry of a hysteresis loop, and fading of the asymmetry (i.e. relaxation of mean stresses) in controlled cyclic strain process preceded by a monotonic straining*).

(ii) Some effects associated with a prior cyclic loading up to a steady state*):

(a) Cycle - dependent hardening (softening). Prior cyclic-loading (controlled symmetric changes in strains) may cause both increase (hardening) and decrease (softening) of the yield limit.

(b) The capacity of a material for a further hardening (softening) - both cycle-dependent and monotonic.

(iii) Some other effects observed during cyclic-loading*):

(a) Relaxation of mean stresses during the cyclic process of controlled asymmetric changes in strains.

(b) "Creep" of a mean strain in a controlled cyclic stress process.

(iv) Some observations concerning the stored energy [39-40]:

(a) Increase-decrease of the efficiency of stored energy (measured by the ratio of stored energy to total plastic work done) with increase of total plastic work done in a monotonic loading.

*) The ample experimental data is available in Refs.[33-38].

(b) A release-storage-release-storage sequence of energy exchange and the cumulation of stored energy during the traversal of a plastic strain hysteresis loop, at cycling loading [21].

4. Throughout the lecture all geometrical effects associated with large strains are neglected, and the following abbreviations are employed

$$\underset{\sim}{AB} \longrightarrow A_{ij}B_j \quad \text{or} \quad A_{ij}B_{jk} \quad \text{or} \quad A_{ijkl}B_{kl}$$

$$A \cdot B \longrightarrow A_iB_i \quad \text{or} \quad A_{ij}B_{ij}$$

$$A^T_{ij} = A_{ji} \;;\; \text{tr}\, \underset{\sim}{A} = A_{ii} \;;\; \underset{\sim}{1} \text{ - unit tensor}$$

## II. COUPLED EFFECTS IN THERMOPLASTICITY

### 1. Gibbs Equation. Energy Dissipation.

(i) We assume that state of a material element of unit mass (and of volume $1/\varrho$) in constrained equilibrium is described by entropy s, tensor of elastic infinitesimal strains $\underset{\sim}{\varepsilon}^e$, one symmetric second-order tensor $\underset{\sim}{\varkappa}$ and two scalar parameters $\varkappa^{(\wedge)}$ and $\varkappa^{(m)}$. The quantities $\underset{\sim}{\varkappa}$, $\varkappa^{(\wedge)}$, $\varkappa^{(m)}$ represent an internal state and we shall call them hardening parameters. Tensor $\underset{\sim}{\varkappa}$ will be used to describe anisotropic (directional) hardening (softening). Thus, $\underset{\sim}{\varkappa}$ may increase or decrease in the course of plastic straining.

Parameters $\varkappa^{(m)}$ and $\varkappa^{(\wedge)}$ are employed to describe monotonic and cycle-dependent indirectional hardening (softening). They do not decrease during the process of plastic deformations. Some additional informations as regards their physical meaning will be given in Sec.6. Let us adopt the following Gibbs equation written in the form of the total differential of the internal energy u

$$du(X^{s\epsilon}) = T\, ds + \frac{1}{\varrho}\, dW^{(o)} \tag{1.1}$$

where

$$\left.\begin{aligned} dW^{(o)} &= dW^{(R)} + dW^{(H)} \\ dW^{(R)} &= \underset{\sim}{\sigma}\cdot d\underset{\sim}{\varepsilon}^{e};\quad dW^{(H)} = \underset{\sim}{\pi}\cdot d\underset{\sim}{\varkappa} + \pi^{(\wedge)} d\varkappa^{(\wedge)} + \pi^{(m)} d\varkappa^{(m)} \\ X^{s\epsilon} &= \left\{ s, \underset{\sim}{\varepsilon}^{e}, \underset{\sim}{\varkappa}, \varkappa^{(\wedge)}, \varkappa^{(m)} \right\} \end{aligned}\right\} \tag{1.2}$$

Here T denotes the absolute temperature, $dW^{(o)}$ is the infinitesimal increment of work in a conceptual reversible process, $W^{(R)}$ represents recoverable external work, whereas $W^{(H)}$ represents the work of internal forces $(\underset{\sim}{\pi}, \pi^{(\wedge)}, \pi^{(m)})$ done on "internal displacements" $(\underset{\sim}{\varkappa}, \varkappa^{(\wedge)}, \varkappa^{(m)})$. Macroscopically the work $W^{(H)}$ displays in the form of strain hardening of solids and can also be detected in the isothermal cyclic process in a stress $(\underset{\sim}{\sigma})$ space.

(ii) During all real infinitesimal homogeneous processes of plastic deformations some agency performs the work on the element of solid only by direct manipulation of external constraints

$$\left.\begin{aligned} &dW = \underset{\sim}{\sigma} \cdot d\underset{\sim}{\varepsilon} = dW^{(e)} + dW^{(p)} \\ &dW^{(e)} = \underset{\sim}{\sigma} \cdot d\underset{\sim}{\varepsilon}^{e} \qquad dW^{(p)} = \underset{\sim}{\sigma} \cdot d\underset{\sim}{\varepsilon}^{p} \end{aligned}\right\} \quad (1.3)$$

i.e. in the course of an actual homogeneous process, hardening parameters may vary though internal forces are unbalanced [14]. Here we have assumed that stress tensor (external force) is totaly balanced so that the elastic work is equal to recoverable external work, $W^{(e)} = W^{(R)}$. During an actual process the element may undergo permanent change of shape represented by an increment of plastic strain $d\underset{\sim}{\varepsilon}^{p} = d\underset{\sim}{\varepsilon} - d\underset{\sim}{\varepsilon}^{e}$ which is not a state variable. The part $\underset{\sim}{\sigma} \cdot d\underset{\sim}{\varepsilon}^{p}$ (increment of plastic work) of total work (which does not occur in a reversible process) together with the fact that internal forces remain unbalanced by external forces, indicate that an actual homogeneous process is irreversible one. For such process, the <u>energy balance law</u> (the first law of thermodynamics) can be written as

$$du = \frac{1}{\varrho}\, dW + dq \qquad (1.4)$$

where dq is an increment of heat absorbed by the material element. We assume that during the homogeneous process the material element is in thermal equilibrium with its surroundings.

(iii) Consider two sufficiently close neighbouring states 1 and 1'. For an actual infinitesimal homogeneous process $1 \rightarrow 1'$ the equation (1.4) is valid, and dq and dW can be measured. Let us compare these quantities with the corresponding quantities of heat $dq^{(o)}$ and work $dW^{(o)}$ which would be exchanged in the course of a conceptual reversible process between the same states. $dq^{(o)} \equiv T ds$ and $dW^{(o)}$ are uniquely determined according to the properties of reversible processes. However, in general $dq^{(o)} \neq dq$ and $dW^{(o)} \neq dW$, but the change of internal energy du is the same. Thus,

$$du = dq^{(o)} + \frac{1}{\varrho} dW^{(o)} = dq + dW \frac{1}{\varrho} \tag{1.5}$$

or

$$dW^{(D)} = dW - dW^{(o)} = (dq^{(o)} - dq)\ \varrho \tag{1.6}$$

where $dW^{(D)}$ is so called "energy dissipation" or "mechanical work dissipation" in an actual infinitesimal homogeneous process.

The counterpart of so called "Inequality of Clausius" [1] (the second part of 2d law of thermodynamics) for infinitesimal homogeneous processes can be written as

$$T\ ds - dq \geqslant 0 \tag{1.7}$$

Hence,

$$\varrho\ T\ ds^{(i)} \equiv dW^{(D)} = dW^{(p)} - dW^{(H)} =$$

$$= \underset{\sim}{\sigma} \cdot d\underset{\sim}{\varepsilon}^{p} - \underset{\sim}{\pi} \cdot d\underset{\sim}{\varkappa} - \pi^{(\wedge)} d\varkappa^{(\wedge)} - \pi^{(m)} d\varkappa^{(m)} \geqslant 0 \tag{1.8}$$

where $ds^{(i)}$ is the "entropy production" due to dissipation in an infinitesimal homogeneous process. Let us introduce two 14-dimensional vectors as follows

$$Z_i = \left\{\sigma_{11}, \sigma_{12}, \dots, \sigma_{33}, -\pi_{11}, -\pi_{12}, \dots, -\pi_{33}, -\pi^{(\Lambda)}, -\pi^{(m)}\right\}$$

$$\dot{z}_i = \left\{\dot{\epsilon}^p_{11}, 2\dot{\epsilon}^p_{12}, \dots, \dot{\epsilon}^p_{33}, \dot{\varkappa}_{11}, 2\dot{\varkappa}_{12}, \dots, \dot{\varkappa}_{33}, \dot{\varkappa}^{(\Lambda)}, \dot{\varkappa}^{(m)}\right\}$$

Then, the rate of energy dissipation may be written in compact form

$$\dot{W}^{(D)} = \underset{\sim}{Z} \cdot \underset{\sim}{\dot{z}} \geqslant 0 \tag{1.9}$$

The forces $Z_i$ are frequently called dissipative thermodynamic forces, whereas $\underset{\sim}{\dot{z}}$, are the conjugate rates.

By eliminating $\dot{u}$ between (1.1) i (1.4) we find

$$T\, ds = \frac{1}{\varrho}\, dW^{(D)} + dq \tag{1.10}$$

## 2. Thermodynamic Potentials. Thermal Equations of State.

The set $X^{s\epsilon}$ (cf.(1.2)$_3$) is the set of natural independent variables for the internal energy function. Once u is known function of $X^{s\epsilon}$ all thermostatic properties of a material element are known, and are described by a set of so called "state equations" (cf.Table 1). For this reason $u(X^{s\epsilon})$ is called thermodynamic potential [15]. Note, that the internal energy when expressed in terms of other independent variables e.g.

$$X^{T\epsilon} = \left\{T, \underset{\sim}{\epsilon}^e, \underset{\sim}{\varkappa}, \varkappa^{(\Lambda)}, \varkappa^{(m)}\right\} \tag{2.1}$$

or

$$X^{T\sigma} = \left\{T, \underset{\sim}{\sigma}, \underset{\sim}{\varkappa}, \varkappa^{(\Lambda)}, \varkappa^{(m)}\right\} \tag{2.2}$$

is the function of thermodynamic state but is not a thermodynamic potential.

Other thermodynamical potential may be derived from u by application of technique of Legendre transformation [15, 41] to the internal energy function. For example Helmholtz free energy $\varphi$ is that partial Legendre transform of u which replaces the entropy by temperature as independent variable, whereas Gibbs function (Gibbs free energy) g is that transform of $\varphi$ which replaces $\underset{\sim}{\varepsilon}^e$ by $\underset{\sim}{\sigma}$. The full relationship between internal energy, free energy and Gibbs function is summarized in the following schematic comparison where corresponding thermal equations of state are also given.

Table 1.

| Internal energy | Helmholtz free energy | Gibbs function | |
|---|---|---|---|
| $u(X^{s\varepsilon})$ | $\varphi(X^{T\varepsilon}) = u - Ts$ | $g(X^{T\sigma}) = \varphi - \frac{1}{\varrho}\underset{\sim}{\sigma}\cdot\underset{\sim}{\varepsilon}^e$ | (2.3) |
| $T = \frac{\partial u(X^{s\varepsilon})}{\partial s}$ | $s = -\frac{\partial\varphi(X^{T\varepsilon})}{\partial T}$ | $s = -\frac{\partial g(X^{T\sigma})}{\partial T}$ | (2.4) |
| $\frac{1}{\varrho}\underset{\sim}{\sigma} = \frac{\partial u(X^{s\varepsilon})}{\partial\underset{\sim}{\varepsilon}^e}$ | $\frac{1}{\varrho}\underset{\sim}{\sigma} = \frac{\partial\varphi(X^{T\varepsilon})}{\partial\underset{\sim}{\varepsilon}^e}$ | $\frac{1}{\varrho}\underset{\sim}{\varepsilon}^e = -\frac{\partial g(X^{T\sigma})}{\partial\underset{\sim}{\sigma}}$ | (2.5) |
| $\frac{1}{\varrho}\underset{\sim}{\pi} = \frac{\partial u(X^{s\varepsilon})}{\partial\underset{\sim}{\varkappa}}$ | $\frac{1}{\varrho}\underset{\sim}{\pi} = \frac{\partial\varphi(X^{T\varepsilon})}{\partial\underset{\sim}{\varkappa}}$ | $\frac{1}{\varrho}\underset{\sim}{\pi} = \frac{\partial g(X^{T\sigma})}{\partial\underset{\sim}{\varkappa}}$ | (2.6) |
| $\frac{1}{\varrho}\pi^{(\wedge)} = \frac{\partial u(X^{s\varepsilon})}{\partial\varkappa^{(\wedge)}}$ | $\frac{1}{\varrho}\pi^{(\wedge)} = \frac{\partial\varphi(X^{T\varepsilon})}{\partial\varkappa^{(\wedge)}}$ | $\frac{1}{\varrho}\pi^{(\wedge)} = \frac{\partial g(X^{T\sigma})}{\partial\varkappa^{(\wedge)}}$ | |
| $\frac{1}{\varrho}\pi^{(m)} = \frac{\partial u(X^{s\varepsilon})}{\partial\varkappa^{(m)}}$ | $\frac{1}{\varrho}\pi^{(m)} = \frac{\partial\varphi(X^{T\varepsilon})}{\partial\varkappa^{(m)}}$ | $\frac{1}{\varrho}\pi^{(m)} = \frac{\partial g(X^{T\sigma})}{\partial\varkappa^{(m)}}$ | |

State equations occuring in different columns of the Table 1 are naturally equivalent. Each column contains set of 15-th independent state equations which in turn constitute the first group of constitutive equations mentioned in the introduction.

## 3. Reference State. Stored Energy.

(i) We shall henceforth assume that the free energy function may be expressed in the form

$$\varphi(X^{T\varepsilon}) = \varphi^{(e)}(X^T_\varepsilon) + \varphi^{(H)}(X^T_\varkappa) \tag{3.1}$$

where

$$X^T_\varepsilon = \{T, \underset{\sim}{\varepsilon}^e\} \quad ; \quad X^T_\varkappa = \{T, \underset{\sim}{\varkappa}, \varkappa^{(\wedge)}, \varkappa^{(m)}\} \tag{3.2}$$

As a consequence of the assumption (3.1), elastic constants and thermal expansion coefficients will not be influenced by prior plastic deformation (cf.Sec.4).

Due to (3.1) we have

$$\left.\begin{aligned} & g(X^{T\sigma}) = g^{(e)}(X^T_\sigma) + \varphi^{(H)}(X^T_\varkappa) \\ & s(X^{T\varepsilon}) = s^{(e)}(X^T_\varepsilon) + s^{(H)}(X^T_\varkappa) ; \quad s(X^{T\sigma}) = s^{(e)}(X^T_\sigma) + s^{(H)}(X^T_\varkappa) \end{aligned}\right\} \tag{3.3}$$

where

$$\left.\begin{aligned} & X^T_\sigma = \{T, \underset{\sim}{\sigma}\} \; ; \quad g^{(e)} = [\varphi^{(e)} - \frac{1}{\varrho}\, \underset{\sim}{\sigma}\cdot\underset{\sim}{\varepsilon}^e]_{\underset{\sim}{\varepsilon}^e = \underset{\sim}{\varepsilon}^e(X^T_\sigma)} \\ & s^{(e)}(X^T_\varepsilon) = -\frac{\partial\varphi^{(e)}}{\partial T} ; \; s^{(e)}(X^T_\sigma) = -\frac{\partial g^{(e)}(X^T_\sigma)}{\partial T} \\ & s^{(H)} = -\frac{\partial\varphi^{(H)}}{\partial T} \end{aligned}\right\} \tag{3.4}$$

Assumption (3.1) makes it possible to present the internal energy function u in the form of a sum provided that u is expressed in terms of $X^{T\epsilon}$ or $X^{T\sigma}$ *), e.g.

$$u(X^{T\epsilon}) = u^{(e)}(X^T_\epsilon) + u^{(H)}(X^T_\varkappa) \tag{3.5}$$

where

$$\left.\begin{aligned} u^{(e)}(X^T_\epsilon) &= \varphi^{(e)}(X^T_\epsilon) + T\, s^{(e)}(X^T_\epsilon) \\ u^{(H)}(X^T_\varkappa) &= \varphi^{(H)}(X^T_\varkappa) + T\, s^{(H)}(X^T_\varkappa) \end{aligned}\right\} \tag{3.6}$$

The function $u^{(H)}$ represents stored (internal) energy due to isothermal cold work.

(ii) Among all possible states of an element we distinguish unloaded states that are characterized by

$$T = T_0 , \qquad \underset{\sim}{\epsilon}^e = 0 , \qquad \underset{\sim}{\sigma} = 0 \tag{3.7}$$

Here $T_0$ is a certain reference temperature. The unloaded state at which a material element is in the unconstrained equilibrium i.e., such that

$$\underset{\sim}{\pi} = \underset{\sim}{\varkappa} = \underset{\sim}{0} , \quad \varkappa^{(\wedge)} = \varkappa^{(m)} = \pi^{(\wedge)} = \pi^{(m)} = 0 , \tag{3.8}$$

we shall call thermodynamic reference state (natural state). A sample of fully annealed plastically unstrained metal at the temperature $T_0$ (at which the metal is metallurgically

*) The function u expressed in terms of those variables is not a thermodynamical potential.

stable) is usually assumed to be in the thermodynamic reference state.

From (3.7) it follows that $[\partial\varphi^{(e)}/\partial\underset{\sim}{\varepsilon}^{e}] = \underset{\sim}{0}$, $[\partial g^{(e)}/\partial\underset{\sim}{\sigma}] = \underset{\sim}{0}$ at an unloaded state. Without any physical restriction we can assume that

$$\varphi^{(H)} = 0 \quad \text{for} \quad \underset{\sim}{\varkappa} = \underset{\sim}{0},\ \varkappa^{(\Lambda)} = \varkappa^{(m)} = 0 \tag{3.9}$$

and arbitrary T. Hence

$$s^{(H)} = u^{(H)} = 0 \quad \text{for } \underset{\sim}{\varkappa} = \underset{\sim}{0},\ \varkappa^{(\Lambda)} = \varkappa^{(m)} = 0 \tag{3.10}$$

At natural state ($T = T_o$, $\underset{\sim}{\varkappa} = \underset{\sim}{0}$, $\varkappa^{(\Lambda)} = 0$, $\varkappa^{(m)} = 0$) we have additionally

$$\frac{\partial\varphi^{(H)}}{\partial\underset{\sim}{\varkappa}} = \underset{\sim}{0}\ ; \quad \frac{\partial\varphi^{(H)}}{\partial\varkappa^{(\Lambda)}} = 0\ ; \quad \frac{\partial\varphi^{(H)}}{\partial\varkappa^{(m)}} = 0 \tag{3.11}$$

(iii) Consider an isothermal (T = const.) cyclic process in $\underset{\sim}{\varepsilon}^{e}$ - space (or $\underset{\sim}{\sigma}$ - space) which begins at some state 1 and ends at another state 2 (Fig.3.1). From (3.5) we get

$$u_2-u_1 = \int_1^2 du^{(e)} + \int_1^2 du^{(H)} = u^{(H)}(\overset{(2)}{X}{}^{T}_{\varkappa}) - u^{(H)}(\overset{(1)}{X}{}^{T}_{\varkappa}) \tag{3.12}$$

$$\overset{(i)}{X}{}^{T}_{\varkappa} = \left\{T, \underset{\sim}{\varkappa}_i,\ \varkappa_i^{(\Lambda)},\ \varkappa_i^{(m)}\right\} \qquad i = 1,\ 2$$

On the other hand from the first law (1.4) it follows that

$$u_2-u_1 = \int_1^2 dq + \int_1^2 \frac{1}{\varrho}\,\underset{\sim}{\sigma}\cdot d\underset{\sim}{\varepsilon}^{e} + \int_1^2 \frac{1}{\varrho}\,\underset{\sim}{\sigma}\cdot d\underset{\sim}{\varepsilon}^{p} = Q(2,1) + W^{(p)}(2,1)\,\frac{1}{\varrho} \tag{3.13}$$

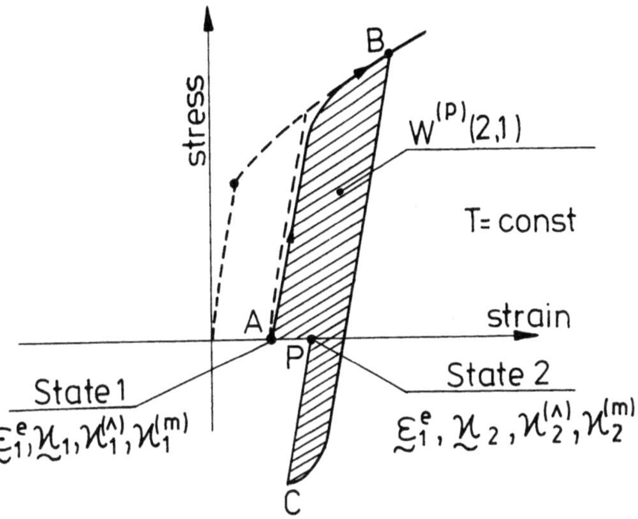


Fig.3.1. The plastic work is not equal to the heat absorbed by a surrounding.

The second integral in (3.13) vanishes since T = const, $\underset{\sim}{\sigma}$ is the state function which due to (3.1) is independent of hardening parameters, and $\underset{\sim}{\varepsilon}_2^e = \underset{\sim}{\varepsilon}_1^e$. Whence

$$u^{(H)}(\overset{(2)}{X}{}^T_{\varkappa}) - u^{(H)}(\overset{(1)}{X}{}^T_{\varkappa}) = Q(2,1) + W^{(p)}(2,1)\,\frac{1}{\varrho} \qquad (3.14)$$

If the state 1 is so choosen that $\underset{\sim}{\varkappa}_1 = \underset{\sim}{0}$ , $\varkappa_1^{(\wedge)} = \varkappa_1^{(m)} = 0$ then $u^{(H)}(\overset{(1)}{X}{}^T_{\varkappa}) = 0$. It is seen that $u^{(H)}(X^T_{\varkappa})$ represents stored energy due to isothermal cold work. It is equal to the difference between work expanded in cold work and heat -Q(2,1) which is absorbed by surrounding. The heat - Q(2,1) which takes part in such kind of experiments, however, is not generally equal to the energy dissipation. For, it can be shown that

$$W^{(D)}(2,1) = W^{(p)}(2,1) - \varrho\, \varphi^{(H)}(2,1) \qquad (3.15)$$

$$\frac{1}{\varrho} W^{(D)}(2,1) = -Q(2,1) + T[s^{(H)}(\overset{(2)}{X}{}^T_{\alpha}) - s^{(H)}(\overset{(1)}{X}{}^T_{\alpha})] \quad (3.16)$$

on account of (1.8), (3.1), $(3.6)_2$ and (3.14).

It is seen that the heat absorbed by the surrounding $-Q(2,1)$ when reduced by the heat due to the change in entropy ("heat of structural hardening") is equal to the total energy dissipation in a cyclic isothermal process in $\underset{\sim}{\sigma}$ - space. Since $\dot{W}^{(D)}$ can not be negative (cf. 1.8), it follows from (3.15) that

$$W^{(p)}(2,1) \geqslant \varrho\, \varphi^{(H)}(2,1) \quad (3.17)$$

or when 2 and 1 are sufficiently close neighbouring state

$$\underset{\sim}{\sigma} \cdot \dot{\underset{\sim}{\varepsilon}}{}^p \geqslant \varrho\, \dot{\varphi}^{(H)} \quad (3.18)$$

and we arrived at the conclusion (Kestin, Rice [11]) that in the absence of stresses the potential $\varphi^{(H)}$ can only decrease. Note that all statements (3.15) - (3.18) are independent of particular form of the rate equations (to be discussed in Sec.5). Finally, it is perhaps worth-while to mention that in the so called "recovery" stage of annealing, the physical properties (e.g. yield stress) that suffered changes as a result of plastic deformations tend to recover their original values. This is done on the expense of stored energy [20].

## 4. Coupled Effects, Equation for Temperature, Structural Entropy.

(i) The isothermal elastic stiffness $\underset{\sim}{L}$ and compliances $\underset{\sim}{M}$ as well as specific heats $C_\epsilon$ and $C_\sigma$, and thermal expansion tensor $\underset{\sim}{\alpha}$ are defined in the same manner as in the usual thermoelasticity

$$\left.\begin{aligned} L_{ijmn}(X^{T\epsilon}) &= \frac{\partial\sigma_{ij}(X^{T\epsilon})}{\partial\epsilon^e_{mn}} = \varrho\,\frac{\partial^2\varphi}{\partial\epsilon^e_{mn}\partial\epsilon^e_{ij}} \\ M_{ijmn}(X^{T\sigma}) &= \frac{\partial\epsilon^e_{ij}(X^{T\sigma})}{\partial\sigma_{mn}} = -\,\varrho\,\frac{\partial^2 g}{\partial\sigma_{ij}\partial\sigma_{mn}} \end{aligned}\right\} \tag{4.1}$$

$$\left.\begin{aligned} C_\epsilon\ (X^{T\epsilon}) &= T\,\frac{\partial s(X^{T\epsilon})}{\partial T} = \frac{\partial u(X^{T\epsilon})}{\partial T} = -\,T\,\frac{\partial^2\varphi(X^{T\epsilon})}{\partial T^2} \\ C_\sigma(X^{T\sigma}) &= T\,\frac{\partial s(X^{T\sigma})}{\partial T} = -\,T\,\frac{\partial^2\varphi(X^{T\sigma})}{\partial T^2} \end{aligned}\right\} \tag{4.2}$$

$$\underset{\sim}{\alpha}(X^{T\sigma}) = \frac{\partial\underset{\sim}{\epsilon}^{(e)}(X^{T\sigma})}{\partial T} = -\,\varrho\,\frac{\partial^2 g}{\partial T\partial\underset{\sim}{\sigma}} \tag{4.3}$$

Note that $\underset{\sim}{L}$, $\underset{\sim}{M}$ and $\underset{\sim}{\alpha}$ are independent of $\underset{\sim}{\varkappa}$, $\varkappa^{(\wedge)}$ and $\varkappa^{(m)}$, whenever (3.1) - (3.3) holds. However, $C_\sigma$ and $C_\epsilon$ can still be influenced by a hardening.

Let us introduce new quantities defined as follows

$$L^{(\pi)}_{ijmn} = \frac{\partial\pi_{ij}}{\partial\varkappa_{mn}} = \varrho\,\frac{\partial^2\varphi^{(H)}}{\partial\varkappa_{mn}\partial\varkappa_{ij}}\ ;\ \underset{\sim}{L}^{(\wedge)} = \frac{\partial\underset{\sim}{\pi}}{\partial\varkappa^{(\wedge)}} = \frac{\varrho\ \partial^2\varphi^{(H)}}{\partial\underset{\sim}{\varkappa}\partial\varkappa^{(\wedge)}} = \frac{\partial\pi^{(\wedge)}}{\partial\underset{\sim}{\varkappa}} \tag{4.4}$$

$$\left.\begin{aligned} &\underset{\sim}{L}^{(m)} = \frac{\partial \underset{\sim}{\pi}}{\partial \varkappa^{(m)}} = \varrho \frac{\partial^2 \varphi^{(H)}}{\partial \underset{\sim}{\varkappa} \partial \varkappa^{(m)}} = \frac{\partial \pi^{(m)}}{\partial \underset{\sim}{\varkappa}} ; \\ &l^{(\wedge)} = \frac{\partial \pi^{(\wedge)}}{\partial \varkappa^{(\wedge)}} ; \quad l^{(m)} = \frac{\partial \pi^{(m)}}{\partial \varkappa^{(m)}} ; \quad l^{(\wedge, m)} = \frac{\partial \pi^{(m)}}{\partial \varkappa^{(\wedge)}} = \frac{\partial \pi^{(\wedge)}}{\partial \varkappa^{(m)}} \end{aligned}\right\} \begin{matrix}(4.4)\\ \text{(cont.)}\end{matrix}$$

They will loosely be called moduli of structural hardening.

Physical quantities (4.1) - (4.3) are not all independent. They satisfy well-known identities

$$\left.\begin{aligned} &2 L_{ijrs} M_{rsmn} = \delta_{im}\delta_{jn} + \delta_{jm}\delta_{in} \\ &C_\sigma(X^{T\sigma}) = \left[C_\epsilon + \frac{T}{\varrho}(\underset{\sim}{\alpha} \cdot \underset{\approx}{L}\underset{\sim}{\alpha})\right]_{\underset{\sim}{\epsilon}^e = \underset{\sim}{\epsilon}^e(X^{T\sigma})} \end{aligned}\right\} \quad (4.5)$$

which can be verified by differentation both sides of

$$\frac{1}{\varrho}\underset{\sim}{\sigma} = \left(\frac{\partial \varphi}{\partial \underset{\sim}{\epsilon}^e}\right)_{\underset{\sim}{\epsilon}^e = \underset{\sim}{\epsilon}^e(X^{T\sigma})} \tag{4.6}$$

and

$$s(X^{T\epsilon})\Big|_{\underset{\sim}{\epsilon}^e = \underset{\sim}{\epsilon}^e(X^{T\sigma})} = s(X^{T\sigma})$$

with respect to $\underset{\sim}{\sigma}$ and T, respectively, and by use of

$$-\varrho \frac{\partial s(X^{T\epsilon})}{\partial \underset{\sim}{\epsilon}^e} = \frac{\partial \underset{\sim}{\sigma}(X^{T\epsilon})}{\partial T} = \underset{\approx}{L}(X^{T\epsilon})\,\underset{\sim}{\alpha} \tag{4.7}$$

The latter equality follows from the calculation of the partial derivative of both sides of (4.6) with respect to T.

One usually deals with so called intrinsic stable materials with respect to thermoelastic properties. For such materials the matrix of second order derivatives of u with respect to s and $\underset{\sim}{\varepsilon}^e$ is positive definite. The conditions of positive-definiteness imply

$$C_\varepsilon > 0 \qquad d\underset{\sim}{\varepsilon}^e \cdot \underset{\sim}{L}\, d\underset{\sim}{\varepsilon}^e > 0 \quad \text{for every } d\underset{\sim}{\varepsilon}^e \neq \underset{\sim}{0} \tag{4.8}$$

(ii) To discuss some coupled effects let us calculate the total differentials of dependent state variables assuming in sequence, the set $X^{T\varepsilon}$, and $X^{T\sigma}$ as a set of independent state variables. Accounting for assumption (3.1) and (4.1)--(4.4) we find

$$Tds = \underbrace{C_\varepsilon dT}_{\text{heat capacity}} + \underbrace{\frac{T}{\varrho}\, d\underset{\sim}{\varepsilon}^e \cdot \underset{\sim}{L}\underset{\sim}{\alpha}}_{\text{heat of elastic deformation}} + \underbrace{T\left[\frac{\partial s^{(H)}}{\partial \underset{\sim}{\varkappa}} \cdot d\underset{\sim}{\varkappa} + \frac{\partial s^{(H)}}{\partial \varkappa^{(\wedge)}} d\varkappa^{(\wedge)} + \frac{\partial s^{(H)}}{\partial \varkappa^{(m)}} d\varkappa^{(m)}\right]}_{\text{heat of structural hardening}} \tag{4.9}$$

$$Tds = \overbrace{C_\sigma dT} + \underbrace{\frac{T}{\varrho}\, \underset{\sim}{\alpha} \cdot d\underset{\sim}{\sigma}}_{\text{piezocaloric effect}} + \overbrace{T\left[\frac{\partial s^{(H)}}{\partial \underset{\sim}{\varkappa}} \cdot d\underset{\sim}{\varkappa} + \frac{\partial s^{(H)}}{\partial \varkappa^{(\wedge)}} d\varkappa^{(\wedge)} + \frac{\partial s^{(H)}}{\partial \varkappa^{(m)}} d\varkappa^{(m)}\right]} \tag{4.10}$$

$$d\underset{\sim}{\sigma} = \underbrace{\underset{\sim}{L}\, d\underset{\sim}{\varepsilon}^e}_{\text{elasticity}} - \underbrace{\underset{\sim}{L}\, \underset{\sim}{\alpha}\, dT}_{\text{elastic thermal stresses}} \tag{4.11}$$

$$d\underset{\sim}{\varepsilon}^e = \overbrace{\underset{\sim}{M}\, d\underset{\sim}{\sigma}} + \underbrace{\underset{\sim}{\alpha}\, dT}_{\text{thermal expansion}} \tag{4.12}$$

$$\left.\begin{aligned} d\underset{\sim}{\pi} &= \underset{\sim}{L}^{(\pi)} \cdot d\underset{\sim}{\varkappa} + \underset{\sim}{L}^{(\wedge)} d\varkappa^{(\wedge)} + \underset{\sim}{L}^{(m)} d\varkappa^{(m)} + \underset{\sim}{\pi}_T\, dT \\ d\pi^{(\wedge)} &= \underset{\sim}{L}^{(\wedge)} \cdot d\underset{\sim}{\varkappa} + l^{(\wedge)} d\varkappa^{(\wedge)} + l^{(\wedge,m)} d\varkappa^{(m)} + \pi_T^{(\wedge)} dT \\ d\pi^{(m)} &= \underbrace{\underset{\sim}{L}^{(m)} \cdot d\underset{\sim}{\varkappa} + l^{(\wedge,m)} d\varkappa^{(\wedge)} + l^{(m)} d\varkappa^{(m)}}_{\text{structural hardening}} + \underbrace{\pi_T^{(m)} dT}_{\text{thermal structural softening}} \end{aligned}\right\} \tag{4.13}$$

where moduli

$$\underset{\sim}{\pi}_T = \frac{\partial \underset{\sim}{\pi}}{\partial T} \quad ; \quad \pi_T^{(\wedge)} = \frac{\partial \pi^{(\wedge)}}{\partial T} \quad ; \quad \pi_T^{(m)} = \frac{\partial \pi^{(m)}}{\partial T} \tag{4.14}$$

represent the "thermal structural softening".

The interpretation of various terms occuring in eqs. (4.9)-(4.12) follows directly from the physical meaning of corresponding partial derivatives [42]. The name associated with last terms occuring in (4.9)-(4.10) and in (4.13) are rather general. One can not give more accurate interpretation until a specific physical meaning is attached to the internal parameters. Nevertheless we have introduced them in order to underline the fact that new thermostatic effects occur when body is plastically strained.

The possible couplings are illustrated by the diagram*) in Fig.(4.1). In three outer corners are temperature T, stress $\underset{\sim}{\sigma}$ and internal forces $\Pi = \{\underset{\sim}{\pi}, \pi^{(\wedge)}, \pi^{(m)}\}$. In the three corresponding inner corners appear entropy s, elastic strain $\underset{\sim}{\epsilon}^e$ and hardening parameters $X_\varkappa = \{\underset{\sim}{\varkappa}, \varkappa^{(\wedge)}, \varkappa^{(m)}\}$. Each of these two sets can be regarded as the set of independent variables e.g., when s, $\underset{\sim}{\epsilon}^e$ and $X_\varkappa$ are independent variables then their changes cause the changes in T, $\underset{\sim}{\sigma}$ and $\underset{\sim}{\Pi}$. It is possible to form a number of sets of independent variables so that every variable may be the member of a set

*) This is a development of a diagram given by Nye [42] for the thermoelasticity.

of independent variables. Hence, there is a number of "coupled effects".

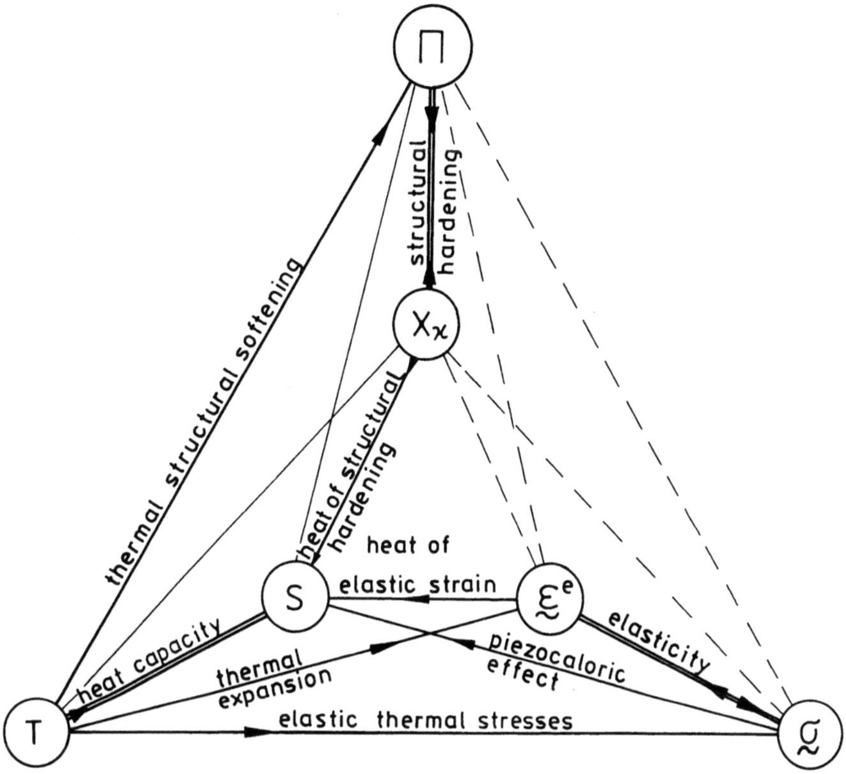


Fig.4.1. The diagram of some thermostatic coupled effects in thermoplasticity

Some of them are illustrated in Fig.4.1 by lines joining pairs of points which are not both at the same corner. The double lines represent the "principal effects". The lines connecting T, s, $\underset{\sim}{\sigma}$ and $\underset{\sim}{\varepsilon}^e$ denote the coupled effects of thermoelastic behaviour, whereas the dotted lines indicate couplings which are excluded from the consideration due to assumption (3.1).

(iii) By eliminating $T\,\dot{s}$ between (1.10) and (4.9) or (4.10) with account of (1.8),(3.1),(3.4),(3.6)$_2$ we obtain two alternative forms of the equation for temperature

$$\begin{aligned} \varrho\, C_\varepsilon \dot{T} &= \underset{\sim}{\sigma}\cdot\underset{\sim}{\dot{\varepsilon}}^p - T\,\underset{\sim}{\dot{\varepsilon}}^e\cdot\underset{\sim}{L}\,\underset{\sim}{\alpha} - \varrho\,\overset{\circ}{u}^{(H)} + \varrho\,\dot{q} \\ \varrho\, C_\sigma \dot{T} &= \underset{\sim}{\sigma}\cdot\underset{\sim}{\dot{\varepsilon}}^p - T\,\underset{\sim}{\alpha}\cdot\underset{\sim}{\dot{\sigma}} - \varrho\,\overset{\circ}{u}^{(H)} + \varrho\,\dot{q} \end{aligned} \tag{4.15}$$

where

$$\overset{\circ}{u}^{(H)} \equiv \frac{\partial u^{(H)}}{\partial \underset{\sim}{\varkappa}}\cdot\underset{\sim}{\dot{\varkappa}} + \frac{\partial u^{(H)}}{\partial \varkappa^{(\wedge)}}\,\dot{\varkappa}^{(\wedge)} + \frac{\partial u^{(H)}}{\partial \varkappa^{(m)}}\,\dot{\varkappa}^{(m)} \tag{4.16}$$

Note that the entropy function $s^{(H)}$ does not occur in (4.15) explicitly.

(iv) Let us finally discuss slightly in a more detail the stored free energy $\varphi^{(H)}$ of cold work and "structural entropy" $s^{(H)}$.

Under the assumption (3.1) (cf.also (3.5)) we can write

$$C_\varepsilon = C'_\varepsilon(X^T_\varepsilon) + \Delta C(X^T_\varkappa) \tag{4.17}$$

where $\Delta C = 0$ for $\underset{\sim}{\varkappa} = \underset{\sim}{0}$, $\varkappa^{(\wedge)} = 0$, $\varkappa^{(m)} = 0$. Similary, the stored energy can be written in the form of sum

$$u^{(H)}(X^T_\varkappa) = \bar{u}^{(H)}(X_\varkappa) + \Delta u^{(H)}(X^T_\varkappa) \tag{4.18}$$

where $\Delta u^{(H)} = \bar{u}^{(H)} = 0$ for $\underset{\sim}{\varkappa} = \underset{\sim}{0}$, $\varkappa^{(\wedge)} = 0$, $\varkappa^{(m)} = 0$ and $\Delta u^{(H)} = 0$ for $T = T_0$, i.e. $\bar{u}^{(H)}$ represents stored internal energy at the reference temperature $T_0$. Suppose that $C_\varepsilon$ and $u^{(H)}$ are known functions (i.e. measured in some experimental programm). It is easy to verify that (cf.(4.2)) $\Delta C$

and $\Delta u^{(H)}$ are not independent

$$\Delta C(X_{\varkappa}^{T}) = \frac{\partial}{\partial T}\,\Delta u^{(H)}$$

and by integrating the equation $(3.6)_2$ we find

$$\varphi^{(H)} = T\int_{T_0}^{T}\Delta u^{(H)}(X_{\varkappa}^{T_1})\,d\Big(\frac{1}{T_1}\Big) - T\,\bar{s}^{(H)}(X_{\varkappa}) + \bar{u}^{(H)}(X_{\varkappa}) \qquad (4.19)$$

$$\left.\begin{aligned} s^{(H)} &= -\frac{\partial\varphi^{(H)}}{\partial T} = \bar{s}^{(H)}(X_{\varkappa}) + \Delta s^{(H)}(X_{\varkappa}^{T}) \\ \Delta s^{(H)} &= \int_{T_0}^{T}\frac{1}{T_1}\,\frac{\partial}{\partial T_1}\,[\Delta u^{(H)}(X_{\varkappa}^{T_1})]\,dT_1 \end{aligned}\right\} \qquad (4.20)$$

where $\bar{s}^{(H)}(X_{\varkappa})$ represents the structural entropy at the reference temperature $T = T_0$; $\bar{s}^{(H)} = 0$ for $\underset{\sim}{\varkappa} = \underset{\sim}{0}$, $\varkappa^{(\wedge)} = \varkappa^{(m)} = 0$. It is seen that the structural entropy $\bar{s}^{(H)}$ at reference temperature can not be determined neither through measurement of specific heats nor energy stored in isothermal process of plastic deformations. The experimental determination of $\bar{s}^{(H)}$ is one of the actual unsolved problems of the phenomenological theory of plasticity. It occurs as a consequence of our inability to contrive a real process which would be close to the assumed hypothetical reversible process. Generally one expects help in this matter from the Physics of Solids, and it is frequently belived [20] that the term $T\,\bar{s}^{(H)}$ is neglegible in comparison to that of $\bar{u}^{(H)}$. Here we promote the following way of the estimation of $\bar{s}^{(H)}$. Suppose that at reference temperature $T_0$ we are able to determine experimentally the values of internal forces $\Pi = \{\underline{\pi}, \pi^{(\wedge)}, \pi^{(m)}\}$ and values of hardening parameters $X_\varkappa$. This can be done in the stage of experimental identification of hardening functions occuring in the theory with the aid of the normality postulate (cf. Sec.5). Having known $\Pi$, we can identify $\varphi^{(H)}$ at $T = T_0$, and since $\bar{u}^{(H)}$ is measurable the structural entropy $\bar{s}^{(H)}$ may be estimated as follows

$$\bar{s}^{(H)} = \frac{1}{T_0} [\varphi^{(H)}\big|_{T=T_0} - \bar{u}^{(H)}] \tag{4.21}$$

## III. RATE EQUATIONS

### 5. Dissipative Potential for Rate-Independent Elastic-Plastic Bodies.

The other group of "constitutive relations" is formed by rate equations that link the dissipative forces $Z_i$ (cf. (1.9)) with rates $\dot{z}_i$ occuring in the expression (1.8) for the entropy production. Here we assume that forces $Z_i$ possess so called dissipative potential $\Psi(\dot{z}_i)$, such that

$$Z_i = \partial\Psi(\dot{\underline{z}})/\partial\dot{z}_i \tag{5.1}$$

The potential $\Psi$ may also be dependent upon some of the state variables from the set $X^{T\sigma}$, what is not explicitly marked in (5.1). When $\Psi$ has continuous second derivatives with respect to all rates the assumption (5.1) is equivalent to so called "Gyarmati postulat" [9]. Edelen [43,44] has shown that this postulat exclude from the consideration only such force which when multiplied by conjugate rates are equal to zero, i.e., so called "nondissipative forces". Some of the writers (see e.g. [12,13]) assume additionally that the dissipative potential is a convex function of $\dot{z}_i$, whereas Vakulenko in his early papers [5,6] applied the relations of the type (5.1) only to the external quantities i.e. $\dot{\underset{\sim}{\varepsilon}}^p$ and $\underset{\sim}{\sigma}$.

For rate independent elastic-plastic materials, equations (5.1) must be invariant under a change in a time scale. Hence $\Psi$ is a homogeneous function of order one with respect to $\dot{z}_i$, i.e.

$$\left.\begin{aligned} &\Psi(\tau\dot{z}_i) = \tau\,\Psi(\dot{z}_i) \qquad \tau > 0 \\ &Z_i\,\dot{z}_i = \partial\Psi/\partial\dot{z}_i \cdot \dot{z}_i = \Psi > 0 \;\; (i = 1,2,\ldots,14) \end{aligned}\right\} \quad (5.2)$$

and surfaces $\Psi$ = const. in $\dot{z}_i$ - space are the surfaces of constant rate of energy dissipation (cf.Eq.1.9). If additionally the region enclosed by a surface $\Psi$ = const. is assumed to be convex then the postulate (5.1) coincides with the "orthogonality principle" developed by Ziegler [7,8].

Let us now discuss briefly some mathematical aspects of the assumptions (5.1),(5.2) [45]. Since $\Psi$ is a homogeneous and positive definite function it has no derivatives in ordinary sense at the point $\dot{\underset{\sim}{z}} = \underset{\sim}{0}$, and $Z_i$ is not uniquely determined by (5.1) at $\dot{\underset{\sim}{z}} = 0$ i.e., at a constrained equilibrium ($\dot{\underset{\sim}{z}} = 0$, $\Psi = \dot{W}^D = 0$)

the dissipative forces do not need to vanish. Suppose that $\dot{\underset{\sim}{z}} \neq \underset{\sim}{0}$. It is not difficult to verify that matrix $|\Psi_{ij}| = |\partial\Psi/\partial\dot{z}_i\partial\dot{z}_j|$ is singular i.e.

$$\det|\Psi_{ij}| = 0 \, .$$

Assume that the rank of the matrix $|\Psi_{ij}|$ is $14 - n$ $(1 \leqslant n < 14)$. It can be proved [45] that among all functions $Z_i$ there are only 14-n independent functions, so that there exist n independent relations of the type

$$f^{(j)}(Z_i) = 0, \qquad j = 1,\dots,n \geqslant 1 \, . \tag{5.3}$$

Moreover, when $\dot{\underset{\sim}{z}} \neq \underset{\sim}{0}$ the set of equations (5.1) can be solved with respect to $\dot{z}_i$. The solution is

$$\dot{\underset{\sim}{z}} = \sum_{j=1}^{n} \lambda^{(j)} \frac{\partial f^{(j)}}{\partial \underset{\sim}{Z}} \tag{5.4}$$

where multipliers $\lambda^{(j)}$ should be determined with the aid of (5.3) and with account for the inequality $\Psi = \dot{W}^D > 0$.

From (5.3) it follows that in 14-dimensional space of dissipative forces $\underset{\sim}{Z}$ there may exists n <u>generalized yield surfaces.</u> When $f^{(j)} \neq 0$ then $\dot{z}_i = 0$, so that the process $\dot{\underset{\sim}{z}} \neq 0$ is possible only when corresponding $\underset{\sim}{Z}$ satisfies (5.3). In particular, when $Z_i$ is such that $f^{(\alpha)}(Z_i) = 0$ and $f^{(\beta)}(Z_i) \neq 0$ for all $\beta \neq \alpha$, the relations (5.4) have simpler form

$$\dot{\underset{\sim}{z}} = \lambda^{(\alpha)} \frac{\partial f^{(\alpha)}}{\partial \underset{\sim}{Z}} \, . \tag{5.5}$$

Substituting (5.4) into (1.9) we obtain the condition

$$\sum_{j=1}^{n} \lambda^{(j)} \frac{\partial f^{(j)}}{\partial Z_i} Z_i \geq 0 \qquad (5.6)$$

that is supposed to be satisfied for all $\underset{\sim}{Z}$ for which (5.3) holds. Under appropriate choice of the functions $f^{(j)}$ the inequality (5.6) fixes the signs of multipliers $\lambda^{(j)}$ to be positive, whenever $\dot{\underset{\sim}{z}} \neq 0$.

## 6. Generalized Plastic Flow Rules.

(i) The mathematical aspects discussed in Sec.5 show that the hypothesis of normal dissipativity (5.1) does not contradict the multisurface formulation of cyclic plasticity [23,24]. We shall henceforth restrict the discussion to the bisurface model of hardening for which the rank of the matrix $|\Psi_{ij}|$ equals 12. To get some correlation with the concepts introduced so far in the theory of plasticity it is expedient to consider two surfaces described by the following general equations written with use of tensorial notation as follows

$$f^{(1)}(\underset{\sim}{S} - \underset{\sim}{\pi},\ \pi^{(\wedge)};\ \varkappa^{(m)}, T) = 0 \qquad (6.1)$$

$$f^{(2)}(\underset{\sim}{\pi};\ \pi^{(m)};\ T) = 0 \qquad (6.2)$$

where $\underset{\sim}{S}$ is deviatoric part of stress tensor $\underset{\sim}{\sigma}$

$$\underset{\sim}{S} = \underset{\sim}{\sigma} - \tfrac{1}{3}\,\underset{\sim}{1}(\mathrm{tr}\,\underset{\sim}{\sigma})$$

and it is assumed that

$$\underset{\sim}{\pi} = \text{dev}\ \underset{\sim}{\pi} \qquad \text{or tr}\ \underset{\sim}{\pi} = 0$$

Here, (6.1) is an equation of the usual yield surface that separates states of overall elastic behaviour $f^{(1)} < 0$ from that of elastic-plastic behaviour ($f^{(1)} = 0$). According to the classical concept of a yield surface, states for which $f^{(1)} > 0$ are inaccessible. Tensor $\underset{\sim}{\pi}$ describes a translatory motion of yield surface in $\underset{\sim}{S}$ - space (directional hardening). With the aid of special models, tensor $\underset{\sim}{\pi}$ is sometimes interpreted [46,47] as a phenomenological measure of "microstresses". The explicit dependence of $f^{(1)}$ on the state variables $\varkappa^{(m)}$ and T, and $f^{(2)}$ on T, reflects the possible alteration in the dissipative potential mode due to the change in thermodynamics state.

Consequently, besides the structural hardening (emphasized in (6.1) by dependence of $f^{(1)}$ on $\pi^{(\wedge)}$) there exists other source of overall indirectional hardening which may be loosely called "frictional hardening". Similary, overall thermal softening can be attributed to the sensitivity of $\Psi$ on temperature T.

For fixed values of $\pi^{(m)}$ and T, eq.(6.2) describes the other family of surfaces in 5-dimensional space. We shall limit our analysis to the case when the surfaces (5.2) possess the centre of symmetry (Fig.6.1).

The variation of internal forces can be divided into 2

groups: those for which $f^{(2)} = 0$ (Π - inelastic process, such as B - C - D, Fig.6.1) and those for which $f^{(2)} < 0$ (Π - elastic process, such as 0 - A - $A_1$, Fig.6.1). Again, states such that $f^{(2)} > 0$ are assumed to be inaccessible.

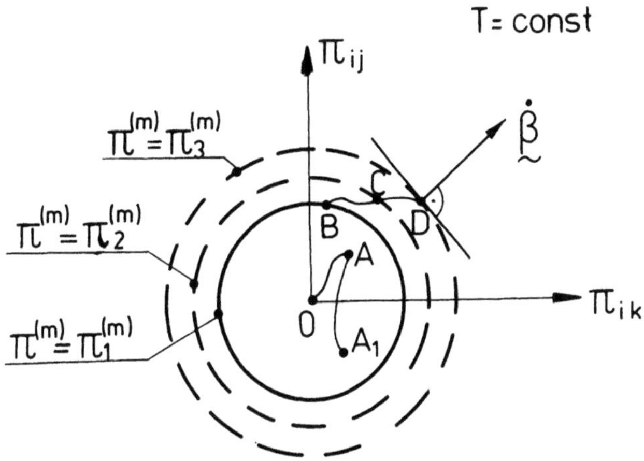


Fig.6.1. Family of surfaces $f^{(2)} = 0$
for different values of $\pi^{(m)}$.

Note that during any isothermal Π - elastic process the surface (6.2) in π - space need not to be fixed, it can both shrink and expand, depending upon the variation of $\pi^{(m)}$ in the course of the process.

The examples of the specific surfaces are that associated with familiar Huber-Mises yield condition

$$f^{(1)} = [(\underset{\sim}{S}-\underset{\sim}{\pi})\cdot(\underset{\sim}{S}-\underset{\sim}{\pi})]^{\frac{1}{2}} - Y^{(1)}(\pi^{(\wedge)}, \varkappa^{(m)}, T) = 0 \qquad (6.3)$$

$$f^{(2)} = (\underset{\sim}{\pi}\cdot\underset{\sim}{\pi})^{\frac{1}{2}} - Y^{(2)}(\pi^{(m)},T) = 0 \tag{6.3 (cont)}$$

(ii) Suppose that $\dot{\underset{\sim}{z}} \neq \underset{\sim}{0}$. Substituting (6.1)-(6.2) into (5.4) we obtain the following flow rules associated with (6.1)-(6.2)

$$\dot{\underset{\sim}{\varepsilon}}^p = \begin{cases} \lambda^{(1)} \dfrac{\partial f^{(1)}}{\partial \underset{\sim}{S}} & \text{if } f^{(1)} = 0,\ \lambda^{(1)} > 0 \\ 0 & \text{if } f^{(1)} < 0 \end{cases} \tag{6.4}$$

$$\dot{\underset{\sim}{\varkappa}} = \begin{cases} \dot{\underset{\sim}{\varepsilon}}^p - \lambda^{(2)} \dfrac{\partial f^{(2)}}{\partial \underset{\sim}{\pi}} & \text{if } f^{(1)}=f^{(2)}=0,\ \lambda^{(1)}>0, \lambda^{(2)}>0 \\ \dot{\underset{\sim}{\varepsilon}}^p & \text{if } f^{(1)}=0,\ f^{(2)}<0,\ \lambda^{(1)}>0 \\ - \lambda^{(2)} \dfrac{\partial f^{(2)}}{\partial \underset{\sim}{\pi}} & \text{if } f^{(2)}=0,\ f^{(1)}<0,\ \lambda^{(2)}>0 \end{cases} \tag{6.5}$$

$$\dot{\varkappa}^{(\wedge)} = \begin{cases} - \lambda^{(1)} \dfrac{\partial f^{(1)}}{\partial \pi^{(\wedge)}} & \text{if } f^{(1)}=0,\ \lambda^{(1)}>0 \\ 0 & \text{if } f^{(1)}<0 \end{cases} \tag{6.6}$$

$$\dot{\varkappa}^{(m)} = \begin{cases} - \lambda^{(2)} \dfrac{\partial f^{(2)}}{\partial \pi^{(m)}} & \text{if } f^{(2)}=0,\ \lambda^{(2)}>0 \\ 0 & \text{if } f^{(2)}<0 \end{cases} \tag{6.7}$$

$$\operatorname{tr} \dot{\underset{\sim}{\varepsilon}}^p = \operatorname{tr} \dot{\underset{\sim}{\varkappa}} = 0$$

From (6.5) it follows that the plastic strain rate tensor $\dot{\underset{\sim}{\varepsilon}}^p$ is made up of two components

$$\dot{\underset{\sim}{\varepsilon}}^{p} = \dot{\underset{\sim}{\varkappa}} + \dot{\underset{\sim}{\beta}} \qquad \mathrm{tr}\,\dot{\underset{\sim}{\beta}} = 0 \tag{6.8}$$

where $\dot{\underset{\sim}{\beta}}$ is related to $\underset{\sim}{\pi}$ in similar manner as $\dot{\underset{\sim}{\varepsilon}}^{p}$ to $\underset{\sim}{S}$

$$\dot{\underset{\sim}{\beta}} = \left\{ \begin{array}{ll} \lambda^{(2)} \dfrac{\partial f^{(2)}}{\partial \underset{\sim}{\pi}} & \text{if } f^{(2)} = 0, \quad \lambda^{(2)} > 0 \\ \underset{\sim}{0} & \text{if } f^{(2)} < 0 \end{array} \right\} \tag{6.9}$$

Tensor $\underset{\sim}{\beta}$ may, therefore, be interpreted as a certain phenomonological measure of inelastic "micro-strains". It may vary only in the course of $\Pi$ - inelastic process, and $\dot{\underset{\sim}{\beta}}$ has the direction normal to the surfaces $f^{(2)} = 0$, such as those shown in Fig.6.1. Thus, surfaces $f^{(2)} = 0$ play the role of "yield surfaces" for the "micro-inelastic flow". The hardening parameter $\underset{\sim}{\varkappa}$ is related to $\underset{\sim}{\pi}$ through the appriopriate equation of state, (2.6). It can be interpreted as a measure of elastic "micro-strains". Tensor $\underset{\sim}{\varkappa}$ is of course the state variable, whereas $\underset{\sim}{\beta}$ is not.

Suppose that $f^{(1)}$ and $f^{(2)}$ are so choosen that

$$\frac{\partial f^{(1)}}{\partial \underset{\sim}{S}} \cdot \frac{\partial f^{(1)}}{\partial \underset{\sim}{S}} = 1 \; ; \qquad \frac{\partial f^{(2)}}{\partial \underset{\sim}{\pi}} \cdot \frac{\partial f^{(2)}}{\partial \underset{\sim}{\pi}} = 1 \tag{6.10}$$

Then, from (6.4) and (6.9) we obtain

$$\lambda^{(1)} = (\dot{\underset{\sim}{\varepsilon}}^{p} \cdot \dot{\underset{\sim}{\varepsilon}}^{p})^{\frac{1}{2}} \; , \quad \lambda^{(2)} = (\dot{\underset{\sim}{\beta}} \cdot \dot{\underset{\sim}{\beta}})^{\frac{1}{2}} \tag{6.11}$$

The parameters $\varkappa^{(\wedge)}$ and $\varkappa^{(m)}$ may, therefore, be called accumulated scalar measures of $\dot{\underset{\sim}{\varepsilon}}^{p}$ and $\dot{\underset{\sim}{\beta}}$, respectively. When the following conditions are satisfied, $\partial f^{(2)}/\partial \pi^{(m)} \leqslant 0$ and

$\partial f^{(1)}/\partial \pi^{(\wedge)} \leqslant 0$, both scalar parameters are nondecreasing functions of time. Accumulated inelastic micro-strain $\varkappa^{(m)}$ is constant during any $\Pi$ - elastic process, whereas accumulated overall plastic strain $\varkappa^{(\wedge)}$ may vary both in a course of $\Pi$ - elastic and $\Pi$ - inelastic process. Note that, in general, the "microelasticity" is coupled with "micro-inelastic flow" because all moduli of structural hardening (4.4) as well as $\pi^{(m)}$ occuring in (6.2) are influenced by $\varkappa^{(\wedge)}$ and $\varkappa^{(m)}$.

(iii) Consider initially $\Pi$ - elastic process of deformation $\underset{\sim}{\varepsilon}^p(t)$ $(0 \leqslant t \leqslant t_f)$ at two different temperatures $T_1$ and $T_2$. From (6.5)-(6.7), (4.14) and (6.1)-(6.2) it follows that the values of hardening parameters at $t = t_f$ will, in general, be different in both cases provided that any of two kinds of thermal softening is accounted for. Exceptional is the case when $f^{(1)}$ and $f^{(2)}$ are linear functions of $\pi^{(\wedge)}$ and $\pi^{(m)}$, respectively, with proportionality coefficients independent of T. <u>Thus, the model in natural manner predicts occurrence of temperature history effects.</u>

At fixed hardening parameters and $\underset{\sim}{\varepsilon}^p$ = const., the variation of the temperature causes changes in material properties (thermal softening) due to the dependence of internal forces $\Pi$ and $f^{(1)}$, $f^{(2)}$ on T. The variable temperature can also cause the variation of hardening parameters $\underset{\sim}{\varkappa}$ and $\varkappa^{(m)}$ <u>in an unloaded specimens</u> ($\underset{\sim}{\sigma} = 0$, $\dot{\underset{\sim}{\varepsilon}}^p = 0$). This may happen most likely in cold-worked specimens (i.e. being not in the thermodynamic reference state) at the moment when a state such that $f^{(2)} = 0$ is met and when varying temperature results in $\lambda^{(2)} > 0$. Then (cf.(6.5)$_3$ and (6.8)) $\dot{\underset{\sim}{\varkappa}} = -\dot{\underset{\sim}{\beta}}$, i.e., there is a mutual redistribution of inelastic and elastic "micro-strains" due to the temperature action itself. Such process may be called "thermally induced Π - inelastic process". In consequence of it, certain cyclic variations of temperature can alter mechanical properties of stress-free cold worked specimens. Thus, within the presented frame--work it is possible to describe some cycle (thermal) - dependent changes in mechanical properties (extent in Bauschinger effect, hardening/softening etc.). Note that isothermal processes such that $\dot{\underset{\sim}{\varkappa}} = -\dot{\underset{\sim}{\beta}}$ can not occur since vanishing $\dot{\underset{\sim}{\varepsilon}}^p$ implies vanishing $\lambda^{(2)}$. This can be justified by considering the "consistency equation" $\dot{f}^{(2)} = 0$. Thus, processes described by (6.5)$_3$ can be activated only by varying temperature.

Suppose that a material element has been proportionally deformed up to $\underset{\sim}{\varepsilon}^p = \underset{\sim}{\varepsilon}^p_o$. Consider subsequent isothermal (T = const.) cyclic process of deformation, e.g.,

$$\underset{\sim}{\varepsilon}^p(t) = \underset{\sim}{\varepsilon}^p_o + \varepsilon^p(t)\, \underset{\sim}{n}_\varepsilon \; ; \; \underset{\sim}{n}_\varepsilon \cdot \underset{\sim}{n}_\varepsilon = 1, \; \mathrm{tr}\, \underset{\sim}{\varepsilon}^p_o = \mathrm{tr}\, \underset{\sim}{n}_\varepsilon = 0 \qquad (6.12)$$

where $\varepsilon^p(t)$ is a periodic function of time whereas $\underset{\sim}{\varepsilon}^p_o$ and $\underset{\sim}{n}_\varepsilon$ are constant tensors. In general, the response of a material element is initially irregular. According to (6.5 - - 6.8) the transitory time period may be divided into two stages. In stage 1 the cyclic deformation is accompanied by a number of consecutive $\Pi$ - elastic and $\Pi$ - inelastic processes. The stage 1 ends when "microstresses $\underset{\sim}{\pi}$ shake down" to the prescribed programm of cyclic plastic deformation (6.12). In stage 2 the deformation process is purely $\Pi$ - - elastic; inelastic "micro-strains" $\underset{\sim}{\beta}$ and parameter $\varkappa^{(m)}$ are constant. When $\varkappa^{(\wedge)} \to \infty$ the steady state of a cyclic process is attained. The amplitudes and mean values of stresses at steady state depend upon the final values of $\underset{\sim}{\beta}$ and $\varkappa^{(m)}$, the latter in turn are influenced by $\underset{\sim}{\varepsilon}^{p}_{o}$. These remarks reflect the essence of the bi-yield surface concept. It enables to take into account the effect of prior plastic straining on the response at steady state.

(iv) Since $\lambda^{(1)}$, $\lambda^{(2)}$ are assumed to be non-negative, the inequality (1.8) implies

$$\left.\begin{aligned} &(\underset{\sim}{S} - \underset{\sim}{\pi}) \cdot \frac{\partial f^{(1)}}{\partial \underset{\sim}{S}} + \pi^{(\wedge)} \frac{\partial f^{(1)}}{\partial \pi^{(\wedge)}} > 0 \quad \text{at} \quad f^{(1)} = 0 \\ &\frac{\partial f^{(2)}}{\partial \underset{\sim}{\pi}} \cdot \underset{\sim}{\pi} + \pi^{(m)} \frac{\partial f^{(2)}}{\partial \pi^{(m)}} > 0 \quad \text{at} \quad f^{(2)} = 0 \end{aligned}\right\} \qquad (6.13)$$

In view of the uncertainity concerning the role that the structural entropy plays, the conditions (6.13) are the part of constitutive assumptions rather then the strict consequence of the second law of thermodynamics.

## 7. Asymptotically Linear Kinematic Hardening.

(i) In the sequel the attention will be focussed on the isothermal processes that start at the thermodynamic reference state.

Let us employ the yield conditions (6.3) where

$$Y^{(1)}\Big|_{T=T_0} = Y^{(1)}(\pi^{(\wedge)},\varkappa^{(m)}), \quad Y^{(2)}\Big|_{T=T_0} = Y^{(2)}(\pi^{(m)}), \tag{7.1}$$

and assume that the free energy function $\varphi^{(H)}$ (cf.3.1) can be written as

$$\varrho\,\varphi^{(H)} = \tfrac{1}{2}C^{(\alpha)}(\varkappa^{(\wedge)},\varkappa^{(m)})\underset{\sim}{\varkappa}\cdot\underset{\sim}{\varkappa} + \varphi_0^{(H)}(\varkappa^{(\wedge)},\varkappa^{(m)}) + \\ + \varphi_1^{(H)}(\varkappa^{(\wedge)}) + \varphi_2^{(H)}(\varkappa^{(m)}) \,. \tag{7.2}$$

Here

$$\left.\begin{aligned} &C^{(\alpha)} \geqslant 0\,, \qquad C^{(\alpha)}(\varkappa^{(\wedge)},\infty) \leqslant \bar{C}_1^{(\alpha)}(\varkappa^{(\wedge)}) \\ &C^{(\alpha)}(\infty,\varkappa^{(m)}) \leqslant \bar{C}_2^{(\alpha)}(\varkappa^{(m)});\ \lim_{\varkappa^{(\wedge)}\to\infty}\frac{\partial C^{(\alpha)}}{\partial\varkappa^{(\wedge)}} \equiv \lim_{\varkappa^{(\wedge)}\to\infty} C_\wedge^{(\alpha)}=0 \\ &\lim_{\varkappa^{(m)}\to\infty}\frac{\partial C^{(\alpha)}}{\partial\varkappa^{(m)}} \equiv \lim_{\varkappa^{(m)}\to\infty} C_m^{(\alpha)} = 0 \end{aligned}\right\} \tag{7.3}$$

For example,

$$C^{(\alpha)} = C_2^{(\alpha)}(a_2\varkappa^{(m)} + 1)^{-n_2} + C_1^{(\alpha)}(a_1\varkappa^{(\wedge)} + 1)^{-n_1} \\ 0 \leqslant n_i \leqslant 1 \qquad (i = 1,2), \tag{7.4}$$

so that

$$C_\wedge^{(\alpha)} \leqslant 0\,, \qquad C_m^{(\alpha)} \leqslant 0\,, \tag{7.5}$$

where $C_i^{(\alpha)}, a_i, n_i$ (i=1,2) are positive constants, but in general case some of them can be influenced by $\varkappa^{(\Lambda)}$ and (or) by $\varkappa^{(m)}$.

It is expedient to assume that $\pi^{(m)}$ attains final value when $\varkappa^{(m)} \to \infty$, and since $\pi^{(\Lambda)}$ and $\pi^{(m)}$ must be bounded when $\varkappa^{(\Lambda)} \to \infty$, we have

$$\left.\begin{aligned} &\lim_{\varkappa^{(\Lambda)}\to\infty} \varphi_0^{(H)} < M_1(\varkappa^{(m)}), \quad \lim_{\varkappa^{(\Lambda)}\to\infty} \frac{\partial\varphi_0^{(H)}}{\partial\varkappa^{(\Lambda)}} = 0 \\ &\lim_{\varkappa^{(\Lambda)}\to\infty} \frac{d\varphi_1^{(H)}}{d\varkappa^{(\Lambda)}} < M_2 = \text{const.}, \quad \lim_{\varkappa^{(m)}\to\infty} \frac{d\varphi_2^{(H)}}{d\varkappa^{(m)}} < M_3 = \text{const.} \end{aligned}\right\} \tag{7.6}$$

For example, the power function $\varphi_0^{(H)}$ that satisfies (7.6) may be written as

$$\varphi_0^{(H)} = 1_0[1-(b_2\varkappa^{(m)}+1)^{m_2}][1-(b_1\varkappa^{(\Lambda)}+1)^{-m_1}] \tag{7.7}$$

where

$$b_i > 0 \quad (i=1,2), \qquad m_1 \geqslant 0, \quad m_2 < 1$$

Under assumption (7.2) the internal forces $\pi^{(\Lambda)}$ and $\pi^{(m)}$ are made up of three components

$$\pi^{(\Lambda)} = \pi_1^{(\Lambda)}(\varkappa^{(\Lambda)}) + \pi_{(1,2)}^{(\Lambda)}(\varkappa^{(\Lambda)},\varkappa^{(m)}) + \pi_{(1,3)}^{(\Lambda)}(\varkappa^{(\Lambda)},\varkappa^{(m)},\underset{\sim}{\varkappa}) \tag*{(7.8)$_1$}$$

$$\pi^{(m)} = \pi_2^{(m)}(\varkappa^{(m)}) + \pi_{(2,1)}^{(m)}(\varkappa^{(\Lambda)},\varkappa^{(m)}) + \pi_{(2,3)}^{(m)}(\varkappa^{(\Lambda)},\varkappa^{(m)},\underset{\sim}{\varkappa}) \tag*{(7.8)$_2$}$$

whereas

$$\underset{\sim}{\pi} = C^{(\alpha)}(\varkappa^{(\Lambda)}, \varkappa^{(m)})\, \underset{\sim}{\varkappa} \tag{7.9}$$

Here

$$\pi_1^{(\Lambda)} = \frac{d\varphi_1^{(H)}}{d\varkappa^{(\Lambda)}} \; ; \; \pi_{(1,2)}^{(\Lambda)} = \frac{\partial\varphi_0^{(H)}}{\partial\varkappa^{(\Lambda)}} \; ; \; \pi_{(1,3)}^{(\Lambda)} = \tfrac{1}{2} \, C_\Lambda^{(\alpha)}(\underset{\sim}{\varkappa}\cdot\underset{\sim}{\varkappa}) \tag{7.10$_1$}$$

$$\pi_2^{(m)} = \frac{d\varphi_2^{(H)}}{d\varkappa^{(m)}} \; ; \; \pi_{(2,1)}^{(m)} = \frac{\partial\varphi_0^{(H)}}{\partial\varkappa^{(m)}} \; ; \; \pi_{(2,3)}^{(m)} = \tfrac{1}{2} \, C_m^{(\alpha)}(\underset{\sim}{\varkappa}\cdot\underset{\sim}{\varkappa}). \tag{7.10$_2$}$$

From $(7.3)_2$ and (7.6) it follows that.

$$\lim_{\varkappa^{(\Lambda)}\to\infty} \pi^{(\Lambda)} = \lim_{\varkappa^{(\Lambda)}\to\infty} \pi_1^{(\Lambda)} = \pi_\infty^{(\Lambda)} = \text{const.} \tag{7.11}$$

i.e. the limit value of $\pi^{(\Lambda)}$ at steady state of a cyclic process is independent of $\varkappa^{(m)}$ and $\underset{\sim}{\varkappa}$.

The limit value of $\pi^{(m)}$ at steady state can be arranged as follows

$$\lim_{\varkappa^{(\Lambda)}\to\infty} \pi^{(m)} = \overset{*}{\pi}{}^{(m)}(\varkappa^{(m)}) + \tfrac{1}{2} \, C_m^{(\alpha)}(\infty,\varkappa^{(m)})\underset{\sim}{\varkappa}\cdot\underset{\sim}{\varkappa} \tag{7.12}$$

where

$$\overset{*}{\pi}{}^{(m)} = \lim_{\varkappa^{(\Lambda)}\to\infty}(\pi_2^{(m)} + \pi_{(2,1)}^{(m)}) \tag{7.13}$$

is independent of $\underset{\sim}{\varkappa}$, and

$$\lim_{\varkappa^{(m)}\to\infty} \lim_{\varkappa^{(\Lambda)}\to\infty} \pi^{(m)} = \lim_{\varkappa^{(m)}\to\infty} \overset{*}{\pi}{}^{(m)} = \pi_\infty^{(m)} = \text{const.} \tag{7.14}$$

on account of (7.3).

At thermodynamic reference state the free energy $\varphi^{(H)}$ and all its derivatives vanish, so that

$$\pi^{(\Lambda)} = \pi^{(m)} = 0 \; , \; \underset{\sim}{\pi} = \underset{\sim}{0} \quad \text{for } \underset{\sim}{\varkappa} = 0, \; \varkappa^{(\Lambda)} = \varkappa^{(m)} = 0 \tag{7.15}$$

$$\pi_1^{(\wedge)}(0) = \pi_2^{(m)}(0) = \pi_{(1,2)}^{(\wedge)}(0,0) = \pi_{(2,1)}^{(m)}(0,0) = 0 \quad (7.15)$$
(cont.)

(ii) Consider the cyclic process of deformation (6.12). Denote by $\underset{\sim}{\beta}^*$ and $\overset{*}{\varkappa}^{(m)}$ the values of micro-strains $\underset{\sim}{\beta}$ and hardening parameter $\varkappa^{(m)}$, respectively, at the instant (say $t_s$) for which the microstresses $\underset{\sim}{\pi}$ shakes down to the given programm of deformations (the end of the first stage of the transitory time period of a cyclic process). For $t > t_s$ the process is entirely $\Pi$ - elastic ($f^{(2)} < 0$, $\dot{\underset{\sim}{\beta}} = 0$, $\dot{\varkappa}^{(m)} = 0$). From (6.3)-(6.8),(7.1) and (7.9) it follows that during second stage ($t > t_s$) of an unsteady cyclic process the following relations hold ($f^{(1)} = 0$)

$$\left.\begin{aligned} \underset{\sim}{\varkappa}(t) &= \underset{\sim}{\Delta} + \epsilon^p(t)\,\underset{\sim}{n}_\epsilon \\ \underset{\sim}{\pi}(t) &= \overline{\underset{\sim}{\pi}}(t) + \pi_c(t)\,\underset{\sim}{n}_\epsilon \\ \underset{\sim}{S}(t) &= \overline{\underset{\sim}{S}}(t) + S_c(t)\,\underset{\sim}{n}_\epsilon \qquad (\dot{\underset{\sim}{\epsilon}}^p \neq 0) \end{aligned}\right\} \quad (7.16)$$

where $\varkappa^{(\wedge)}$ varies and

$$\left.\begin{aligned} \underset{\sim}{\Delta} = \underset{\sim}{\epsilon}_o^p - \underset{\sim}{\beta}^* = \text{const.};\ \overline{\underset{\sim}{S}} &= \overline{\underset{\sim}{\pi}}(t) = C^{(\alpha)}(\varkappa^{(\wedge)}, \overset{*}{\varkappa}^{(m)})\underset{\sim}{\Delta} \\ \pi_c(t) &= C^{(\alpha)}(\varkappa^{(\wedge)}, \overset{*}{\varkappa}^{(m)})\ \epsilon^p(t) \\ S_c(t) &= \pi_c(t) + Y^{(1)}(\operatorname{sgn} \dot{\epsilon}^p) \end{aligned}\right\} \quad (7.17)$$

The response functions consist of two terms. The first terms occuring in (7.16) represent the "mean" values of $\underset{\sim}{\varkappa}$,

$\underset{\sim}{\pi}$ and $\underset{\sim}{S}$, respectively. It is seen that the mean value $\bar{\underset{\sim}{\pi}}$ of the elastic micro-stress tensor $\underset{\sim}{\pi}$ is at every instant of the second stage equal to the mean value of stresses. Since $C^{(\alpha)}$ is monotonic and decreasing function of $\varkappa^{(\wedge)}$ mean stresses relax in this stage, and tend toward a limit value of a steady state

$$\bar{\underset{\sim}{S}}^{\infty} = \bar{\underset{\sim}{\pi}}^{\infty} = C^{(\alpha)}(\infty, \overset{*}{\varkappa}^{(m)})\ \underset{\sim}{\Delta} \tag{7.18}$$

The other terms occuring in (7.16) represent vibrational changes in $\underset{\sim}{\varkappa}$, $\underset{\sim}{\pi}$ and $\underset{\sim}{S}$. The decreasing vibration is the characteristic of $\pi_c(t)$ i.e. an element undergoes "cyclic - -microelastic softening". As a result, cyclic variation of overall stresses $S_c(t)$ may be either decreasing or increasing, depending upon which of the two terms occuring in $(7.17)_3$ is dominating, provided that $Y^{(1)}$ is increasing. When $\pi_c$ is dominating there is a cyclic softening. Otherwise, cyclic hardening occurs during 2-d stage of a transitory time period. When $Y^{(1)}$ is not increasing, cyclic hardening is precluded.

At steady state ($\varkappa^{(\wedge)} \to \infty$) both $\pi_c$ and $S_c(t)$ are periodic functions that have the same time period as $\epsilon^p(t)$ and the following amplitudes

$$\begin{aligned} a_\pi &= C^{(\alpha)}(\infty, \overset{*}{\varkappa}^{(m)})\ a_\epsilon \\ a_s &= a_\pi + Y^{(1)}(\pi_\infty^{(\wedge)}, \overset{*}{\varkappa}^{(m)}) \end{aligned} \tag{7.19}$$

where $a_\epsilon$ is the amplitude of $\epsilon^p(t)$. When $\varkappa^{(\wedge)} \to \infty$, the relation between $\underset{\sim}{S} = \underset{\sim}{S}^{\infty}(t)$ and $\underset{\sim}{\varepsilon}^{p}(t)$ is piecewise-linear

$$\underset{\sim}{S}^{\infty}(t) = C^{(\alpha)}(\infty,\overset{*}{\varkappa}^{(m)})(\underset{\sim}{\varepsilon}^{p}(t) - \underset{\sim}{\beta}^{*}) + Y^{(1)}(\pi_{\infty}^{(\wedge)},\overset{*}{\varkappa}^{(m)}) \times \\ \times (\mathrm{sgn}\ \dot{\varepsilon}^{p})\cdot\underset{\sim}{n}_{\varepsilon} \qquad (7.20)$$

i.e. steady states are described by a family (parameters are $\overset{*}{\underset{\sim}{\beta}}$ and $\overset{*}{\varkappa}^{(m)}$) of <u>Prager's linear kinematical hardening laws</u>. Note that for fixed $\underset{\sim}{\beta}$ and $\varkappa^{(m)}$, the behaviour of the model is essentially the same as the combined kinematical model of hardening.

The possible $\underset{\sim}{\pi}$ - path during 2-d stage of the transitory time and at steady state is illustrated in Fig.(7.1) in 5-dimensional space where the trajectory of $\underset{\sim}{S}^{\infty}$ is also presented.

The response function $\underset{\sim}{S}(t)$ can always be decomposed in accordance with $(7.16)_3$ *) i.e. $\bar{\underset{\sim}{S}}(t)$ and $S_c$ can be evaluated when $\underset{\sim}{S}$ is experimentally determined. At instants, say $t_1, t_2, .., t_i, ..,$ when $\dot{\varepsilon}^p = 0$, stress $\underset{\sim}{S}$ and $S_c$ are discontinuous. Introduce the notation

$$\left.\begin{aligned} \underset{\sim}{S}^{+}(t_i) &= \underset{\sim}{S}(t_i - 0) \ ; \ S_c^{+}(t_i) = S_c(t_i - 0) \\ \underset{\sim}{S}^{-}(t_i) &= \underset{\sim}{S}(t_i + 0) \ ; \ S_c^{-}(t_i) = S_c(t_i + 0) \end{aligned}\right\} \qquad (7.21)_1$$

when $\dot{\varepsilon}^p(t_i + 0) < 0$ and $\dot{\varepsilon}^p(t_i - 0) > 0$, and

*) Decompose $\underset{\sim}{S}$ in the direction $\underset{\sim}{n}_{\varepsilon}$ and the direction $\underset{\sim}{n}_{\varepsilon}^{\perp}$ orthogonal to $\underset{\sim}{n}_{\varepsilon}(\underset{\sim}{n}_{\varepsilon}\cdot n_{\varepsilon}^{\perp}=0)$; $\underset{\sim}{S} = S\,\underset{\sim}{n}_{\varepsilon} + S^{\perp}\underset{\sim}{n}_{\varepsilon}^{\perp}$. Subsequent decomposition of $S = \underset{\sim}{S}\cdot\underset{\sim}{n}_{\varepsilon}$ into the periodic (mean value equals zero) function $S_c$ and a monotonic function $S_{(m)}$ enables to determine $\bar{\underset{\sim}{S}}(t)$; $\bar{\underset{\sim}{S}}(t) = S^{\perp}\underset{\sim}{n}_{\varepsilon}^{\perp} + S_{(m)}\underset{\sim}{n}_{\varepsilon}$.

$$\left.\begin{aligned} \underset{\sim}{S}^+(t_i) &= \underset{\sim}{S}(t_i+0) \;;\; S_c^+(t_i) = S_c(t_i+0) \\ \underset{\sim}{S}^-(t_i) &= \underset{\sim}{S}(t_i-0) \;;\; S_c^-(t_i) = S_c(t_i-0) \end{aligned}\right\} \qquad (7.21)_2$$

when $\dot{\epsilon}^p(t_i+0) > 0$ and $\dot{\epsilon}^p(t_i-0) < 0$, so that $S_c^+(t_i) > 0$ and $S_c^-(t_i) < 0$.

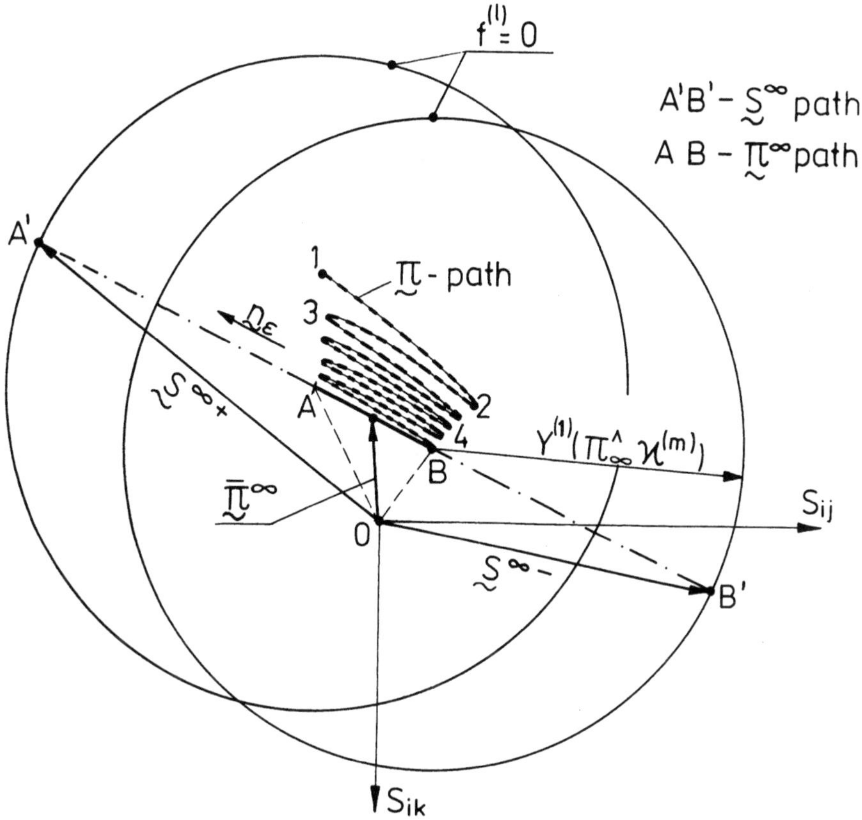


Fig. 7.1.

Since $\underset{\sim}{\kappa}$, $\kappa^{(\wedge)}$, $\epsilon^p$ are continuous at $t_i$ it follows from (7.16) and (7.17) that

$$\underset{\sim}{S}^+ + \underset{\sim}{S}^- = 2\underset{\sim}{\pi}(t_i) \;;\; \underset{\sim}{S}^+ - \underset{\sim}{S}^- = 2Y^{(1)}\Big|_{t=t_i} \underset{\sim}{n}_\epsilon \qquad (7.22)$$

$$S_c^+ + S_c^- = 2\pi_c(t_i) \;\; ; \;\; S_c^+ - S_c^- = 2Y^{(1)}\Big|_{t=t_i} \qquad (7.22)\ (\text{cont.})$$

Thus, through $(7.17)_2$, (7.22) one can evaluate $\pi_c$, $Y^{(1)}$ and $C^{(\alpha)}$ at every instant $t_i$. Having known $C^{(\alpha)}$ and $\underset{\sim}{\bar{\pi}} = \underset{\sim}{\bar{S}}$ one can determinate $\underset{\sim}{\Delta}$ and prior inelastic "micro-strains" $\overset{*}{\underset{\sim}{\beta}}$ using $(7.17)_1$.

(iii) Finally let us briefly discuss the variation of free (stored) energy $\varphi^{(H)}$ at steady state and its relation with the plastic work done. Consider a single cycle. The relation between $\varphi^{(H)}$ and $\epsilon^p$ during two consecutive half cycles may be written in the following dimensionless form

$$\Delta\bar{\varphi}^{(H)} = \bar{\varphi}^{(H)} - \bar{\varphi}_o^{(H)} = \bar{\varphi}_n^{(H)} + \bar{\varphi}_c^{(H)} \qquad (7.23)$$

$$\left.\begin{aligned} \bar{\varphi}_n^{(H)} &= -\delta(1-e) - \gamma(1-e^2) \\ \bar{\varphi}_c^{(H)} &= \begin{cases} P_\infty\, y_1^\infty(1-e) & \text{- first half cycle} \\ P_\infty\, y_1^\infty(e+3) & \text{- other half cycle} \end{cases} \end{aligned}\right\} \qquad (7.24)$$

where

$$\left.\begin{aligned} \bar{\varphi}^{(H)} &= \frac{\varphi^{(H)}}{2a_\epsilon\, Y^{(1)}(\pi_\infty^{(\wedge)}, \varkappa^{(m)})} \;\; ; \;\; e = \frac{\epsilon^p(t)}{a_\epsilon} \\ \gamma &= \frac{a_\epsilon\, C^{(\alpha)}(\infty,\varkappa^{(m)})}{4Y^{(1)}(\pi_\infty^{(\wedge)},\varkappa^{(m)})} = 0.25\left[\frac{a_s}{a_\epsilon\, C^{(\alpha)}(\infty,\varkappa^{(m)})} - 1\right]^{-1} \\ \delta &= \frac{C^{(\alpha)}(\infty,\varkappa^{(m)})\,|\underset{\sim}{\Delta}\cdot\underset{\sim}{n}_\epsilon|}{2Y^{(1)}(\pi_\infty^{(\wedge)},\varkappa^{(m)})} = 0.5\,|\underset{\sim}{\Delta}\cdot\underset{\sim}{n}_\epsilon|\,C^{(\alpha)}(\infty,\varkappa^{(m)}) \times \\ &\qquad \times [a_s - C^{(\alpha)}(\infty,\varkappa^{(m)})a_\epsilon]^{-1} \end{aligned}\right\} \qquad (7.25)$$

$$P_\infty = \frac{\pi_\infty^{(\wedge)}}{2Y^{(1)}(\pi_\infty^{(\wedge)}, \varkappa^{(m)})} \quad ; \quad y_1^\infty = \left.\frac{\partial Y^{(1)}}{\partial \pi^{(\wedge)}}\right|_{\pi^{(\wedge)} = \pi_\infty^{(\wedge)}} \qquad \begin{matrix}(7.25)\\ \text{(cont.)}\end{matrix}$$

$\bar{\varphi}_n^{(H)}$ and $\bar{\varphi}_c^{(H)}$ represent noncumulative and possible cumulative components of $\Delta\varphi^{(H)}$, respectively. Here for simplicity we have assumed that the first cycle starts when $e = 1$.Note that the parameter $\gamma$ is expressed in terms of measurable at steady state quantities: the amplitude of plastic strain $a_\epsilon$, the amplitude of stress $a_s$ and linear hardening modulus $C^{(\alpha)}(\infty, \varkappa^{(m)})$ of hystheresis loop. Both $C^{(\alpha)}$ and $a_s$ depend on prior plastic strain history, through dependence on accumulated inelastic micro-strain $\varkappa^{(m)}$.

When $y_1^\infty \approx 0$, the cumulative component vanishes and there is a cyclic "saturation" level of free (stored) energy. It is reached at steady state. The saturation level of stored internal energy $u^{(H)}$ has been experimentally observed by Halford [21].

The relation between the plastic work done and $\epsilon^p$ is

$$\left.\begin{aligned} \Delta\omega^p &= \omega^p - \omega_o^p = (1-e)[\tfrac{1}{2} - \delta - (1+e)\gamma] && \text{(first half cycle)} \\ \Delta\omega^p &= \omega^p - \omega_o^p = \frac{e+3}{2} - (1-e)[\delta + \gamma(1+e)] && \text{(other half cycle)} \end{aligned}\right\} \quad (7.26)$$

where

$$\omega^p = \frac{W^{(pl)}}{2a_\epsilon \, Y^{(1)}(\pi_\infty^{(\wedge)}, \varkappa^{(m)})}$$

Regarding e as a parameter one can deduce from (7.24)-

-(7.26) the relation between $\Delta\varphi^{(H)}$ and $\Delta\omega^{p}$. The noncumulative component of $\Delta\varphi^{(H)}$ is plotted in Fig.7.2 versus $\Delta\omega^{p}$ for one complete cycle of $\epsilon^{p}$ and for the following values of $\gamma$ and $\delta$

$$\gamma = 0.08\ ,\quad \delta = 0\ ,\quad \delta = 0.03$$

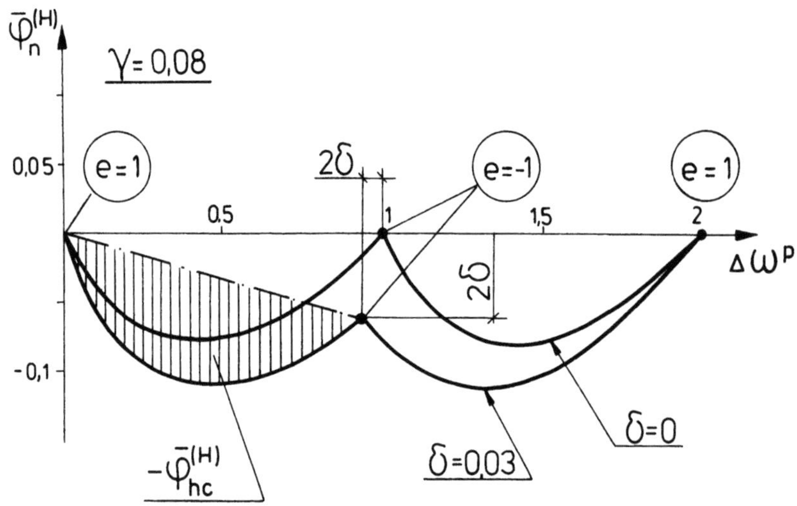


Fig. 7.2.

It is seen that during traversal of a plastic strain hysteresis loop there is a release-storage-release-storage sequence of the free energy exchange. When there is no prior plastic strains and when an element shakes down (during the first stage of a transitory time) to zero inelastic "micro-strains" ($\overset{*}{\beta} = 0$) then $\delta = 0$, and the nature of the free energy exchange during half cycle is repetitious. Otherwise, the repetitious is only the nature of the energy exchange during a complete cycle. The work done during one of the half cycle is smaller (by amount $2\delta$) than the work done during the other half cycle*). The former we shall call "s-half cycle". This statement indicates the possible manner of experimental determination of a parameter $\delta$.

Let us define the noncumulative stored free energy of "s-half cycle" $\bar{\varphi}_{hc}^{(H)}$ (cf. lined area in Fig.7.2)

$$\bar{\varphi}_{hc}^{(H)} = \bar{\varphi}_{n}^{(H)} + \frac{2\delta}{1-2\delta}\,\Delta\omega^{p} \tag{7.27}$$

Consider a fraction of work stored as noncumulative free energy during the s - half cycle

$$\chi^{(H)} = \frac{\bar{\varphi}_{hc}^{(H)}}{\Delta\omega^{p}} \tag{7.28}$$

While elaborating experimental data it is expedient to represent $\chi^{(H)}$ as a function of fraction $\chi^{(p)}$ of plastic work done per s - half cycle

$$\chi^{(p)} = \frac{\Delta\omega^{p}}{1-2\delta} \tag{7.29}$$

Using $(7.24)_1$ - $(7.26)_1$ we find that

$$\chi^{(H)} = \frac{1}{1-2\delta}\left[1 - \frac{\sqrt{\frac{1-2\delta}{\gamma}\chi^{(p)} + \left(\frac{1-2\delta-4\gamma}{4\gamma}\right)^{2}} - \left(\frac{1-2\delta-4\gamma}{4\gamma}\right)}{2\chi^{(p)}}\right] \tag{7.30}$$

for the present model.

*) Note that there may occure some asymmetry of a hysteresis loop even in an annealed ($\underset{\sim}{\varepsilon}_{o}^{p}=0$) tost specimen. This happens when $\underset{\sim}{\beta}$ generated during the first stage of a transitory time does not vanish at the end of this stage.

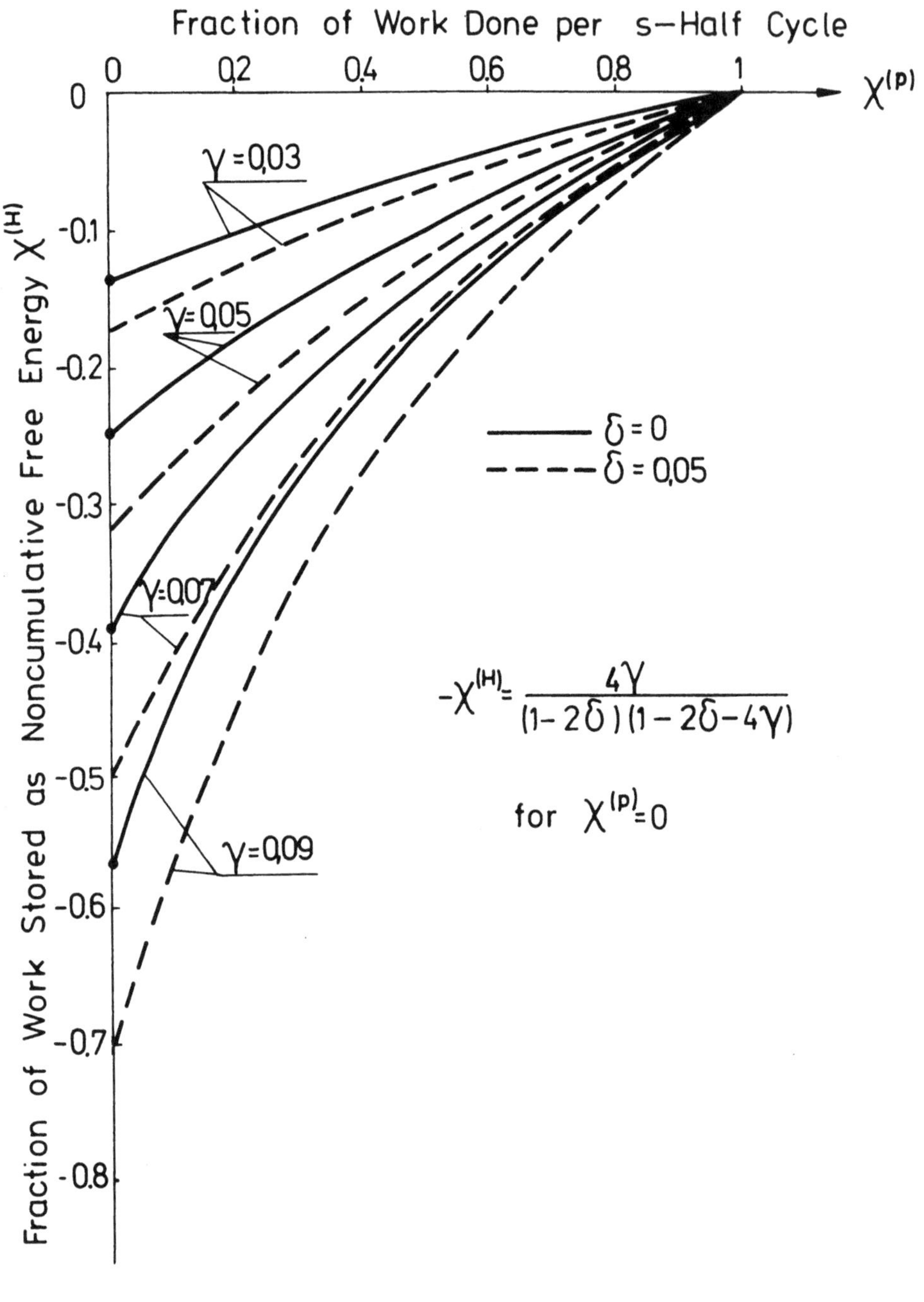
Fraction of Work Done per s-Half Cycle
0
0,2
0,4
0,6
0,8
1
$\chi^{(p)}$
0
-0.1
-0.2
-0.3
-0.4
-0.5
-0.6
-0.7
-0.8
Fraction of Work Stored as Noncumulative Free Energy $\chi^{(H)}$
$\gamma$ = 0,03
$\gamma$ = 0,05
$\gamma$ = 0,07
$\gamma$ = 0,09
$\delta$ = 0
$\delta$ = 0,05
$-\chi^{(H)} = \frac{4\gamma}{(1-2\delta)(1-2\delta-4\gamma)}$
for $\chi^{(p)} = 0$

Fig. 7.3.

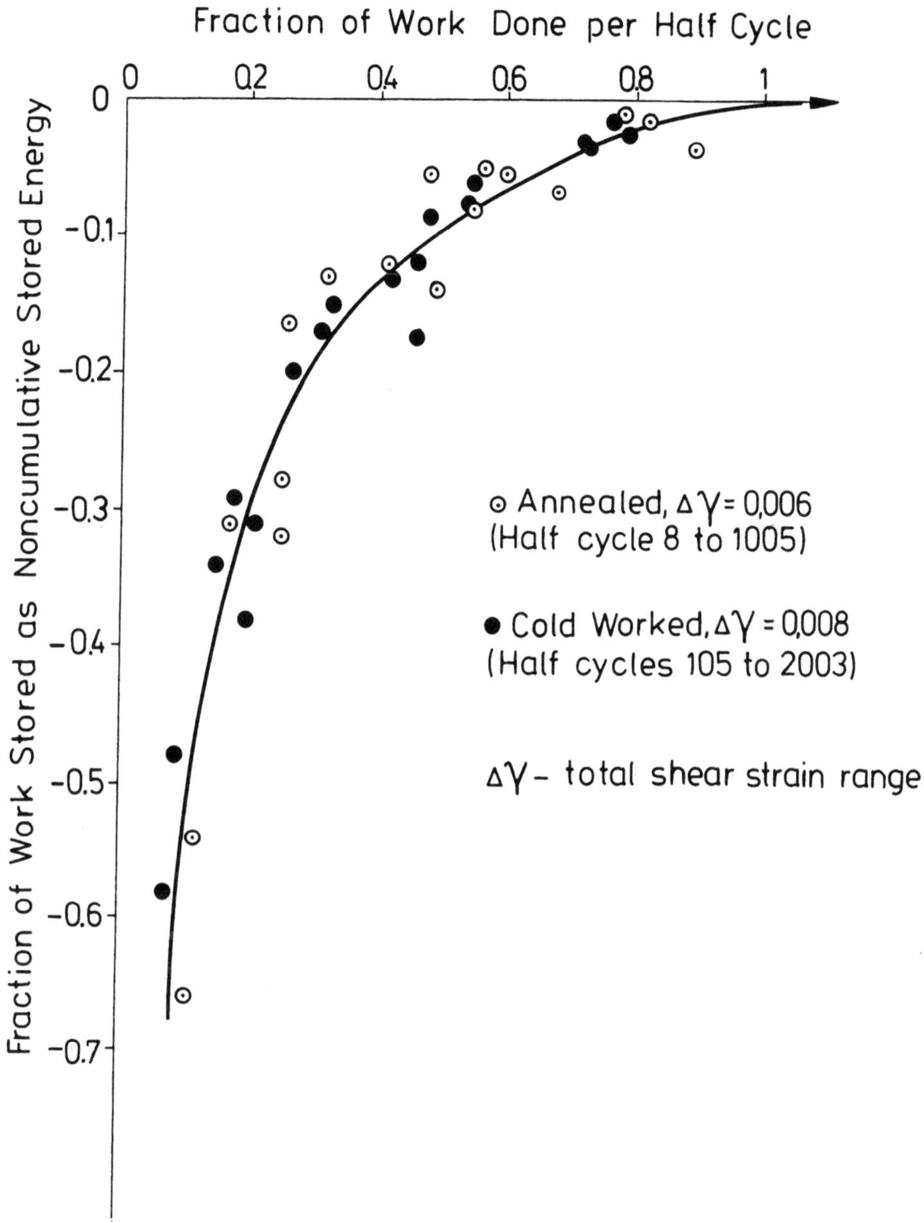


Fig.7.4. Dimensionless representation of noncumulative stored energy of cold work changes during completely reversed half cycles of shear strained OFHC copper (after Halford [21]).

The theoretical curves $\chi^{(H)} = \chi^{(H)}(\chi^{(p)})$ for a few values of $\gamma$ and $\delta$ are shown in Fig.7.3. In Fig.7.4 one of Halford's experimental curve (shear strained OFHC Copper) of the same kind is reproduced. Although it concerns stored internal energy (and not stored free energy) its character is similar. It is interesting to note (Fig.7.3) that cold work ($\delta \neq 0$) does not affect the monotone character of the $\chi^{(H)}(\chi^{(p)})$ - curve. The rate of energy release is the highest at the instant when the direction of plastic strain is reversed ($\chi^{(p)} = 0$). This is in agreement with Halford's [21] observation (Fig.7.4) that there is the indiscernible noncumulative stored (internal) energy changes of the annealed and cold worked materials when comparison is made on the basis of approximately equal total strain ranges. The present model, however, does not predict so high maximum rates of energy release as observed by Halford (at $\chi^{(p)} = 0$) due to inherent linearity of hardening at a steady state.

The above mentioned observations suggest that the energy exchange associated with structural entropy change $T_0\bar{s}^{(H)}$ may be indeed insignificant at steady state. At least, it seems to be plaussible to assume that

$$T_0\bar{s}^{(H)} = \nu(\varkappa^{(m)})\,\bar{u}^{(H)}(X_\varkappa) \qquad (7.31)$$

where $|\nu| \ll 1$.

# IV. PROPORTIONAL LOADINGS

## 8. Basic Set of Equations.

(i) In this part we shall discuss in more detail the response of the asymptotically linear hardening model (Sec.7) to a proportional variation of $\underset{\sim}{\varepsilon}^p$, i.e. we shall assume that $\underset{\sim}{\varepsilon}^p$ vary according to (6.12), where $\underset{\sim}{\varepsilon}^p_o = \underset{\sim}{0}$ and $\varepsilon^p(t)$ is not necessarily a periodic function of time.

From (6.3)-(6.5), (6.8)-(6.9) and (7.9) it follows that

$$\left.\begin{array}{c} \underset{\sim}{S} = S(t)\,\underset{\sim}{n}_\varepsilon \; ; \; \underset{\sim}{\varkappa} = \varkappa(t)\,\underset{\sim}{n}_\varepsilon \; ; \; \underset{\sim}{\pi} = \pi(t)\,\underset{\sim}{n}_\varepsilon \\ \underset{\sim}{\beta} = \beta(t)\,\underset{\sim}{n}_\varepsilon \end{array}\right\} \tag{8.1}$$

i.e. all tensorial quantities vary in proportional manner provided that the initial state is the thermodynamic reference state

$$\varepsilon^p = S = \pi = \pi^{(\wedge)} = \pi^{(m)} = \varkappa = \varkappa^{(\wedge)} = \varkappa^{(m)} = 0 \quad \text{at } t=0. \tag{8.2}$$

(ii) At those time periods at which $\dot{\underset{\sim}{\varepsilon}}^p = \underset{\sim}{0}$ the process is entirely elastic, $\dot{\underset{\sim}{\varkappa}} = \underset{\sim}{0}$, $\dot{\varkappa}^{(\wedge)} = 0$, $\dot{\varkappa}^{(m)} = 0$. Let $f^{(1)} = 0$ and $\dot{\underset{\sim}{\varepsilon}}^p \neq 0$. Accounting for (8.1) the yield condition $(6.3)_1$ may be written as

$$\bar{f}^{(1)} = S - \pi - \operatorname{sgn}(S-\pi)\, Y^{(1)} = 0 \tag{8.3}$$

and from $(6.4)_1$ we get

$$\operatorname{sgn}\,(S-\pi) = \operatorname{sgn}\,\dot{\varepsilon}^p \tag{8.4}$$

The other yield condition $(6.3)_2$ in the case of proportional loadings (8.1) takes the form

$$\bar{f}^{(2)} = \pi - \operatorname{sgn}\pi\; Y^{(2)} = 0 \tag{8.5}$$

where

$$\operatorname{sgn}\pi = \operatorname{sgn}\dot{\beta} = \operatorname{sgn}(\dot{\varepsilon}^p - \dot{\varkappa}) \tag{8.6}$$

on account of (6.9) and provided that $\dot{\beta} \neq 0$.

Equations (8.3) and (8.4) enable to determine S when hardening parameters $\varkappa$, $\varkappa^{(\wedge)}$, $\varkappa^{(m)}$ are known functions of time.

(iii) The parameter $\lambda^{(2)}$ occuring in (6.5),(6.7) and (6.9) may be expressed in terms of the parameter $\lambda^{(1)}$ with the aid of the consistency relation $\dot{f}^{(2)} = 0$. Using (6.3)-(6.7),(7.9), (8.1) and (4.4), and accounting for $\lambda^{(2)} \geq 0$ and $\lambda^{(1)} = |\dot{\varepsilon}^p|$ we eventually find that

$$\lambda^{(2)} = |\dot{\varepsilon}^p| < \mu_o(\operatorname{sgn}\pi)(\operatorname{sgn}\dot{\varepsilon}^p) - \nu_o > \tag{8.7}$$

where (cf.(4.4))

$$< z > = \left\{ \begin{array}{ll} z & \text{when } z \geq 0 \\ 0 & \text{when } z < 0 \end{array} \right\} \tag{8.8}$$

$$\mu_o = \frac{1}{\eta_o}\left[1 - Y^{(2)}\, y_2\, C_m^{(\alpha)}\, C^{(\alpha)^{-2}}\right] \tag{8.9}$$

$$\left.\begin{aligned}
\nu_o &= \frac{y_1}{\eta_o}\left[y_2\, l^{(\wedge,m)}\, c^{(\alpha)^{-1}} - Y^{(2)}\, c_{\wedge}^{(\alpha)}\, c^{(\alpha)^{-2}}\right] \\
\eta_o &= 1 - 2Y^{(2)}\, y_2\, c_m^{(\alpha)}\, c^{(\alpha)^{-2}} + l^{(m)}\, c^{(\alpha)^{-1}}\, y_2^2 \\
y_1 &= \frac{\partial Y^{(1)}}{\partial \pi^{(\wedge)}} \quad ; \qquad y_2 = \frac{dY^{(2)}}{d\pi^{(m)}}
\end{aligned}\right\} \quad \begin{matrix}(8.9)\\ \text{(cont.)}\end{matrix}$$

Here $\eta_o$ represents "the hardening in an inelastic process of microdeformations". We shall assume henceforth that

$$\begin{aligned}
&1 \geqslant \mu_o \geqslant 0 \quad , \quad \eta_o > 0 \qquad (c^{(\alpha)} \neq 0) \\
&\frac{|\nu_o|}{\mu_o} \leqslant 1
\end{aligned} \qquad (8.10)$$

(iv) Under assumption (8.10) the <u>basic differential equations</u> for the hardening parameters are ($\bar{f}^{(1)} = 0$)

$$\left.\begin{aligned}
\dot{\varkappa} &= \begin{cases} (1-\mu_o+\nu_o)\dot{\varepsilon}^p & \text{if } \bar{f}^{(2)}=0 \text{ and sgn } \dot{\varepsilon}^p = \text{sgn } \varkappa \\ \dot{\varepsilon}^p & \text{if } \bar{f}^{(2)}<0 \text{ or } \bar{f}^{(2)}=0 \text{ and sgn } \dot{\varepsilon}^p \neq \text{sgn } \varkappa \end{cases} \\
\dot{\beta} &= \dot{\varepsilon}^p - \dot{\varkappa}
\end{aligned}\right\} \quad (8.11)$$

$$\left.\begin{aligned}
\dot{\varkappa}^{(m)} &= \begin{cases} (\mu_o-\nu_o)y_2|\dot{\varepsilon}^p| & \text{if } \bar{f}^{(2)}=0 \text{ and sgn } \dot{\varepsilon}^p = \text{sgn } \varkappa \\ 0 & \text{if } \bar{f}^{(2)}<0 \text{ or } \bar{f}^{(2)}=0 \text{ and sgn } \dot{\varepsilon}^p \neq \text{sgn } \varkappa \end{cases} \\
\dot{\varkappa}^{(\wedge)} &= y_1|\dot{\varepsilon}^p|
\end{aligned}\right\} \quad (8.12)$$

The set of equations (8.11)-(8.12) together with the

thermal equations of state (7.8)-(7.10)*) and (8.5) enable to determine the variation of hardening parameters and internal forces $\pi$, $\pi^{(\wedge)}$ and $\pi^{(m)}$, provided the function $\epsilon^p(t)$ is prescribed. The variation of stress S can be found from (8.3)-(8.4).

## 9. Mean Stress. Cyclic Stress-Plastic Strain Curve.

(i) Consider stress-plastic strain S - $\epsilon^p$ curve shown in

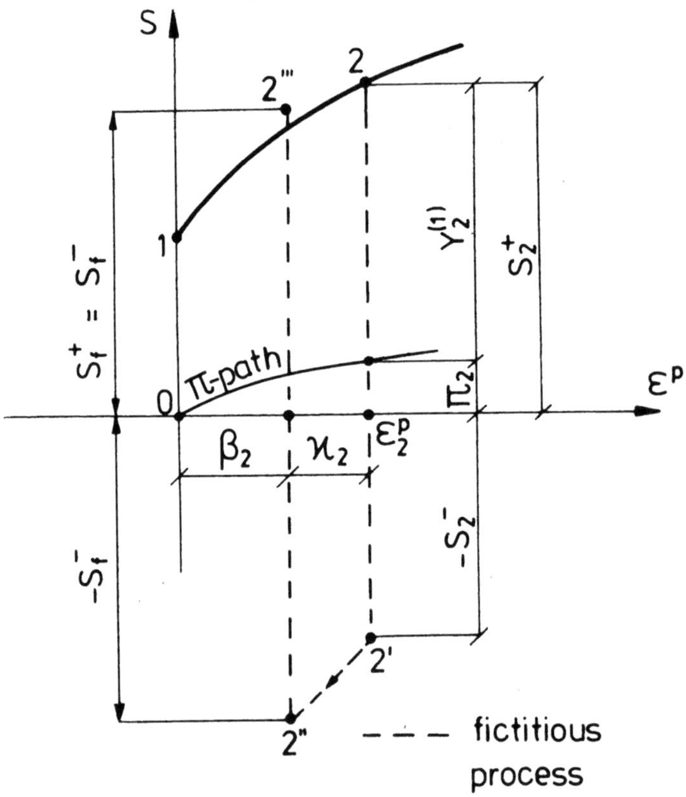


Fig. 9.1.

*) In the case of proportional loadings $\underset{\sim}{\varkappa}\cdot\underset{\sim}{\varkappa}$ occuring in (7.2),(7.8) and (7.9) may be replaced by $\varkappa^2$, whereas the equation (7.9) becomes $\pi = C^{(\alpha)}\varkappa$.

Fig.9.1. The value of internal force $\pi$ corresponding to a point, say point 2, of the curve equals half of the difference between the actual yield stress ($S_2^+$ in Fig.9.1) in the direction $\underset{\sim}{n}_\epsilon$ and the yield stress ($-S_2^-$ in Fig.9.1) in the direction $-\underset{\sim}{n}_\epsilon$

$$\pi = \tfrac{1}{2}(S^+ + S^-), \tag{9.1}$$

whereas the "active" stress $Y^{(1)}$ can be evaluated from (cf. (7.22),(8.3)-(8.4))

$$Y^{(1)} = \tfrac{1}{2}(S^+ - S^-). \tag{9.2}$$

The value of the parameter $\varkappa$ assigned to given point of the curve is the difference between the actual plastic strain ($\epsilon_2^p$ in Fig.9.1) and the plastic strain ($\beta_2$ in Fig. 9.1) which would have obtained had the specimen in sequence been unloaded and reloaded in reversed direction up to the level where the yield stresses ($S_f^+$ and $-S_f^-$ in Fig.9.1) in the directions $\underset{\sim}{n}_\epsilon$ and $-\underset{\sim}{n}_\epsilon$ are equal. Thus, conceptually both elastic and inelastic "micro-strains" are measurable. Hence, the measurable is also $Y^{(2)}$ that equals $|\pi|$ (cf. (8.5)) provided that $\beta$ is varying. The possible $\pi$-path corresponding to $S - \epsilon^p$ curve is also illustrated in Fig.9.1.

(ii) Consider the cyclic process of plastic straining shown in Fig.9.2 a

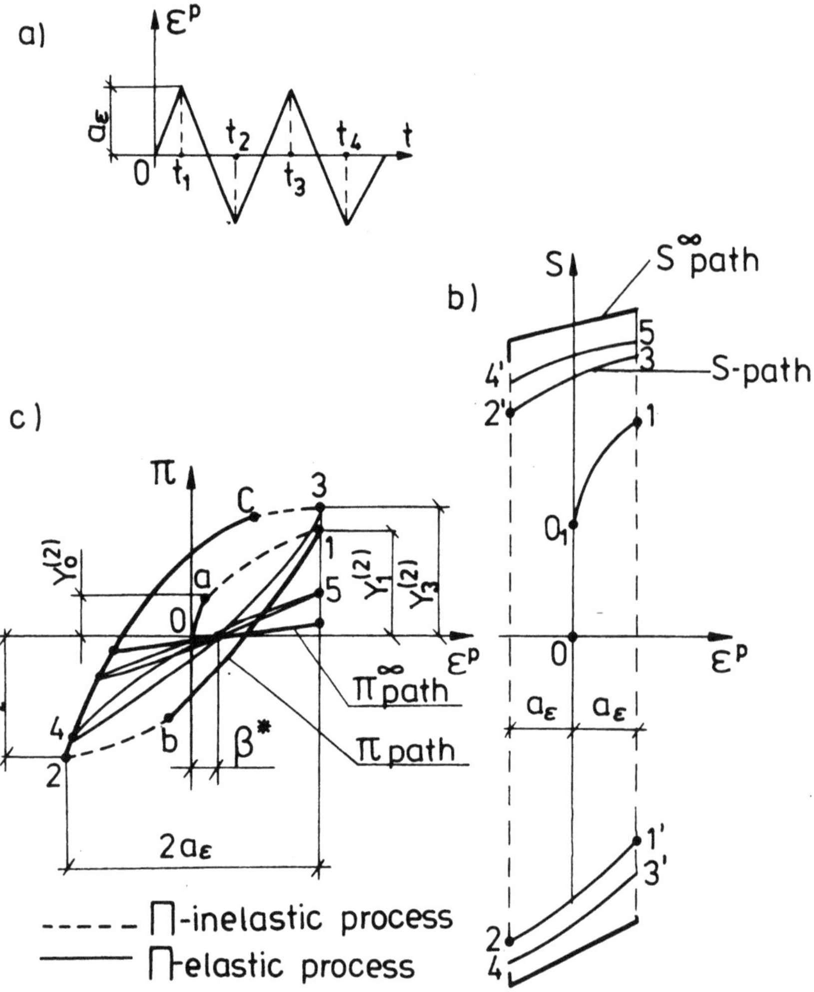


Fig. 9.2.

At instants $t_i$ $(i = 1,2,\dots)$ $\dot{\varepsilon}^p$ changes sign and S is discontinuous. Simultaneously $\pi$ and $\varkappa$ attain local extremes. The internal forces and hardening parameters are, of course, continuous functions of time. The possible S-path and $\pi$-path are illustrated in Fig.9.2 b and 9.2 c. During the first stage of transitory time, a number of consecutive $\Pi$ - inelastic processes occur. They are illustrated by dotted lines in Fig.9.2 c (a-1, b-2, c-3). $\Pi$ - inelastic processes must end at time $t_i$ since then sgn $\dot{\varepsilon}^p \neq$ sgn $\pi$, provided that $\bar{f}^{(2)} = 0$. At times $t > t_k$, where $t_k$ is certain instant ($t_3$ in Fig.9.2 c) at which microstress $\pi$ shakes down to the given programm of variation of $\varepsilon^p$, $\Pi$ - inelastic process does not occur.

Assume the inequalities (7.5) to be satisfied and

$$y_2 = \frac{dY^{(2)}}{d\pi^{(m)}} = \text{const. i.e. } Y^{(2)} = Y_0^{(2)} + y_2\, \pi^{(m)} \qquad (9.3)$$

If a shakedown of microstresses occur then there must exist two numbers $\overset{*}{\beta}$, $\overset{*}{\varkappa}{}^{(m)}$ such that

$$C^{(\alpha)}(\infty, \overset{*}{\varkappa}{}^{(m)})[a_3 + |\overset{*}{\beta}|] \leqslant \bar{Y}^{(2)}(\overset{*}{\varkappa}{}^{(m)}) \qquad (9.4)$$

where (cf. 7.3 and 7.13)

$$\bar{Y}^{(2)}(\varkappa^{(m)}) = z(\varkappa^{(m)}) \sqrt{z^2 + 2\mathfrak{M}(\varkappa^{(m)})} - z^2$$

$$z(\varkappa^{(m)}) = C^{(\alpha)}(\infty, \varkappa^{(m)})[-y_2\, C_m^{(\alpha)}(\infty, \varkappa^{(m)})]^{-\frac{1}{2}} > 0 \qquad (9.5)$$

$$\mathfrak{M} = Y_0^{(2)} + y_2\, \overset{*}{\pi}{}^{(m)}(\varkappa^{(m)})$$

The instant $t_k$ defines the beginning of 2d stage of transitory time during which only $\Pi$ - elastic process occurs (continuous lines in Fig.9.2 c) and $\beta = \overset{*}{\beta} = \text{const.}$, $\varkappa^{(m)} = \overset{*}{\varkappa}{}^{(m)} = \text{const.}$

The most frequent response of actual metals is characterized by concave $S - \epsilon^p$ curves when $\dot{\epsilon}^p > 0$, and convex $S - \epsilon^p$ curves when $\dot{\epsilon}^p < 0$, i.e.,

$$\operatorname{sgn} \frac{d^2 S}{d^2 \epsilon^p} = - \operatorname{sgn} \dot{\epsilon}^p \tag{9.6}$$

The functions $Y^{(1)}$ and $\pi_1^{(\wedge)}$, $\pi_{(1,2)}^{(\wedge)}$ occuring in $(7.8)_1$ can always be so choosen to satisfy this condition. Therefore, it is sufficient to consider the case when $\pi_1^{(\wedge)} = \pi_{(1,2)}^{(\wedge)} = 0$ (this may happen when $\varphi_0^{(H)} \approx \varphi_1^{(H)} \approx 0$ - cf.(7.2)) so that (cf.(7.8))

$$\pi^{(\wedge)} = \pi_{(1,3)}^{(\wedge)} \tag{9.7}$$

Assuming the power law (7.4) and $y_1 = \partial Y^{(1)}/\partial \pi^{(\wedge)} = \text{const.} > 0$ it can be proved (the proof is omitted here) that even in this "degenerate" situation the condition (9.6) is satisfied. Note that in this case during second stage of transitory time there are two "stationary" points on hysteresis loops with coordinates $\epsilon^p = - \overset{*}{\beta}$ and $S = \pm Y^{(1)}(0, \overset{*}{\varkappa}{}^{(m)})$ i.e., all $S - \epsilon^p$ curves intersect each other at $\epsilon^p = - \beta^*$. This observation may be usefull when elaborating the general scheme for identification of constitutive functions, even though such stationary points are not detected in actual experiments.

Let us decompose the stress S into mean stress $\bar{S}$ and cyclic stress $S_c$. According to (7.16) for 2d stage of transitory time we may write

$$S = \bar{S}(t) + S_c(t) \tag{9.8}$$

where

$$\bar{S}(t) = -C^{(\alpha)}(\varkappa^{(\Lambda)}, \overset{*}{\varkappa}^{(m)}) \overset{*}{\beta} \tag{9.9}$$

and $S_c$ is determined by $(7.17)_3$. The decomposition (9.8) is not unique. More frequently mean stresses are defined as

$$\bar{\bar{S}} = \tfrac{1}{2}(S^{max} - S^{min}) \tag{9.10}$$

where $S^{max}$ and $S^{min}$ are maximum and minimum values of S attained during the traversal of single hysteresis loop. The variable stress amplitude is then defined as

$$\bar{a}_s = \tfrac{1}{2}(S^{max} + S^{min}). \tag{9.11}$$

Assuming that

$$Y^{(1)} = Y_o^{(1)} + y_1\, \pi^{(\Lambda)}, \tag{9.12}$$

where $Y_o^{(1)}$ and $y_1$ depend on $\varkappa^{(m)}$, we may derive the approximate formula for $\bar{\bar{S}}$ and $\bar{a}_s$

$$\bar{a}_s \approx Y_o + \tfrac{1}{2}y_1(a_\varepsilon^2 + \overset{*}{\beta}{}^2)\, C_\Lambda^{(\alpha)}(\varkappa^{(\Lambda)}, \overset{*}{\varkappa}^{(m)}) +$$

$$+ y_1(\pi_1^{(\Lambda)} + \pi_{(1,2)}^{(\Lambda)}) + C^{(\alpha)}(\varkappa^{(\Lambda)}, \overset{*}{\varkappa}^{(m)})a_\varepsilon \tag{9.13}$$

$$\bar{\bar{S}} \approx [y_1\, C_\Lambda^{(\alpha)}(\varkappa^{(\Lambda)}, \overset{*}{\varkappa}^{(m)})a_\varepsilon - C^{(\alpha)}(\varkappa^{(\Lambda)}, \overset{*}{\varkappa}^{(m)})]\, \overset{*}{\beta} \tag{9.14}$$

which are sufficiently accurate when $(y_1 a_\epsilon/\varkappa^{(\Lambda)}) \approx 0$, and are exact when $\varkappa^{(\Lambda)} \to \infty$.

Both $|\bar{S}|$ and $|\bar{\bar{S}}|$ decrease during 2d stage of transitory time, i.e., self-induced mean stress relaxes, provided that $C_\Lambda^{(\alpha)} \leqslant 0$. Note that mean stresses are essentially the result of inelastic "micro-strains" developed during the first stage of transitory time period. In the course of this stage $\beta$ varies cyclicly and thus, mean stress are just as likely to change (during 1th stage) in cycle dependent fashion as $\beta$.

At steady state $(\varkappa^{(\Lambda)} \to \infty)$ (9.9) and (9.14) coincide so that

$$\bar{S} = \bar{\bar{S}} = \bar{S}^\infty = - C^{(\alpha)}(\infty, \overset{*}{\varkappa}^{(m)}) \overset{*}{\beta} \tag{9.15}$$

and $\bar{a}_s = a_s$ (cf.$(7.19)_2$). The steady hysteresis loop is shown in Fig.9.3. For small plastic strain amplitudes that satisfy the condition

$$a_\epsilon \leqslant Y^{(2)}(0)/C^{(\alpha)}(0,0) , \tag{9.16}$$

we have $\beta^* = 0$, and the loop is symmetric. The inelastic micro-strain $\beta^*$ may vanish also for larger amplitudes provided that microstresses shake down to the zero value of $\beta$ during the first stage of transitory time.

The equation of cyclic stress-strain curve $S^\infty - a_\epsilon$ is (Fig. 9.3)

$$S^\infty = C^{(\alpha)}(\infty, \overset{*}{\varkappa}^{(m)})(a_\epsilon - \beta^*) + Y^{(1)}(\pi_\infty^{(\Lambda)}, \overset{*}{\varkappa}^{(m)}) \tag{9.17}$$

Here $\overset{*}{\varkappa}^{(m)}$ and $\beta^*$ are functionals of $\epsilon^p(t)$, i.e., their values depend on given programm of plastic strain variation during first stage of transitory time period. For the case of the simple programm shown in Fig.9.2 a $\overset{*}{\varkappa}^{(m)}$ and $\beta^*$ become complex functions of $a_\epsilon$. They may be found analytically through detailed analysis of the solution of basic set of equations (8.11-8.12) when constitutive functions occurring there are specified

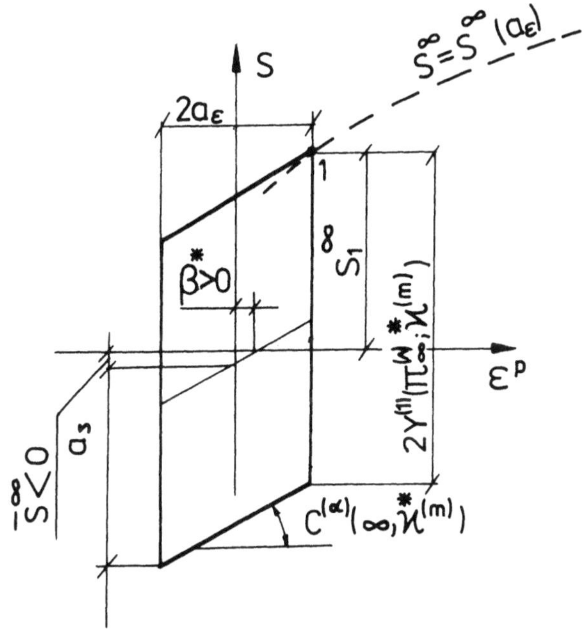


Fig. 9.3.

The important advantage of the model, however, is that $C^{(\alpha)}(\infty, \overset{*}{\varkappa}^{(m)})$, $\beta^*$ and $Y^{(1)}$ can be experimentally determined in terms of $a_\varepsilon$ (cf. Fig.9.3).

The $S^{\infty} - a_{\varepsilon}$ curve is entirely different from static (monotonic) stress-plastic strain curve, for the stress S attained in monotonic process and corresponding to $\varepsilon^{p} = a_{\varepsilon}$ is $(\bar{f}^{(2)}= 0,\ \bar{f}^{(1)}= 0)$

$$S = C^{(\alpha)}(\varkappa_0^{(\Lambda)},\varkappa_0^{(m)})(a_{\varepsilon} - \beta_0) + Y^{(1)}(\pi_0^{(\Lambda)},\varkappa_0^{(m)}) \qquad (9.18)$$

As a rule $\overset{*}{\varkappa}^{(m)} > \varkappa_0^{(m)}$, and when $C_{\Lambda}^{(\alpha)} \leqslant 0$ and $C_{m}^{(\alpha)} \leqslant 0$, $S^{\infty}$ can be smaller than S . Thus, the model is able to reproduce cyclic softening. Note that $\sqrt{3/2}\ Y^{(1)}(\pi_{\infty}^{(\Lambda)},0)$ is the cyclic "initial yield limit" in tension-compression. It differs from conventional yield limit (of annealed specimen) that equals $\sqrt{3/2}\ Y^{(1)}(0,0)$. Note also that (9.17) is valid provided that the cyclic process of amplitude $a_{\varepsilon}$ is the <u>first cyclic process</u> that undergoes a test specimen, i.e., to determine (9.17) experimentally, it is necessary to use as many specimens as experimental points to be recorded.

## 10. Two More Complex Programms of $\varepsilon^{p}$.

(i) $C^{(1)}- C^{(2)}\ldots C^{(n)}$ processes.

Consider the sequence $C^{(1)}- C^{(2)}\ldots C^{(n)}$ of cyclic processes. Suppose that the steady state has been attained in the course of the first ($C^{(1)}$) cyclic process so that $\varkappa^{(\Lambda)}\to\infty$, and that the subsequent cyclic processes have progressively increasing amplitudes. If the amplitude $\overset{(2)}{a}_{\varepsilon} > \overset{(1)}{a}_{\varepsilon}$ of the process $C^{(2)}$ is such that the current "microelasticity" limit $\bar{Y}^{(2)}(\overset{*}{\varkappa}{}_1^{(m)})$ (cf.(9.5)) is not exceeded during $C^{(2)}$ (cf. point $A_1$ in Fig.10.1), i.e.,

$$|\overset{(2)}{a}_{\epsilon} - \overset{*}{\beta}_1| \leq \bar{Y}^{(2)}(\overset{*}{\varkappa}{}_1^{(m)})/C^{(\alpha)}(\infty,\varkappa_1^{(m)}) \qquad (10.1)$$

then the steady state of the process $C^{(2)}$ is attained just after the first cycle. Here $\overset{*}{\varkappa}{}_1^{(m)}$ and $\overset{*}{\beta}_1$ are the values of $\varkappa^{(m)}$ and $\beta$ reached during the process $C^{(1)}$. If, however, the inequality (10.1) is violated then a development of inelastic micro-strains acompanies the process $C^{(2)}$. During transitory time $\beta$ varies and parameter $\varkappa^{(m)}$ increases from $\overset{*}{\varkappa}{}_1^{(m)}$ to $\overset{*}{\varkappa}{}_2^{(m)}$. The latter value being attained at the instant of a shakedown of microstresses.

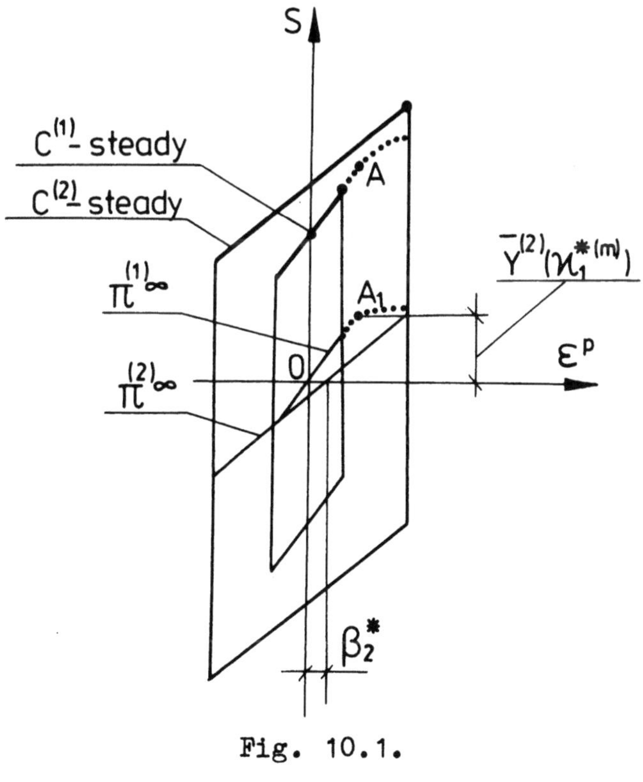


Fig. 10.1.

At the instant of a shakedown, simultaneously the steady state of the process $C^{(2)}$ is attained. That is, the model does not predicts the occurence of the second stage of transitory time of any cyclic process following $C^{(1)}$ - process. Each subsequent cyclic process $C^{(k)}$ proceeds similarly as $C^{(2)}$. In the course of certain, say $C^{(n)}$ $(n \geq k)$, process of large amplitude, shakedown of microstresses may not occur. The associated alternating microinelasticity will probably accelerate the cyclic damage. The asymptotic behaviour of the model when $\varkappa^{(m)} \to \infty$ requires, however, individual carefull investigation.

The cyclic stress-plastic strain curve determined in the course of $C^{(1)}-\ldots C^{(n)}$ - processes is also described by Eq.(9.17). However, the values of $\overset{*}{\varkappa}^{(m)}$ and $\beta^*$ corresponding to given amplitude $a_\varepsilon$ may be different due to different history of loading of a specimen.

(ii) $M^{(1)}-C^{(1)}-M^{(2)}$ processes.

Consider the sequence $M^{(1)}-C^{(1)}-M^{(2)}$ of monotonic and cyclic plastic strain variation. Let at the end of monotonic loading $M^{(1)}$ (0-A-0' - Fig.10.2) the values of $\varepsilon^p$, $\varkappa^{(m)}$, $\varkappa^{(\wedge)}$, $\beta$, $\pi^{(\wedge)}$ equal

$$\varepsilon^p_o\ ,\quad \varkappa^{(m)}_o,\ \varkappa^{(\wedge)}_o,\ \beta_o\ ,\ \pi^{(\wedge)}_o$$

The subsequent cyclic loading $C^{(2)}$ induces further changes of hardening parameters so that at the end of the first stage of transitory time

$$\varkappa^{(m)} = \overset{*}{\varkappa}^{(m)} \;, \quad \beta = \overset{*}{\beta} \;, \quad \pi^{(\wedge)} = \overset{*}{\pi}^{(\wedge)}$$

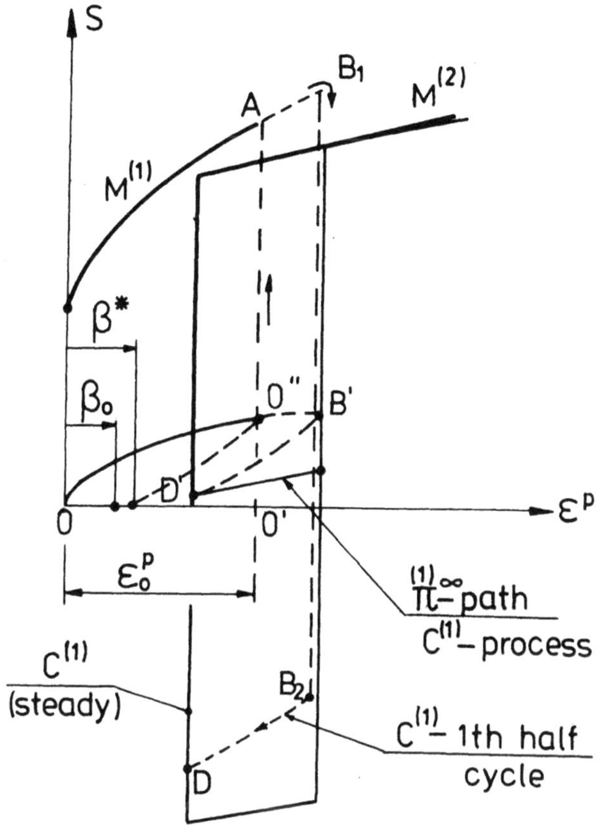


Fig.10.2. 0-0" π-path during $M^{(1)}$
0"-B'-D' π-path during 1th half cycle of $C^{(1)}$
0'-A-$B_1$-$B_2$-D S-path (1th half cycle of $C^{(1)}$)

If $|\varepsilon_0^p| >> a_\varepsilon$, the microstresses most probably shake down after completing the first cycle (the part of the first cycle is illustrated by dotted line 0'-A-$B_1$-$B_2$-D in Fig. 10.2), so that

$$\overset{*}{\varkappa}^{(m)} \approx \varkappa_0 \;, \quad \overset{*}{\beta} \approx \beta_0 \;, \quad \overset{*}{\pi}^{(\wedge)} \approx \pi_0^{(\wedge)}$$

The prior monotonic loading $M^{(1)}$ induces, due to Bauschinger effect, high mean stresses (higher then self--induced mean stresses) that relax during the transitory time of $C^{(1)}$. At the same time active stress $Y^{(1)}$ may increase. As a result, if

$$[C^{(\alpha)}(\varkappa_o^{(\wedge)},\varkappa_o^{(m)}) - C^{(\alpha)}(\infty,\varkappa_o^{(m)})] \times$$
$$\times\ (\varepsilon_o^p - \beta_o) > Y^{(1)}(\pi_\infty^{(\wedge)},\varkappa_o^{(m)}) - Y^{(1)}(\pi_o^{(\wedge)},\varkappa_o^{(m)}) \qquad (10.2)$$

the relaxation of mean stresses dominates and there is softening during $C^{(2)}$ process; the hysteresis loops move downward in the Fig.10.2. Otherwise cyclic hardening will acompany $C^{(2)}$ process. The model predicts that the cyclic softening is most likely to occur in the solids with the relatively small ratio of ultimate stress to initial yield stress at annealed state, and in those solids that display remarkable Bauschinger effect. This is in agreement with the experimental evidence (cf.[48]).

The monotonic process $M^{(2)}$ following the steady state of $C^{(1)}$ induces further hardening. The stress-strain curve corresponding to $M^{(2)}$ is initially linear (with hardening modulus $C^{(\alpha)}(\infty,\overset{*}{\varkappa}{}^{(m)})$ depending upon prior monotonic straining $M^{(1)}$ and cyclic straining $C^{(1)}$) until the actual yield limit $\bar{Y}^{(2)}(\overset{*}{\varkappa}{}^{(m)})$ of microstress is attained.

## References


1. H.A.Buchdahl: The Concepts of Classical Thermodynamics, Cambridge University Press 1966.
2. E.W.Hart: In Constitutive Equations in Plasticity, Ed. Argon, MIT Press, 1975, 149-197.
3. E.W.Hart: Acta Met., 18, June 1970, 599-610.
4. B.Raniecki, A.Sawczuk: ZAMM, 55, 6, 1975, 333-341.
5. A.A.Vakulenko: Lectures presented 22-25 XII, 1954, Izd. A.N. SSSR (in Russian).
6. A.A.Vakulenko: Dokl.A.N. SSSR, 126, 1959, 736-739.
7. H.Ziegler: Progress in Solid Mechanics, North Holland, Amsterdam, 1963, 613-619.
8. H.Ziegler: ZAMP, 21, 1970, 853-863.
9. J.Gyarmati: Non-equilibrium Thermodynamics, Field Theory and Variational Principles, Springer, Berlin, N-Y, 1970.
10. J.Mandel: Plasticite Classique et Viscoplasticite, Lecture course at CISM, Udine, Springer, 1972.
11. J.Kestin, J.R.Rice: In "A Critical Review of Thermodynamics", Mono Book Corp. 1970.
12. B.Halphen, Nguyen Quoc Son: Journal de Mécanique 14, 1975, No.1.
13. B.Halphen: Int.J.Solids Structures 11 (1975), 947-960.
14. J.Kestin: In "Thermoplasticity" Ed. "Ossolineum", Wroclaw, 1973, Lecture Notes for Summer School in Jablonna (in Polish).
15. H.B.Callen: Thermodynamics, Ed. I.Wiley and Sons, N-Y, London, 1960.
16. B.Raniecki, K.Thermann: Mitteilungen aus dem Institut für Mechanik, 2, Juni 1978.
17. G.A.Kluitenberg: Physica 28, 1962, pp.217,561,1173.
18. G.A.Kluitenberg: Physica 30, 1964, 1945-1972.
19. Th.Lehmann: Arch.Mech.Stos., 24, 1972, 975-984.
20. R.E.Reed Hill: Physical Metallurgy Principles, D.Van Nostrand Comp.Jnc., Toronto, London, 1964.
21. G.R.Halford: Stored Energy of Cold Work Changes Induced by Cyclic Deformation, Ph.D.-thesis Dept.of Theoretical and Appl.Mech., University of Illinois, 1966.

22. Z.Mróz, B.Raniecki; Thermodynamic Aspects of Cyclic and Monotonic Plasticity (in progress).

23. Z.Mróz: J.Mech.Phys.Sol., 15, 1967, 163-175.

24. Z.Mróz: J.de Mécanique Appliquée, 5, 3, 1981, 241-259.

25. R.D.Krieg: J.Appl.Mech., 42, 1975, 641-646.

26. Y.F.Dafalias, E.P.Popov: Acta Mechanica, 21, 1975, 173--192.

27. J.F.Besseling: Nat.Aera Res.Inst.Rep. S-410, 1953.

28. J.E.Dorn, A.Goldberg, T.E.Tietz: Metal Technology T.P., No 2445, September, 1948.

29. W.D.Sylwestrowicz: AIME Trans. 212, 1958, 617.

30. A.C.Willhelm, J.R.Kattus, Proc.Americ.Soc.Testing Mat., 63, 1963, 613.

31. M.Conrad: in Mechanical Behaviour of Materials at Elevated Temperatures. Ed.J.E.Dorn, McGraw-Hill, 1961, 149.

32. J.Klepaczko, J.Duffy: Engineering Transactions, 25, 1, 1977.

33. L.F.Coffin: Low cycle fatigue: a review, Appl.Mat.Res., 1, 1962, 129-141.

34. L.F.Coffin: J.Basic Eng., 82 D, 1960, 671-682.

35. A.P.Gusenkov: in "Low cycle failure and deformation", Moscow, 1967 (in Russian).

36. S.S.Manson: Thermal stresses and law-cycle fatigue, Mc Graw-Hill Book Co., 1966.

37. B.I.Sandor: Fundamentals of Cyclic Stress and Strain, Univ.of Wisconsin Press, USA, 1972.

38. V.V.Moskvitin: Plasticity under varying loads (in Russian), Moscow Univ., 1965.

39. M.B.Bever, D.L.Holt, A.L.Titchener: The stored energy of cold work, in: Progress in Material Sciences 17, Pergamon Press, 1973.

40. M.A.Bolshanyna, W.E.Panyn: Stored Energy of Deformation in: Issledovanija po Physice Tvierdovo Tiela, Izd.A.N.SSSR, 1957, 193-233 (in Russian).

41. B.Noble, M.J.Sewell: J.Inst.Math.Applies 9, 1972, 123--193.

42. J.F.Nye; Physical properties of crystals, Clarendon Press Oxford, 1957.

43. D.G.Edelen: Arch.Rational Mech.Anal., 51, 1973, 218.

44. D.G.Edelen: Int.J.Engn.Sci., 12,1974, 397-411.

45. L.I.Sedov: Mechanics of Continuous Medium, vol.2,Nauka, Moscow, 1970 (in Russian, cf.English translation).

46. K.Kadeshevith, V.V.Novozhilov; P.M.M., 22, 1958, 104--118.

47. V.V.Novozhilov: P.M.M., 28, 1964, 393-401.

48. R.Smith, M.Hirschberg, S.Manson: NASA TN-D-1574, 1963.

# FOUNDATIONS OF LARGE DEFORMATIONS

**KLAUS THERMANN**
**Universität Dortmund**
**D–4600 Dortmund 50**

## 6.1 Introduction

In these lecture notes we will recapitulate some basic facts from the mechanics of a continuous medium at finite deformations, which are needed for the development of thermomechanical constitutive theory, presented in separate articles of this course on thermoplasticity. Our topic are thus material independent quantities like strain, strain-rate, stress or mechanical work. Whereas in the older literature index notation of tensorial quantities is almost exclusively used, either in spatial coordinate systems, e.g. ERINGEN (1962), or in convected coordinate systems, e.g. SEDOV (1965), GREEN and ZERNA (1954), there is a tendency towards a direct or coordinate-free notation, e.g. TRUESDELL and NOLL (1965). Since in textbooks on continuum mechanics usually the one or the other approach is preferred, it is difficult for the student to interpret results stated in a "language" he has not learned. The main purpose of these notes is therefore to develop the various concepts partly parallel, and to work out in some detail the existing relations. Special emphasis is laid on the transition from a description in convected coordinates to the coordinate-free notation and vice versa. To introduce the notation used here, we list a few defining relations on vectors and second order tensors.

If $\underline{g}_i$ (i = 1,2,3) is a basis in the vector space associated with the three dimensional Euclidean point space, its dual basis is denoted as $\underline{g}^i$ and defined as

$$\underline{g}^i \cdot \underline{g}_k = \delta^i_k = \begin{cases} 1, & i = k \\ 0, & i \neq k \end{cases} , \qquad (1)$$

where the dot denotes the inner product. Given a vector $\underline{v}$,

$$\underline{v} = v_i \, \underline{g}^i = v^i \, \underline{g}_i , \qquad (2)$$

its co- and contravariant components are obtained as

$$v_i = \underline{v} \cdot \underline{g}_i , \qquad v^i = \underline{v} \cdot \underline{g}^i \qquad (3)$$

and we have the identities

$$\underline{v} = (\underline{v} \cdot \underline{g}_i)\underline{g}^i = (\underline{v} \cdot \underline{g}^i)\underline{g}_i . \qquad (4)$$

Summation over repeated indices from 1 to 3 is always implied. A second order tensor $\underline{S}$ is most conveniently looked upon as a linear transformation that assigns to each vector $\underline{v}$ another vector $\underline{S}\underline{v}$. Obviously, the tensor product $\underline{a} \otimes \underline{b}$ of two vectors, defined as

$$(\underline{a} \otimes \underline{b})\underline{v} = \underline{a}(\underline{b} \cdot \underline{v}) \text{ for every } \underline{v}, \qquad (5)$$

is a tensor. Moreover, any tensor $\underline{S}$, mapping $\underline{v}$ linearly into $\underline{S}\underline{v}$, is expressible as a linear combination of the tensor products of basis vectors $\underline{g}_i$,

$$\underline{\underline{S}}\underline{v} = \underline{\underline{S}}\underline{g}^k(\underline{v}\cdot\underline{g}_k) = (\underline{g}^i\cdot\underline{\underline{S}}\underline{g}^k)\underline{g}_i(\underline{g}_k\cdot\underline{v}) = (\underline{g}^i\cdot\underline{\underline{S}}\underline{g}^k)(\underline{g}_i\otimes\underline{g}_k)\underline{v}. \tag{6}$$

Thus,

$$\underline{S} = S^{ik}\underline{g}_i\otimes\underline{g}_k \text{ with contravariant components } S^{ik} = \underline{g}^i\cdot\underline{\underline{S}}\underline{g}^k, \tag{7}$$

and the set of all second order tensors is a vector space of dimension nine with the nine tensors $\underline{g}_i\otimes\underline{g}_k$ as a basis. Alternatively,

$$\underline{S} = S_{ik}\underline{g}^i\otimes\underline{g}^k = S^i{}_k\underline{g}_i\otimes\underline{g}^k = S_i{}^k\underline{g}^i\otimes\underline{g}_k, \tag{8}$$

with covariant and mixed components

$$S_{ik} = \underline{g}_i\cdot\underline{\underline{S}}\underline{g}_k\ , \quad S^i{}_k = \underline{g}^i\cdot\underline{\underline{S}}\underline{g}_k\ , \quad S_i{}^k = \underline{g}_i\cdot\underline{\underline{S}}\underline{g}^k. \tag{9}$$

The transpose $\underline{S}^T$ of a tensor $\underline{S}$ is the unique tensor with the property

$$\underline{u}\cdot(\underline{\underline{S}}\underline{v}) = (\underline{S}^T\underline{u})\cdot\underline{v} \quad \text{for all vectors } \underline{u} \text{ and } \underline{v}, \tag{10}$$

as a consequence

$$(\underline{a}\otimes\underline{b})^T = \underline{b}\otimes\underline{a}. \tag{11}$$

$\underline{S}$ is called symmetric if $\underline{S} = \underline{S}^T$, skew if $\underline{S} = -\underline{S}^T$. It follows for the components of a symmetric tensor

$$S_{ik} = S_{ki}\ , \quad S^{ik} = S^{ki}\ , \quad S^i{}_k = S_k{}^i =: S^i_k \tag{12}$$

however, $S^i{}_k \neq S^k{}_i$. Trace and determinant are denoted and defined as

$$\operatorname{tr}\underline{S} = S^i{}_i = S_i{}^i\ , \quad \det\underline{S} = \det(S^i{}_k) = \det(S_i{}^k). \tag{13}$$

These definitions make sence, since they are independent of the basis to which the mixed components are referred. This would not be the case if the matrices of the co- or contravariant components had been used in (13).

In the vector space of second order tensors an additional operation, besides the usual affine operations of addition and multiplication with scalars, has meaning: the product $\underline{S}\underline{T}$ of two tensors $\underline{S}$ and $\underline{T}$, which is the linear transformation mapping first any vector $\underline{v}$ into $\underline{T}\underline{v}$ and then $\underline{T}\underline{v}$ into $\underline{S}(\underline{T}\underline{v})$, i. e.

$$(\underline{S}\underline{T})\underline{v} = \underline{S}(\underline{T}\underline{v}) \text{ for every } \underline{v}. \tag{14}$$

Using this operation the space of tensors can be endowed with an inner product $\underline{S}.\underline{T}$,

$$\underline{S}.\underline{T} = \text{tr } (\underline{S}^T\underline{T}) = S^{ik}T_{ik} = S^i{}_k T_i{}^k = S_{ik}T^{ik}, \tag{15}$$

and the magnitude $|\underline{S}|$ of a tensor $\underline{S}$ be defined as

$$|\underline{S}| = \sqrt{(\underline{S}.\underline{S})} = \sqrt{(\text{tr } \underline{S}^T\underline{S})} = \sqrt{(S^{ik}S_{ik})}. \tag{16}$$

From the above definitions the usual transformation rules for vector and tensor components under a change of basis follow, which could have been used as an alternative starting point for the definition of tensors.

## 6.2 Kinematics

During the motion of a continuous medium its material particles are carried to various positions in space. A region in space, occupied by the body, is called a configuration. The description of the motion requires to specify the positions in space of all material particles as function of time t. We therefore need means (i) to identify points in space and (ii) to identify material particles. The identification of points in space is achieved by introduction of a coordinate system, which is fixed in space. By choosing any configuration of the body as a reference configuration, material particles can be identified by their spatial coordinates in this particular configuration. Through the choice of the reference configuration as well as the choice of generally curvilinear coordinate systems considerable freedom in the quantitative description of motion is offered. We will discuss some possibilities. A natural way, conceptually general and simple, is the description in Cartesian coordinates.

In a Cartesian frame of reference, consisting of a fixed origin in space together with three orthonormal basis vectors $\{\underline{e}_1,\underline{e}_2,\underline{e}_3\}$, the motion of the body is given by

$$x^\alpha = x^\alpha(\overset{o}{x}{}^\beta, t), \tag{17}$$

where $x^\alpha$ are the Cartesian coordinates of a material particle at time t, whose Cartesian coordinates in the reference configuration are $\overset{o}{x}{}^\alpha$. The reference configuration is taken as that region in space, which was occupied by the body during its motion at some initial instant of time $t_o$, and denoted as initial configuration $B^o$. The actual configuration B at time t is then the set of points in space $\{x^\alpha(\overset{o}{x}{}^\beta,t) \mid \overset{o}{x}{}^\beta \in B^o\}$.

To obtain the local behaviour of the mapping (17) from $B^o$ onto B, we consider infinitesimal vectors $d\overset{o}{\underline{x}}$ and $d\underline{x}$, connecting the same particles in the two configurations. From (17) $d\underline{x}$ is calculated as

$$\begin{aligned} d\underline{x} &= dx^\alpha\, \underline{e}_\alpha = \frac{\partial x^\alpha}{\partial \overset{o}{x}{}^\beta}\, d\overset{o}{x}{}^\beta\, \underline{e}_\alpha = \frac{\partial x^\alpha}{\partial \overset{o}{x}{}^\beta}\, (\underline{e}^\beta \cdot d\overset{o}{\underline{x}})\, \underline{e}_\alpha \\ &= \frac{\partial x^\alpha}{\partial \overset{o}{x}{}^\beta}\, (\underline{e}_\alpha \otimes \underline{e}_\beta)\, d\overset{o}{\underline{x}} =: \underline{F}\, d\overset{o}{\underline{x}}, \end{aligned} \tag{18}$$

thus, the linear transformation of the infinitesimal neighbourhood of a material particle from the initial to the actual configuration is obtained with the deformation gradient $\underline{F}$

$$\underline{F} = \frac{\partial x^\alpha}{\partial \overset{o}{x}{}^\beta} \underline{e}_\alpha \otimes \underline{e}_\beta \quad \text{through} \quad d\underline{x} = \underline{F}\, d\underline{\overset{o}{x}}. \tag{19}$$

Due to the linearity of the last relation, the image of the whole neighbourhood of a particle is given as soon as the images of three linearly independent vectors $d\overset{o}{\underline{x}}_\alpha$ are known.

If the motion of a body with a particular geometric shape has to be studied, it is often convenient to use a <u>description in spatial curvilinear coordinate systems</u>. If we express the Cartesian coordinates $x^\alpha$ of points in space through invertible functions of three other variables $y^1, y^2, y^3$, i.e.

$$x^\alpha = x^\alpha(y^\beta), \tag{20}$$

we can use $y^\beta$ for the identification of points in space as well. (20) defines a line in space, the coordinate line $y^\beta$, when $y^\beta$ is varying and the two other parameters kept fixed. In every point of space a natural basis is given by

$$\underline{G}_\beta = \frac{\partial x^\alpha}{\partial y^\beta} \underline{e}_\alpha, \tag{21}$$

where the vectors $\underline{G}_\beta$ are tangential to the coordinate lines $y^\beta$. Now, in a similar notation as before, we denote by $\overset{o}{y}{}^\alpha$ the spatial curvilinear coordinates of a particle in the initial configuration $B^o$, and by $y^\alpha$ its coordinates in the actual configuration B. $\overset{o}{\underline{G}}_\alpha$ and $\underline{G}_\alpha$ are basis vectors of the same coordinate system at the points $\overset{o}{y}{}^\alpha$ and $y^\alpha$, respectively.

The description of motion (17) is then transformed into

$$y^{\alpha} = y^{\alpha}(\overset{o}{y}^{\beta},t), \tag{22}$$

and the deformation gradient obtained from

$$d\overset{o}{\underline{x}} = d\overset{o}{x}^{\alpha}\,\underline{e}_{\alpha} = \frac{\partial \overset{o}{x}^{\alpha}}{\partial \overset{o}{y}^{\beta}}\, d\overset{o}{y}^{\beta}\,\underline{e}_{\alpha} = d\overset{o}{y}^{\beta}\,\overset{o}{\underline{G}}_{\beta} \tag{23}$$

and

$$d\underline{x} = dx^{\beta}\,\underline{e}_{\beta} = dy^{\alpha}\,\underline{G}_{\alpha} = \frac{\partial y^{\alpha}}{\partial \overset{o}{y}^{\beta}}\, d\overset{o}{y}^{\beta}\,\underline{G}_{\alpha} = \frac{\partial y^{\alpha}}{\partial \overset{o}{y}^{\beta}}\,\underline{G}_{\alpha}(\overset{o}{\underline{G}}^{\beta}\cdot d\overset{o}{\underline{x}}) \tag{24}$$

as

$$\underline{F} = \frac{\partial y^{\alpha}}{\partial \overset{o}{y}^{\beta}}\,\underline{G}_{\alpha} \otimes \overset{o}{\underline{G}}^{\beta}. \tag{25}$$

The matrix $\partial y^{\alpha}/\partial \overset{o}{y}^{\beta}$ of partial derivatives has to be related to basis vectors at the material point in the initial and actual configuration to obtain the tensor $\underline{F}$. For simplicity, we used the same spatial curvilinear coordinate system for the identification of particles in $B^{o}$ and to describe their positions in B. Obviously, $\overset{o}{y}^{\beta}$ interpreted as coordinates of a second spatial coordinate net with basis $\overset{o}{G}_{\beta}$ in $B^{o}$ leaves all relations (22)-(25) unaltered.

The use of spatial curvilinear coordinates is conceptually no more general than the Cartesian description, and can always be avoided, unless there is a definite need in the solution of specific problems. In contrast, a <u>description in convected coordinates</u> leads always to curvilinear coordinates (unless the motion is very special).

For the identification of particles we choose three numbers $\xi^{i}$ in one-to-one correspondence with their Cartesian coordinates in $B^{o}$,

$$\overset{o}{x}^{\alpha} = \overset{o}{x}^{\alpha}(\xi^{i}). \tag{26}$$

The motion (17) is then given through

$$x^{\alpha} = x^{\alpha}(\xi^{i},t). \tag{27}$$

The transformation (26) defines a net of coordinate lines $\xi^{i}$ in the initial configuration with basis vectors

$$\overset{o}{\underline{g}}_{i} = \frac{\partial \overset{o}{x}^{\alpha}}{\partial \xi^{i}} \underline{e}_{\alpha}. \tag{28}$$

If we look upon this net as beeing embedded in the body, it deforms together with the body during its motion, according to (27). With this interpretation the $\xi^{i}$ are called convected or embedded coordinates. Note, that the curvilinear net $\xi^{i}$ of material coordinate lines is only defined in those regions of space occupied by the body. The basis vectors, tangential to the coordinate lines $\xi^{i}$, are now time dependent and given by

$$\underline{g}_{i} = \frac{\partial x^{\alpha}(\xi^{i},t)}{\partial \xi^{i}} \underline{e}_{\alpha}. \tag{29}$$

Again, from

$$d\overset{o}{\underline{x}} = \frac{\partial \overset{o}{x}^{\alpha}}{\partial \xi^{i}} d\xi^{i} \underline{e}_{\alpha} = d\xi^{i} \overset{o}{\underline{g}}_{i} \tag{30}$$

$$d\underline{x} = dx^{\alpha} \underline{e}_{\alpha} = \frac{\partial x^{\alpha}}{\partial \xi^{i}} d\xi^{i} \underline{e}_{\alpha} = d\xi^{i} \underline{g}_{i} = (d\overset{o}{\underline{x}}\cdot\overset{o}{\underline{g}}^{i}) \underline{g}_{i}, \tag{31}$$

the deformation gradient is obtained and represented as

$$\underline{F} = \underline{g}_{i} \otimes \overset{o}{\underline{g}}^{i}. \tag{32}$$

Hence, $\underline{F}$ transforms the initial basis $\overset{o}{\underline{g}}_{i}$ into $\underline{g}_{i}$,

$$\underline{g}_{i} = \underline{F}\, \overset{o}{\underline{g}}_{i}. \tag{33}$$

For future reference we list the properties and representations of the transpose and inverse of $\underline{F}$, which follow immediately from (32) and (33),

$$\overset{o}{\underline{g}}_i = \underline{F}^{-1} \underline{g}_i , \qquad \underline{F}^{-1} = \overset{o}{\underline{g}}_i \otimes \underline{g}^i \tag{34}$$

$$\overset{o}{\underline{g}}{}^i = \underline{F}^T \underline{g}^i , \qquad \underline{F}^T = \overset{o}{\underline{g}}{}^i \otimes \underline{g}_i \tag{35}$$

$$\underline{g}^i = \underline{F}^{-T} \overset{o}{\underline{g}}{}^i , \qquad \underline{F}^{-T} = \underline{g}^i \otimes \overset{o}{\underline{g}}_i . \tag{36}$$

The matrices of the tensors $\underline{F}$, $\underline{F}^{-1}$, $\underline{F}^T$, $\underline{F}^{-T} := (\underline{F}^T)^{-1} = (\underline{F}^{-1})^T$ are all unity, when referred to co- and contravariant basis vectors of a convected coordinate sytem in $B^o$ and B, and are therefore so uninteresting that these quantities are usually not considered at all by users of convected coordinates.

Let us now turn to the <u>definition of strain measures</u>. A clear geometrical picture is obtained by using the polar decomposition of invertible tensors and the spectral decomposition of symmetric tensors. According to the first, the invertible deformation gradient is uniquely decomposed into

$$\underline{F} = \underline{R}\underline{U} = \underline{V}\underline{R}, \tag{37}$$

where $$\underline{U} = \underline{U}^T , \quad \underline{V} = \underline{V}^T , \quad \underline{R}^T = \underline{R}^{-1} , \quad \det \underline{R} = +1 , \tag{38}$$

according to the second, there exist (at least) three orthonormal vectors $\overset{o}{\underline{p}}_1, \overset{o}{\underline{p}}_2, \overset{o}{\underline{p}}_3$, which preserve their directions under the transformation $\underline{U}$

$$\underline{U} \overset{o}{\underline{p}}_i = u_i \overset{o}{\underline{p}}_i \quad \text{(no summation)} , \tag{39}$$

and allow the spectral decomposition of $\underline{U}$ in the form

$$\underline{U} = \sum u_i \overset{o}{\underline{p}}_i \otimes \overset{o}{\underline{p}}_i . \tag{40}$$

If $u_i$, $\overset{o}{p}_i$ are the principal values and directions of $U$, then from (37) and (40)

$$V = RUR^T = R(\sum u_i \overset{o}{p}_i \otimes \overset{o}{p}_i)R^T = \sum u_i R\overset{o}{p}_i \otimes R\overset{o}{p}_i, \tag{41}$$

hence, $V$ has the same principal values as $U$ but principal directions $p_i = R\,\overset{o}{p}_i$, obtained from those of $U$ through rotation by $R$.

The physical interpretation is, that we can always find three mutually perpendicular material directions, passing through the considered particle, which remain mutually perpendicular - in the infinitesimal neighbourhood of that particle - in the motion from the initial to the actual configuration. The orientation in space of this orthogonal material triad is given by $\overset{o}{p}_i$ in $B^o$ and by $p_i$ in B. The material element undergoes stretches of amount $u_i$ along this triad, defined, according to (39), as the ratio of actual to initial length of the material line element. If all principal stretches $u_i$ are unity, then

$$U = \sum \overset{o}{p}_i \otimes \overset{o}{p}_i = 1 \quad , \quad V = 1 \quad , \quad F = R \quad , \tag{42}$$

and the material element is rigidly rotated, otherwise it is strained.

Instead of using the principal stretches $u_i$, we may take any sufficiently smooth monotone function of one variable $f(u_i)$, and measure the principal values of strain by $\varepsilon_i$,

$$\varepsilon_i = f(u_i). \tag{43}$$

This generates two families of strain tensors, which have as their principal directions either those of $U$ or $V$,

$$\overset{o}{\varepsilon} = f(U) := \sum f(u_i)\, \overset{o}{p}_i \otimes \overset{o}{p}_i \quad \text{or} \quad \varepsilon = f(V) := \sum f(u_i)\, p_i \otimes p_i. \tag{44}$$

In both cases the principal directions are given by the same material orthogonal triad, taken in the initial or actual configuration, respectively. The corresponding strain tensors are sometimes called Lagrangean and Eulerian measures of strain.

The tensor valued functions $\underline{f}(\underline{U})$ and $\underline{f}(\underline{V})$ are defined in (44) via principal values and directions of their arguments. If the argument $\underline{U}$ is replaced by $\underline{Q}\underline{U}\underline{Q}^T$ then, similar to (41), its principal values are unaltered, but the principal directions changed into $\underline{Q}\,\overset{o}{\underline{p}}_i$. It follows

$$\underline{f}(\underline{Q}\underline{U}\underline{Q}^T) = \underline{Q}\underline{f}(\underline{U})\underline{Q}^T \text{ for all orthogonal } \underline{Q}, \tag{45}$$

the equivalent, frequently met, definition of an isotropic tensor-valued function of one tensor variable.

It is convenient to define the strain measure in such a way that it coincides with the usual strain measure for small deformations after linearization. This is achieved with the further restrictions on f,

$$f(1) = 0 \quad , \quad f'(1) = 1 \quad , \tag{46}$$

HILL (1968). Some strain tensors of practical interest with this property are contained in the family

$$f(u_i) = \begin{cases} \dfrac{(u_i)^{\varkappa} - 1}{\varkappa} & , \quad \varkappa \neq 0 \\[2ex] \log(u_i) & , \quad \varkappa = 0 \end{cases} \tag{47}$$

e.g.,

$$\overset{o}{\underline{\varepsilon}}_{(-2)} = \sum \frac{1-(u_i)^{-2}}{2}\, \overset{o}{\underline{p}}_i \otimes \overset{o}{\underline{p}}_i = \frac{1}{2}(\underline{1}-\underline{U}^{-2}) \tag{48}$$

$$\overset{o}{\underline{\varepsilon}}_{(-1)} = \sum \frac{1-(u_i)^{-1}}{1}\, \overset{o}{\underline{p}}_i \otimes \overset{o}{\underline{p}}_i = \underline{1}-\underline{U}^{-1} \tag{49}$$

$$\overset{o}{\underline{\varepsilon}}_{(0)} = \sum \log(u_i)\, \overset{o}{\underline{p}}_i \otimes \overset{o}{\underline{p}}_i = \log(\underline{U}) \tag{50}$$

$$\overset{\circ}{\underline{\varepsilon}}_{(1)} = \sum \frac{u_i - 1}{1} \overset{\circ}{\underline{p}}_i \otimes \overset{\circ}{\underline{p}}_i = \underline{U} - \underline{1} \tag{51}$$

$$\overset{\circ}{\underline{\varepsilon}}_{(2)} = \sum \frac{(u_i)^2 - 1}{2} \overset{\circ}{\underline{p}}_i \otimes \overset{\circ}{\underline{p}}_i = \frac{1}{2}(\underline{U}^2 - \underline{1}) \, . \tag{52}$$

The corresponding Eulerian strain tensors are obtained with $\underline{V}$ instead of $\underline{U}$, for example,

$$\underline{\varepsilon}_{(2)} = \sum \frac{(u_i)^2 - 1}{2} \underline{p}_i \otimes \underline{p}_i = \frac{1}{2}(V^2 - 1). \tag{53}$$

A particular property of the logarithmic strain tensors is the simple relation between their traces and the volume ratio $dV/d\overset{\circ}{V}$ of the strained element

$$\mathrm{tr}\,\overset{\circ}{\underline{\varepsilon}}_{(0)} = \mathrm{tr}\;\underline{\varepsilon}_{(0)} = \sum \log(u_i) = \log(u_1 u_2 u_3) = \log \frac{dV}{d\overset{\circ}{V}}. \tag{54}$$

From a practical point of view, the disadvantage of these measures - and the members of the family (44), in general - is the necessity to perform a transformation to principal axes for the actual calculation of their components. This is avoided, if the strain tensors $\overset{\circ}{\underline{\varepsilon}}_{(2)}$, $\underline{\varepsilon}_{(2)}$, $\overset{\circ}{\underline{\varepsilon}}_{(-2)}$, $\underline{\varepsilon}_{(-2)}$ are used, which can be directly calculated from the right and left Cauchy-Green tensor,

$$\underline{F}^T\underline{F} = (\underline{RU})^T\underline{RU} = \underline{U}^T\underline{U} = \underline{U}^2 \tag{55}$$

$$\underline{FF}^T = (\underline{VR})(\underline{VR})^T = \underline{VV}^T = V^2, \tag{56}$$

and the inverse of theses quantities. These tensors appear immediately, if we calculate the square of material line elements,

$$(ds)^2 = d\underline{x}.d\underline{x} = \underline{F}d\overset{\circ}{\underline{x}}.\underline{F}d\overset{\circ}{\underline{x}} = d\overset{\circ}{\underline{x}}.\underline{F}^T\underline{F}d\overset{\circ}{\underline{x}} \tag{57}$$

$$(d\overset{\circ}{s})^2 = d\overset{\circ}{\underline{x}}.d\overset{\circ}{\underline{x}} = \underline{F}^{-1}d\underline{x}.\underline{F}^{-1}d\underline{x} = d\underline{x}.(\underline{FF}^T)^{-1}d\underline{x} \, . \tag{58}$$

In convected coordinates we have from (30) and (31)

$$(ds)^2 = d\xi^i \underline{g}_i \cdot d\xi^k \underline{g}_k = g_{ik} d\xi^i d\xi^k \tag{59}$$

$$(d\mathring{s})^2 = d\xi^i \mathring{\underline{g}}_i \cdot d\xi^k \mathring{\underline{g}}_k = \mathring{g}_{ik} d\xi^i d\xi^k \,, \tag{60}$$

and the metric coefficients

$$g_{ik} = \underline{g}_i \cdot \underline{g}_k \quad , \quad g^{ik} = \underline{g}^i \cdot \underline{g}^k \tag{61}$$

$$\mathring{g}_{ik} = \mathring{\underline{g}}_i \cdot \mathring{\underline{g}}_k \quad , \quad \mathring{g}^{ik} = \mathring{\underline{g}}^i \cdot \mathring{\underline{g}}^k \tag{62}$$

will enter the definition of strain. The representation of the strain tensors in convected coordinates are obtained easily with the help of (32)-(36). The Cauchy-Green tensors are expressed as

$$\underline{U}^2 = \underline{F}^T\underline{F} = (\mathring{\underline{g}}^i \otimes \underline{g}_i)(\underline{g}_k \otimes \mathring{\underline{g}}^k) = g_{ik}\, \mathring{\underline{g}}^i \otimes \mathring{\underline{g}}^k = \mathring{g}^{ir} g_{rk}\, \mathring{\underline{g}}_i \otimes \mathring{\underline{g}}^k \tag{63}$$

$$\underline{V}^2 = \underline{F}\underline{F}^T = (\underline{g}_i \otimes \mathring{\underline{g}}^i)(\mathring{\underline{g}}^k \otimes \underline{g}_k) = \mathring{g}^{ik}\, \underline{g}_i \otimes \underline{g}_k = \mathring{g}^{ir} g_{rk}\, \underline{g}_i \otimes \underline{g}^k \,, \tag{64}$$

hence, the same matrix

$$q^i_k = \mathring{g}^{ir} g_{rk} \,, \tag{65}$$

represents the mixed components of $\underline{U}^2$ and $\underline{V}^2$, when the basis to the material coordinate net $\xi^i$ is taken in $B^o$ and B, respectively. $q^i_k$ transforms the metric from the initial to the actual configuration, according to

$$g_{ik} = \mathring{g}_{ir} q^r_k = q^r_i \mathring{g}_{rk} \,, \tag{66}$$

and components of the two families of strain measures in $B^o$ and B are generated with the help of it. For example, those members of (47), which can be expressed without performing the transformation to principal axes, assume the form

$$\underline{\mathring{\varepsilon}}_{(-2)} = \frac{1}{2}\,(\underline{1} - \underline{U}^{-2}) = \frac{1}{2}\,(\mathring{g}^{ik} - g^{ik})\; \underline{\mathring{g}}_i \otimes \underline{\mathring{g}}_k \tag{67}$$

$$= \frac{1}{2}\,(\delta^i_k - g^{ir}\mathring{g}_{rk})\; \underline{\mathring{g}}_i \otimes \underline{\mathring{g}}^k = \frac{1}{2}\,(\delta^i_k - (q^{-1})^i_k)\; \underline{\mathring{g}}_i \otimes \underline{\mathring{g}}^k$$

$$\underline{\mathring{\varepsilon}}_{(2)} = \frac{1}{2}\,(\underline{U}^2 - \underline{1}) = \frac{1}{2}\,(g_{ik} - \mathring{g}_{ik})\; \underline{\mathring{g}}^i \otimes \underline{\mathring{g}}^k \tag{68}$$

$$= \frac{1}{2}\,(\mathring{g}^{ir} g_{rk} - \delta^i_k)\; \underline{\mathring{g}}_i \otimes \underline{\mathring{g}}^k = \frac{1}{2}\,(q^i_k - \delta^i_k)\; \underline{\mathring{g}}_i \otimes \underline{\mathring{g}}^k$$

$$\underline{\varepsilon}_{(-2)} = \frac{1}{2}\,(\underline{1} - \underline{V}^{-2}) = \frac{1}{2}\,(g_{ik} - \mathring{g}_{ik})\; \underline{g}^i \otimes \underline{g}^k \tag{69}$$

$$= \frac{1}{2}\,(\delta^i_k - g^{ir}\mathring{g}_{rk})\; \underline{g}_i \otimes \underline{g}^k = \frac{1}{2}\,(\delta^i_k - (q^{-1})^i_k)\; \underline{g}_i \otimes \underline{g}^k$$

$$\underline{\varepsilon}_{(2)} = \frac{1}{2}\,(\underline{V}^2 - \underline{1}) = \frac{1}{2}\,(\mathring{g}^{ik} - g^{ik})\; \underline{g}_i \otimes \underline{g}_k \tag{70}$$

$$= \frac{1}{2}\,(\mathring{g}^{ir} g_{rk} - \delta^i_k)\; \underline{g}_i \otimes \underline{g}^k = \frac{1}{2}\,(q^i_k - \delta^i_k)\; \underline{g}_i \otimes \underline{g}^k ,$$

where $$(q^{-1})^i_k = g^{ir}\mathring{g}_{rk} \tag{71}$$

is the matrix inverse to $q^i_k$. Note, that the unit tensor can be expressed as

$$\underline{1} = g^{ik}\, \underline{g}_i \otimes \underline{g}_k = g_{ik}\, \underline{g}^i \otimes \underline{g}^k = \delta^i_k\, \underline{g}_i \otimes \underline{g}^k = \mathring{g}^{ik}\, \underline{\mathring{g}}_i \otimes \underline{\mathring{g}}_k , \tag{72}$$

and that raising and lowering of indices has to be done with the metric tensor corresponding to the basis in which the components are written.

Furthermore, when calculating principal values and directions the starting point is, of course, the tensorial definition. E.g.,

$$(\underline{\mathring{\varepsilon}}_{(2)} - \lambda \underline{1})\, \underline{\mathring{p}} = \underline{0} \tag{73}$$

yields the equivalent algebraic equations for $\lambda$ and $\mathring{p}^k = \underline{\mathring{g}}^k \cdot \underline{\mathring{p}}$ in covariant components as

$$\{ \frac{1}{2} (g_{ik} - \overset{o}{g}_{ik}) - \lambda \overset{o}{g}_{ik} \} \overset{ok}{p} = 0 \ , \tag{74}$$

and in mixed components as

$$\{ \frac{1}{2} (\overset{oir}{g} g_{rk} - \delta^i_k) - \lambda \delta^i_k \} \overset{ok}{p} = \{ \frac{1}{2} (q^i_k - \delta^i_k) - \lambda \delta^i_k \} \overset{ok}{p} = 0. \tag{75}$$

Hence, the fact that $\overset{o}{\underset{-}{\varepsilon}}_{(\varkappa)}$ and $\underset{-}{\varepsilon}_{(\varkappa)}$ have the same principal values is readily read off from their mixed components, but not from their co- or contravariant components.

We conclude the discussion of kinematics with a consideration of rates. The velocity, as the rate at which material particles change their position in space, is obtained for the three descriptions of motion as

$$\underset{-}{v} = \frac{\partial x^\alpha(\overset{o}{x}^\beta, t)}{\partial t} \underset{-\alpha}{e} = \dot{x}^\alpha \underset{-\alpha}{e} = v^\alpha(x^\beta, t) \underset{-\alpha}{e} \tag{76}$$

$$= \frac{\partial x^\alpha}{\partial y^\beta} \frac{\partial y^\beta(\overset{o}{y}^\mu, t)}{\partial t} \underset{-\alpha}{e} = \dot{y}^\beta \underset{-\beta}{G} = V^\beta(y^\mu, t) \underset{-\beta}{G} \tag{77}$$

$$= v^\alpha \frac{\partial \xi^i}{\partial x^\alpha} \underset{-i}{g} = v^i(x^\alpha, t) \underset{-i}{g} \ . \tag{78}$$

Since one usually wishes to calculate a neighbouring configuration of the body at time t+dt from its actual configuration, the velocity is considered as a field of the actual particle positions and time, their initial coordinates beeing replaced through the motion. This has to be observed when taking the material time derivative, e.g. of the velocity to get the accelaration,

$$\underset{-}{a} = \dot{\underset{-}{v}} = \dot{v}^\alpha \underset{-\alpha}{e} = \{ \frac{\partial v^\alpha(x^\beta, t)}{\partial t} + \frac{\partial v^\alpha}{\partial x^\beta} \frac{\partial x^\beta(\overset{o}{x}^\mu, t)}{\partial t} \} \underset{-\alpha}{e}$$

$$= \{ \frac{\partial v^\alpha}{\partial t} + \frac{\partial v^\alpha}{\partial x^\beta} v^\beta \} \underset{-\alpha}{e} \ . \tag{79}$$

By definition, the spatial gradient grad $\underline{v}$ of the velocity field is the second order tensor, which gives the change in velocity, when the field point is changed by $d\underline{x}$,

$$d\underline{v} = (\text{grad}\ \underline{v})\ d\underline{x}\ . \tag{80}$$

In Cartesian coordinates,

$$\text{grad}\ \underline{v} = \frac{\partial v^{\alpha}}{\partial x^{\beta}}\ \underline{e}_{\alpha} \otimes \underline{e}^{\beta}\ , \tag{81}$$

however, in spatial curvilinear coordinates, the change of basis $\underline{G}_{\alpha}$ under a change of field point has to be accounted for, and the partial derivatives have to be replaced by covariant derivatives of components,

$$\text{grad}\ \underline{v} = \left( \frac{\partial V^{\alpha}}{\partial y^{\beta}} + \Gamma^{\alpha}_{\beta\mu} V^{\mu} \right) \underline{G}_{\alpha} \otimes \underline{G}^{\beta} = V^{\alpha}|_{\beta}\ \underline{G}_{\alpha} \otimes \underline{G}^{\beta}\ . \tag{82}$$

The infinitesimal vector $d\underline{x}$ connects two particles in the actual configuration and its rate of change, following these particles, is

$$(d\underline{x})^{\cdot} = d\underline{v}\ . \tag{83}$$

From $d\underline{x} = \underline{F}\ d\overset{o}{\underline{x}}$ we have

$$(d\underline{x})^{\cdot} = \dot{\underline{F}}\ d\overset{o}{\underline{x}} = \dot{\underline{F}}\underline{F}^{-1}\ d\underline{x}\ , \tag{84}$$

and the velocity gradient can be expressed through the deformation gradient and its material time derivative,

$$\text{grad}\ \underline{v} = \dot{\underline{F}}\underline{F}^{-1} =: \underline{L}\ . \tag{85}$$

The rate of change of a material line element of length ds at time t,

$$(ds^2)^{\cdot} = (d\underline{x} . d\underline{x})^{\cdot} = d\underline{x} . (\underline{L} + \underline{L}^T)\ d\underline{x}, \tag{86}$$

depends on $\underline{L}$ only through its symmetric part

$$\underline{D} = \frac{1}{2}\ (\underline{L} + \underline{L}^T)\ . \tag{87}$$

$\underset{-}{D}$ is therefore called <u>rate of deformation</u>*), whereas the skew part of $\underset{-}{L}$,

$$\underset{-}{W} = \frac{1}{2} (\underset{-}{L} - \underset{-}{L}^T) \ , \tag{88}$$

denoted as <u>spin</u>, describes the angular velocity of an instantaneously rigid rotation of the material element in B. As should be expected, the rate of deformation is expressible through the metric coefficients of an embedded coordinate system and their material time derivatives. From (86) with $d\underset{-}{x} = d\xi^i \underset{=}{g}_i$ ,

$$\begin{aligned} (ds^2)^{\cdot} &= (d\xi^i \underset{=}{g}_i \cdot d\xi^k \underset{=}{g}_k)^{\cdot} = \dot{g}_{ik} \, d\xi^i d\xi^k \\ &= d\underset{-}{x} \cdot 2\underset{-}{D} \, d\underset{-}{x} \qquad = \underset{=}{g}_i \cdot 2\underset{-}{D} \, \underset{=}{g}_k \, d\xi^i d\xi^k \ , \end{aligned} \tag{89}$$

hence, $\dot{g}_{ik}$ are the components of $2\underset{-}{D}$ in the basis $\underset{=}{g}^i \otimes \underset{=}{g}^k$, and

$$\underset{-}{D} = \frac{1}{2} \dot{g}_{ik} \, \underset{=}{g}^i \otimes \underset{=}{g}^k = \frac{1}{2} g^{ir} \dot{g}_{rk} \, \underset{=}{g}_i \otimes \underset{=}{g}^k = -\frac{1}{2} \dot{g}^{ik} \, \underset{=}{g}_i \otimes \underset{=}{g}_k \ , \tag{90}$$

where the covariant components follow with the help of

$$g^{ir} g_{rk} = \delta^i_k \quad , \quad \dot{g}^{ir} g_{rk} = - g^{ir} \dot{g}_{rk} \quad . \tag{91}$$

The mixed components can be rewritten as

$$d^i_k = \frac{1}{2} g^{ir} \dot{g}_{rk} = \frac{1}{2} g^{ir\!\!\!\circ} g_{rs} g^{\circ sl} \dot{g}_{lk} = \frac{1}{2} (q^{-1})^i_s \, (q^s_k)^{\cdot} \ , \tag{92}$$

this shows again the importance of the metric transformation matrix $q^i_k$ for the description of kinematic quantities in the convected coordinate system.

---

*) The nomenclature is not unique in the literature. Stretching, velocity of deformation, Eulerian strain rate are alternative names.

Note, that the two tensors which can be constructed with the numbers $(q^i_k)^\cdot$ in the basis corresponding to the actual configuration are different,

$$(q^i_k)^\cdot \, \underline{g}_i \otimes \underline{g}^k \neq (q^i_k)^\cdot \, \underline{g}^k \otimes \underline{g}_i \; , \tag{93}$$

in contrast to

$$\begin{aligned} q^i_k \, \underline{g}_i \otimes \underline{g}^k &= q^i_k \, \underline{g}^k \otimes \underline{g}_i \; , \\ q^i_k \, \overset{o}{\underline{g}}_i \otimes \overset{ok}{\underline{g}} &= q^i_k \, \overset{ok}{\underline{g}} \otimes \overset{o}{\underline{g}}_i \; , \\ (q^i_k)^\cdot \, \overset{o}{\underline{g}}_i \otimes \overset{ok}{\underline{g}} &= (q^i_k)^\cdot \, \overset{ok}{\underline{g}} \otimes \overset{o}{\underline{g}}_i \; . \end{aligned} \tag{94}$$

## 6.3 Stress and mechanical work

With the help of the symmetric Cauchy or "true" stress tensor $\underline{\sigma}$, the stress vector $\underline{t}$, acting on a plane element with unit normal $\underline{n}$ in the actual configuration, is obtained as

$$\underline{t} = \underline{\sigma}\underline{n} \; . \tag{95}$$

If $\underline{\sigma}$ is the actual state of stress at a particle in the configuration B, then under a velocity field $\underline{v}$ the power of the stresses on the velocities is simply,

$$\text{per unit volume in B} \quad \sigma^{\alpha\beta} \frac{\partial v^\alpha}{\partial x^\beta} = \underline{\sigma}.\text{grad}\, \underline{v} = \underline{\sigma}.\underline{L} \; , \tag{96}$$

$$\text{or per unit mass} \quad \dot{w} = \frac{1}{\rho}\, \underline{\sigma}.\underline{\mathbf{L}} = \frac{1}{\rho}\, \underline{\sigma}.\underline{D} \; . \tag{97}$$

A symmetric stress tensor $\underline{\tau}$ is called work conjugate to the Lagrangian strain measure $\overset{o}{\underline{\varepsilon}}$, if the specific rate of work is expressible as

$$\dot{w} = \frac{1}{\rho_o} \underline{\tau} . (\overset{o}{\underline{\varepsilon}})^{\bullet} \; . \tag{98}$$

Hence, $\underline{\tau}$ is defined through

$$\underline{\tau} . (\overset{o}{\underline{\varepsilon}})^{\bullet} = \frac{\rho_o}{\rho} \underline{\sigma} . \underline{D} = \underline{s} . \underline{D} \; , \tag{99}$$

where the Cauchy stress, multiplied with the ratio of the mass densities in the reference and actual configuration, is introduced as the Kirchhoff stress tensor

$$\underline{s} = \frac{\rho_o}{\rho} \underline{\sigma} \; . \tag{100}$$

In the balance equation of energy, the specific rate of mechanical work is one part encountered. In order to obtain the restrictions on constitutive equations from thermodynamical arguments, it is useful to have such conjugate pairs, which have been studied e.g. by MACVEAN (1968), HILL (1968) and RICE (1975). Their discussion allows at the same time to introduce some frequently met stress measures.

As can be seen by straightforward calculation after substitution in (99), the symmetric Piola-Kirchhoff stress

$$\underline{\tau}_{(2)} = \underline{F}^{-1} \underline{s} \underline{F}^{-T} \tag{101 a}$$

is conjugate to the Green strain tensor

$$\overset{o}{\underline{\varepsilon}}_{(2)} = \frac{1}{2} (\underline{F}^T \underline{F} - \underline{1}) \; , \tag{101 b}$$

and

$$\underline{\tau}_{(-2)} = \underline{F}^T \underline{s} \underline{F} \tag{102 a}$$

is conjugate to the Almansi strain tensor

$$\overset{o}{\underline{\varepsilon}}_{(-2)} = \frac{1}{2} \{ \underline{1} - (\underline{F}^T \underline{F})^{-1} \} \; . \tag{102 b}$$

HILL (1968) has shown that

$$\underline{\tau}_{(0)} = \underline{R}^T \underline{s} \underline{R} \tag{103 a}$$

is conjugate to the logarithmic strain

$$\overset{o}{\underline{\varepsilon}}_{(0)} = \frac{1}{2} \log(\underline{F}^T \underline{F}) \ , \tag{103 b}$$

if the principal axes of stress $\underline{\sigma}$ coincide with the principal axes of strain in the actual configuration. This would be the case, if the material is isotropic elastic with respect to the reference configuration. In order to obtain the pair (103), one has to use the fact that the normal components of the rate of deformation tensor $\underline{D}$ on the principal axes of the left stretch tensor $\underline{V}$ are equal to the normal components of the logarithmic strain rate on the principal axes of $\underline{U}$.

In a convected coordinate system the Cauchy or Kirchhoff stress would be written in components with respect to basis vectors in the actual configuration,

$$\underline{\sigma} = \sigma^{ik} \, \underline{g}_i \otimes \underline{g}_k = \sigma_{ik} \, \underline{g}^i \otimes \underline{g}^k = \sigma^i_k \, \underline{g}_i \otimes \underline{g}^k \ , \tag{104}$$

$$\underline{s} = s^{ik} \, \underline{g}_i \otimes \underline{g}_k = s_{ik} \, \underline{g}^i \otimes \underline{g}^k = s^i_k \, \underline{g}_i \otimes \underline{g}^k \ . \tag{105}$$

If the contravariant representation for $\underline{s}$ is used in (101 a), then with the help of (33)-(36)

$$\underline{\tau}_{(2)} = \underline{F}^{-1}(s^{ik} \, \underline{g}_i \otimes \underline{g}_k)\underline{F}^{-T} = s^{ik} \underline{F}^{-1} \underline{g}_i \otimes \underline{F}^{-1} \underline{g}_k = s^{ik} \, \overset{o}{\underline{g}}_i \otimes \overset{o}{\underline{g}}_k \ , \tag{106}$$

hence, the Piola-Kirchhoff stress, conjugate to Green strain, has contravariant components with respect to the fixed basis $\overset{o}{\underline{g}}_i$ in the reference configuration, which are equal to the contravariant components of

Kirchhoff stress with respect to the convected basis in the actual configuration. From (68) we have the representation

$$\overset{o}{\underline{\varepsilon}}_{(2)} = \frac{1}{2}\,(g_{ik} - \overset{o}{g}_{ik})\, \overset{oi}{\underline{g}} \otimes \overset{ok}{\underline{g}} \tag{107}$$

and after construction of the conjugate $\underline{\tau}_{(2)}$ we finally arrive at an expression for the specific rate of work

$$\dot{w} = \frac{1}{2\rho_o}\, s^{ik}\dot{g}_{ik}\ , \tag{108}$$

which on account of (90), (97) and (100) would be obvious for the user of convected coordinates.

In the same way,

$$\underline{\tau}_{(-2)} = \underline{F}^T\underline{s}\underline{F} = \underline{F}^T(s_{ik}\ \underline{g}^i \otimes \underline{g}^k)\underline{F} = s_{ik}\ \overset{oi}{\underline{g}} \otimes \overset{ok}{\underline{g}}\ , \tag{109}$$

and the stress conjugate to Almansi strain has covariant components in $\overset{oi}{\underline{g}}$ equal to the covariant components of Kirchhoff stress in $\underline{g}^i$. With the pair (102) the specific rate of work is simply

$$\dot{w} = \frac{1}{2\rho_o}\, s_{ik}\dot{g}^{ik}\ . \tag{110}$$

For the stress conjugate to logarithmic strain the interpretation in convected coordinates seems to be more difficult. However, taking into account that this pair is conjugate under arbitrary velocity fields only if the principal axes of stress coincide with those of the left stretch $\underline{V}$, it follows that $\underline{s}$ commutes with $\underline{V}$,

$$\underline{s}\underline{V} = \underline{V}\underline{s}\ . \tag{111}$$

We therefore have the identity

$$\underline{s} = \underline{V}^{-1}\underline{s}\underline{V}\ , \tag{112}$$

and observing $\underline{R} = \underline{V}^{-1}\underline{F}$ and $\underline{V}^{-2} = (\underline{F}\underline{F}^T)^{-1}$ it follows

$$\underline{\tau}_{(0)} = \underline{R}^T\underline{s}\underline{R} = \underline{R}^T\underline{V}^{-1}\underline{s}\underline{V}\underline{R} = (\underline{F}^T\underline{V}^{-1})\underline{V}^{-1}\underline{s}\underline{V}(\underline{V}^{-1}\underline{F}) = \underline{F}^{-1}\underline{s}\underline{F}\ , \tag{113}$$

and after substitution of

$$\underline{s} = s^i_k\ \underline{g}_i \otimes \underline{g}^k\ , \tag{114}$$

$$\underline{\tau}_{(0)} = s^i_k\ \underline{F}^{-1}\underline{g}_i \otimes \underline{F}^T\underline{g}^k = s^i_k\ \overset{o}{\underline{g}}_i \otimes \overset{o}{\underline{g}}^k . \tag{115}$$

The stress tensor $\underline{\tau}_{(0)}$, conjugate to logarithmic strain and the Kirchhoff stress have the same mixed components in the initial and actual basis, respectively. The result (113) seems to contradict the symmetry of $\underline{\tau}_{(0)}$, what on account of (111) is, however, not the case and we may alternatively write

$$\underline{\tau}_{(0)} = \underline{R}^T\underline{s}\underline{R} = \underline{F}^{-1}\underline{s}\underline{F} = \underline{F}^T\underline{s}\underline{F}^{-T}\ . \tag{116}$$

For the three strain measures considered, the conjugate stresses have particular simple interpretations, their components in the fixed reference basis beeing just the components of Kirchhoff stress on the convected actual basis. In the general case the interpretation of conjugate stress in terms of the mixed components $s^i_k$ and strain is also possible. What is happening to the body is properly described by these components, the metric $g_{ik}$ of the body and its rate of change $\dot{g}_{ik}$.

Finally we mention that the unsymmetrical Piola-Kirchhoff stress $\overset{o}{\underline{S}}$, obtained from (101 a) by multiplication with $\underline{F}$ from the left,

$$\overset{o}{\underline{S}} = \underline{s}\underline{F}^{-T} = \frac{\rho_o}{\rho}\ \sigma\underline{F}^{-1} = \det(\underline{F})\underline{\sigma}\ \underline{F}^{-1} \tag{117}$$

has some importance, since it allows to calculate the force $\underline{t}$ da, acting on a plane element in the actual configuration,

$$\underline{t}\,da = \overset{o}{\underline{S}}\,\overset{o}{\underline{n}}\,d\overset{o}{a} = \overset{o}{\underline{t}}\,d\overset{o}{a} \tag{118}$$

from the normal $\overset{o}{\underline{n}}$ and area $d\overset{o}{a}$ of this element in the reference configuration. Hence, the nominal stress vector $\overset{o}{\underline{t}}$ - "force referred to the initial area" - is obtained as

$$\overset{o}{\underline{t}} = \overset{o}{\underline{S}}\,\overset{o}{\underline{n}}\ . \tag{119}$$

Some authors prefer to write this as $\overset{o}{\underline{t}} = \underline{S}^T\,\overset{o}{\underline{n}}$, with the usual interpretation of stress indices in engineering, and call $\underline{S} = \overset{o}{\underline{S}}{}^T$ nominal stress tensor. Obviously, $\overset{o}{\underline{S}}$ and $\underline{F}$ are conjugate, however, $\underline{F}$ is not a suitable measure of strain.

## 6.4 Decomposition of rate of deformation and mechanical work

A typical feature of an elastic plastic material is its different behaviour in loading and unloading. The plastic part of deformation of a material element, which has been deformed into the plastic range, would be revealed in an elastic unloading process, where the stresses are removed and the temperature is brought back to its reference value. We may therefore associate with each configuration (B at time t) of a material element, assumed in the course of its motion, a virtual unloaded configuration $(B^*, t)$ from which its actual configuration would be obtained in a purely elastic transition. Such an intermediate configuration proves to be useful in a description of the material behaviour through internal variables, in which the past history of observable variables like strain and temperature is condensed. The internal variables may also be viewed as thermodynamic state variables, which describe on the macrolevel some average of microstructural rearrangements, due to plastic deformation. It is obvious, that the precise definition of suitable (hidden and observable) variables for the description of the elastic plastic material behaviour is part of the constitutive problem as a whole. The following discussion of kinematic quantities, on kinematic grounds alone, must therefore be considered as preliminary. What is attributed as "elastic" or "plastic" part can find its justification only within the framework of the complete theory.

We will use $^*$ as a descriptive mark for quantities related to the intermediate configuration $\overset{*}{B}$, e.g. $\overset{*}{\underline{g}}_i$ is the basis vector to the material coordinate line $\xi^i$, $d\overset{*}{\underline{x}}$ a vector connecting two particles in $\overset{*}{B}$.

If $\underline{F}_p$ denotes the linear mapping from $\overset{o}{B}$ to $\overset{*}{B}$, $\underline{F}_e$ the mapping from $\overset{*}{B}$ to B, we have

$$d\overset{*}{\underline{x}} = \underline{F}_p \, d\overset{o}{\underline{x}}, \quad d\underline{x} = \underline{F}_e \, d\overset{*}{\underline{x}} = \underline{F}_e \underline{F}_p \, d\overset{o}{\underline{x}} , \tag{120}$$

and the deformation gradient is multiplicatively decomposed in the form

$$\underline{F} = \underline{F}_e \underline{F}_p . \tag{121}$$

This decomposition is not unique. In fact, with $\overset{*}{\underline{Q}}$ orthogonal,

$$\underline{F}'_e = \underline{F}_e \overset{*}{\underline{Q}}{}^T, \quad \underline{F}'_p = \overset{*}{\underline{Q}} \underline{F}_p , \quad \underline{F} = \underline{F}'_e \underline{F}'_p \tag{122}$$

is another decomposition with the same principal stretches, corresponding to an intermediate configuration rotated by $\overset{*}{\underline{Q}}$.

The concept of an intermediate configuration implies that elastic strain is measured by comparison of the actual and the intermediate configuration. If $B^o$ is replaced by $\overset{*}{B}$, two families of strain measures follow in the same way as (43) and (44),

$$\overset{*}{\underline{\varepsilon}}_e = \underline{f}(\underline{U}_e) = f^i_k(q_{es}{}^r)\overset{*}{\underline{g}}_i \otimes \overset{*k}{\underline{g}} \quad \text{and} \quad \underline{\varepsilon}_e = \underline{f}(\underline{V}_e) = f^i_k(q_{es}{}^r)\underline{g}_i \otimes \underline{g}^k, \tag{123}$$

where

$$\underline{U}_e^2 = \underline{F}_e^T \underline{F}_e \qquad \text{and} \quad \underline{V}_e^2 = \underline{F}_e \underline{F}_e^T , \tag{124}$$

$$\underset{(e)}{q}{}^i{}_k = \overset{*ir}{g} g_{rk} , \tag{125}$$

$$\underline{g}_i = \underline{F}_e \overset{*}{\underline{g}}_i . \tag{126}$$

If the intermediate configuration is rotated by $\overset{*}{\underline{Q}}$, then

$$\overset{*}{\underline{\varepsilon}}{}'_e = \overset{*}{\underline{Q}}\overset{*}{\underline{\varepsilon}}_e\overset{*T}{\underline{Q}} \qquad \text{but} \quad \underline{\varepsilon}'_e = \underline{\varepsilon}_e . \tag{127}$$

The rate of change of $d\overset{*}{\underline{x}}$ is from (120)

$$(d\overset{*}{\underline{x}})^{\cdot} = \dot{\underline{F}}_p \, d\overset{o}{\underline{x}} = \dot{\underline{F}}_p \underline{F}_p^{-1} \, d\overset{*}{\underline{x}} = \overset{*}{\underline{L}} d\overset{*}{\underline{x}} , \tag{128}$$

and

$$\overset{*}{\underline{D}} = \frac{1}{2}(\overset{*}{\underline{L}} + \overset{*T}{\underline{L}}) = \{\overset{*}{\underline{L}}\} \text{ \#)} \tag{129}$$

$$\overset{*}{\underline{W}} = \frac{1}{2}(\overset{*}{\underline{L}} - \overset{*T}{\underline{L}}) \tag{130}$$

are rate of deformation and spin in $\overset{*}{B}$, respectively. A material line element changes its length in $\overset{*}{B}$ according to

$$(d\overset{*}{s}{}^2)^{\cdot} = (d\overset{*}{\underline{x}} \cdot d\overset{*}{\underline{x}})^{\cdot} = d\overset{*}{\underline{x}} . 2\overset{*}{\underline{D}} d\overset{*}{\underline{x}} . \tag{131}$$

With (129) and (130) the gradient of the velocity field is expressed as

---

#) $\{\underline{A}\}$ denotes the symmetric part of tensor $\underline{A}$ in the following

$$L = F^{\bullet}F^{-1} = (F_e\,F_p)^{\bullet}\,(F_e\,F_p)^{-1} = F_e^{\bullet}\,F_e^{-1} + F_e\,F_p^{\bullet}\,F_p^{-1}\,F_e^{-1}$$

$$= F_e^{\bullet}\,F_e^{-1} + F_e\,(D^{*} + W^{*})\,F_e^{-1}\,, \tag{132}$$

and its symmetric part, the total rate of deformation, reads

$$D = \{F_e^{\bullet}\,F_e^{-1}\} + \{F_e(D^{*} + W^{*})F_e^{-1}\}\,. \tag{133}$$

If the first and second member are tentatively considered as elastic and plastic part of the total rate of deformation, i.e.

$$D_e^{(1)} = \{F_e^{\bullet}\,F_e^{-1}\} \tag{134}$$

$$D_p^{(1)} = \{F_e\,L^{*}\,F_e^{-1}\} = \{F_e\,W^{*}\,F_e^{-1}\} + \{F_e\,D^{*}\,F_e^{-1}\} \tag{135}$$

then also $W^{*}$ has to be specified through the constitutive law, in order to make the decomposition (121) unique. In fact, if arbitrary rotations of the intermediate configuration are admitted, then the three parts of $L$ are changed into

$$(F_e^{\bullet}\,F_e^{-1})' = F_e^{\bullet}\,F_e^{-1} + F_e\,Q^{*T\bullet}\,Q^{*}\,F_e^{-1} \tag{136}$$

$$(F_e\,W^{*}\,F_e^{-1})' = F_e\,W^{*}\,F_e^{-1} + F_e\,Q^{*\bullet}\,Q^{*T}\,F_e^{-1} \tag{137}$$

$$(F_e\,D^{*}\,F_e^{-1})' = F_e\,D^{*}\,F_e^{-1}\,, \tag{138}$$

and the relative angular velocity, given by $Q^{*\bullet}Q^{*T}$, of one intermediate configuration with respect to another, enters the definition of elastic and plastic rate of deformation (134)-(135). This is avoided, when

$$D_p^{(2)} = \{F_e\,D^{*}\,F_e^{-1}\} \tag{139}$$

is considered as the plastic part of the total rate of deformation in accordance with the fact that the rate of length change of material line elements in $B^*$ is measured by $\underline{D}^*$. Since $\underline{D}_p^{(2)}$ and $\underline{D}$ are unaffected by the spin of the intermediate configuration, so is

$$\underline{D}_e^{(2)} = \underline{D} - \underline{D}_p^{(2)} = \{\dot{\underline{F}}_e \underline{F}_e^{-1} + \underline{F}_e \underline{W}^* \underline{F}_e^{-1}\} , \tag{140}$$

what is also obvious from (136)-(138), taking into account $(\underline{Q}^* \underline{Q}^{*T})^\cdot = 0$.

With a look upon the specific rate of mechanical work,

$$\rho\dot{w} = \underline{\sigma}.\underline{L} = \underline{\sigma}.(\dot{\underline{F}}_e \underline{F}_e^{-1}) + \underline{\sigma}.(\underline{F}_e \underline{W}^* \underline{F}_e^{-1}) + \underline{\sigma}.(\underline{F}_e \underline{D}^* \underline{F}_e^{-1}) , \tag{141}$$

it is in fact not necessary to assume symmetric parts in (139). As has been recently suggested by LEHMANN (1982), the measures

$$\underline{D}_p = \underline{F}_e^{-T} \underline{D}^* \underline{F}_e^{T} \tag{142}$$

$$\underline{D}_e = \underline{D} - \underline{D}_p \tag{143}$$

allow the decomposition of the work rate into elastic and inelastic parts

$$\rho\dot{w} = \underline{\sigma}.\underline{D} = \underline{\sigma}.\underline{D}_e + \underline{\sigma}.\underline{D}_p = \dot{w}_e + \dot{w}_p , \tag{144}$$

since their skew parts do not contribute in the inner product with the symmetric Cauchy stress tensor. With the same argument $\underline{F}_e \underline{D}^* \underline{F}_e^{-1}$ as plastic part is equally well possible. We took (142) in order to obtain the component forms as in LEHMANN (1982).

From

$$2\underline{D}^* = \overset{*\cdot}{g}_{ik}\, \overset{*}{\underline{g}}{}^{i} \otimes \overset{*}{\underline{g}}{}^{k} = \overset{*\cdot}{g}_{il}\, \overset{*}{g}{}^{lk}\, \overset{*}{\underline{g}}{}^{i} \otimes \overset{*}{\underline{g}}_{k} \tag{145}$$

we obtain the representations of $\underline{D}_p$ and $\underline{D}_e$ in the basis $\underline{g}_i$ of embedded coordinates in the actual configuration as follows.

$$2\underline{D}_p = 2\underline{F}_e^{-T} \underline{D}^* \underline{F}_e^{T} = \overset{*\cdot}{g}_{il}\, \overset{*}{g}{}^{lk} (\underline{F}_e^{-T} \overset{*}{\underline{g}}{}^{i} \otimes \underline{F}_e \overset{*}{\underline{g}}_{k}) = \overset{*\cdot}{g}_{il}\, \overset{*}{g}{}^{lk}\, \underline{g}^{i} \otimes \underline{g}_{k} =$$

$$= g^{ir} \dot{\overset{*}{g}}_{rl} \overset{*}{g}^{lm} g_{mk} \mathbf{g}_i \otimes \mathbf{g}^k = - g^{ir} \overset{*}{g}_{rl} \dot{\overset{*}{g}}^{lm} g_{mk} \mathbf{g}_i \otimes \mathbf{g}^k , \tag{146}$$

observing (126) and the relations analogous to (34)-(36). Now,

$$\begin{aligned} \mathbf{D}_e = \mathbf{D} - \mathbf{D}_p &= \frac{1}{2} \left( g^{ir} \dot{g}_{rk} + g^{ir} \overset{*}{g}_{rl} \dot{\overset{*}{g}}^{lm} g_{mk} \right) \mathbf{g}_i \otimes \mathbf{g}^k \\ &= \frac{1}{2} g^{ir} \overset{*}{g}_{rl} \left( \overset{*}{g}^{lm} \dot{g}_{mk} + \dot{\overset{*}{g}}^{lm} g_{mk} \right) \mathbf{g}_i \otimes \mathbf{g}^k \\ &= \frac{1}{2} g^{ir} \overset{*}{g}_{rl} \left( \overset{*}{g}^{lm} g_{mk} \right)^{\cdot} \mathbf{g}_i \otimes \mathbf{g}^k = \underset{(e)}{d}{}^i{}_k \, \mathbf{g}_i \otimes \mathbf{g}^k . \end{aligned} \tag{147}$$

From (125)

$$\underset{(e)}{q}{}^i{}_k = \overset{*}{g}^{ir} g_{rk} , \qquad \left( \underset{(e)}{q}^{-1} \right)^i{}_k = g^{ir} \overset{*}{g}_{rk} , \tag{148}$$

hence

$$\underset{(e)}{d}{}^i{}_k = \frac{1}{2} \left( \underset{(e)}{q}^{-1} \right)^i{}_r \left( \underset{(e)}{\dot{q}} \right)^r{}_k . \tag{149}$$

These component forms may be established directly within the framework of a description in convected coordinates. Concerning this point and further details, we refer to the work of Th. Lehmann and his lectures.

## References

ERINGEN, A.C., Nonlinear theory of continuous media, McGraw-Hill, New York 1962

GREEN, A.E. and W.ZERNA, Theoretical elasticity, Oxford University Press New York 1954

HILL, R., On constitutive inequalities for simple materials. J.Mech. Physics Solids, 16(1968), pp. 229-242

LEHMANN, Th., Some remarks on the decomposition of deformations and mechanical work. Intern.J. Engng. Sci. 20(1982), pp. 281-288

MACVEAN, B., Die Elementararbeit in einem Kontinuum und die Zuordnung von Spannungs- und Verzerrungstensoren. Z. angew. Math. Phys. (ZAMP) 19(1968), pp. 157-185

RICE, J.R., Continuum mechanics and thermodynamics of plasticity in relation to microscale deformation mechanism. In: Constitutive Equations in Plasticity, ed. A.S. Argon, MIT press (1975)

SEDOV, L.I., Introduction to the mechanics of a continuous medium, Addison-Wesley, Reading 1965

TRUESDELL, C. and W. NOLL, The non-linear field theories of mechanics, in: Encyclopedia of Physics, Vol. III/3, ed. Flügge, Springer Berlin (1965)

HILL, R., Aspects of invariance in solid mechanics. In: Advances in Applied Mechanics, Vol.18, Acad. Press, New York 1978, pp. 1-75 *)

*) Not cited in the text, since I was not aware of this highly recommended reference when writing these notes

# WELL POSEDNESS OF CONSTITUTIVE EQUATIONS OF THE KINEMATICAL HARDENING TYPE

G. DEL PIERO
Istituto di Meccanica Teorica e Applicata
Università degli Studi di Udine

## PART ONE - THE KINEMATICAL HARDENING MODEL

### 1 - Introduction

The purpose of the first part of these Lectures is to characterize the class of elastic-plastic materials starting from the more general class of the *materials with elastic range* and introducing a small number of ad hoc assumptions. This is done in the general framework of NOLL's *new theory of simple materials* [6] After some preliminary work intended to re-define in this new context the class of materials with elastic range, a supplementary assumption on the nature of the inelastic behaviour defines a subclass of materials with elastic range of the rate type, which exhibits a number of material properties considered as typical of elastic-plastic materials.

Elastic-plastic materials are obtained by the final assumption that any state of the material be completely determined by actual strain and actual stress; under this assumption, the material admits a differential characterization in terms of stresses, whose basic elements are the yield surface, the flow rule and the hardening rule.

For the same material, an alternative characterization in terms of strains, founded on the concept of plastic deformation, is discussed. It comes out that, while the former characterization is appropriate for describing the most elementary hardening schemes, such as isotropic hardening, kinematical hardening or perfect plasticity, the latter is more convenient when more complicated hardening schemes are involved.

## 2 - An outline of the classical theory

Before introducing the new theory of simple materials, it is convenient to recall some fundamental concepts of the classical theory [1]. Consider a continuous body, and let any particle of it be characterized by the point x occupied in a fixed reference configuration. A motion $\hat{y}$ is a function which to any particle x and to any time t associates the point

$$y = \hat{y}(x,t)$$

occupied by x at the time t. The past history of $\hat{y}$ at the time t is the function $\hat{y}^t$ which to any x and to any non-negative real s associates the point occupied by x at the time t-s:

$$\hat{y}^t(x,s) = \hat{y}(x,t-s) \quad , \quad 0 \leqq s < +\infty .$$

Let $T = \hat{T}(x,t)$ be the (Cauchy) stress at x at the time t. The starting point of the classical theory is the principle of determinism for the stress, which states that T is determined by the past history of the motion at the time t. The functional relation which to any $\hat{y}^t$ associates the corresponding stress is called a constitutive equation. A constitutive equation depends on the nature of the material, or rather characterizes a particular material. Any constitutive equation is subject to two main general restrictions:

(i) the principle of local action: let I be a neighborhood of x, and let $\hat{y}^t$ and $\hat{y}'^t$ be two past histories whose restrictions to $I\times[0,\infty)$ are the same. Then the stresses at (x,t) determined by the two past histories are the same.

(ii) the principle of material frame-indifference: the constitutive equations are indifferent under changes in observer preserving distances and time intervals.

The first principle leads to the following local form of the general constitutive equation

$$\hat{T}(x,t) = \mathscr{F}_x(\hat{y}^t_I) \quad ,$$

(1) For precise definitions, notation and terminology we refer to TRUESDELL & NOLL [2].

where $\mathscr{F}_x$ is the response functional of the material at x, and $\hat{y}^t_I$ is the restriction of $\hat{y}^t$ to $I\times[0,\infty)$. The principle of material frame-indifference imposes to all response functionals the restriction

$$Q(0)\,\mathscr{F}_x\left(\hat{y}^t_I\right)\,(\hat{Q}(0))^T = \mathscr{F}_x\left(\hat{Q}\,\hat{y}^t_I\right),$$

where $\hat{Q}$ is any smooth one-parameter family of orthogonal tensors

$$Q(s)\quad,\quad 0\leqq s<+\infty.$$

The simplest dependence of $\hat{T}(x,t)$ upon $\hat{y}^t_I$ is the dependence upon the histories of the successive gradients of $\hat{y}$ with respect to x, evaluated at (x,t). Accordingly, a material of grade N is a material depending upon the history of the first N gradients. A material of grade one:

$$\hat{T}(x,t) = \mathscr{F}_x\left(\partial_x\hat{y}^t(x,s)\right)\quad,\quad 0\leqq s<+\infty$$

is also called a simple material. The class of simple materials is the framework in which develop the most important theories of material behaviour.

## 3 - The New Theory of simple materials

A theory of material behaviour founded on the concept of constitutive equation presents a severe defect: the assumption that the stress be determined by the whole past history of the motion is hardly reconcilable with the impossibility of knowing, even in principle, events occurred in the infinite past.

The answer of the classical theory to this objection is a further general restriction on the constitutive equation. This restriction, the principle of fading memory, assumes that the response functional be affected only by a limited part of the past history of the motion.

Unfortunately, this does not agree with some universally accepted models of material behaviour. Consider, for instance, those materials whose response functional is indifferent to time re-scaling, the so-called rate-independent materials; because time re-scaling can transform recent events into remote ones and vice-versa, the principle of fading memory does not apply to such materials [1].

Originated from the declared purpose of eliminating the drawbacks of the classical theory, the new theory resulted in a deep upheaval of some well-established views on material behaviour, involving the concept of constitutive equation itself. Basic in the construction of the new theory is the primitive concept of the state of a material element, corresponding to the intuitive idea that "...... when a material element is given in a concrete physical situation, it is given in a definite state. The state determines everything about the element: its configuration, its stress, and, most importantly, the response of the element in every possible test" [2].

Let $\Sigma$ be the set of all states which are admissible for a given material element. We assume that $\Sigma$ is a subset of a finite-dimensional space, called the state space. Let, further, $\hat{G}$ and $\hat{T}$ be the maps which to any state $\sigma$ associate the deformation G and the stress T determined by $\sigma$:

$$G = \hat{G}(\sigma) \quad , \quad T = \hat{T}(\sigma). \tag{3.1}$$

The sets $\hat{G}(\Sigma)$ and $\hat{T}(\Sigma)$ are the sets of all deformations and of all stresses admissible for the given material element. They are assumed to be subsets of two finite-dimensional spaces, called the deformation space and the stress space, respectively.

Consider a state $\sigma_0$ and a deformation process P, defined as a piecewise smooth one-parameter family in $\hat{G}(\Sigma)$. We say that $\sigma_0$ and P are compatible with each other if P starts from the deformation $\hat{G}(\sigma_0)$ determined by $\sigma_0$.

We are ready to make one of the fundamental assumptions of the new theory. We assume that for any material element there exists a function $\hat{\rho}$, called the evolution function, which to any compatible pair $(\sigma_0, P)$ associates the state

$$\sigma = \hat{\rho}(\sigma_0, P). \tag{3.2}$$

---

(1) This is the main reason of the difficulties encountered in fitting to the framework of the classical theory the theories of Plasticity. On the contrary, they fit in a quite natural way to the new theory (see e.g. PERZYNA [7], DEL PIERO [8] ).

(2) NOLL [6], Sect. 7.

$\sigma$ is the state reached by the material element when subject to the deformation process P starting from the initial state $\sigma_0$. Roughly, we can say that the evolution function is the counterpart of the con stitutive equation of the classical theory. The fact that $\sigma$, and the refore the stress $T = \hat{T}(\sigma)$, is determined by a process of finite du= ration rather than by an infinite past history seems to be very rea= sonable from a physical viewpoint, and rules out any need of conside ring fading memory as a general material property.

This essential feature of the new theory will be used in the following Sections as a starting point for a rational definition of elastic-plastic materials.

## 4 - Materials with elastic range

It is a common point of view of many theories of Plasticity that an elastic-plastic material is essentially a material endowed with elastic regions. Roughly, an elastic region is a region in the deformation space in which the material behaves as if it were ela= stic; in the classical theory of simple materials, elastic regions are associated with past histories of motion (1). Even if the ela= stic region associated with a given history is not necessarily uni= que, it is usually assumed that there exists a maximal elastic re= gion containing as subsets all other elastic regions associated with the same past history. This maximal region is called the <u>elastic range</u> of the given past history. It is also generally assumed that the elastic range does not change under elastic continuations, i.e., under continuations contained in the elastic range itself (2).

---

(1) This is, for example, the point of view developed in PIPKIN and RIVLIN's theory of rate-independent materials [3]; the same ideas were used later by OWEN [5] for constructing a more gene= ral theory of materials with elastic range.

(2) The first assumption is made by OWEN, and both assumptions are made by PIPKIN and RIVLIN. A noticeable progress is made in a recent paper by MATSUMOTO [12] who, after defining a (maximal) elastic range in terms of the values taken by a yield functio= nal, is able to prove that this rules out any possibility of change of the elastic range during elastic continuations.

Here we consider materials with elastic range in the new theory of simple materials. In the new theory, an elastic material is a material for which states are identified with deformations. Thus, it seems to be reasonable to call elastic region a region of the state space equipped with a one-one correspondence with its image under $\hat{G}$ in the deformation space.

Definition. Let $\sigma$ be any member of the set $\Sigma$ of admissible states. We say that a subset E of $\Sigma$ is an elastic region for $\sigma$ if it enjoys the following three properties:

(P1) E contains $\sigma$;

(P2) The restriction of the mapping $\hat{G}$ to E is invertible;

(P3) For any deformation process P compatible with $\sigma$ and with values in $\hat{G}(E)$, the final state $\hat{\rho}(\sigma,P)$ belongs to E.

Denote by g the inverse of the restriction of $\hat{G}$ to E; g is the map from $\hat{G}(E)$ onto E which to any G in $\hat{G}(E)$ associates the (unique) state in E whose deformation is G.

Consider any deformation process P compatible with $\sigma$ and with values in $\hat{G}(E)$. Then, the final value G of P belongs to $\hat{G}(E)$ and g(G) is the only state in E whose deformation is G. But G is the deformation associated to the state $\hat{\rho}(\sigma,P)$ and, by (P3), $\hat{\rho}(\sigma,P)$ belongs to E. Therefore, $\hat{\rho}(\sigma,P)$ is the same state as g(G):

$$\hat{\rho}(\sigma,P) = g(G). \tag{4.1}$$

What comes out is that the state reached starting from $\sigma$ via the deformation process P depends on P only through the final value G. In other words, the evolution function is path-independent for all deformation processes contained in $\hat{G}(E)$; we can say that for such processes the material behaves as if it were elastic.

This gives a sound physical motivation to the definition of elastic region given above. Moreover, it can be seen that two very desirable properties are implicit in that definition: the maximality of the elastic range, and its invariance under elastic continuations. Indeed, it is not difficult to prove, starting from (P1), (P2) and (P3), that [(1)]:

(1) Considering as "elastic continuations" the processes starting from $\hat{G}(\sigma)$ and contained in $\hat{G}(E)$, (P5) is equivalent to saying that the elastic range does not change under elastic continuations. For the proof of (P4) and (P5) see DEL PIERO [13].

(P4) Any state admits a maximal elastic region (elastic range);

(P5) The elastic range is the same for all states contained in the same elastic range.

In the following, we denote by $E_\sigma$ the elastic range at $\sigma$, by $\hat{G}_\sigma$ the restriction of $\hat{G}$ to $E_\sigma$ and by $g_\sigma$ its inverse, which will be called the elastic response function at $E_\sigma$. In these notations, the property (P5) can be written in the form

$$\tau \in E_\sigma \Rightarrow E_\tau = E_\sigma .$$

In particular, this implies

$$\tau \in E_\sigma \Rightarrow \sigma \in E_\tau \quad ; \quad \tau \in E_\sigma \text{ and } \pi \in E_\tau \Rightarrow \pi \in E_\sigma ,$$

i.e., the symmetry and transitivity of the relation $\tau \in E_\sigma$. As this relation is also reflexive because of (P1), we conclude that it is an equivalence relation.

Thus, the property (P5) induces on the set $\Sigma$ a partition whose equivalence classes are the elastic ranges. In other words, any state belongs to one and only one elastic range. The idea developed in the following Sections is that yielding is a transition from one elastic range to another one.

## 5 - Materials with elastic range of the rate type

In this Section we discuss a special class of materials with elastic range, obtained by making three supplementary assumptions. The first one concerns the topological nature of the elastic ranges:

(A1) Each elastic range $E_\sigma$ is a closed connected region of $\Sigma$ endowed with a non-empty set of interior points, and such that every boundary point of $E_\sigma$ is also a boundary point of $\Sigma$ (1). Moreover, $\hat{G}$ maps the interior of $E_\sigma$ onto the interior, and the boundary onto the boundary, of $\hat{G}(E_\sigma)$.

(1) The latter requirement is not strictly necessary for the present theory. It complies with the physical idea that yielding occurs at the boundary of the set of admissible states. This occurs, for example, for all situations depicted in Fig. 2 in the next Section.

Let $\sigma$ be an admissible state and let P be a deformation process compatible with $\sigma$. We say that P determines an <u>elastic initial response</u> if there exists a positive real a such that all values taken by P in the time interval $[0,a]$ belong to $\hat{G}(E_\sigma)$. If, on the contrary, all values taken in some semi-open[1] interval $(0,a]$ do not belong to $G(E_\sigma)$, we say that P determines an <u>anelastic initial response</u>. As a consequence of (A1), if $\sigma$ is an interior point of $E_\sigma$ then $\hat{G}(\sigma)$ is an interior point of $\hat{G}(E_\sigma)$, so that any deformation process compatible with $\sigma$ determines an elastic initial response. On the contrary, if $\sigma$ is a boundary point, both elastic and anelastic initial responses are possible. More precisely, let N be the exterior unit normal to $\hat{G}(E_\sigma)$ at $\hat{G}(\sigma)$, and let $\dot{G}$ be the time derivative of G(t) at t=0. Then we have elastic initial response if

$$(5.1) \qquad \dot{G}\cdot N < 0 ,$$

because this condition ensures that P takes values in $\hat{G}(E_\sigma)$ for some finite time interval. Similarly, if the inner product $\dot{G}\cdot N$ is positive, anelastic initial response occurs. More complicated is the case $\dot{G}\cdot N=0$, in which the initial response can be elastic, anelastic, or neither [2].

Denote by G(t) the value of P at time t, and by $\sigma(t)$ the state reached at time t starting from $\sigma$ with P. The following characterization of elastic response follows readily from the definition of elastic range:

$$(5.2) \qquad \sigma(t) = g_\sigma(G(t)) \quad .$$

Our second assumption is a regularity hypothesis for the elastic response function $g_\sigma$:

(A2) The function $g_\sigma$ is continuously differentiable at any interior point of $\hat{G}(E_\sigma)$, and its gradient admits a continuous extension to the boundary of $\hat{G}(E_\sigma)$.

---

(1) Because the property (P1) of the elastic ranges, the initial value $\hat{G}(\sigma)$ of P always belongs to $\hat{G}(E_\sigma)$.

(2) It may well happen that there exists no time interval $(0,a]$ in which P takes all values either inside or outside $\hat{G}(E_\sigma)$. As discussed in [8], this is essentially the origin of the difficulties met in proving the existence and uniqueness of solutions for the differential equation (5.5) below.

This assumption renders possible a differential description of elastic response. Indeed, after denoting by $H(\sigma(t))$ [1] the gradient of $g_\sigma$ at $G(t)$, time differentiation of (5.1) yields

(5.3) $$\dot{\sigma}(t) = H(\sigma(t))\,\dot{G}(t) \quad .$$

The third assumption imposes very severe restrictions on the nature of anelastic response. It induces on the material element the structure of a rate-type material, and the property of rate-independence [2].

(A3) For any compatible pair $(\sigma,P)$, the derivative $\dot{\sigma}(t)$ is determined by $\sigma(t)$ and $\dot{G}(t)$. At anelastic response, the dependence of $\dot{\sigma}(t)$ upon $\dot{G}(t)$ is linear for any fixed $\sigma(t)$.

Denote by $K(\sigma(t))$ the linear function mapping strain rates into state rates at $\sigma(t)$ at anelastic response. Then, the second part of (A3) results in assuming the following differential description of anelastic response:

(5.4) $$\dot{\sigma}(t) = K(\sigma(t))\,\dot{G}(t)$$

We have seen that both elastic and anelastic response can occur when $\dot{G}(t)$ is orthogonal to the normal N. As, by the first part of (A3), $\dot{\sigma}(t)$ is determined by $\dot{G}(t)$, the right-hand sides of (5.3) and (5.4) must coincide for any $\dot{G}(t)$ orthogonal to N. Consider now any $\dot{G}(t)$; after observing that the vector

$$\dot{G}(t) - (\dot{G}(t)\cdot N)N$$

is orthogonal to N, we have that

$$\Big(K(\sigma(t)) - H(\sigma(t))\Big)\Big((\dot{G}(t) - (\dot{G}(t)\cdot N)N\Big) = 0 \quad ,$$

and, therefore,

$$K(\sigma(t))\,\dot{G}(t) = H(\sigma(t))\,\dot{G}(t) + (\dot{G}(t)\cdot N\,\Big(K(\sigma(t)) - H(\sigma(t))\Big)\,N \;.$$

After substituting into (5.4), the differential descriptions of ela=

(1) We adopt the simplified notation $H(\sigma(t))$ instead of the more precise one $H_\sigma(G(t))$, taking advantage of the fact that both $G(t)$ and the elastic range are determined by $\sigma(t)$.

(2) The rate-type structure is explicit in the first part of (A3). Rate-independence can be easily recognized in the differential expression (5.5) below, which is insensitive to time re-scaling.

stic and plastic responses can be synthesized in the single expression [1]

$$(5.5)\qquad \dot{\sigma}(t) = H(\sigma(t))\dot{G}(t) - (\dot{G}(t)\cdot N)^{+}\, m(\sigma(t)) \quad ,$$

where

$$m(\sigma(t)) = \Big(H(\sigma(t)) - K(\sigma(t))\Big)N \quad .$$

This expression holds if $\sigma(t)$ is a boundary point of $E_\sigma$. It can be extended to any $\sigma(t)$ if we agree that $m(\sigma(t))$ be zero at interior points of $E_\sigma$.

The material element described so far has some properties typical of elastic plastic materials: rate-independence, elastic unloading [2]. Moreover, as it will be seen in the next Section, it is possible to recognize in it elastic ranges, flow rules, hardening rules. The problem arises, whether the assumptions made so far give a complete and correct characterization of the material element, in the sense that they determine unambiguously the evolution function or, what is the same, the final state corresponding to any compatible pair $(\sigma,P)$.

This occurs if the differential equation (5.5) subject to the initial condition

$$\sigma(0) = \sigma$$

has a unique solution for any P compatible with $\sigma$. The existence proof is non-trivial, because the right-hand side of (5.5) exhibits a discontinuity at any transition from elastic to plastic response or vice-versa. Moreover, this transition can occur infinitely often at any finite time-interval, so that it becomes impossible to construct a solution of (5.5) as a finite sequence of solutions of the equations (5.2) and (5.4) [3].

---

(1) Here the symbol $(\bullet)^{+}$ denotes the positive part of $\dot{G}(t)\cdot N$. Its presence is explained by observing that when $\dot{G}(t)\cdot N$ is positive then (5.5) must reduce to (5.4) and when it is negative then (5.5) must reduce to (5.3).

(2) Elastic unloading is an elastic response starting from a boundary point of $E_\sigma$.

(3) This circumstance was pointed out by DEL PIERO [8], who worked out an existence proof founded on an ad hoc approximation scheme; other approximation schemes were studied by BUHITE [9].

A systematical study of differential equations with discontinuous right-hand side exhibiting this particular type of discontinuity is due to FILIPPOV [1] . His results have been applied to Plasticity by BUHITE and OWEN [10], to give an existence proof for the equation of the one-dimensional elastic-plastic oscillator. The application of FILIPPOV's technique to more complicated differential equations, such as (5.5), is still open.

## 6 - Elastic-plastic materials

We define an elastic-plastic material as a material with elastic range of the rate type obeying the assumptions (A1), (A2), (A3), for which states are identified with deformation-stress pairs:

$$\sigma = (G,T) \quad .$$

For an elastic-plastic material, the state space is the Cartesian product of the deformation space and the stress space. The maps $\hat{G}$, $\hat{T}$ defined by (3.1) become the canonical projections of the state space onto the deformation and stress space respectively, and the sets $\hat{G}(E_\sigma)$ and $\hat{T}(E_\sigma)$ become the corresponding projections of the elastic range $E_\sigma$, as it is roughly represented in Fig. 1.

In the classical theories of Plasticity, the differential equation (5.5) is replaced by its projection in the stress space:

$$(6.1) \qquad \dot{T}(t) = A(\sigma(t))\,\dot{G}(t) - \left(A(\sigma(t))\dot{G}(t)\cdot\bar{N}\right)^{+} M(\sigma(t)) \quad ,$$

where

$$(6.2) \qquad A(\sigma(t)) = \hat{T}\circ H(\sigma(t)) \quad , \quad M(\sigma(t)) = \hat{T}m(\sigma(t)) \quad ,$$

and where

$$(6.3) \qquad \bar{N} = \left(A^{T}(\sigma(t))\right)^{-1} N$$

is the exterior normal to $\hat{T}(E_\sigma)$ at $T(t)$ (1). Yielding occurs when the stress $T(t)$ belongs to the boundary of $\hat{T}(E_\sigma)$ (yield surface) and the strain rate $\dot{G}(t)$ points outwards from $\hat{T}(E_\sigma)$; the vector $M(\sigma(t))$

(1) It follows from the definition of elastic range that A maps vectors tangent to $\hat{G}(E_\sigma)$ into vectors tangent to $\hat{T}(E_\sigma)$. Therefore, $\dot{G}(t)\cdot N = 0$ implies $A\dot{G}(t)\cdot\bar{N} = 0$ , i.e., $A^{T}\bar{N}$ parallel to N. Note that defining $\bar{N}$ as in (6.3) implies that $\bar{N}$ is not, in general, a unit vector.

determines the stress rate $\dot{T}(t)$ both in direction (flow rule) and in magnitude (hardening rule).

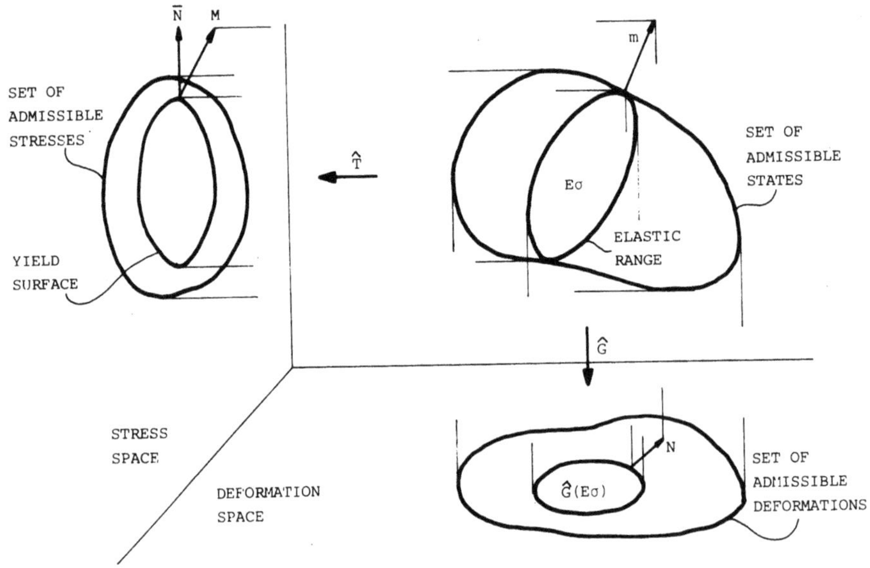


Fig. 1

What is not specified in Eq. (6.1) is how the yield surface changes during yielding. In classical theories, this is done by re= curring to a supplementary assumption: for instance, it is assumed that the yield surface grows homotetically (isotropic hardening), or that it undergoes rigid translations (kinematical hardening). A lar= ger variety of models can be described by recurring to the partition of the state space into elastic ranges introduced in the preceding Sections. Some examples are shown in Fig. 2 where, for simplicity, the stress space and the deformation space are represented as one-di mensional.

Figures 2.1 and 2.2 show two examples in which the sets $\hat{G}(E_\sigma)$ are all nested into each other. Yielding occurs when $\dot{G}\cdot N>0$ , i.e. , in the directions marked by the arrows. In both cases, it co= mes out that it is never possible to go back from a given elastic range $E_\sigma$ to another one whose projection is contained in $\hat{G}(E_\sigma)$. We denote this situation as irreversible yielding.

If, in addition, the yield surfaces $\hat{T}(E_\sigma)$ are also nested as in Fig. 2.1, then the yield surface can only grow in size; we refer to this behaviour as isotropic-like hardening. A kinematical-like hardening is shown in Fig. 2.2, where the elastic ranges exhibit translations accompanied by slight changes in shape.

Figures 2.3 and 2.4 show <u>reversible yielding</u>, i.e. , situati ons in which it is possible to reach any state from any other one. Reversible yielding can be associated to both isotropic-like (Fig. 2.3) and kinematical-like hardening (Fig. 2.4). Finally, Fig. 2.5 shows a model exhibiting a mixed behaviour: reversible yielding in the central zone and irreversible yielding in the two lateral zones.

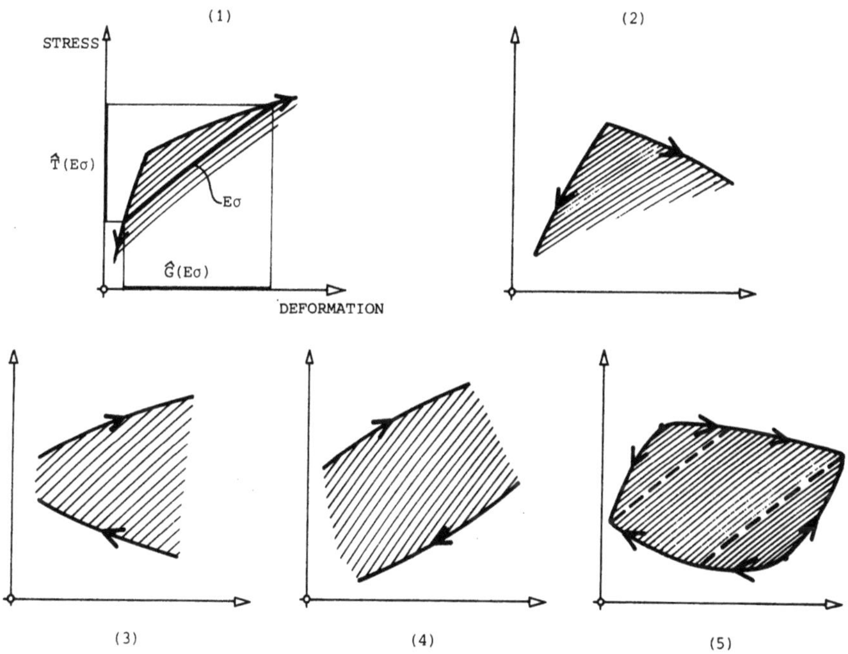


One-dimensional elastic-plastic models:

(1) Irreversible yielding and isotropic-like hardening,
(2) Irreversible yielding and kinematic-like hardening,
(3) Reversible yielding and isotropic-like hardening,
(4) Reversible yielding and kinematic-like hardening,
(5) Mixed.

Fig. 2

## 7 - A description of elastic-plastic materials in terms of strains

In the preceding Section we have established a connection between the definition of an elastic-plastic material given in the framework of the new theory of simple materials, and the description adopted in the classical theories of Plasticity, which is essentially a description in terms of stresses. Here we discuss a description in terms of strains, related to some extent to the classical decomposition of strain into an elastic and a plastic part.

Consider an elastic range $E_\sigma$, and let $g_\sigma$ be the elastic response function at $E_\sigma$. To any deformation G in $\hat{G}(E_\sigma)$, $g_\sigma$ associates the state of $E_\sigma$ whose deformation is G. Thus, any state is determined by its deformation and by its elastic range.

Suppose to associate to any elastic range $E_\sigma$ a deformation $G_\sigma$. If the mapping $E_\sigma \to G_\sigma$ is one-one, then any state is determined by a pair of deformations: the deformation corresponding to the state and the one corresponding to its elastic range. $G_\sigma$ will be called the <u>plastic deformation</u> at $E_\sigma$.

Denote by $\hat{g}(G,G_\sigma)$ the state determined by G and $G_\sigma$. For any fixed $G_\sigma$, the function $\hat{g}(\cdot,G_\sigma)$ reduces to the elastic response function $g_\sigma$ at $E_\sigma$. Therefore, $\hat{g}$ is differentiable with respect to its first argument, and its gradient is the linear mapping H defined in Sect. 5. We assume that $\hat{g}$ be also differentiable with respect to its second argument [1], and denote by L the corresponding gradient.

Consider a compatible state-process pair $(\sigma,P)$, and let $\sigma(t)$, $G(t)$, $G_\sigma(t)$ be the state, the deformation and the plastic deformation at the time t. Then

$$\sigma(t) = \hat{g}(G(t), G_\sigma(t)) \quad ,$$

and, after time differentiation and by the use of the chain rule,

$$\dot{\sigma}(t) = H(\sigma(t))\dot{G}(t) + L(\sigma(t))\, \dot{G}_\sigma(t) \quad .$$

Comparison with (5.5) shows that

$$(7.1) \qquad \dot{G}_\sigma(t) = - (\dot{G}(t)\cdot N)^+ \; L^{-1}(\sigma(t))\; m(\sigma(t)) \quad .$$

---

(1) This assumption involves severe topological restrictions on the set of all admissible deformations.

Assume that, instead of the set of all admissible states partitioned into the elastic ranges, we know the set of all admissible plastic deformations, and that for each of them we know the corresponding region $\hat{G}(E_\sigma)$ of the deformation space and the vector field $L^{-1}m$ defined on the boundary of $\hat{G}(E_\sigma)$. Then, (7.1) determines the change of plastic deformation corresponding to any change in the deformation G. Moreover, if we are able to prove an existence and uniqueness theorem for the solutions of (7.1), then we are able to determine the plastic deformation, and therefore the state, reached at the end of any given deformation process. After observing that all quantities appearing in (7.1) belong to the deformation space, we may conclude that (7.1) supplies a description of the elastic-plastic material element in terms of deformations, equivalent to the description in terms of states introduced in the preceding Sections.

Till now, no physical meaning has been attributed to the plastic deformation. Also, no mention has been made of the decomposition of deformation into an elastic and a plastic part. In fact, this is not essential for the present theory; we introduce here the decomposition of strain only for showing the correlation with the classical theories.

Assume that any deformation G results from the composition of a plastic deformation $G_\sigma$ and an elastic deformation $G_\varepsilon$. Because the operation of composition is not commutative, there are two possible decompositions of G:

$$G = G_\varepsilon G_\sigma \quad , \qquad G = G_\sigma G_\varepsilon \quad . \tag{7.2}$$

The first possibility is the most natural one when the plastic deformation is defined as a permanent deformation, i.e., as the deformation obtained by unloading elastically up to the zero stress. We shall adopt this choice, even if the agreement on it is not general (1).

In infinitesimal Plasticity, the operation of composition is replaced by the sum, and the decomposition of G into an elastic and a plastic part is given unambiguously by

$$G = G_\varepsilon + G_\sigma \quad . \tag{7.3}$$

---

(1) The first decomposition has been introduced by LEE and LIU [4]. The second one has been proposed by NEMAT NASSER [11], who claims that it fits better with incremental calculations because it supplies a more physically meaningful expression of the elastic strain rate.

We have assumed that G and $G_\sigma$ be known for any state of the materi= al; therefore, the expressions (7.2), (7.3) must be considered as al ternative definitions of $G_\varepsilon$.

A different approach is followed in the classical, infinite= simal theory of Plasticity. In it, it is assumed that the stress is not affected by a change in the plastic deformation. Consequently, Eq. (6.1) reduces to

$$\dot{T} = A\dot{G}_\varepsilon \quad , \tag{7.4}$$

and this relation, integrated with respect to time starting from the natural state, determines the actual elastic deformation $G_\varepsilon$. The pla stic deformation comes by difference from (7.3). Again from (7.3) we deduce for (7.4) the equivalent form

$$\dot{T} = A\dot{G} - A\dot{G}_\sigma \quad , \tag{7.5}$$

which, compared with (6.1), gives the following expression of the plastic strain rate:

$$\dot{G}_\sigma = (\dot{G}\cdot N)^+ \, A^{-1} \, M \quad . \tag{7.6}$$

This can be considered as a restriction on the plastic strain rate coming from the assumption that the stress be not affected by a chan ge in plastic deformation. Comparison of (7.6) with the general ex= pression (7.1) of the plastic strain rate shows that this restricti= on results in determining the vector $L^{-1}m$ as a function of A and M.

PART TWO - THE ONE-DIMENSIONAL ELASTIC-PLASTIC OSCILLATOR

## 8 - Introduction

The second part of these Notes is devoted to a short account of the systematic study made by BUHITE & OWEN [10] on the differential equation of the one-dimensional elastic-perfectly plastic oscillator. This study is an important step towards the mathematical settlement of the equations governing the behaviour of elastic-plastic materials, and towards a unification of the theory of shakedown with that of the boundary-value problems of equilibrium and of motion.

## 9 - The differential equation

Consider a mass m retained by an elastic-perfectly plastic spring, governed by the constitutive equation

$$\dot{s} = \begin{cases} \mu\dot{u} & \text{if } |s| < s_0 \text{ or if } |s| = s_0 \text{ and } s\dot{u} < 0 , \\ 0 & \text{if } |s| = s_0 \text{ and } s\dot{u} \geq 0 . \end{cases}$$

Here u is the displacement of the mass, s the tension of the spring, $s_0$ the yield limit, and $\mu$ the elastic modulus of the spring. Under an external excitation f(t), the differential equation of motion is

$$m\ddot{u}(t) = f(t) - s(t) .$$

After setting

$$x_1 = \frac{m\dot{u}}{s_0} , \quad x_2 = \frac{s}{s_0} , \quad h = \frac{f}{s_0} , \quad \beta = \frac{\mu}{m} ,$$

we are led to the sistem of two ordinary differential equations

$$(9.1) \quad \begin{cases} \dot{x}_1(t) = h(t) - x_2(t) , \\ \dot{x}_2(t) = \begin{cases} \beta x_1(t) & \text{if } |x_2(t)| < 1 \\ & \text{or if } |x_2(t)| = 1 \text{ and } x_1(t)x_2(t) < 0, \\ 0 & \text{if } |x_2(t)| = 1 \text{ and } x_1(t)x_2(t) \geq 0 , \end{cases} \end{cases}$$

which, together with the initial conditions

(9.2) $x_1(0) = x_1^\circ \quad , \quad x_2(0) = x_2^\circ \quad ,$

determine the initial-value problem governing the motion of the o= scillator.

The two possibilities appearing in Eq. (9.1.2) are the elastic response and the plastic response, respectively. The transition from one type of response to the other one determines a discontinuity whi= ch, as already discussed in Sect. 5, is the main technical difficulty in the proof of the existence of solutions.

In a geometrical representation, solutions can be represented by trajectories in the $(x_1, x_2)$ plane. Referring to Eq. (9.1.2), we see that plastic response occurs at the half-lines Y,Y' marked in bold-face in Fig. 3, and that elastic response occurs at all remai= ning point of the strip

$$Z = \{(x_1,x_2) \quad : \quad |x_2| \leq 1\} \quad .$$

Transition from elastic to plastic response can occur at any point of Y,Y', whereas that from plastic to elastic response occurs only if the product $x_1x_2$ changes sign, i.e., only when the trajectory re= presenting the solution attains either of the points A,A' in fig. 3.

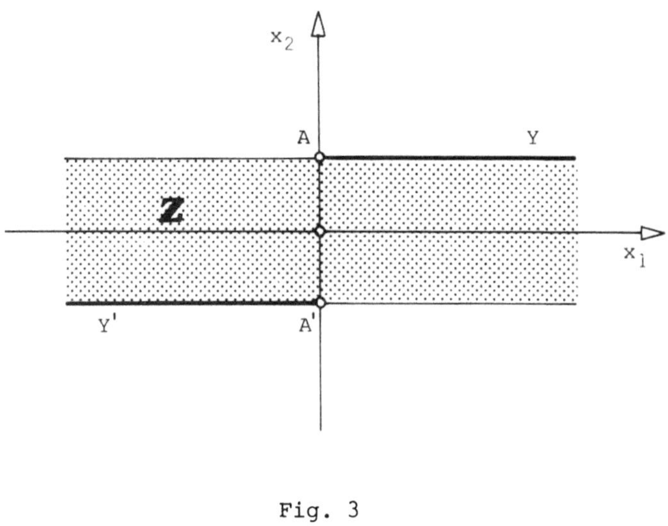


Fig. 3

## 10 - Existence of solutions

Denote by $x(t)$ the vector $(x_1(t),x_2(t))$, by $x^o$ the set $(x_1^o,x_2^o)$ of the initial conditions, and by

$$\phi(x^o,h)$$

the solution, if any, of (9.1) corresponding to the initial conditi= on $x^o$ and to the external force h. In what follows, we take as un= derstood that $x^o$ belongs to Z. A very simple preliminary existence result concerns the case of h constant in time.

Theorem 10.1. Let h be constant in time. Then there exists a soluti= on $\phi(x^o,h)$ for any $x^o$ in Z.

The central idea of the proof is to consider the two initi= al-value problems

$$\begin{cases} \dot{x}_1(t) = h - x_2(t), \\ \dot{x}_2(t) = \beta x_1(t), \\ x(0) = x^o; \end{cases} \qquad \begin{cases} \dot{\bar{x}}_1(t) = h - \bar{x}_2(t), \\ \dot{\bar{x}}_2(t) = 0, \\ \bar{x}(0) = \bar{x}^o, \end{cases}$$

corresponding to a purely elastic and to a purely plastic response, respectively. Here $x^o$ is any point of Z and $\bar{x}^o$ belongs either to Y or to Y'. The trajectory representing the solution of the first pro= blem is the ellipse

$$\beta x_1^2 + (x_2 - h)^2 = \beta x_1^{o\,2} + (x_2^o - h)^2 , \tag{10.1}$$

centered at (0,h), with the diameters proportional to $(\beta^{-1/2},1)$, tra versed counterclockwise. The trajectory corresponding to the second problem is the straight half-line

$$\bar{x}_1 = \bar{x}_1^o + (h - \bar{x}_2^o)t \quad , \quad \bar{x}_2 = \bar{x}_2^o \quad , \quad t \geq 0 \tag{10.2}$$

originating at $\bar{x}^o$ and pointing to a direction depending on h. More precisely, the half-line points towards the $x_2$-axis if $h\bar{x}_2^o < 1$, and to the opposite direction if $h\bar{x}_2^o > 1$.

Assume, for instance, that $x^o$ be an interior point of Z, as in Fig. 4. Then the trajectory starts with the arc of the ellipse (10.1) beginning at $x^o$ and terminating at the intersection $\bar{x}^o$ of the ellipse with either Y or Y', where an elastic-plastic transition oc= curs.

At this point, the ellipse is replaced by the half-line

(10.2); as we have seen before, if $h\overline{x}^{o}_{2} < 1$ then the trajectory crosses either of the points A,A' where a plastic-elastic transition occurs; otherwise, no further transition occurs, and we do have plastic response forever.

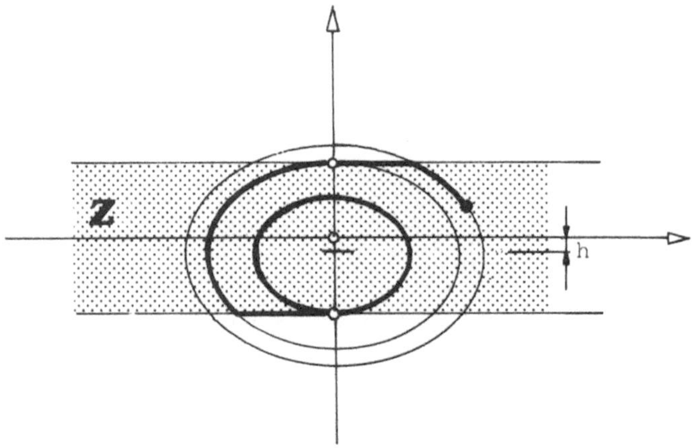


Fig. 4

If plastic-elastic transition occurs as in Fig. 4, we have a new ellipse with the same centre but smaller diameters. By iterating the procedure and after a number of transitions, which is never greater than four, we get an ellipse tangent to either A or A' and entirely contained in Z, corresponding to indefinite elastic response.

By this way, we have constructed a continuous and piecewise differentiable solution $\phi(x^{o},h)$, made of a finite number of pieces of the type (10.1), (10.2). The same technique can be extended without any difficulty from a constant force h to a step function, i.e., to a function taking constant values over each of a finite number of intervals spanning the real positive half-line.

Far from trivial is the extension to more general forces. Indeed, as already pointed out in Sect. 5, it is possible to have an infinite number of elastic-plastic transitions in a finite time interval; this renders vain the hope of constructing the function $\phi(x^{o},h)$ directly, by patching together a finite number of solutions for purely elastic and purely plastic responses.

It seems a very reasonable idea that of approximating h by a sequence $\{h^k\}$ of step functions, and looking whether the correspon=

ding sequence $\{\phi(x^\circ,h^k)\}$ converges to a solution of (9.1). The result which we state here without proof [1] shows that this is a successful idea.

Theorem 10.2 - Let h be $L^1(0,\infty)$, let $\{h^k\}$ be a sequence of step functions converging to h in the norm of $L^1$, and let $\phi(x^\circ,h^k)$ be the solutions of (9.1) for the initial condition $x^\circ$ and the forces $h^k$. Then there exists a time T > 0 such that the restrictions of $\phi(x^\circ,h^k)$ to $[0,T]$ converge uniformly to an absolutely continuous function satisfying (9.1) almost everywhere in $[0,T]$.

## 11 - Uniqueness and Continuous Dependence

In this Section we prove the uniqueness of the solution of the initial-value problem (9.1), (9.2), and the continuous dependence on the data, the continuity being referred to the energy norm

$$\text{(11.1)} \qquad \|x\| = \left(\frac{1}{2}x_1^2 + \frac{1}{2\beta}x_2^2\right)^{1/2} .$$

Let us begin with a preliminary result.

Lemma 11.1 - Let $x = \phi(x^\circ,h)$ and $x' = \phi(x^{\circ\prime},h')$. Then

$$\text{(11.2)} \qquad \|x(t) - x'(t)\|^{\bullet} \le \frac{1}{\sqrt{2}} |h(t) - h'(t)| .$$

This inequality is a consequence of the identity

$$\text{(11.3)} \qquad (\|x(t) - x'(t)\|^2)^{\bullet} = (x_1(t) - x_1'(t))\,(\dot{x}_1(t) - \dot{x}_1'(t)) +$$

$$+ \frac{1}{\beta}(x_2(t) - x_2'(t))\,(\dot{x}_2(t) - \dot{x}_2'(t)) .$$

In it, the derivatives appearing in the right-hand side can be substituted by their expressions (9.1), whose explicit form is different for elastic an plastic response. Assume, for instance, that the response be elastic for x and plastic for x'. Then we have from (9.1) that $\dot{x}_2 = \beta x_1$ and $\dot{x}_2' = 0$. Thus

---

(1) See FILIPPOV [1], BUHITE & OWEN [10].

$$(11.4) \qquad (\| x(t) - x'(t) \|^2)^{\bullet} = (x_1(t) - x_1'(t))\,(h(t) - h'(t)) +$$

$$+ x_1'(t)\,(x_2(t) - x_2'(t)).$$

Again from (9.1) we see that $x_2'$ has absolute value one and has the same sign as $x_1'(t)$. Therefore, not only the product $x_1'(t)x_2'(t)$ is non-negative, but also it is not less than $x_1'(t)x_2$ for any $x_2$ in $[-1,1]$. Consequently,

$$(11.5) \qquad (\| x(t) - x'(t) \|^2)^{\bullet} \leq (x_1(t) - x_1'(t))(h(t) - h'(t)).$$

The same inequality can be proved for any other combination of ela= stic and plastic response. Then, in any case we get

$$2 \| x(t)-x'(t) \| \; \| x(t)-x'(t) \|^{\bullet} \leq \sqrt{2}\, |h(t)-h'(t)| \left(\frac{1}{2}(x_1(t)-x_1'(t))^2\right)^{1/2} \leq$$

$$\leq \sqrt{2}\, |h(t)-h'(t)| \; \| x(t)-x'(t) \| ,$$

and, therefore, (11.2).

We are now in position of proving the theorem of uniqueness and continuous dependence.

Theorem 11.2 - Let $x = \phi(x^{\circ},h)$ and $x' = \phi(x^{\circ\prime},h')$. Then

$$(11.6) \qquad \| x(t)-x'(t) \| \leq \| x^{\circ}-x^{\circ\prime} \| + \frac{1}{\sqrt{2}} \int_0^t |h(s)-h'(s)|\,ds.$$

This inequality, whose proof comes easily from integration of (11.2) over (0,t), establishes a continuous dependence of the "distance" between the solutions $x$ and $x'$ at any time $t$ upon the "distances" between the initial data and between the external forces appearing in right-hand side of (11.6). In particular, the uniqueness theorem follows after observing that $x^{\circ} = x^{\circ\prime}$ and $h = h'$ imply $x = x'$.

## 12 - A stability property related to shakedown theory

Consider two solutions corresponding to two different initial conditions $x^o, x^{o'}$, but to the same external force h. For them, the inequality (11.2) results in the following important stability property: the distance between x and $x'$ is non-increasing in time.

Unfortunately, we have no general a priori condition on the data ensuring that the distance between x and $x'$ vanishes after a finite time [(1)]. This problem can be considered as one of the important open questions in the mathematical theory of Plasticity.

Of particular interest is the case in which one solution, say x, is <u>strictly elastic</u>, in the sense that

$$(12.1) \qquad \varepsilon = \sup_{t \in (0,\infty)} |x_2(t)| < 1 \quad .$$

In this case, it is possible to establish an upper bound[(2)] for the <u>plastic dissipation</u> of x', i.e., for the power expended at plastic response by $x'$ in the time interval (0,t)

$$(12.2) \qquad D(x',t) = \int_{P(t)} x_1'(s)\; x_2'(s)\; ds \quad .$$

Here P(t) represents the subset of (0,t) in which $x'$ is plastic. To establish the upper bound, consider first a time s in P(t). As x is elastic at any time, we are in the situation examined in the proof of Lemma 11.1. Thus, after setting $h = h'$ in (11.4), using (12.1) and the identity

$$(12.3) \qquad x_1'(t)\; x_2'(t) = |x_1'(t)|$$

---

(1) A partial result, restricted to periodical solutions, is given in BUHITE & OWEN [10]

(2) BUHITE & OWEN [10]. In shakedown theory, the existence of an elastic solution for the given external force with suitable initial conditions is considered as a sufficient condition in order $x'$ to become definitively elastic after time.

coming from (9.1) at plastic response, we get

$$(12.4)\qquad (\|x(s)-x'(s)\|^2)^{\bullet} = x_1'(s)(x_2(s)-x_2'(s)) \le$$

$$\le |x_1'(s)|\ (|x_2(s)| - 1) \le -(1-\varepsilon)\ |x_1'(s)| .$$

If s is a time in which x' is elastic, then proceding as in the pro= of of Lemma 11.1 it is easy to prove that

$$(\|x(s) - x'(s)\|^2)^{\bullet} = 0 .$$

Consequently,

$$\int_{P(t)} (\|x(s)-x'(s)\|^2)^{\bullet}\, ds = \int_0^t (\|x(s)-x'(s)\|^2)^{\bullet}\, ds =$$

$$= \|x(t)-x'(t)\|^2 - \|x^o-x^{o\prime}\|^2 .$$

The desired bound for the plastic dissipation follows from integra= tion of (12.4) over P(t), after applying (12.3) to (12.2):

$$D(x',t) = \int_{P(t)} |x_1'(s)|\, ds \le -\frac{1}{1-\varepsilon} \int_{P(t)} (\|x(s)-x'(s)\|^2)^{\bullet}\, ds \le$$

$$\le \frac{1}{1-\varepsilon}\ \|x^o - x^{o\prime}\|^2 .$$

As we see, this bound depends only on the initial data and on the con stant $\varepsilon$ .

## References

[1] FILIPPOV, A.F., Differential Equations with Discontinuous Right-Hand Side (in Russian), Mat. Sbornik 51, 1960. AMS Tran= slations 42, 199-231, 1964.

[2] TRUESDELL, C., & W. NOLL, The Non-Linear Field Theories of Me= chanics, Handbuch der Physik, Vol III/3, Springer 1965.

[3] PIPKIN, A.C., & R.S. RIVLIN, Mechanics of Rate-Independent Ma= terials, Z.A.M.P., 16, 313-327, 1965.

[4] LEE, E.H., & D.T. LIU, Finite-Strain Elastic-Plastic Theory with Application to Plane-Wave Analysis, Jl. of Appl. Physics, 38, 19-27, 1967.

[5] OWEN, D.R., A Mechanical Theory of Materials with Elastic Ran= ge, Arch. Rational Mech. Anal. 37, 85-110, 1970.

[6] NOLL, W., A New Mathematical Theory of Simple Materials, Arch. Rational Mech. Anal. 48, 1-50, 1972.

[7] PERZYNA, P., On material isomorphism in description of dynamic plasticity, Arch. Mech. Stosow. 27, 473-484, 1975.

[8] DEL PIERO, G., On the Elastic-Plastic Material Element, Arch. Rational Mech. Anal. 59, 111-129, 1975.

[9] BUHITE, J.L., Ph. D. Thesis - Dept. of Mathematics, Carnegie-Mellon University, 1978.

[10] BUHITE, J.L., & D.R. OWEN, An Ordinary Differential Equation from the Theory of Plasticity, Arch. Rational Mech. Anal. 71, 357-383, 1979.

[11] NEMAT-NASSER, S., Decomposition of Strain Measures and their Rates in Finite Deformation Elastoplasticity, Int. J. Solids Structures 15, 155-166, 1979.

[12] MATSUMOTO, E., A Mathematical Theory of Elastic-Plastic Mate= rials with Memory with General Work-Hardening, Q. Jl. Mech. Appl. Math. 35, 197-218, 1982.

[13] DEL PIERO, G., Materials with Elastic Range and the New Theory of Simple Materials (in preparation).

# GENERAL FRAME FOR THE DEFINITION OF CONSTITUTIVE LAWS FOR LARGE NON-ISOTHERMIC ELASTIC-PLASTIC AND ELASTIC-VISCO-PLASTIC DEFORMATIONS

**THEODOR LEHMANN**
**Ruhr-University Bochum**
**D–4630 Bochum**

## 8.1 Introduction

The deformations of a body considered as a classical continuum can be derived from the description of the motion of the material points against a suitably defined space of observation furnished with a space-fixed rigid coordinate system. We can, however, also describe these deformations by considering the changes of the metric of a body-fixed coordinate system which is co-moving and co-deforming with the body. Both methods are in principle equivalent. We shall see, however, that the use of a body-fixed coordinate system may have some advantages at least for the definition of constitutive laws. Therefore we shall begin with some considerations concerning the use of space-fixed and body-fixed coordinate systems.

From some phenomena which can be observed in inelastic deformations of solid bodies we derive a material model which seems to be suitable to reflect the elastic-plastic or elastic-viscoplastic behaviour particularly of polycrystalline metals. Many features of this model, however, may apply to the inelastic behaviour of other materials, too.

The next step consists in an extended discussion of the thermodynamical foundations for the definition of constitutive laws in thermoplasticity. This discussion will be based on the assumption of local thermodynamical equilibrium on the macroscale even in such cases where the body as a whole is not in thermodynamical equilibrium. This assumption, however, restricts our considerations to processes which are running not too far from thermodynamical equilibrium.

From the introduced material model on the one hand and the thermodynamical foundations on the other hand we shall derive a general frame for the formulation of the constitutive law for large, non-isothermic, elastic-plastic, or elastic-viscoplastic deformations. This frame shall not only comprehend hardening and softening processes due to inelastic deformations, but also internal processes like solid phase transformations, recrystallization, and other annealing processes. Within this frame some possible formulations of the constitutive law for elastic-plastic and elastic-viscoplastic bodies will be discussed with respect to some consequences.

Finally the total scheme of the description of thermo-mechanical processes will be given. This scheme enables us to make some statements about certain coupling-effects in thermo-mechanical processes during inelastic deformations.

## 8.2 The use of space-fixed and body-fixed coordinate systems

The following considerations may serve as an introduction into the notation used in this chapter on the one hand. On the other hand these considerations may help to clarify some questions arising from the use of different coordinate systems. For more details we relate to the lectures given by K. Thermann and to [1, 2].

8.2.1 Space-fixed coordinate systems. Our space of observation is represented by a space-fixed, rigid coordinate system $x^\alpha$ $(\alpha = 1,2,3)$ with an Euclidian metric $g_{\alpha\beta}(x^\rho)$ (s. fig. 1). The initial position of a material particle at time $\overset{o}{t}$ (in the initial configuration of the body) is given by

$$\overset{o}{x}{}^\alpha = x^\alpha(\overset{o}{t}). \tag{2.1}$$

The corresponding base vectors and the metric of the coordinate system at this position are denoted by

$$\overset{o}{\underset{\sim}{g}}{}_\alpha = \underset{\sim}{g}_\alpha(\overset{o}{x}{}^\rho) \tag{2.2}$$

and

$$\overset{o}{g}_{\alpha\beta} = g_{\alpha\beta}(\overset{o}{x}^{\varrho}). \tag{2.3}$$

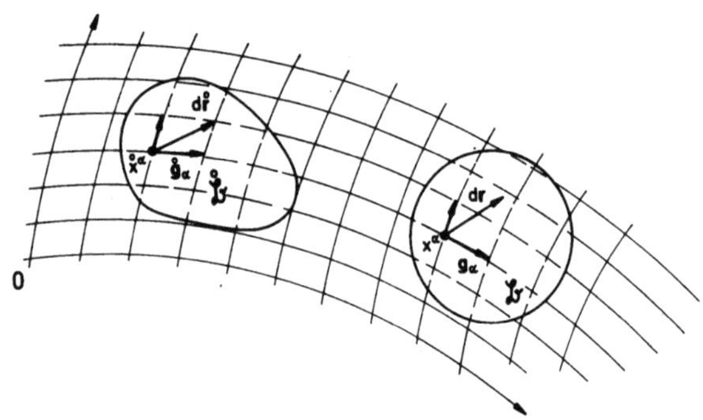


fig. 1: motion in the space-fixed coordinate system

The current position of the same material particle at time t (in the actual configuration of the body) is

$$x^{\alpha} = x^{\alpha}(\overset{o}{x}^{\varrho}, t). \tag{2.4}$$

(2.4) describes the motion of the material particles depending on the initial position and time t. The corresponding base vectors and the metric in the actual position are

$$\underset{\sim}{g}_\alpha = \underset{\sim}{g}_\alpha(x^\rho) = \underset{\sim}{g}_\alpha(\overset{o}{x}{}^\rho, t) \tag{2.5}$$

and

$$g_{\alpha\beta} = g_{\alpha\beta}(x^\rho) = g_{\alpha\beta}(\overset{o}{x}{}^\rho, t) \; . \tag{2.6}$$

A material line element defined at time $\overset{o}{t}$ by

$$d\overset{o}{\underset{\sim}{r}} = d\overset{o}{x}{}^\alpha \, \overset{o}{\underset{\sim}{g}}_\alpha \tag{2.7}$$

transforms by the motion into

$$d\underset{\sim}{r} = dx^\alpha \, \underset{\sim}{g}_\alpha \; . \tag{2.8}$$

This transition can be written in the form

$$d\underset{\sim}{r} = \underset{\sim}{F} \cdot d\overset{o}{\underset{\sim}{r}} \; , \tag{2.9}$$

where

$$\underset{\sim}{F} = \frac{\partial x^\alpha}{\partial \overset{o}{x}{}^\beta} \underset{\sim}{g}_\alpha \, \overset{o}{\underset{\sim}{g}}{}^\beta = \underset{\sim}{r} \, \overset{o}{\underset{\sim}{\nabla}} \tag{2.10}$$

denotes the so-called deformation gradient, which represents a double-field tensor. $\overset{o}{\underset{\sim}{\nabla}}$ means the Nabla-Operator corresponding to the initial configuration.

The inverse transformation is given by

$$d\overset{o}{\underset{\sim}{r}} = \underset{\sim}{F}^{-1} \cdot d\underset{\sim}{r} \tag{2.11}$$

with

$$\underset{\sim}{F}^{-1} = \frac{\partial \overset{o}{x}^{\alpha}}{\partial x^{\beta}} \overset{o}{\underset{\sim}{g}}_{\alpha} \underset{\sim}{g}^{\beta} = \overset{o}{\underset{\sim}{r}} \underset{\sim}{\nabla} . \tag{2.12}$$

In this formula $\underset{\sim}{\nabla}$ denotes the Nabla-Operator corresponding to the actual configuration.

The respective transposed tensors are

$$\underset{\sim}{F}^{T} = \frac{\partial x^{\alpha}}{\partial \overset{o}{x}^{\beta}} \overset{o}{\underset{\sim}{g}}^{\beta} \underset{\sim}{g}_{\alpha} = \overset{o}{\underset{\sim}{\nabla}} \underset{\sim}{r}, \quad (\underset{\sim}{F}^{-1})^{T} = \frac{\partial \overset{o}{x}^{\alpha}}{\partial x^{\beta}} \underset{\sim}{g}^{\beta} \overset{o}{\underset{\sim}{g}}_{\alpha} = \underset{\sim}{\nabla} \overset{o}{\underset{\sim}{r}} = (\underset{\sim}{F}^{T})^{-1} . \tag{2.13}$$

The deformation gradient can be decomposed according to

$$\underset{\sim}{F} = \underset{\sim}{R} \cdot \underset{\sim}{U} = \underset{\sim}{V} \cdot \underset{\sim}{R} \tag{2.14}$$

into a rotation $\underset{\sim}{R}$ and into the right stretch tensor $\underset{\sim}{U}$ or the left stretch tensor $\underset{\sim}{V}$ , respectively. In order to eliminate the rigid body rotations of the material elements from $\underset{\sim}{F}$ , which must not enter the constitutive laws, we multiply $\underset{\sim}{F}$ from the left hand side or from the right hand side, respectively, by $\underset{\sim}{F}^T$. Then we obtain

$$\underset{\sim}{F}^T \cdot \underset{\sim}{F} = \underset{\sim}{U} \cdot \underset{\sim}{U} = \frac{\partial x^\rho}{\partial \overset{\circ}{x}^\alpha} g_{\rho\sigma} \frac{\partial x^\sigma}{\partial \overset{\circ}{x}^\beta} \overset{\circ}{\underset{\sim}{g}}{}^\alpha \overset{\circ}{\underset{\sim}{g}}{}^\beta$$

$$= \underset{(o)}{C}_{\alpha\beta} \overset{\circ}{\underset{\sim}{g}}{}^\alpha \overset{\circ}{\underset{\sim}{g}}{}^\beta = \underset{(o)}{\underset{\sim}{C}} , \qquad (2.15)$$

$$\underset{\sim}{F} \cdot \underset{\sim}{F}^T = \underset{\sim}{V} \cdot \underset{\sim}{V} = \frac{\partial x^\alpha}{\partial \overset{\circ}{x}^\rho} \overset{\circ}{g}{}^{\rho\sigma} \frac{\partial x^\beta}{\partial \overset{\circ}{x}^\sigma} \underset{\sim}{g}_\alpha \underset{\sim}{g}_\beta$$

$$= B^{\alpha\beta} \underset{\sim}{g}_\alpha \underset{\sim}{g}_\beta = \underset{\sim}{B} . \qquad (2.16)$$

$\underset{(o)}{\underset{\sim}{C}}$ denotes the right Cauchy tensor which is related to the initial position $\overset{\circ}{x}{}^\rho$ , whereas $\underset{\sim}{B}$ means the left Cauchy tensor, which is related to the actual position. These tensors can be used for the description of the (total) deformation of the body on account of the relation

$$\frac{1}{2}\left\{ds^2 - d\overset{\circ}{s}{}^2\right\} = \frac{1}{2}\left\{d\underset{\sim}{r} \cdot d\underset{\sim}{r} - d\underset{\sim}{\overset{\circ}{r}} \circ d\underset{\sim}{\overset{\circ}{r}}\right\}$$

$$= \frac{1}{2}\left\{ \underset{(\circ)}{C}_{\alpha\beta} - \overset{\circ}{g}_{\alpha\beta}\right\} d\overset{\circ}{x}{}^{\alpha}\, d\overset{\circ}{x}{}^{\beta} = \underset{(\circ)}{E}_{\alpha\beta}\, d\overset{\circ}{x}{}^{\alpha}\, d\overset{\circ}{x}{}^{\beta} \qquad (2.17)$$

$$= \frac{1}{2}\left\{g_{\alpha\beta} - (B^{-1})_{\alpha\beta}\right\} dx^{\alpha}\, dx^{\beta} = E_{\alpha\beta}\, dx^{\alpha}\, dx^{\beta}. \qquad (2.18)$$

$\underset{(\circ)}{E}_{\alpha\beta}$ represents the so-called Cauchy-Green strain tensor which is related to the initial configuration $\overset{\circ}{\mathfrak{L}}$ whereas $E_{\alpha\beta}$ denotes the so-called Almansi-Hamel tensor which is related to the actual configuration $\mathfrak{L}$. The decomposition of the total deformation into its reversible part and its remaining part will be discussed later in connection with thermodynamical considerations.

Analogously to the strain measures we may also relate the acting stresses to the actual configuration $\mathfrak{L}$ or to the initial configuration $\overset{\circ}{\mathfrak{L}}$ of the body. This leads in the first case to the definition of the so-called Cauchy stress tensor

$$\underset{\sim}{\Sigma} = \sigma^{\alpha\beta} \, \underset{\sim}{g}_\alpha \, \underset{\sim}{g}_\beta \tag{2.19}$$

or, for physical reasons, to the introduction of the weihted Cauchy stress tensor

$$\underset{\sim}{S} = \frac{\overset{\circ}{\rho}}{\rho} \sigma^{\alpha\beta} \, \underset{\sim}{g}_\alpha \, \underset{\sim}{g}_\beta = S^{\alpha\beta} \, \underset{\sim}{g}_\alpha \, \underset{\sim}{g}_\beta \tag{2.20}$$

where $\rho$ denotes the mass density. In the second case we obtain the so-called second Piola-Kirchhoff stress-tensor which is defined by

$$\underset{(\circ)}{\underset{\sim}{S}} = \underset{\sim}{F}^{-1} \cdot \underset{\sim}{S} \cdot (\underset{\sim}{F}^{-1})^T \tag{2.21}$$

$$= \frac{\overset{\circ}{\rho}}{\rho} \frac{\partial \overset{\circ}{x}^\alpha}{\partial x^\rho} \sigma^{\rho\mu} \frac{\partial \overset{\circ}{x}^\beta}{\partial x^\mu} \, \overset{\circ}{\underset{\sim}{g}}_\alpha \, \overset{\circ}{\underset{\sim}{g}}_\beta = \underset{(\circ)}{S}^{\alpha\beta} \, \overset{\circ}{\underset{\sim}{g}}_\alpha \, \overset{\circ}{\underset{\sim}{g}}_\beta \, .$$

Analogously we may relate any other physical quantity either to the actual or to the initial configuration of the body. Doing so, however, we have to observe the physical meaning of the respective quantities.

Concerning the derivation of quantities which are related to the space-fixed coordinate system with respect to the space-fixed coordinates we have to distinguish between the derivation in the initial configuration $\overset{\circ}{\mathfrak{L}}$ and in the actual configuration $\mathfrak{L}$ . In our considerations we can restrict ourselves to the two following simple cases which are given by

$$\underset{(\circ)}{\underset{\sim}{a}}\overset{\circ}{\underset{\sim}{\nabla}} = \frac{\partial}{\partial \overset{\circ}{x}^{\gamma}}\left\{ \underset{(\circ)}{a}{}^{\alpha}_{\cdot\beta}(\overset{\circ}{x}^{\rho}, t)\, \overset{\circ}{\underset{\sim}{g}}_{\alpha}\, \overset{\circ}{\underset{\sim}{g}}{}^{\beta} \right\} \overset{\circ}{\underset{\sim}{g}}{}^{\gamma} = \underset{(\circ)}{a}{}^{\alpha}_{\cdot\beta}\Big|_{\gamma} \overset{\circ}{\underset{\sim}{g}}_{\alpha}\, \overset{\circ}{\underset{\sim}{g}}{}^{\beta}\, \overset{\circ}{\underset{\sim}{g}}{}^{\gamma} \tag{2.22}$$

and

$$\underset{\sim}{a}\,\underset{\sim}{\nabla} = \frac{\partial}{\partial x^{\gamma}}\left\{ a^{\alpha}_{\cdot\beta}(x^{\rho}, t)\, \underset{\sim}{g}_{\alpha}\, \underset{\sim}{g}^{\beta} \right\} \underset{\sim}{g} = a^{\alpha}_{\cdot\beta}\Big|_{\gamma} \underset{\sim}{g}_{\alpha}\, \underset{\sim}{g}^{\beta}\, \underset{\sim}{g}^{\gamma}. \tag{2.23}$$

We disregard other possible cases, since we do not need them in the following. The derivations with respect to time will be considered in a separate section 8.2.3. This will enable us to obtain a better comparison between the different respective situations for space-fixed and body-fixed coordinate systems.

8.2.2 Body-fixed coordinate system. We furnish the body with a co-moving and co-deforming body-fixed coordinate system $\xi^{i}$ (i = 1,2,3). Therefore each material point keeps its coordinates during the whole process. In the initial configuration $\overset{\circ}{\mathfrak{L}}$ at time $\overset{\circ}{t}$ the corresponding base vectors and metric are (see fig. 2)

$$\overset{\circ}{g}_i (\xi^r) = g_i (\xi^r, \overset{\circ}{t}), \tag{2.24}$$

$$\overset{\circ}{g}_{ik} (\xi^r) = g_{ik} (\xi^r, \overset{\circ}{t}). \tag{2.25}$$

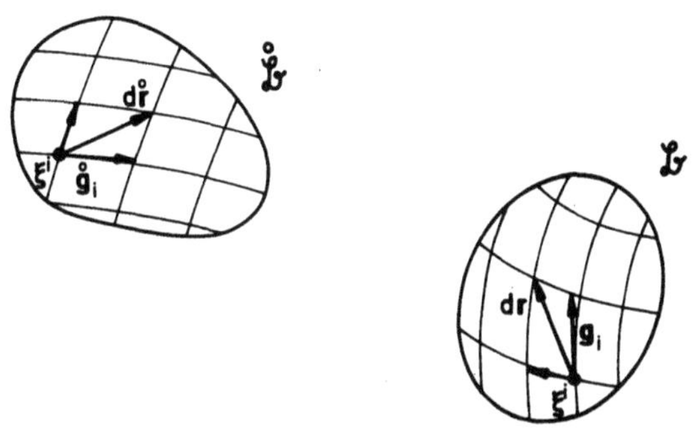


fig. 2: body-fixed coordinate system

It may be simplifying to assume that in the initial configuration the body-fixed and the space-fixed coordinates coincide:

$$\xi^i = \delta^i_\alpha \overset{\circ}{x}^\alpha. \tag{2.26}$$

This assumption, however, is not important.

In the actual configuration the corresponding base vectors and metric are

$$g_i = g_i (\xi^r, t), \tag{2.27}$$

$$g_{ik} = g_{ik}\,(\xi^r,\ t). \tag{2.28}$$

The total deformations of the material elements can be expressed by the difference of the metric in the actual configuration and in the initial configuration based on the relation

$$\frac{1}{2}\left\{ds^2 - d\mathring{s}^2\right\} = \frac{1}{2}\left\{g_{ik} - \mathring{g}_{ik}\right\} d\xi^i\, d\xi^k \tag{2.29}$$

$$= e_{ik}\,(\xi^r,\ t)\ d\xi^i\, d\xi^k. \tag{2.30}$$

In doing so it is still open whether the quantity $e_{ik}$ is related either to the initial configuration $\mathring{\mathfrak{L}}$ or to the actual configuration $\mathfrak{L}$. In the first case we obtain a strain tensor

$$\underset{(\circ)}{\tilde{e}} = \frac{1}{2}\left\{g_{ik} - \mathring{g}_{ik}\right\} \tilde{\mathring{g}}^i\, \tilde{\mathring{g}}^k = \underset{(\circ)}{e}{}_{ik}\, \tilde{\mathring{g}}^i\, \tilde{\mathring{g}}^k \tag{2.31}$$

which corresponds to the Cauchy-Green strain tensor defined by (2.17). In the second case we are led to a strain tensor

$$\tilde{e} = \frac{1}{2}\left\{g_{ik} - \mathring{g}_{ik}\right\} \tilde{g}^i\, \tilde{g}^k = e_{ik}\, \tilde{g}^i\, \tilde{g}^k \tag{2.32}$$

which corresponds in a certain way to the Almansi-Hamel strain tensor defined by (2.18) (for more details see [2]).

Another possibility for the description of deformations starts from the relation

$$g_{ik} = q_i^r \overset{\circ}{g}_{rk} = \overset{\circ}{g}_{ir} q_k^r \tag{2.33}$$

with

$$q_k^i = \overset{\circ}{g}{}^{ir} g_{rk} \tag{2.34}$$

or its inverse

$$\overset{\circ}{g}_{ik} = (q^{-1})_i^r g_{rk} = g_{ir} (q^{-1})_k^r \tag{2.35}$$

with

$$(q^{-1})_k^i = g^{ir} \overset{\circ}{g}_{rk}, \tag{2.36}$$

respectively. The relations (2.33) and (2.35) can be considered as metric transformation. Accordingly the quantitiy $q_k^i$ can be called metric transformation tensor. $q_k^i$ can be related either to the initial configuration or to the actual configuration:

$$\underset{(o)}{\underset{\sim}{q}} = \overset{\circ}{g}{}^{ir} g_{rk} \underset{\sim}{\overset{\circ}{g}}_i \underset{\sim}{\overset{\circ}{g}}{}^k, \tag{2.37}$$

$$\underset{\sim}{q} = \overset{\circ}{g}{}^{ir} g_{rk} \underset{\sim}{g}_i \underset{\sim}{g}^k . \tag{2.38}$$

(2.37) and (2.38) define different tensors. From both definitions, however, we may derive an unlimited number of equivalent strain tensors by means of isotropic tensor functions.

Examples are:

$$\underset{(\circ)}{\underset{\sim}{e}} = \frac{1}{2} \left\{ q^i_k - \delta^i_k \right\} \overset{\circ}{\underset{\sim}{g}}_i \overset{\circ}{\underset{\sim}{g}}{}^k = \underset{(\circ)}{e}{}^i_k \overset{\circ}{\underset{\sim}{g}}_i \overset{\circ}{\underset{\sim}{g}}{}^k , \tag{2.39}$$

$$\underset{\sim}{e} = \frac{1}{2} \left\{ \delta^i_k - (q^{-1})^i_k \right\} \underset{\sim}{g}_i \underset{\sim}{g}^k = e^i_k \underset{\sim}{g}_i \underset{\sim}{g}^k , \tag{2.40}$$

$$\underset{(\circ)}{\underset{\sim}{\varepsilon}} = \frac{1}{2} (\ln q)^i_k \overset{\circ}{\underset{\sim}{g}}_i \overset{\circ}{\underset{\sim}{g}}{}^k = \underset{(\circ)}{\varepsilon}{}^i_k \overset{\circ}{\underset{\sim}{g}}_i \overset{\circ}{\underset{\sim}{g}}{}^k , \tag{2.41}$$

$$\underset{\sim}{\varepsilon} = \frac{1}{2} (\ln q)^i_k \underset{\sim}{g}_i \underset{\sim}{g}^k = \varepsilon^i_k \underset{\sim}{g}_i \underset{\sim}{g}^k . \tag{2.42}$$

(2.39) coincides with the tensor defined by (2.31), (2.40) with (2.32). The strain tensors (2.41) and (2.42) are attributed to Hencky.

From the foregoing considerations we see, that by the use of body-fixed coordinates the rigid body rotations of the material elements against the space of observation are eliminated from the very beginning. Therefore we obtain immediately an objective representation of the processes. This means an important advantage particularly in the formulation of constitutive laws. We shall find this fact corroborated also in many subsequent considerations. It has to be mentioned, however, that the use of body-fixed coordinates is restricted to processes without (material) diffusion.

Analogously to the strain measures we may relate the acting stresses to the actual configuration $\mathcal{L}$ of the body or to the initial configuration $\overset{\circ}{\mathcal{L}}$. This procedure leads (among others) to the definition of the following stress tensors:

Cauchy-stress tensor (2.19):

$$\underset{\sim}{\Sigma} = \sigma^{\alpha\beta} \, \underset{\sim}{g}_\alpha \, \underset{\sim}{g}_\beta = \sigma^{ik} \, \underset{\sim}{g}_i \, \underset{\sim}{g}_k = \sigma^i_k \, \underset{\sim}{g}_i \, \underset{\sim}{g}^k , \tag{2.43}$$

weighted Cauchy-stress tensor (2.20)

$$\underset{\sim}{S} = \frac{\overset{\circ}{\rho}}{\rho} \, \sigma^{\alpha\beta} \, \underset{\sim}{g}_\alpha \, \underset{\sim}{g}_\beta = \frac{\overset{\circ}{\rho}}{\rho} \, \sigma^{ik} \, \underset{\sim}{g}_i \, \underset{\sim}{g}_k = S^{ik} \, \underset{\sim}{g}_i \, \underset{\sim}{g}_k = S^i_k \, \underset{\sim}{g}_i \, \underset{\sim}{g}^k , \tag{2.44}$$

second Piola-Kirchhoff stress tensor (2.21)

$$\underset{(\circ)}{S} = \underset{(\circ)}{S}^{\alpha\beta} \, \overset{\circ}{\underset{\sim}{g}}_\alpha \, \overset{\circ}{\underset{\sim}{g}}_\beta = \underset{(\circ)}{S}^{ik} \, \overset{\circ}{\underset{\sim}{g}}_i \, \overset{\circ}{\underset{\sim}{g}}_k . \tag{2.45}$$

It may be emphasized, that

$$S^{ik} = \underset{(\circ)}{S}^{ik} \tag{2.46}$$

This relation, however, is only true with respect to the contravariant notation of these stress tensors. This corresponds to the fact that

$$e_{ik} = \underset{(o)}{e}_{ik} \tag{2.47}$$

is only true with respect to the covariant notation of these strain tensors.

Concerning the derivation of quantities which are related to the body-fixed coordinate system with respect to these coordinates we have again to distinguish between the derivation in the initial configuration $\mathring{\mathfrak{L}}$ and in the actual configuration $\mathfrak{L}$ . Analogously to (2.22) and (2.23) we write

$$\frac{\partial}{\partial \xi^l} \{ \underset{(o)}{a}^i_{\cdot k} (\xi^r, t) \, \mathring{g}_i \, \mathring{g}^k \} \, \mathring{g}^l = \underset{(o)}{a}^i_{\cdot k}|_l \, \mathring{g}_i \, \mathring{g}^k \, \mathring{g}^l , \tag{2.48}$$

$$\frac{\partial}{\partial \xi^l} \{ a^i_{\cdot k} (\xi^r, t) \, g_i \, g^k \} \, g^l = a^i_{\cdot k}|_l \, g_i \, g^k \, g^l . \tag{2.49}$$

The transformation or mapping of physical quantities defined in the body-fixed coordinate system into the space-fixed coordinate system at an arbitrary but fixed time t obeys the usual rules. We can skip these procedures. Instead we want to focus our attention to the different definitions of time derivatives in the space-fixed and in the body-fixed coordinate system and their mutual relations.

8.2.3 Material time derivatives. Reflecting on material processes we can restrict our considerations of derivation with respect to time to such derivatives where the material coordinates $\overset{\circ}{x}^{\alpha}$ or $\xi^{i}$, respectively, are held constant. We call such derivatives summarizing material derivatives. We have, however, to distinguish different definitions of such time derivatives.

To begin with we consider material derivatives of quantities which are related to the space-fixed coordinate system in the actual position of the body. The definition of the velocity

$$\frac{D\underset{\sim}{r}}{dt} = \dot{\underset{\sim}{r}} = \underset{\sim}{v} = v^{\alpha}\,(x^{\rho},\ t)\ \underset{\sim}{g}_{\alpha} = \frac{\partial x^{\alpha}(\overset{\circ}{x}^{\rho},\ t)}{\partial t}\,\underset{\sim}{g}_{\alpha} = \dot{x}^{\alpha}\,\underset{\sim}{g}_{\alpha} \tag{2.50}$$

represents a particular case since the coordinates $x^{\alpha}$ do not form a vector in arbitrary coordinate systems. Similar the material time derivative of the deformation gradient represents an exceptional case, since $\underset{\sim}{F}$ is a double field tensor. We define

$$\begin{aligned}\dot{\underset{\sim}{F}} &= \frac{\partial}{\partial t}\,\underset{\sim}{F}\,(\overset{\circ}{x}^{\rho},\ t) = \frac{\partial}{\partial t}\,\Big(\frac{\partial x^{\alpha}(\overset{\circ}{x}^{\rho},\ t)}{\partial \overset{\circ}{x}^{\beta}}\,\underset{\sim}{g}_{\alpha}\Big)\ \overset{\circ}{\underset{\sim}{g}}{}^{\beta}\\ &= \frac{\partial}{\partial \overset{\circ}{x}^{\beta}}\,(v^{\alpha}\,\underset{\sim}{g}_{\alpha})\ \overset{\circ}{\underset{\sim}{g}}{}^{\beta} = \underset{\sim}{v}\,\overset{\circ}{\underset{\sim}{\nabla}}.\end{aligned} \tag{2.51}$$

Multiplying $\dot{\underset{\sim}{F}}$ from the right hand side by $\underset{\sim}{F}^{-1}$ we obtain

$$\dot{\underset{\sim}{F}} \circ \underset{\sim}{F}^{-1} = \frac{\partial}{\partial \overset{\circ}{x}^{\mu}} (v^{\alpha} \, \underset{\sim}{g}_{\alpha}) \, \overset{\circ}{\underset{\sim}{g}}{}^{\mu} \circ \frac{\partial \overset{\circ}{x}^{\gamma}}{\partial x^{\beta}} \, \overset{\circ}{\underset{\sim}{g}}_{\gamma} \, \underset{\sim}{g}^{\beta}$$

$$= v^{\alpha}\big|_{\beta} \, \underset{\sim}{g}_{\alpha} \, \underset{\sim}{g}^{\beta}, \qquad (2.52)$$

i.e. the velocity gradient. Its symmetrical part

$$\underset{\sim}{d} = d^{\alpha}_{\beta} \, \underset{\sim}{g}_{\alpha} \, \underset{\sim}{g}^{\beta} = \frac{1}{2} \left\{ v^{\alpha}\big|_{\beta} + v_{\beta}\big|^{\alpha} \right\} \underset{\sim}{g}_{\alpha} \, \underset{\sim}{g}^{\beta} \qquad (2.53)$$

represents the rate of deformation whereas its skew-symmetric part

$$\underset{\sim}{\omega} = \omega^{\alpha}_{\cdot\beta} \, \underset{\sim}{g}_{\alpha} \, \underset{\sim}{g}^{\beta} = \frac{1}{2} \left\{ v^{\alpha}\big|_{\beta} - v_{\beta}\big|^{\alpha} \right\} \underset{\sim}{g}_{\alpha} \, \underset{\sim}{g}^{\beta} \qquad (2.54)$$

describes the rate of (rigid body) rotation of the material elements.

The material time derivative of an arbitrary second rank tensor which is related to the space-fixed coordinate system in the actual position is given by

$$\frac{D \underset{\sim}{a}}{dt} = \frac{\partial}{\partial t}\bigg|_{\overset{\circ}{x}^{\mu}} \left\{ a^{\alpha}_{\cdot\beta}(x^{\rho}, t) \, \underset{\sim}{g}_{\alpha}(x^{\rho}) \, \underset{\sim}{g}^{\beta}(x^{\rho}) \right\}$$

$$= \frac{\partial}{\partial t}\bigg|_{x^{\mu}} (a^{\alpha}_{\cdot\beta}) \, \underset{\sim}{g}_{\alpha} \, \underset{\sim}{g}^{\beta} + \frac{\partial}{\partial x^{\rho}} \left\{ a^{\alpha}_{\cdot\beta} \, \underset{\sim}{g}_{\alpha} \, \underset{\sim}{g}^{\beta} \right\} \frac{\partial x^{\rho}}{\partial t}\bigg|_{\overset{\circ}{x}^{\mu}}$$

$$= \left\{ \frac{\partial}{\partial t}\bigg|_{x^{\mu}} (a^{\alpha}_{\cdot\beta}) + v^{\rho} \, a^{\alpha}_{\cdot\beta}\big|_{\rho} \right\} \underset{\sim}{g}_{\alpha} \, \underset{\sim}{g}^{\beta} = \dot{a}^{\alpha}_{\cdot\beta} \, \underset{\sim}{g}_{\alpha} \, \underset{\sim}{g}^{\beta}. \qquad (2.55)$$

This material time derivative is commutative with the procedure of lowering or raising of an index, i. e.

$$\overset{\circ}{a}{}^{\alpha}_{\cdot\beta} = g^{\alpha\varrho}\,\overset{\circ}{a}_{\varrho\beta} = g_{\beta\varrho}\,\overset{\circ}{a}{}^{\alpha\varrho}. \tag{2.56}$$

The definition (2.55) can be extended easily to arbitrary other tensors.

The preceding definition of a material time derivative is not objective with respect to rigid body rotations of the material elements. In order to eliminate the effects of these rotations we have to introduce the so-called objective material derivation with respect to time defined by

$$\overset{\nabla}{\underset{\sim}{a}} = \left\{ \overset{\circ}{a}{}^{\alpha}_{\cdot\beta} - \omega^{\alpha}_{\cdot\rho}\, a^{\rho}_{\cdot\beta} + a^{\alpha}_{\cdot\rho}\,\omega^{\rho}_{\cdot\beta} \right\} \underset{\sim}{g}_{\alpha}\, \underset{\sim}{g}^{\beta}$$

$$= \overset{\nabla}{a}{}^{\alpha}_{\cdot\beta}\, \underset{\sim}{g}_{\alpha}\, \underset{\sim}{g}^{\beta}. \tag{2.57}$$

This time derivative is due to Zaremba [3] and Jaumann [4] . Its definition can be extended easily to other tensors of arbitrary rank. It is also commutative with lowering and raising of indices.

The material time derivative of a quantity, for instance of a second rank tensor, which is related to the initial position of the body in the space-fixed coordinate system is given by

$$\frac{D\, \underset{(o)}{a}}{dt} = \frac{\partial}{\partial t} \left\{ \underset{(o)}{a}^{\alpha}{}_{\circ\beta}(\overset{\circ}{x}{}^{\rho},t)\, \overset{\circ}{g}_{\alpha}(\overset{\circ}{x}{}^{\rho})\, \overset{\circ}{g}{}^{\beta}(\overset{\circ}{x}{}^{\rho}) \right\} = \frac{\partial}{\partial t} \left( \underset{(o)}{a}^{\alpha}{}_{\circ\beta}(\overset{\circ}{x}{}^{\rho},t) \right) \overset{\circ}{g}_{\alpha}\, \overset{\circ}{g}{}^{\beta}$$

$$= \underset{(o)}{\dot{a}}^{\alpha}{}_{\cdot\beta}\, \overset{\circ}{g}_{\alpha}\, \overset{\circ}{g}{}^{\beta}. \tag{2.58}$$

This time derivative is objective and commutative with lowering and raising of indices.

Concerning the material time derivatives of quantities which are related to the body-fixed coordinate system in the actual configuration we have to distinguish between the derivative of the numerical values of the tensor components and the derivative of the tensor itself. In the first case we obtain, for instance,

$$\frac{\bar{D}}{dt}(a^{i}{}_{\cdot k})\, g_i\, g^k = \frac{\partial}{\partial t} \left\{ a^{i}{}_{\circ k}(\xi^r,t) \right\} g_i\, g^k = \dot{a}^{i}{}_{\cdot k}\, g_i\, g^k. \tag{2.59}$$

$\bar{D}$ means that the changes of the respective quantities shall be considered with respect to the body-fixed coordinate system. This material time derivative of the tensor components is not communicative with lowering or raising of indices, since

$$\dot{g}_{ik} = \frac{\partial}{\partial t} \left\{ g_{ik}(\xi^r,t) \right\} = 2\, d_{ik}(\xi^r,t), \tag{2.60}$$

$$\dot{g}^{ik} = \frac{\partial}{\partial t}\left\{ g^{ik}\,(\xi^r,t)\right\} = -2\, d^{ik}\,(\xi^r,t) \tag{2.61}$$

where $d_{ik}$ denotes the deformation rate of the body.

Therefore

$$\dot{a}^{\cdot i}_{\cdot k} \neq g^{ir}\,\dot{a}_{rk} + \dot{a}^{\cdot ir}\, g_{rk} \tag{2.62}$$

if

$$d_{ik} \neq 0.$$

Furthermore we have to observe that the application of this time derivative to a symmetric tensor in the mixed variant notation $a^i_k$ yields an unsymmetric tensor in general. Therefore we have to write

$$\frac{\bar{D}}{dt}\,(a^i_k)\,\mathbf{g}_i\,\mathbf{g}^k = \frac{\partial}{\partial t}\,a^i_k\,(\xi^r,t)\,\mathbf{g}_i\,\mathbf{g}^k = \dot{a}^{\cdot i}_{\cdot k}\,\mathbf{g}_i\mathbf{g}^k$$

$$+\; \dot{a}_k^{\cdot\cdot i}\,\mathbf{g}^k\,\mathbf{g}_i = \frac{\partial}{\partial t}\,a^i_k\,(\xi^r,\ t)\,\mathbf{g}^k\,\mathbf{g}_i \tag{2.63}$$

where

$$\dot{a}_k^{\cdot\cdot i} = g_{kr}\,\dot{a}^{\cdot r}_{\cdot m}\,g^{mi}\ .$$

The derivation of the covariant or contravariant components of such a symmetric tensor, however, leads again to a symmetric tensor. This means

$$\dot{a}^{ik} = \dot{a}^{ki} \quad \text{if} \quad a^{ik} = a^{ki} ,$$
$$\dot{a}_{ik} = \dot{a}_{ki} \quad \text{if} \quad a_{ik} = a_{ki} . \tag{2.64}$$

The preceding defined material time derivatives of tensor components are not affected by rigid body rotations which are eliminated by introduction of the body-fixed coordinate system from the very beginning.

The material time derivative of a tensor itself which is related to the body-fixed coordinate system in its actual configuration is (in the case of a second rank tensor) defined by

$$\frac{\bar{D}}{dt} \underset{\sim}{a} = \overset{\nabla}{\underset{\sim}{a}} = \frac{\partial}{\partial t} \left\{ a^{i}_{\cdot k} (\xi^r, t) \, \underset{\sim}{g}_i (\xi^r, t) \, \underset{\sim}{g}^k (\xi^r, t) \right\}$$
$$= \left\{ \dot{a}^{i}_{\cdot k} + d^{i}_{r} a^{r}_{\cdot k} - a^{i}_{\cdot r} d^{r}_{k} \right\} \underset{\sim}{g}_i \, \underset{\sim}{g}^k = \overset{\nabla}{a}{}^{i}_{\cdot k} \, \underset{\sim}{g}_i \, \underset{\sim}{g}^k . \tag{2.65}$$

This result can be obtained by means of the relations

$$\frac{\bar{D}}{dt} \underset{\sim}{g}_i (\xi^r, t) = d^{r}_{i} \underset{\sim}{g}_r , \quad \frac{\bar{D}}{dt} \underset{\sim}{g}^k (\xi^r, t) = - d^{k}_{r} \underset{\sim}{g}^r , \tag{2.66}$$

where $d^{i}_{k}$ means again the deformation rate of the body. Equivalent definitions of this time derivative are

$$\frac{\bar{D}\underset{\sim}{a}}{dt} = \overset{\nabla}{\underset{\sim}{a}} = \left\{ \dot{a}^{ik} + d^{i}_{r}\, a^{rk} + a^{ir}\, d^{k}_{r} \right\} \mathbf{g}_i\, \mathbf{g}_k = \overset{\nabla}{a}{}^{ik}\, \mathbf{g}_i \mathbf{g}_k \tag{2.67a}$$

$$= \left\{ \dot{a}_{ik} - d^{r}_{i}\, a_{rk} - a_{ir}\, d^{r}_{k} \right\} \mathbf{g}^i\, \mathbf{g}^k = \overset{\nabla}{a}_{ik}\, \mathbf{g}^i\, \mathbf{g}^k. \tag{2.67b}$$

This time derivative corresponds to the objective material time derivative of a tensor which is related to the space-fixed coordinate system in the actual position of the body (Zaremba-Jaumann time derivative) defined by (2.57). Its definition can be extended easily to tensors of arbitrary rank.

This time derivation is commutative with the procedure of lowering or raising of indices since

$$\overset{\nabla}{g}_{ik} = \overset{\nabla}{g}{}^{ik} = 0. \tag{2.68}$$

From the definitions (2.65) and (2.67) we can furthermore conclude with respect to symmetric second rank tensors that

$$\overset{\nabla}{a}{}^{i}_{k} = \frac{1}{2} \left\{ g^{ir}\, \dot{a}_{rk} + \dot{a}^{ir}\, g_{rk} \right\} = \frac{1}{2} \left\{ \dot{a}^{i}_{\cdot k} + \dot{a}_{k}^{\cdot i} \right\} \tag{2.69}$$

if $a^{ik} = a^{ki}$.

This means that $\overset{\nabla i}{a}_{k}$ represents the symmetric part of $\dot{a}^{i}_{\cdot k}$ for a symmetric tensor $a^{i}_{k}$ .

Finally it may be mentioned that the definition of this material time derivative and its attributes can be gained quite naturally by introduction of a four-dimensional body-fixed coordinate system containing time as fourth (or zeroth) coordinate. In this case the preceding defined material time derivative turns out to be the covariant derivation of tensors with respect to time [5]. Abbreviating therefore we shall call it covariant time derivative.

The material time derivative of quantities which are related to the body-fixed coordinate system in the initial configuration shows the same aspects as the material time derivative of quantities related to the space-fixed coordinate system in the initial position of the body given by (2.58). It needs no further discussion.

8.2.4 Some applications to the definition of deformation rate, specific mechanical work rate and strain rate. Using different material time derivatives we can define the deformation rate of the body in different ways. Some possibilities are:

a) $$\underset{\sim}{d} = \text{sym.}\left\{ \dot{\underset{\sim}{F}} \cdot \underset{\sim}{F}^{-1} \right\} = \frac{1}{2}\left\{ v^{\alpha}\big|_{\beta} + v_{\beta}\big|^{\alpha} \right\} \underset{\sim}{g}_{\alpha}\, \underset{\sim}{g}^{\beta} = d^{\alpha}_{\beta}\; \underset{\sim}{g}_{\alpha}\, \underset{\sim}{g}^{\beta} \qquad (2.70)$$

with $\underset{\sim}{F}^{-1}$ according to (2.12) and $\dot{\underset{\sim}{F}}$ according to (2.51),

b) $$\underset{(o)}{\underset{\sim}{d}} = \underset{(o)}{\dot{E}}_{\alpha\beta} \overset{o}{\underset{\sim}{g}}{}^{\alpha} \overset{o}{\underset{\sim}{g}}{}^{\beta} = \frac{1}{2} \underset{(o)}{\dot{C}}_{\alpha\beta} \overset{o}{\underset{\sim}{g}}{}^{\alpha} \overset{o}{\underset{\sim}{g}}{}^{\beta} = \underset{(o)}{d}_{\alpha\beta} \overset{o}{\underset{\sim}{g}}{}^{\alpha} \overset{o}{\underset{\sim}{g}}{}^{\beta} \tag{2.71}$$

with $\underset{(o)}{C}_{\alpha\beta}$ according to (2.15) and $\underset{(o)}{E}_{\alpha\beta}$ according to (2.17),

c) $$\underset{(\tilde{o})}{d} = \underset{(o)}{\dot{e}}_{ik} \overset{o}{\underset{\sim}{g}}{}^{i} \overset{o}{\underset{\sim}{g}}{}^{k} = \frac{1}{2} \dot{g}_{ik} \overset{o}{\underset{\sim}{g}}{}^{i} \overset{o}{\underset{\sim}{g}}{}^{k} = \underset{(o)}{d}_{ik} \overset{o}{\underset{\sim}{g}}{}^{i} \overset{o}{\underset{\sim}{g}}{}^{k} \tag{2.72}$$

with $\underset{(o)}{e}_{ik}$ according to (2.31),

d) $$\underset{\sim}{d} = \dot{e}_{ik} \underset{\sim}{g}{}^{i} \underset{\sim}{g}{}^{k} = \frac{1}{2} \dot{g}_{ik} \underset{\sim}{g}{}^{i} \underset{\sim}{g}{}^{k} = d_{ik} \underset{\sim}{g}{}^{i} \underset{\sim}{g}{}^{k} \tag{2.73}$$

with $e_{ik}$ according to (2.32),

e) $$\underset{\sim}{d} = \frac{1}{2} (q^{-1})^{i}_{r} \dot{q}^{\cdot r}_{\cdot k} \underset{\sim}{g}_{i} \underset{\sim}{g}{}^{k} = \frac{1}{2} g^{im} \dot{g}_{mk} \underset{\sim}{g}_{i} \underset{\sim}{g}{}^{k} = d^{i}_{k} \underset{\sim}{g}_{i} \underset{\sim}{g}{}^{k} \tag{2.74}$$

with $q^{i}_{k}$ according to (2.34).

All these definitions (and some others more) of the deformation rate are equivalent.

This results from the fact that the double scalar product of the deformation rate and the correspondingly defined stress tensor yields the specific mechanical work rate

$$\dot{w} = \frac{1}{\rho} \sigma^{\alpha\beta} d_{\alpha\beta} = \frac{1}{\rho} \underset{\sim}{\Sigma} \cdot\cdot \underset{\sim}{d} = \frac{1}{\bar{\rho}} S^{\alpha\beta} d_{\alpha\beta} = \frac{1}{\bar{\rho}} \underset{\sim}{S} \cdot\cdot \underset{\sim}{d} \,. \tag{2.75}$$

Other expressions for this work rate are

$$\overset{\circ}{w} = \frac{1}{\overset{\circ}{\rho}} \underset{(\tilde{o})}{S} \cdot\cdot \underset{(\tilde{o})}{d} = \frac{1}{\overset{\circ}{\rho}} \underset{(o)}{S}^{\alpha\beta} \underset{(o)}{\dot{E}}_{\alpha\beta}. \tag{2.76}$$

with $\underset{(\tilde{o})}{S}^{\alpha\beta}$ according to (2.21),

$$\overset{\circ}{w} = \frac{1}{\overset{\circ}{\rho}} \underset{(\tilde{o})}{S} \cdot\cdot \underset{(\tilde{o})}{d} = \frac{1}{\overset{\circ}{\rho}} \underset{(o)}{S}^{ik} \underset{(o)}{\dot{e}}_{ik} \tag{2.77}$$

with $\underset{(o)}{S}^{ik}$ according to (2.45),

$$\overset{\circ}{w} = \frac{1}{\overset{\circ}{\rho}} \underset{\sim}{S} \cdot\cdot \underset{\sim}{d} = \frac{1}{\overset{\circ}{\rho}} S^{ik} \dot{e}_{ik} \tag{2.78}$$

with $S^{ik}$ according to (2.44),

$$\dot{w} = \frac{1}{\overset{\circ}{\rho}} \underset{\sim}{S} \cdot\cdot \underset{\sim}{d} = \frac{1}{\overset{\circ}{\rho}} S^{i}_{k} (q^{-1})^{k}_{r} \frac{1}{2} \dot{q}^{r}_{\cdot i} \tag{2.79}$$

with $S^{i}_{k}$ according to (2.44).

In the formulas (2.76) to (2.78) the deformation rate is expressed by material time derivatives of strain tensors, i.e. by strain rates. In this cases we call

$$\underset{(o)}{S}^{\alpha\beta}, \; \underset{(o)}{E}_{\alpha\beta} \quad ; \quad \underset{(o)}{S}^{ik}, \; \underset{(o)}{e}_{ik} \quad ; \quad S^{ik}, \; e_{ik}$$

conjugated pairs of stress and strain. It must be emphasized, however, that $S^{ik}$ and $e_{ik}$ are only conjugated if the stress is represented contravariant and the strain covariant.

According to equation (2.79) we may also consider

$$\tilde{S}^{i}_{\circ k} = S^{i}_{r} \, (q^{-1})^{r}_{\cdot k} \qquad \text{and} \qquad \frac{1}{2}\, q^{i}_{k} = \frac{1}{2}\, \overset{\circ}{g}^{ir} \, g_{rk} \tag{2.80}$$

as an conjugated pair of (modified) stress and strain. Later on we shall see that this combination has an important meaning with respect to the decomposition of mechanical work and deformations into a reversible part and into a remaining part.

## 8.3. Some observed phenomena in thermoplasticity and a corresponding material model

In a simple quasistatic, isothermic tension test, for instance, with a steel specimen at room temperature we observe a relation between stress and strain as sketched in fig. 3a.

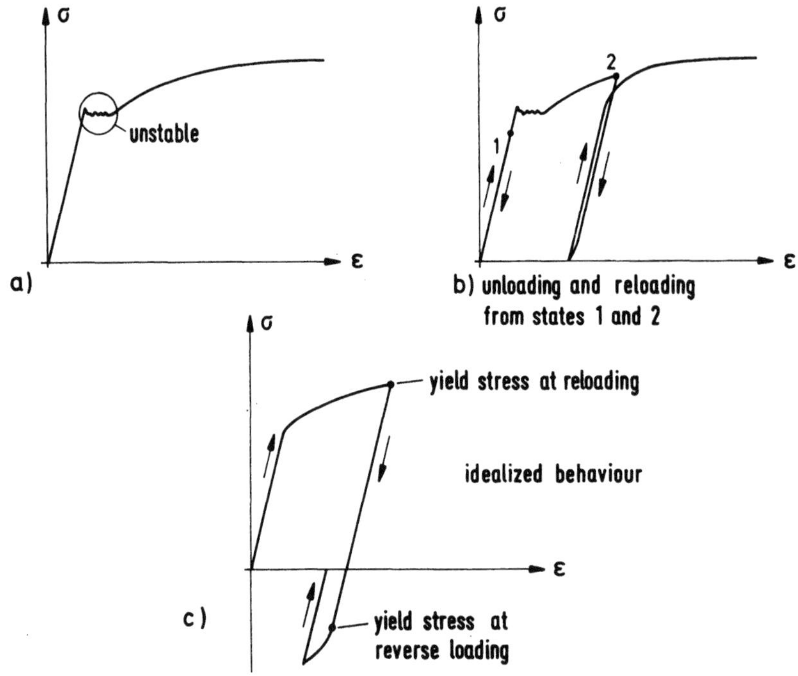


fig. 3: uniaxial quasistatic isothermic loading

Extending the test to processes including also unloading, reloading (fig. 3b), and reverse loading (fig. 3c) we find:

1. at loading beyond a certain limit (yield stress) we obtain besides reversible (elastic) also irreversible deformations;
2. at reloading after the occurence of irreversible deformations we observe an increased yield stress in comparison with the first loading;

3. at least at the first reverse loading after the occurence of irreversible deformations the absolute value of the yield stress is lower than at reloading (Bauschinger effect);
4. the elastic behaviour is rather unaffected by accompanying or preceding irreversible deformations.

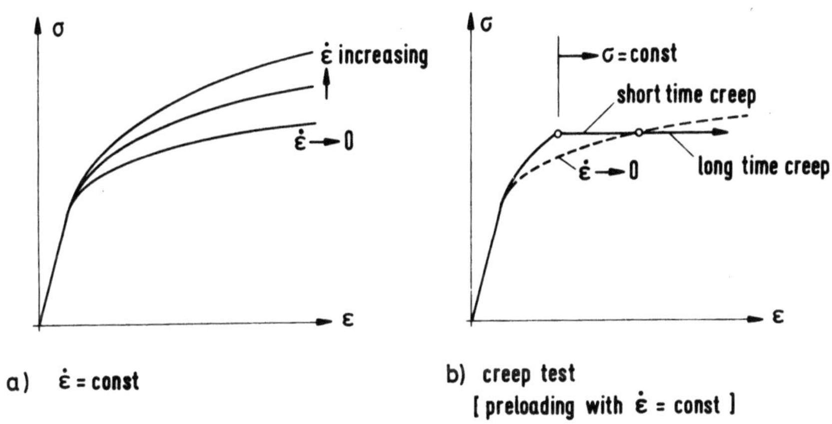


fig. 4: uniaxial isothermic loading

With increasing deformation rate higher stresses are required as fig. 4a indicates. Altering the experiment in such a way that from a certain time we do not prescribe the deformation rate any more but hold the stress constant (see fig. 4b), we observe a creep process leading in the short time range to a strain which about corresponds to the strain reached in an analogous quasistatic experiment. In the long time range (days, years) we observe further creep processes but indeed with essential lower creep rate. This long time creep shows in general three different stages as indicated in fig. 5.

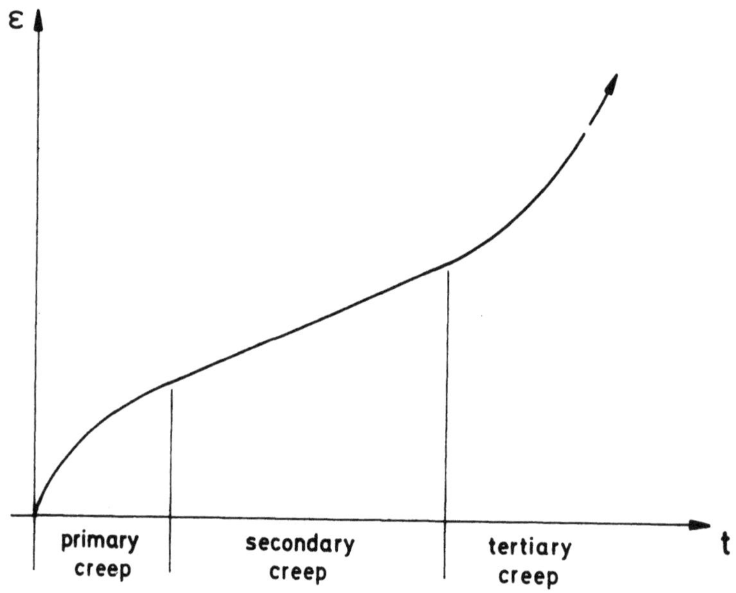


fig. 5: long time creep

Taking into account different temperature levels or even non-isothermic processes the situation becomes still much more complicated as fig. 6 shows. Solid phase transformations, recrystallization, and other annealing processes may occur interacting with hardening or softening processes induced by inelastic deformations.

stress-temperature diagramm

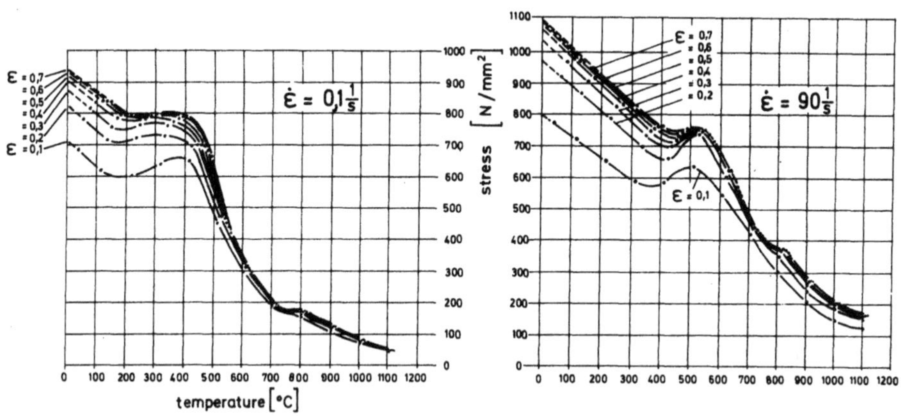


$$\varepsilon = \ln \frac{l}{l_o} \qquad \dot{\varepsilon} = \frac{\dot{l}}{l}$$

fig. 6: uniaxial loading depending on $\dot{\varepsilon}$ and T steel C45 (Fritsch-Siegel [6] )

A simple (uniaxial) material model which about renders the observed phenomena is sketched in fig. 7.

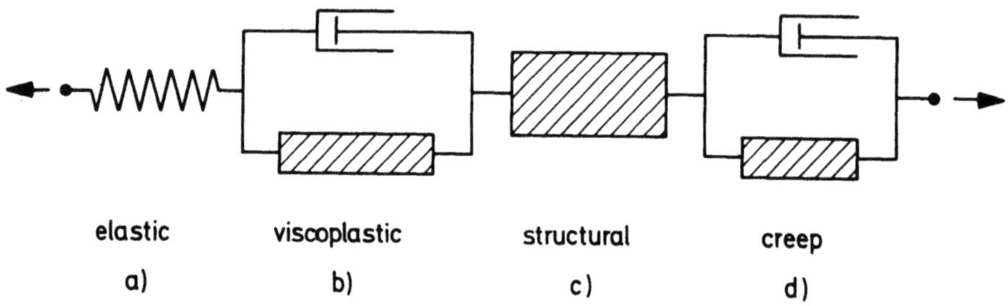


fig. 7: material model

In this model four different mechanisms are arranged in one row:

a) an elastic element;

b) a complex viscoplastic element consisting of a rate insensitive (plastic) element (involving likewise certain changes of the internal structure) and a parallel arranged rate sensitive (viscous) element responsible also for short time creep and relaxation phenomena;

c) an element rendering certain changes of the internal structure of the material independent of viscoplastic and creep processes;

d) a complex viscous element representing unlimited long time creep and relaxation phenomena connected with corresponding structural changes.

The elastic element stands for reversible processes governed by thermo-dynamical state equations, whereas the viscous subelements reflect pure irreversible dissipative processes.

Concerning the processes connected with changes of the internal structure the situation is more complex. We have to distinguish between different kinds of such internal processes.

On the one hand we know strictly reversible internal processes like quasistatic solid phase transformations governed by state equations. Such processes (sometimes called pseudo-elastic) have to be included into the elastic element which comprehends all strictly reversible processes. On the other hand we obtain also internal processes like recrystallization which are pure dissipative. They can be treated within the separate "black box" representing the changes of internal structure.

We know still a third kind of such internal processes, where the applied mechanical work is not dissipated immediately but stored in the produced lattice defects like dislocations and disclinations. This stored energy, however, is not governed by thermodynamical state equations at the macroscale. Storing and regaining of this energy (corresponding to hardening and softening of the material due to inelastic deformations) depends on the history of the thermomechanical process. These internal processes are to attach partially to the viscoplastic and to the creep processes, partially to the separate mechanism.

The material model may be extended in several directions. One possibility, for instance, consists in replacing the elastic element by a viscoelastic element. This can become necessary when very fast deformations shall be covered by the model. Another possibility consists in a parallel arrangement of several such material models in so-called fractional models. We shall restrict ourselves, however, to simple models as sketched in fig. 7. They offer already much freedom in the formulation of constitutive laws in thermoplasticity.

## 8.4. Thermodynamical frame for the formulation of the constitutive law

8.4.1 Some general preliminary remarks. We have to distinguish between the description of a thermo-mechanical process in the space of process variables and the accompanying description of such processes in the space of thermodynamical state.

Concerning the description of the thermodynamical state we assume that each material element has the properties of a local thermodynamical system whose state is uniquely defined by the values of a suitable set of well defined external and internal state variables. This means we adopt the assumption of local thermodynamical equilibrium even in irreversible processes due to non-equilibrium states of the body as the whole. A theory based on such an assumption, of course, is limited to processes running not too far from thermodynamical equilibrium.

External state variables are the stresses and the temperature. They can be replaced by their conjugated state variables, i.e. the reversible strain and the entropy, respectively. The internal state variables reflect the actual internal structure of the material elements. They represent the distribution of solid phase fractions, lattice defects etc. by means of statistical methods. Theoretically this requires an infinite set of internal state variables of even and odd rank. Within the frame of a phenomenological theory we have to restrict ourselves to a finite set of internal state variables containing only even rank tensors (including scalars). For this reason each phenomenological theory of thermoplasticity can only represent an approximation.

Within the frame of our phenomenological considerations a thermo-mechanical process starts in the initial state $\dot{\mathcal{L}}$ of the body defined by the initial geometrical configuration of the body and by the initial values of the state variables of each material element.

The thermo-mechanical process is governed by the history of the independent process variables. These are

a) the prescribed thermo-mechanical boundary conditions,

b) the prescribed specific body forces and energy sources.

A theoretical process description intends to deliver (explicitly or implicitly) the values of the dependent process variables in the actual state $\mathcal{L}$ of the body. These dependent process variables are

a) the non-prescribed thermo-mechanical quantities at the boundary,

b) the displacements (strains), the stresses and the temperature as well as the non-prescribed specific body forces and energy sources inside the body,

c) the internal state variables, and

d) all other interesting quantities as, for instance, the interchanged energy.

The link between the initial state and the history of the independent process variables on the one hand and the resulting actual state on the other hand is formed by

1. the general (material independent) field equations, consisting of balance equations for mass, compatibility conditions for deformations, balance equations for momentum and moment of momentum, balance equations for energy and entropy,
2. the constitutive law.

We shall focus our considerations to the constitutive law.

The real thermo-mechanical process carries the body from the initial configuration and state $\overset{\circ}{\mathcal{L}}$ into the actual configuration and state $\mathcal{L}$. All physical quantities are acting in the respective current configuration of the body. Therefore we prefer in the following to relate all physical quantities to the body-fixed coordinate system in its actual configuration.

The real thermo-mechanical process contains strictly reversible and other parts. We attach to the actual state $\mathcal{L}$ of the body an accompanying fictitious reference state $\overset{\bullet}{\mathcal{L}}$ by a fictitious reversible process. This fictitious process carries each material element from its actual thermodynamical state into an unstressed state at reference temperature $\overset{*}{T}$ . During this fictitious process the internal variables are kept constant (see fig. 8).

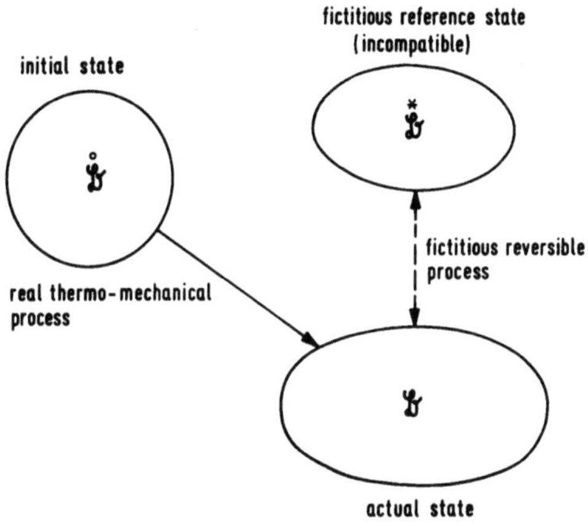


fig. 8: thermo-mechanical process

A realization of this fictitious reversible process is, in general, impossible for two reasons:

1. the resulting reference state $\overset{*}{\mathfrak{S}}$ is in general characterized by an incompatible configuration requiring a tearing of the body.
2. A real process carrying a material element from the actual stress and temperature into the reference state of stress and temperature may lead to changes of the internal variables which are excluded by our definition of the fictitious reversible process. We can desist from the necessity of "freezing" the internal variable during the fictitious process only if the

respective internal variables are governed by state functions depending uniquely on stress and temperature.

The defined fictitious process serves for an unique definition of that what we call reversible energy. The energy really gained back depends on the further history of the thermo-mechanical process.

From the preceding remarks it follows immediately that we can not define a (real or fictitious) process leading from the initial state $\overset{\bullet}{\mathcal{L}}$ to the fictitious reference state $\overset{*}{\mathcal{L}}$. For the same reason it becomes unnecessary to introduce a strain tensor defining the non-reversible deformations uniquely. We need, however, an unique decomposition of the mechanical work into its reversible and its remaining part. Furthermore we require an unique measure for the reversible strains, i.e. a definition of an elastic strain tensor which enters the expression for the reversible energy. This will be discussed later on.

8.4.2. Balance equations for energy and entropy. The first law of thermodynamics states

$$\dot{u} = \dot{w} - \frac{1}{\rho} q^i\Big|_i + r. \qquad (4.1)$$

In this relation mean

u specific internal energy,

w specific mechanical work,

$q^i$ energy flux,

$r$ specific energy sources.

The energy flux does not only comprehend the heat flux but also other energy fluxes. Such other kinds of energy fluxes may due, for instance, to the migration of lattice defects as far as it is not connected with macroscopic mechanical work (diffusion of lattice defects). These energy fluxes different from heat may be small in solid bodies and negligible in many cases. However, they have to be taken into account in general considerations.

The mechanical work rate can be decomposed into its strictly reversible part $\underset{(r)}{\dot{w}}$ and into its remaining part $\underset{(i)}{\dot{w}}$ according to

$$\dot{w} = \underset{(r)}{\dot{w}} + \underset{(i)}{\dot{w}} . \tag{4.2}$$

Within the frame of our assumptions the specific internal energy must be expressible as an unique function of a suitable set of thermodynamical state variables. We introduce as state variables:

$\underset{(r)}{\tilde{\varepsilon}^i_k}$ suitable chosen elastic strain tensor,

$s$ specific entropy,

$b$, $\beta^i_k$ representative set of internal state variables.

It has to be emphasized that the macroscopic inelastic deformations can not be used as thermodynamic state variables. This can, for instance, be concluded from the fact that the migration of a lattice defect through a crystal grain leaves a macroscopic inelastic deformation but not a remaining change of the internal structure of the crystal. We can also argue that an annealing process like recrystallization lets the material forget the preceding inelastic deformations.

The number of internal state variables is, as already stated, unlimited. We only introduce a representative set consisting of one scalar valued variable $b$ and one tensor valued variable $\beta^{i}_{k}$ standing for an arbitrary but finite set of such internal variables.

On account of the preceding remarks we can write

$$u = u\,(\underset{(r)}{\tilde{\varepsilon}}{}^{i}_{k},\ s,\ b,\ \beta^{i}_{k}). \tag{4.3}$$

For some reasons it is advantageous to replace $\underset{(r)}{\tilde{\varepsilon}}{}^{i}_{k}$ and $s$ by their conjugated state variables, i.e. the stress $\tilde{S}^{i}_{\cdot k}$ and the temperature $T$. This requires that $\tilde{S}^{i}_{\cdot k}$ and $\underset{(r)}{\tilde{\varepsilon}}{}^{i}_{k}$ form an conjugated pair of stress and strain with respect to the reversible work, i.e.

$$\underset{(r)}{\dot{w}} = \frac{1}{\overset{\circ}{\rho}}\,\tilde{S}^{i}_{\cdot k}\,(\dot{\tilde{\varepsilon}})^{k}_{\cdot i}. \tag{4.4}$$

The stress tensor $\tilde{S}^{i}_{\cdot k}$ which is conjugated to the symmetric strain tensor $\underset{(r)}{\tilde{\varepsilon}}{}^{i}_{k}$ is not necessary symmetric as indicated by the notation. The introduction of $\tilde{S}^{i}_{\cdot k}$ and $T$ as thermodynamical state variables is achieved by a double Legendre transformation leading to the specific free enthalpy

$$\psi = \dot{u} - \frac{1}{\overset{\circ}{\rho}} \tilde{S}^{i}_{\cdot k} \underset{(r)}{\tilde{\varepsilon}}{}^{k}_{\dot{i}} - Ts = \psi(\tilde{S}^{i}_{k}, T, b, \beta^{i}_{k}). \qquad (4.5)$$

Since the unsymmetric part of $\tilde{S}^{i}_{\circ k}$ vanishes in the double scalar product with the symmetric strain tensor only the symmetric part of $\tilde{S}^{i}_{k}$ enters the definition of the free enthalpy.

From (4.5) we obtain for the evolution of the specific free enthalpy on the one hand

$$\dot{\psi} = \dot{u} - \frac{1}{\overset{\circ}{\rho}} \dot{\tilde{S}}^{i}_{\cdot k} \underset{(r)}{\tilde{\varepsilon}}{}^{k}_{\dot{i}} - \frac{1}{\overset{\circ}{\rho}} \tilde{S}^{i}_{\cdot k} (\underset{(r)}{\dot{\tilde{\varepsilon}}})^{k}_{\cdot i} - \dot{T}s - T\dot{s} \qquad (4.6)$$

$$= \underset{(r)}{\dot{w}} + \underset{(i)}{\dot{w}} - \frac{1}{\rho} q^{i}\Big|_{i} + r - \frac{1}{\overset{\circ}{\rho}} \dot{\tilde{S}}^{i}_{\cdot k} \underset{(r)}{\tilde{\varepsilon}}{}^{k}_{\dot{i}} - \frac{1}{\overset{\circ}{\rho}} \tilde{S}^{i}_{\cdot k} (\underset{(r)}{\dot{\tilde{\varepsilon}}})^{k}_{\cdot i} - \dot{T}s - T\dot{s}$$

and on the other hand

$$\psi = \frac{\partial\psi}{\partial \tilde{S}^{i}_{k}} (\dot{\tilde{S}})^{i}_{\cdot k} + \frac{\partial\psi}{\partial T} \dot{T} + \frac{\partial\psi}{\partial b} \dot{b} + \frac{\partial\psi}{\partial \beta^{i}_{k}} (\dot{\beta})^{i}_{\circ k}. \qquad (4.7)$$

From the properties of the Legendre transformations it follows immediately:

thermic state equation: $$\underset{(r)}{\tilde{\varepsilon}}{}^{i}_{k} = - \rho \frac{\partial \psi}{\partial \tilde{S}^{k}_{i}} = \underset{(r)}{\tilde{\varepsilon}}{}^{i}_{k} (\tilde{S}^{i}_{k}, T, b, \beta^{i}_{k}), \tag{4.8}$$

caloric state equation: $$s = - \frac{\partial \psi}{\partial T} = s(\tilde{S}^{i}_{k}, T, b, \beta^{i}_{k}). \tag{4.9}$$

Substituting (4.8) and (4.9) into (4.6) and equating (4.6) with (4.7) we obtain the Gibbs equation

$$T \dot{s} = \underset{(i)}{\dot{w}} - \frac{1}{\rho} q^{i}\Big|_{i} + r - \frac{\partial \psi}{\partial b} \mathring{b} - \frac{\partial \psi}{\partial \beta^{i}_{k}} (\dot{\beta})^{i}_{\cdot k}. \tag{4.10}$$

This relation represents the balance equation for the specific entropy s.

On the other hand we derive from the caloric state equation (4.9)

$$T \dot{s} = - T \left\{ \frac{\partial^{2} \psi}{\partial \tilde{S}^{i}_{k} \partial T} (\dot{\tilde{S}})^{i}_{\cdot k} + \frac{\partial^{2} \psi}{\partial T^{2}} \dot{T} + \frac{\partial^{2} \psi}{\partial b \, \partial T} \dot{b} + \frac{\partial^{2} \psi}{\partial \beta^{i}_{k} \, \partial T} (\dot{\beta})^{i}_{\cdot k} \right\}. \tag{4.11}$$

From the thermic state equation (4.8) we conclude

$$\underset{(r)}{\dot{w}} = \frac{1}{\rho} \tilde{S}^{i}_{\circ k} \, (\underset{(r)}{\dot{\tilde{\varepsilon}}})^{k}_{\cdot i} = - \tilde{S}^{i}_{\cdot k} \left\{ \frac{\partial^{2} \psi}{\partial \tilde{S}^{r}_{s} \, \partial \tilde{S}^{i}_{k}} (\dot{\tilde{S}})^{r}_{\cdot s} + \frac{\partial^{2} \psi}{\partial T \, \partial \tilde{S}^{i}_{k}} \mathring{T} \right.$$
$$\left. + \frac{\partial^{2} \psi}{\partial b \, \partial \tilde{S}^{i}_{k}} \dot{b} + \frac{\partial^{2} \psi}{\partial \beta^{r}_{s} \, \partial \tilde{S}^{i}_{k}} (\dot{\beta})^{r}_{\circ s} \right\} . \tag{4.12}$$

This relation balances the supplied reversible work with the changes of the state variables. On the other hand we obtain by equation (4.10) and (4.11)

$$\underset{(i)}{\dot{w}} - \frac{1}{\rho} q^i\Big|_i + r = - T \left\{ \frac{\partial^2 \psi}{\partial \tilde{S}^i_k \partial T} (\dot{\tilde{S}})^i_{\circ k} + \frac{\partial^2 \psi}{\partial T^2} \dot{T} \right\}$$

$$+ \frac{\partial}{\partial b} \left\{ \psi - T \frac{\partial \psi}{\partial T} \right\} \dot{b} + \frac{\partial}{\partial \beta^i_k} \left\{ \psi - T \frac{\partial \psi}{\partial T} \right\} (\dot{\beta})^i_{\cdot k} . \quad (4.13)$$

where

$$\psi - T \frac{\partial \psi}{\partial T} = \psi + T s = h (\tilde{S}^i_k, T, b, \beta^i_k) \quad (4.14)$$

means the specific enthalpy (Gibbs function).

The relation (4.13) balances the remaining energy supply with the changes of the state variables. The sum of (4.12) and (4.13) represents the total energy balance containing on the left hand side the energy supply and on the right hand side the changes of the thermodynamical state.

In the above balance equations for the energy the coefficients of the material time derivatives of the state variables are derivable from the specific free enthalpy. From this fact some important restrictions arise with respect to the formulations of the constitutive law. This will be discussed later.

Other restrictions result from the second law of thermodynamics. In the balance equation (4.10) for the specific entropy the evolution of entropy has to be decomposed into its reversible part $\underset{(r)}{\dot{s}}$ and into its irreversible part $\underset{(irr)}{\dot{s}}$ according to

$$T \dot{s} = T \underset{(r)}{\dot{s}} + T \underset{(irr)}{\dot{s}} . \tag{4.15}$$

In order to define the irreversible part of $\dot{s}$ we have to decompose the terms on the right hand side of equation (4.10) into those parts which are dissipated immediately and into the remaining parts. The immediately dissipated parts which contributes to the entropy production are

1. the immediately dissipated mechanical work

$$\underset{(d)}{\dot{w}} = \underset{(i)}{\dot{w}} - \underset{(h)}{\dot{w}}$$

where $\underset{(h)}{w}$ denotes the specific mechanical work stored in changes of the internal structure of the material,

2. the irreversible part of the heat flux $\underset{(T)}{q^i}$ , i.e.

$$- \frac{1}{T} \underset{(T)}{q^i} T|_i = - \frac{1}{T} (q^i - \underset{(h)}{q^i}) T|_i \tag{4.17}$$

where $\underset{(h)}{q^i}$ represents the energy fluxes apart from heat,

3. the entropy production (summarized in $\dot{\eta}$ ) due to dissipative

processes in remaining energy fluxes $\underset{(h)}{q}^{i}$ , in energy supply by sources $r$ , and in changes of the internal material structure.

Using these notations the second law of thermodynamics reads

$$T \underset{(irr)}{\dot{s}} = \underset{(d)}{\dot{w}} - \frac{1}{\rho T} \underset{(T)}{q}^{i} T\Big|_{i} + T \dot{\eta} \geq 0. \qquad (4.18)$$

$\underset{(d)}{\overset{\circ}{w}}$ and $\dot{\eta}$ have to be defined within the constitutive law.

8.5. Reversible work and deformations

According to our material model (fig. 7) the decomposition of mechanical work (4.2) can also be written

$$\dot{w} = \underset{(r)}{\dot{w}} + \underset{(i)}{\dot{w}}$$

$$= \frac{1}{\rho} S^{i}_{k} \underset{(r)}{d}^{k}_{i} + \frac{1}{\rho} S^{i}_{k} \underset{(i)}{d}^{k}_{i} . \qquad (5.1)$$

This leads to an additive decomposition of the deformation rate into one part $\underset{(r)}{d}^{i}_{k}$ corresponding to the reversible deformations and into a remaining part $\underset{(i)}{d}^{i}_{k}$ :

$$d^i_k = \underset{(r)}{d}{}^i_k + \underset{(i)}{d}{}^i_k \ . \tag{5.2}$$

$\underset{(r)}{d}{}^i_k$ and $\underset{(i)}{d}{}^i_k$ are not necessary symmetric tensors but their sum is symmetric. Therefore we keep the symmetric notations of $\underset{(r)}{d}{}^i_k$ and $\underset{(i)}{d}{}^i_k$ apart from such cases where it becomes necessary to emphasize the unsymmetry.

From thermodynamical considerations we concluded on the other hand that $\underset{(r)}{\dot{w}}$ must be expressible in the form (4.4). Thus we obtain

$$\underset{(r)}{\dot{w}} = \frac{1}{\overset{o}{\rho}} S^i_k \underset{(r)}{d}{}^k_i = \frac{1}{\overset{o}{\rho}} \tilde{S}^i_{\cdot k} \underset{(r)}{\dot{\tilde{\varepsilon}}}{}^k_{\cdot i} \ . \tag{5.3}$$

$\underset{(r)}{d}{}^i_k$, $\tilde{S}^i_{\cdot k}$, and $\underset{(r)}{\tilde{\varepsilon}}{}^i_k$ have still to be defined in a suitable manner meeting the following physical requirements:

1. The tensor $\underset{(r)}{\tilde{\varepsilon}}{}^i_k$ has to describe the reversible deformations of the material elements independent of possible accompanying inelastic deformations. This implies among others that the invariants of the tensor $\underset{(r)}{\tilde{\varepsilon}}{}^i_k$ must only depend on the ratios $ds/\overset{*}{ds}$ measured in the three main axes of $\underset{(r)}{\tilde{\varepsilon}}{}^i_k$ where $\overset{*}{ds}$ relates to the fictitious reference configuration $\overset{*}{\mathfrak{L}}$ of the body.

2. Vanishing invariants of $\underset{(r)}{d}{}^i_k$ imply vanishing material time derivatives of the invariants of $\underset{(r)}{\tilde{\varepsilon}}{}^i_k$.

3. $\tilde{S}^{i}_{\cdot k}$ and $\tilde{\varepsilon}^{i}_{k}$ have to represent an conjugated pair of stress and strain.

According to the different ways concerning the description of total deformation and total deformation rate as discussed in section 8.2.4 we may follow different lines in decomposition of total deformation and deformation rate. Some possibilities are:

1. We decompose the deformation gradient (2.10) multiplicatively according to [7] to [10].

$$\underset{\sim}{F} = \frac{\partial x^{\alpha}}{\partial \overset{o}{x}^{\beta}} \underset{\sim}{g}_{\alpha} \overset{o}{\underset{\sim}{g}}^{\beta} = \underset{(\tilde{r})}{F} \cdot \underset{(\tilde{i})}{F} = \frac{\partial x^{\alpha}}{\partial \overset{*}{x}^{\rho}} \underset{\sim}{g}_{\alpha} \overset{*}{\underset{\sim}{g}}^{\rho} \cdot \frac{\partial \overset{*}{x}^{\sigma}}{\partial x^{\beta}} \overset{*}{\underset{\sim}{g}}_{\sigma} \underset{\sim}{g}^{\beta} . \tag{5.4}$$

2. We split the changes of the metric (2.32) of the body-fixed coordinate system additively obtaining [11], [12].

$$e_{ik} = \frac{1}{2} \left\{ g_{ik} - \overset{o}{g}_{ik} \right\} = \frac{1}{2} \left\{ g_{ik} - \overset{*}{g}_{ik} \right\} + \frac{1}{2} \left\{ \overset{*}{g}_{ik} - \overset{\bullet}{g}_{ik} \right\}$$

$$= \underset{(r)}{e}_{ik} + \underset{(i)}{e}_{ik} . \tag{5.5}$$

3. We decompose the metric transformation (2.34) multiplicatively corresponding to [13], [14] .

$$q^{i}_{k} = \overset{o}{g}^{ir} g_{rk} = \overset{o}{g}^{im} \overset{*}{g}_{mr} \overset{*}{g}^{rs} g_{sk} = \underset{(i)}{q}^{i}_{\cdot r} \underset{(r)}{q}^{r}_{k} . \tag{5.6}$$

It can be easily shown that the decomposition (5.5) does not meet our physical requirements since

$$\underset{(r)}{e}_{ik}\, d\xi^i\, d\xi^k = \frac{1}{2}\left\{ds^2 - \overset{*}{ds}{}^2\right\} = \frac{1}{2}\overset{*}{ds}{}^2\left\{\left(\frac{ds}{\overset{*}{ds}}\right)^2 - 1\right\} \tag{5.7}$$

does not only depend on $ds/\overset{*}{ds}$ but also on $\overset{*}{ds}$ itself. Some difficulties arise also from the decomposition (5.4). The elimination of the involved rotation from $\underset{(r)}{\tilde{F}}$ and the establishment of an unique connection between the time derivative of an objective strain tensor based on $\underset{(r)}{\tilde{F}}$ and the corresponding reversible deformation rate requires additional considerations [9, 10].

Therefore we prefer the decomposition (5.6). It leads to an additive decomposition of the total deformation rate according to

$$d^i_k = \frac{1}{2}(q^{-1})^i_r (\dot{q})^r_{\cdot k} = \frac{1}{2}\underset{(r)}{(q^{-1})}{}^i_r \underset{(r)}{(\dot{q})}{}^r_{\cdot k} + \frac{1}{2}(q^{-1})^i_r \underset{(i)}{(\dot{q})}{}^r_{\cdot m} \underset{(r)}{q}{}^m_k$$

$$= \underset{(r)}{d}{}^i_{\cdot k} + \underset{(i)}{d}{}^i_{\cdot k} \tag{5.8}$$

$$\text{with} \quad \underset{(r)}{d}{}^i_{\cdot k} = \frac{1}{2}\underset{(r)}{(q^{-1})}{}^i_r \underset{(r)}{(\dot{q})}{}^r_{\cdot k}\,. \tag{5.9}$$

As already mentioned the components $\underset{(r)}{d}{}^i_{\cdot k}$ and $\underset{(i)}{d}{}^i_{\cdot k}$ are not necessary symmetric tensors but their sum is symmetric. Furthermore only their symmetric parts enter the expressions for the respectively decomposed work rate.

Inserting (5.9) into equation (5.3) we obtain

$$\underset{(r)}{\dot{w}} = \frac{1}{\overset{\circ}{\rho}} s^{i}_{r} \, (\underset{(r)}{q}^{-1})^{r}_{k} \, \frac{1}{2} \, (\underset{(r)}{\dot{q}})^{r}_{\cdot k} = \frac{1}{\overset{\circ}{\rho}} \tilde{S}^{i}_{\cdot k} \, \overset{\circ}{(\tilde{\varepsilon})}{}^{k}_{\cdot i} . \tag{5.10}$$

This leads to the conclusion that

$$\tilde{S}^{i}_{\cdot k} = s^{i}_{r} \, (\underset{(r)}{q}^{-1})^{r}_{k} \quad \text{and} \quad \tilde{\varepsilon}^{i}_{k} = \frac{1}{2} \underset{(r)}{q}{}^{i}_{k} \tag{5.11}$$

represent an conjugated pair of stress and strain.

Therefore the thermic state equation (4.8) can also be written

$$\underset{(r)}{\tilde{\varepsilon}}{}^{i}_{k} = \frac{1}{2} \underset{(r)}{q}{}^{i}_{k} = - \overset{\circ}{\rho} \frac{\partial \psi}{\partial \tilde{S}^{k}_{i}} . \tag{5.12}$$

From this relation we derive the following expression for the reversible deformation rate, i.e. the incremental law for the reversible deformations:

$$\underset{(r)}{d}{}^{i}_{\cdot k} = \frac{1}{2} \, (\underset{(r)}{q}^{-1})^{i}_{r} \, (\underset{(r)}{\dot{q}})^{r}_{\cdot k}$$

$$= - \overset{\circ}{\rho} \, (\underset{(r)}{q}^{-1})^{i}_{r} \left\{ \frac{\partial^2 \psi}{\partial \tilde{S}^{1}_{m} \partial \tilde{S}^{k}_{r}} \overset{\circ}{(\tilde{S})}{}^{1}_{\cdot m} + \frac{\partial^2 \psi}{\partial T \partial \tilde{S}^{k}_{r}} \dot{T} + \frac{\partial^2 \psi}{\partial b \, \partial \tilde{S}^{k}_{r}} \dot{b} + \frac{\partial^2 \psi}{\partial \beta^{1}_{m} \partial \tilde{S}^{k}_{r}} (\dot{\beta})^{1}_{m} \right\}$$

$$= -\rho \left( \underset{(r)}{q}^{-1} \right)^{i}_{r} \left\{ A^{rm}_{kl} (\overset{\circ}{\tilde{S}})^{l}_{\cdot m} + B^{r}_{k} \dot{T} + C^{r}_{k} \dot{b} + D^{rm}_{kl} (\dot{\beta})^{l}_{\cdot m} \right\} . \qquad (5.13)$$

Only the symmetric part of $\underset{(r)}{d}^{i}_{\circ k}$, however, is needed.

The coefficients $A^{rm}_{kl}$ etc. have to fulfil certain reciprocity conditions following immediately from their definitions. They have the meaning of integrability conditions securing that the constitutive law for the reversible deformations is derivable from a thermodynamical potential. The reciprocity conditions are

$$\frac{\partial A^{ir}_{ks}}{\partial T} = \frac{\partial B^{i}_{k}}{\partial \tilde{S}^{s}_{r}} \; ; \qquad \frac{\partial A^{ir}_{ks}}{\partial b} = \frac{\partial C^{i}_{k}}{\partial \tilde{S}^{s}_{r}} \; ;$$

$$\frac{\partial A^{ir}_{ks}}{\partial \beta^{m}_{n}} = \frac{\partial D^{in}_{km}}{\partial \tilde{S}^{s}_{r}} \; ; \qquad \frac{\partial B^{i}_{k}}{\partial b} = \frac{\partial C^{i}_{k}}{\partial T} \; ; \qquad (5.14)$$

$$\frac{\partial B^{i}_{k}}{\partial \beta^{s}_{r}} = \frac{\partial D^{ir}_{ks}}{\partial T} \; ; \qquad \frac{\partial C^{i}_{k}}{\partial \beta^{s}_{r}} = \frac{\partial D^{ir}_{ks}}{\partial b} \; .$$

In the incremental law for the reversible deformations the increments of the internal variables have to be expressed by the respective evolution laws. If dissipative internal processes are not involved the incremental law for the reversible deformations can be reduced to an expression of the form

$$\underset{(r)}{d}{}^{i}_{k} = E^{ir}_{ks} \overset{\nabla}{\tilde{S}}{}^{s}_{r} + F^{i}_{k} \dot{T} \tag{5.15}$$

where the coefficients $E^{ir}_{ks}$ and $F^{i}_{k}$ may still depend on the whole set of state variables. It may be mentioned that in the formulation of (5.15) we are only interested in the symmetric part of $\underset{(r)}{d}{}^{i}_{\cdot k}$ since only this part enters the respective work rate.

The situation becomes much simpler if the reversible processes behave isotropically. In this case $\underset{(r)}{q}{}^{i}_{k}$ and $S^{i}_{k}$ become co-axial. This enables us to introduce $S^{i}_{k}$ itself and the logarithmic strain tensor

$$\underset{(r)}{\varepsilon}{}^{i}_{k} = \frac{1}{2} (\ln \underset{(r)}{q})^{i}_{k} \tag{5.16}$$

as conjugated pair of stress and strain, since in the isotropic case

$$\underset{(r)}{\dot{w}} = \frac{1}{\overset{\circ}{\rho}} S^{i}_{k} \underset{(r)}{\dot{\varepsilon}}{}^{k}_{\cdot i} = \frac{1}{\overset{\circ}{\rho}} S^{i}_{k} \underset{(r)}{\overset{\nabla}{\varepsilon}}{}^{k}_{\cdot i} \ . \tag{5.17}$$

It can be shown [13] that in many cases the reversible (hyperelastic) behaviour which is derivable from a thermodynamical potential can be replaced by a simplified linear incremental law of the form

$$\underset{(r)}{d}{}^{i}_{k} = E^{ir}_{ks} \overset{\nabla}{S}{}^{s}_{r} + F^{i}_{k} \dot{T} \tag{5.18}$$

with constant tensors $E^{ir}_{ks}$ and $F^{i}_{k}$. In the isotropic case this reduces to

$$\underset{(r)}{d}{}^{i}_{k} = \frac{1}{2G} \overset{\nabla}{t}{}^{i}_{k} + \left\{ \frac{1}{9K} \dot{S}^{r}_{r} + \alpha \dot{T} \right\} \delta^{i}_{k} \tag{5.19}$$

with $G$ shear modulus,

$K$ bulk modulus,

$\alpha$ coefficient of thermal expansion,

$t^{i}_{k}$ (weithted) stress deviator.

For simplicity we assume in the following isotropic reversible behaviour. For this reason we omit in the following the tilda on $S^{i}_{k}$ and $\underset{(r)}{\varepsilon}{}^{i}_{k}$. We keep, however, in mind that in general cases we always have to start with conjugated pairs of stress and strain. The subsequent transition to other measures of stress and strain is always possible and can be easily achieved [14].

## 8.6 Some remarks on the evolution laws for internal variables.

When we have to take into account additional energy fluxes apart from heat flux the evolution laws for the internal variables have the form of balance equations containing also spatial gradients of the state variables. They represent in this case partial differential equations in space and time. If, however, such energy fluxes can be disregarded the balance equations for the internal variables are reduced to simple evolution laws representing ordinary first order differential equations in time. Their general form is

$$\dot{b} = \dot{b}\left\{\overset{i}{S}_k,\ T,\ b,\ \overset{i}{\beta}_k;\ \overset{\nabla i}{S}_k,\ \mathring{T}\right\},$$

$$\overset{\nabla i}{\beta}_k = \overset{\nabla i}{\beta}_k\left\{\overset{i}{S}_k,\ T,\ b,\ \overset{i}{\beta}_k;\ \overset{\nabla i}{S}_k,\ \dot{T}\right\}. \qquad (6.1)$$

In the following we shall assume that the additional energy fluxes can be disregarded since in solid bodies these fluxes remain small in most cases. This means we adopt the general form (6.1) for the evolution laws of the internal variables. Concerning the particular form of the evolution laws we can distinguish three types:

(A) Equilibrium type.

It is characterized by the fact that the right hand side of the evolution laws (6.1) becomes zero only if $\overset{\nabla i}{S}_k$ and $\dot{T}$ are vanishing at the same time. The changes of the respective internal variables represent in this case a sequence of equilibrium states.

(B) Non-equilibrium type.

It is characterized by the fact that $\overset{\nabla i}{S}_k$ and $\dot{T}$ do not enter the evolution law. The changes of the respective internal variable run through non-equilibrium states.

(C) Mixed type.

Strictly reversible changes of internal variables like, for instance, certain quasistatic solid phase transformations are governed by additional state functions allowing for eliminating of supernumerary variables. Such processes fall into the scope of section 8.5 independent whether or not they are connected with macroscopic deformations. They need no further considerations in this context.

We know, however, other processes of the equilibrium type (A) which are not governed by state equations. This is, for instance, true for certain hardening and softening processes due to inelastic deformations. The mechanical work involved in such processes is not dissipated immediately but stored inside the body. Within the energy balance this work has to be balanced separately with the corresponding changes of the internal energy due to changes of the internal variables.

Dissipative internal processes are described by evolution laws of non-equilibrium type (B). They require likewise a separate treatment within the balance equations of energy and entropy. The respective details of balancing have to be specified within the constitutive law.

Using the decomposition (4.16) the balance equation (4.13) for the remaining energy supply reads

$$\underset{(h)}{\dot{w}} + \underset{(d)}{\dot{w}} - \frac{1}{\rho} q^i\Big|_i + r = - T \left\{ \frac{\partial^2 \psi}{\partial s^i_k \partial T} (\overset{\circ}{S})^i_{\cdot k} + \frac{\partial^2 \psi}{\partial T^2} \overset{\circ}{T} \right\}$$
$$+ \frac{\partial h}{\partial b} \dot{b} + \frac{\partial h}{\partial \beta^i_k} (\dot{\beta})^i_{\cdot k}$$

$$= T \left\{ \frac{c_p}{T} \dot{T} - B^k_{i} (\dot{S})^i_{\cdot k} \right\} + G\dot{b} + H^k_{i} (\dot{\beta})^i_{\cdot k}. \tag{6.2}$$

where

$$c_p = - T \frac{\partial^2 \psi}{\partial T^2} \tag{6.3}$$

means the heat capacity at constant stress and constant internal variables. The coefficient $B^i_k$ in (6.2) is by the way identical with the coefficient $B^i_k$ in (5.13).

The coefficient on the right hand side of (6.2) have again to obey certain reciprocity conditions. These are

$$- \frac{\partial}{\partial S^i_k} \left( \frac{c_p}{T} \right) = \frac{\partial B^k_{i}}{\partial T} , \quad \frac{\partial G}{\partial \beta^i_k} = \frac{\partial H^k_{i}}{\partial b} . \tag{6.4}$$

They are less restrictive than the corresponding conditions (5.14) holding for reversible processes.

For demonstration let us consider an isotropic hardening law of any material. The hardening state may be reflected by a scalar valued internal variable b. The internal process may represent an interaction between a hardening or softening process due to inelastic deformations and an annealing process caused by recrystallization. This interaction leads to an evolution law for b of the form

$$\overset{\circ}{b} = \zeta \underset{(h)}{\dot{w}} - (b - \overset{\circ}{b})\vartheta$$

$$\text{with} \quad \underset{(h)}{\dot{w}} = \underset{(h)}{\overset{\circ}{w}} \{S^i_k, T, b, \beta^i_k, \overset{\nabla}{S}{}^i_k, \overset{\circ}{T}\} \gtrless 0 \quad \text{depending on certain side conditions,}$$

$$\vartheta(S^i_k, T, b, \beta^i_k) \begin{cases} > 0 & \text{for } T > T_R(S^i_k, b, \beta^i_k) \\ = 0 & \text{for } T \leq T_R(S^i_k, b, \beta^i_k). \end{cases} \tag{6.5}$$

$T_R$ denotes the recrystallization temperature. Above $T_R$ the recrystallization starts.

$\underset{(h)}{\dot{w}}$ has still to be specified by the constitutive law. This will be discussed in connection with the other laws for inelastic deformations.

The evolution law for the internal variable b represents a mixed type. The first term is of the equilibrium type (A); the second is of the non-equilibrium type (B) and describes a dissipative process. It is obvious that the first term has to be balanced with one part of $\underset{(h)}{\dot{w}}$ on the left hand side of the balance equation (6.2), whereas the second term has to be treated as entropy production.

From the preceding considerations it can be already seen how the different parts of the constitutive law are intermeshing into one another. This makes the task of formulation of a consistent constitutive law difficult.

8.7. Inelastic work and deformations

Corresponding to our material model (fig. 7) we can decompose the total work and the total deformation rate for the present according to

$$\dot{w} = \underset{(r)}{\dot{w}} + \underset{(s)}{\dot{w}} + \underset{(p)}{\dot{w}} + \underset{(c)}{\dot{w}}$$

$$= \frac{1}{\overset{\circ}{\rho}} S^i_k \underset{(r)}{d^k_{\cdot i}} + \frac{1}{\overset{\circ}{\rho}} S^i_k \underset{(s)}{d^k_{\cdot i}} + \frac{1}{\overset{\circ}{\rho}} S^i_k \underset{(p)}{d^k_{\cdot i}} + \frac{1}{\overset{\circ}{\rho}} S^i_k \underset{(c)}{d^k_{\cdot i}} \qquad (7.1)$$

The corresponding decomposition of the total deformation rate is

$$d^i_k = \underset{(r)}{d^i_k} + \underset{(s)}{d^i_k} + \underset{(p)}{d^i_k} + \underset{(c)}{d^i_k} \qquad (7.2)$$

The introduced indices relate to the following processes

(r): reversible process

(p): viscoplastic processes

(s): processes leading to certain changes of the internal structure (hardening and softening)

(c): dissipative long time creep or relaxation processes.

$\underset{(r)}{\dot{w}}$ and $\underset{(r)}{d^i_k}$ are already discussed in section 4.

From the thermodynamical point of view the remaining part of the mechanical work rate has to be split into the dissipated part $\underset{(d)}{\dot{w}}$ and into that part $\underset{(h)}{\dot{w}}$ which is connected with changes of the stored energy. This leads to

$$\dot{w}_{(i)} = \dot{w}_{(s)} + \dot{w}_{(p)} + \dot{w}_{(c)} = \dot{w}_{(h)} + \dot{w}_{(d)} . \qquad (7.3)$$

Concerning $\dot{w}_{(s)}$ and $\dot{w}_{(c)}$ we may assume that $\overset{\circ}{w}_{(s)}$ contributes to the changes of the stored energy whereas $\overset{\circ}{w}_{(c)}$ characterizes a pure dissipative process. The viscoplastic work, however, requires a separate discussion. Indeed the mainpart of this work is dissipated. This part may be denoted by $\overset{\circ}{w}_{(p/d)}$ .

However, one small part

$$\dot{w}_{(p/h)} = \xi \overset{\circ}{w}_{(p)} \qquad (\xi \ll 1) \qquad (7.4)$$

is involved in changes of the internal structure.

Therefore we obtain

$$\dot{w}_{(d)} = \dot{w}_{(p/d)} + \dot{w}_{(c)} = (1 - \xi) \dot{w}_{(p)} + \overset{\circ}{w}_{(c)} \qquad (7.5)$$

and

$$\dot{w}_{(h)} = \overset{\circ}{w}_{(s)} + \xi \dot{w}_{(p)} . \qquad (7.6)$$

Usually it is assumed that viscoplastic deformations $\underset{(p)}{d^i_k}$ and deformations $\underset{(s)}{d^i_k}$ connected with changes of the internal structure occur only under certain constraints concerning the actual state. The constraints (yield conditions) for these two kinds of processes, however, are not to be identical. The introduction of different yield conditions rather allows for a better description of certain phenomena. This is, for instance, true of the hysteresis at unloading and reloading in tension tests (see fig. 9) and at other cyclic processes.

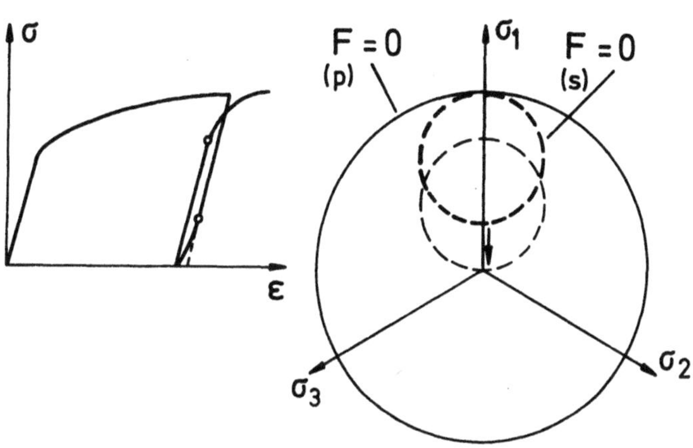


fig. 9: hysteresis in unloading and reloading

From the balance equation (4.13) for the supplied energy apart from strictly reversible processes and from the fact that $\underset{(s)}{\dot{w}}$ represents a non-dissipative part of $\underset{(i)}{\overset{\circ}{w}}$ we conclude that $\underset{(s)}{\dot{w}}$ must be expressible in the general form

$$\underset{(s)}{\dot{w}} = \frac{1}{\overset{\circ}{\rho}} S^k_i \left\{ a^{is}_{kr} \overset{\nabla}{S}{}^r_s + b^i_k \overset{\circ}{T} + c^i_k \dot{b} + d^{is}_{kr} \overset{\nabla}{\beta}{}^r_s \right\} . \qquad (7.7)$$

This leads for $\underset{(s)}{d}{}^i_k$ to an expression of the general form

$$\underset{(s)}{d}{}^i_k = a^{is}_{kr}\,\overset{\nabla}{S}{}^r_s + b^i_k\,\dot{T} + c^i_k\,\mathring{b} + d^{is}_{kr}\,\overset{\nabla}{\beta}{}^r_s \,. \tag{7.8}$$

$\dot{b}$ and $\overset{\nabla}{\beta}{}^i_k$ do not necessarily enter these expressions. If they do we can express $\dot{b}$ and $\overset{\nabla}{\beta}{}^i_k$ by the respective evolution laws (6.1). Finally we obtain

$$\underset{(s)}{d}{}^i_k = \underset{(s)}{d}{}^i_k \left\{ S^i_k,\ T,\ b,\ \beta^i_k,\ \overset{\nabla}{S}{}^i_k,\ \dot{T} \right\} . \tag{7.9}$$

Only if the evolution laws for the internal variables are of the equilibrium type or if they do not at all enter (7.8) this reduces to an expression of the form

$$\underset{(s)}{d}{}^i_k = \chi^{ir}_{ks}\,\overset{\nabla}{S}{}^s_r + \lambda^i_k\,\dot{T} \,. \tag{7.10}$$

The (necessary but not sufficient) yield condition for the appearance of separate deformations $\underset{(s)}{d}{}^i_k$ apart from $\underset{(p)}{d}{}^i_k$ can be written in the form

$$\underset{(s)}{F}\,(S^i_k,\ T,\ b,\ \beta^i_k) = 0 \tag{7.11}$$

The continuation of such processes requires additionally

$$\dot{F}_{(s)} = 0 = \frac{\partial F_{(s)}}{\partial S^i_k} \overset{\nabla}{S}{}^i_k + \frac{\partial F_{(s)}}{\partial T} \dot{T} + \frac{\partial F_{(s)}}{\partial b} \overset{\circ}{b} + \frac{\partial F_{(s)}}{\partial \beta^i_k} \overset{\nabla}{\beta}{}^i_k \; . \tag{7.12}$$

Expressing again $\dot{b}$ and $\overset{\nabla}{\beta}{}^i_k$ by the respective evolution laws (6.1) we can derive from (7.12) a sufficient condition for the existence of $\underset{(s)}{d}{}^i_k \neq 0$.

From the preceding considerations we see that the yield condition (7.11), the evolution laws for the internal variables (6.1), and the law (7.9) for $\underset{(s)}{d}{}^i_k$ are interacting within the frame given by the balance equation (4.13).

A similar situation exists with respect to the viscoplastic deformations. We assume again that such deformations occur only if a yield condition of the form

$$\underset{(p)}{F} \, (S^i_k, T, b, \beta^i_k) \geq 0 \tag{7.13}$$

is fulfilled. In (7.13) the equality holds only when the viscous effect is vanishing. When viscous effects have to be taken into account according to our material model (fig. 7) the stresses $S^i_k$ are split into one part $\bar{S}^i_k$ acting in the rate independent mechanism and into another part $S^i_k - \bar{S}^i_k$ working in the rate dependent mechanism:

$$S^i_k = \bar{S}^i_k + (S^i_k - \bar{S}^i_k) \; . \tag{7.14}$$

The stresses $\bar{S}^{i}_{k}$ obey during viscoplastic deformations the condition

$$\underset{(p)}{\bar{F}}\,(\bar{S}^{i}_{k},\ T,\ b,\ \beta^{i}_{k}) = 0 \tag{7.15}$$

This function $\underset{(p)}{\bar{F}}$ differs from $\underset{(p)}{F}$ only by the fact that $S^{i}_{k}$ is replaced by $\bar{S}^{i}_{k}$. $\underset{(p)}{\bar{F}}$ characterizes the equilibrium state whom the viscoplastic process tends to.

As usually we assume for $\underset{(p)}{F} > 0$

$$\underset{(p)}{d^{i}_{k}} = \gamma\,\Phi\,(\underset{(p)}{F})\,\frac{\partial \underset{(p)}{F}}{\partial S^{k}_{i}} \tag{7.16a}$$

$$= \lambda\,\frac{\partial \underset{(p)}{\bar{F}}}{\partial \bar{S}^{k}_{i}}\,. \tag{7.16b}$$

$\gamma$ has the meaning of a constant reference quantity.

By constructing the double scalar products of both relations (7.16a) and (7.16b) we derive

$$\lambda = \gamma\,\Phi\,(\underset{(p)}{F})\,\sqrt{\frac{\dfrac{\partial \underset{(p)}{F}}{\partial S^{i}_{k}}\,\dfrac{\partial \underset{(p)}{F}}{\partial S^{k}_{i}}}{\dfrac{\partial \underset{(p)}{\bar{F}}}{\partial \bar{S}^{r}_{s}}\,\dfrac{\partial \underset{(p)}{\bar{F}}}{\partial \bar{S}^{s}_{r}}}} \tag{7.17}$$

Using this result we can also express $\bar{S}^{i}_{k}$ by $S^{i}_{k}$ and the other state variables [15]. Furthermore we can prove that during viscoplastic deformations the consistency condition $\bar{\underset{(p)}{F}} = 0$ is identically fulfilled. We obtain finally for the viscoplastic deformation rate an expression of the form

$$\underset{(p)}{d}{}_k^i = \underset{(p)}{d}{}_k^i \left\{ S_k^i, T, b, \beta_k^i \right\}. \tag{7.18}$$

The law of rate independent plastic deformations cannot be achieved by a transition to the limit case of quasistatic viscoplastic deformations. It requires a separate treatment.

Starting with $S_k^i = \bar{S}_k^i$ the yield condition now reads

$$\bar{\underset{(p)}{F}} = \underset{(p)}{F} (S_k^i, T, b, \beta_k^i). \tag{7.19}$$

Assuming as usually that the so-called normality condition holds we put

$$\underset{(p)}{d}{}_k^i = \lambda \frac{\partial \underset{(p)}{F}}{\partial S_i^k}. \tag{7.20}$$

$\lambda$ has to be determined from the consistency condition for plastic deformations

$$\dot{\underset{(p)}{F}} = \frac{\partial \underset{(p)}{F}}{\partial S_k^i} \overset{\nabla}{S}{}_k^i + \frac{\partial \underset{(p)}{F}}{\partial T} \dot{T} + \frac{\partial \underset{(p)}{F}}{\partial b} \dot{b} + \frac{\partial \underset{(p)}{F}}{\partial \beta_k^i} \overset{\nabla}{\beta}{}_k^i = 0 \tag{7.21}$$

where $\dot{b}$ and $\overset{\nabla}{\beta}{}_k^i$ again have to be expressed by the respective evolution laws (6.1). This leads to a law for the plastic deformation rate in the form

$$\underset{(p)}{d}{}^i_k = \underset{(p)}{d}{}^i_k \left\{ S^i_k,\ T,\ b,\ \beta^i_k,\ \overset{\nabla}{S}{}^i_k,\ \dot{T} \right\} . \qquad (7.22)$$

In comparison with the law for viscoplastic deformations now the increments of $S^i_k$ and $T$ enter the relation. However only if dissipative, non-equilibrium processes in connection with changes of the internal material structure are absent this relation reduces to an expression of the form

$$\underset{(p)}{d}{}^i_k = K^{ir}_{ks} \overset{\nabla}{S}{}^s_r + L^i_k \dot{T} \qquad (7.23)$$

In many considerations we don't find a clear distinction between the viscoplastic deformation which also can lead to creep and relaxation processes on the one hand and the long time creep and relaxation processes on the other hand. The difference is the following. Creep and relaxation processes involved in viscoplastic deformations are limited and tend to an equilibrium state. Theoretically this equilibrium state is reached at infinite time asymptotically. Practically these creep and relaxation processes are only essential for a certain short time period. Long time creep and relaxation processes, however, are unlimited. They lead to damage in the long time range. Therefore they are important among others for certain safety problems, for instance, at pressure vessels, or at prestressed concrete structures.

In most technical problems we can separate viscoplastic deformations and long time creep or relaxation processes on account of the different time scales. Systematical investigations on possible coupling effects between these two different kinds of inelastic processes are not known as yet. Concerning long time creep and relaxation processes we relate to the special literatur in the different fields of application [22] to [27]. We shall focus our consideration to the problem of viscoplasticity.

## 8.8. The total scheme of the constitutive laws in thermoplasticity

The constitutive law for a material undergoing non-isothermic, elastic-viscoplastic deformations (including long time creep and relaxation) comprehends

(a) the state function for the specific free enthalpy determining at the same time the reversible processes (see section 8.5),

(b) the evolution laws for the different components of the inelastic deformations (see section 8.7),

(c) the evolution laws for the internal energy (see section 8.6),

(d) the flux laws for the different kinds of energy (heat flux and others),

(e) the laws of entropy production specifying $\dot{w}_{(d)}$ and $\dot{\eta}$ .

Due to the occurence of energy fluxes the constitutive law forms a system of partial differential equations in space and time completed by some side conditions. Only if all energy fluxes are vanishing the constitutive law is reducible to a system of first order ordinary differential equations in time with some side conditions. We obtain, however, already a simplified constitutive law, if we can at least disregard the energy fluxes apart from heat. Then the evolution laws for the strain and the internal variables as well as the laws for the entropy production are reduced to first order ordinary differential equations in time. In this case we obtain the following simplified scheme of the constitutive law.

specific free enthalpy: $\psi = \psi(S^i_k, T, b, \beta^i_k)$, (8.1)

evolution laws (with side conditions)

total strain: $d^i_k = d^i_k (S^i_k, T, b, \beta^i_k; \overset{\nabla}{S}{}^i_k, \dot{T})$, (8.2)

internal variables $\dot{b} = \breve{b} (S^i_k, T, b, \beta^i_k; \overset{\nabla}{S}{}^i_k, \dot{T})$, (8.3)

$\overset{\nabla}{\beta}{}^i_k = \overset{\nabla}{\beta}{}^i_k (S^i_k, T, b, \beta^i_k; \overset{\nabla}{S}{}^i_k, \dot{T})$, (8.4)

flux law of heat $q^i = q^i (S^i_k, T, b, \beta^i_k, S^i_k|_r, T|_r)$, (8.5)

laws of entropy production:

$$\underset{(d)}{\dot{w}} = \underset{(d)}{\dot{w}} \; (S^i_k, \; T, \; b, \; \beta^i_k; \; \overset{\nabla i}{S}_k, \; \dot{T}), \tag{8.6}$$

$$T\dot{\eta} = T\dot{\eta} \; (S^i_k, \; T, \; b, \; \beta^i_k; \; \overset{\nabla i}{S}_k, \; \dot{T}). \tag{8.7}$$

The evolution laws for the inelastic parts of the deformations and for the internal variables on the one hand and the laws of energy flux and entropy production on the other hand have to be defined in accordance with the balance equations of energy and entropy. One example is given in the next section.

## 8.9. An example for the constitutive law of an elastic-viscoplastic and an elastic-plastic material

We consider the constitutive law of an elastic-viscoplastic material and of an corresponding elastic-plastic material. Long time creep or relaxation processes are disregarded. Likewise solid phase transformation are not taken into account.

The specific free enthalpy may be given in the form

$$\psi(S^i_k, \; T, \; b, \; \beta^i_k) = \psi^* \; (S^i_k, \; T) + \psi^{**} \; (T, \; b) + \psi^{***} \; (T, \; B)$$

$$\text{with } B - \beta^i_k \; \beta^k_i. \tag{9.1}$$

This means that the reversible deformations are not influenced by changes of the internal variables.

With (9.1) the balance equation (6.2) reads

$$\underset{(h)}{\dot{w}} + \underset{(d)}{\dot{w}} - \frac{1}{\rho} q^i\Big|_i + r = c_p \dot{T} - T B^i_k \overset{\nabla k}{S_i} + h\dot{b} + 2g\beta^i_k \overset{\nabla k}{\beta_i} \tag{9.2}$$

$$\text{with:} \quad h(b, T) = \frac{\partial}{\partial b}\left\{ \psi^{**} - T \frac{\partial \psi^{**}}{\partial T} \right\}$$

$$g(T, B) = \frac{\partial}{\partial B}\left\{ \psi^{***} - T \frac{\partial \psi^{***}}{\partial T} \right\} .$$

If $\psi^* (S^i_k, T)$ is known we can derive the exact evolution law for the reversible deformations. We replace this (hyperelastic) law by an approximating (hypoelastic) law according to (5.19) setting

$$\underset{(r)}{d^i_k} = \frac{1}{2G} \overset{\nabla i}{t_k} + \left\{ \frac{1}{9K} \dot{S}^r_r + \alpha \dot{T} \right\} \delta^i_k . \tag{9.3}$$

Now we consider at first the viscoplastic material.

For simplicity we assume that $\underset{(s)}{d^i_k}$ and $\underset{(p)}{d^i_k}$ depend on corresponding yield conditions, i.e. (s. equ. (7.11), (7.13) and (7.15)).

$$\underset{(p)}{\bar{F}} = \underset{(s)}{F} = \bar{F}, \quad \underset{(p)}{F} = F . \tag{9.4}$$

In order to reflect as well isotropic as anisotropic hardening we suppose a yield condition of the form

$$F = (t^i_k - \beta^i_k)\,(t^k_i - \beta^k_i) - k^2\,(b,\,T) = f^2 - k^2 \;>\; 0. \tag{9.5}$$

Analogously we obtain for the equilibrium stresses the condition

$$\bar{F} = (\bar{t}^i_k - \beta^i_k)\,(\bar{t}^k_i - \beta^k_i) - k^2\,(b,\,T) = \bar{f}^2 - k^2 = 0. \tag{9.6}$$

Concerning the law for the viscoplastic deformations we assume according to (7.16a) and (7.16b)

$$\underset{(p)}{d^i_k} = \gamma \frac{1}{\sqrt{k^2}} \left\{ e^{a(\frac{F}{k^2})} - 1 \right\} 2\,(t^i_k - \beta^i_k) \tag{9.7a}$$

$$= \lambda \frac{1}{\sqrt{k^2}}\, 2\,(\bar{t}^i_k - \beta^i_k)\;. \tag{9.7b}$$

Using the relation (7.17) we derive from (9.7a) and (9.7b)

$$\lambda = \gamma \sqrt{\frac{f^2}{k^2}} \left\{ e^{a(\frac{F}{k^2})} - 1 \right\} = \gamma \sqrt{\frac{f^2}{k^2}}\, \Phi\,(F) \tag{9.8}$$

and furthermore

$$\bar{t}^i_k - \beta^i_k = \sqrt{\frac{k^2}{f^2}}\,(t^i_k - \beta^i_k) \Rightarrow \bar{t}^i_k = \bar{t}^i_k\,(t^i_k,\,T,\,b,\,\beta^i_k)\;. \tag{9.9}$$

Concerning the law for $\underset{(s)}{d}{}^i_k$ we assume

$$\underset{(s)}{d}{}^i_k = \varkappa(\overset{\nabla}{t}{}^i_k - \overset{\nabla}{\beta}{}^i_k) \; . \tag{9.10}$$

This means $\underset{(s)}{d}{}^i_k$ may depend on the changes of the active stress $\dot{t}^i_k - \dot{\beta}^i_k$. It may be emphasized, however, that also other approaches for $\underset{(s)}{d}{}^i_k$ are imaginable.

With respect to the evolution laws for the internal variables $b$ and $\beta^i_k$ we suppose (with $\overset{\circ}{b} = 0$, $\overset{\circ}{\beta}{}^i_k = 0$)

$$\dot{b} = \frac{\delta}{\overset{\circ}{\rho}} (t^i_k - \beta^i_k) \underset{(s)}{d}{}^k_i + \frac{\mu}{\overset{\circ}{\rho}} (\bar{t}^i_k - \beta^i_k) \underset{(p)}{d}{}^k_i - b \underset{(b)}{\vartheta} \tag{9.11}$$

$$\overset{\nabla}{\beta}{}^i_k = \zeta \underset{(s)}{d}{}^i_k + \nu \underset{(p)}{d}{}^i_k - \beta^i_k \underset{(\beta)}{\vartheta} \tag{9.12}$$

with side conditions of the form

$$\underset{(b,\beta)}{\vartheta} (S^i_k, T, b\ \beta^i_k) \begin{cases} > 0 \text{ for } T > T_R\,(S^i_k, b, \beta^i_k) \\ = 0 \text{ for } T \leq T_R\,(S^i_k, b, \beta^i_k). \end{cases} \tag{9.13}$$

These approaches reflect the interaction of changes of the internal state due to inelastic deformations on the one hand and changes caused by dissipative annealing processes on the other hand.

The balance equation (6.3) for the energy supply apart from reversible work reads (see 9.2)

$$\underset{(h)}{\dot{w}} + \underset{(d)}{\dot{w}} - \frac{1}{\rho} q^i\Big|_i + r = c_p \dot{T} - TB^k_{\cdot i} \overset{\nabla}{S}{}^i_k + h\dot{b} + 2g\, \beta^k_{\cdot i} \overset{\nabla}{\beta}{}^i_k \quad . \qquad (9.14)$$

In this equation

$$\underset{(h)}{\dot{w}} = \underset{(s)}{\overset{\circ}{w}} + \underset{(p/h)}{\dot{w}}$$

$$= \frac{1}{\overset{\circ}{\rho}} t^i_k \underset{(s)}{d^k_{\cdot i}} + \xi t^i_k \underset{(p)}{d^k_{\cdot i}} \qquad (9.15)$$

has to balanced with the non-dissipative terms of $h\overset{\circ}{b} + 2\, g\beta^k_{\cdot i} \overset{\nabla}{\beta}{}^i_k$.

This leads to the equation

$$\frac{1}{\overset{\circ}{\rho}} t^i_k \underset{(s)}{d^k_{\cdot i}} + \frac{1}{\overset{\circ}{\rho}} t^i_k \underset{(p)}{d^k_{\cdot i}} = \left\{ \frac{\delta h}{\overset{\circ}{\rho}} (t^i_k - \beta^i_k) + \zeta 2g \;\; \beta^i_k \right\} \underset{(s)}{d^k_{\cdot i}}$$

$$+ \left\{ \frac{\mu h}{\overset{\circ}{\rho}} (\bar{t}^i_k - \beta^i_k) + \nu 2\, g \;\; \beta^i_k \right\} \underset{(p)}{d^k_{\cdot i}} \; . \qquad (9.16)$$

From (9.16) we conclude

$$\delta h = 1 \quad ; \qquad \zeta = \frac{1}{2g\overset{\circ}{\rho}}$$

$$0 < \mu h \ll 1 \quad ; \qquad \nu = \frac{\mu h}{2g\overset{\circ}{\rho}} \qquad (9.17)$$

It may be mentioned that the relation for $\nu$ is not a necessary condition but simplifying. With (9.17) we obtain on the one hand

$$\overset{\circ}{\underset{(h)}{w}} = \frac{1}{\overset{\circ}{\rho}} t^{i}_{k} \underset{(s)}{d^{k}_{i}} + \frac{\mu h}{\overset{\circ}{\rho}} \bar{t}^{i}_{k} \underset{(p)}{d^{k}_{i}} \quad \text{with: } \mu h = \xi \frac{t^{i}_{k} \underset{(p)}{d^{k}_{i}}}{\bar{t}^{r}_{s} \underset{(p)}{d^{s}_{r}}} \tag{9.18}$$

$$\underset{(d)}{\dot{w}} = \frac{1}{\overset{\circ}{\rho}} \left\{ t^{i}_{k} - \mu h \, \bar{t}^{i}_{k} \right\} \underset{(p)}{d^{k}_{i}} \geq 0 \tag{9.19}$$

and on the other hand according to (9.11) and (9.12)

$$\overset{\circ}{b} = \frac{\varkappa}{\overset{\circ}{\rho} h} (t^{i}_{k} - \beta^{i}_{k}) (\overset{\nabla k}{t_{i}} - \overset{\nabla k}{\beta_{i}}) + 2 \frac{\mu}{\overset{\circ}{\rho}} \gamma \sqrt{f^{2}} \, \Phi(F) - b \underset{(b)}{\vartheta} \tag{9.20}$$

$$\overset{\nabla i}{\beta_{k}} = \varkappa \zeta \, (\overset{\nabla i}{t_{k}} - \overset{\nabla i}{\beta_{k}}) + \mu h \zeta \frac{\gamma}{\sqrt{k^{2}}} \Phi(\dot{F}) \, 2 \, (t^{i}_{k} - \beta^{i}_{k}) - \beta^{i}_{k} \underset{(\beta)}{\vartheta} \, . \tag{9.21}$$

From (9.21), (9.20) we derive eventually the final form of the evolution laws for the internal variables

$$\overset{\nabla i}{\beta_{k}} = \frac{1}{1+\varkappa\zeta} \left\{ \varkappa\zeta \overset{\nabla i}{t_{k}} + \mu h \zeta \frac{\gamma}{\sqrt{k^{2}}} \Phi(F) \, 2 \, (t^{i}_{k} - \beta^{i}_{k}) - \beta^{i}_{k} \underset{(\beta)}{\vartheta} \right\} \tag{9.22}$$

$$\text{with} \quad \zeta = \frac{1}{2 \overset{\circ}{\rho} \, g(B,T)} \tag{9.23}$$

$$\overset{\bullet}{b} = \frac{1}{1+\varkappa\zeta} \frac{\varkappa}{\overset{\circ}{\rho} h} \left\{ (t^{i}_{k} - \beta^{i}_{k}) \overset{\nabla k}{t_{i}} - \mu h \zeta \frac{\gamma}{\sqrt{k^{2}}} 2 f^{2} \, \Phi(F) + (t^{i}_{k} - \beta^{i}_{k}) \underset{(\beta)}{\vartheta} \beta^{k}_{i} \right\}$$

$$+ 2 \frac{\mu}{\overset{\circ}{\rho}} \gamma \sqrt{f^{2}} \, \Phi(F) - b \underset{(b)}{\vartheta} \tag{9.24}$$

$$= \frac{1}{2} \frac{\varkappa}{\overset{\circ}{\rho} h} (\dot{f}^2) + 2 \frac{\mu}{\overset{\circ}{\rho}} \gamma \sqrt{f^2} \; \Phi (F) - b \underset{(b)}{\vartheta} . \qquad (9.25)$$

We see, the evolution laws for the internal variables depend on the set of the state variables $S^i_k$, T, b, $\beta^i_k$ and on $\overset{\nabla}{t}{}^i_k$. They are of mixed type.

We recognize, that only the parameters (or functions, respectively)

$$g (B, T) = \frac{\partial}{\partial B} \left\{ \psi^{***} - T \frac{\partial \psi^{***}}{\partial T} \right\} = \frac{1}{2 \overset{\circ}{\rho} \zeta (B,T)} \qquad (9.26)$$

$$h (b, T) = \frac{\partial}{\partial b} \left\{ \psi^{**} - T \frac{\partial \psi^{**}}{\partial T} \right\} \qquad (9.27)$$

$$k^2 = k^2 (b, T) \qquad (9.28)$$

$$\varkappa = \varkappa (S^i_k, T, b, \beta^i_k) \qquad (9.29)$$

$$\mu h = \mu h (S^i_k, T, b, \beta^i_k) \qquad (9.30)$$

$$\underset{(b,\beta)}{\vartheta} = \underset{(b,\beta)}{\vartheta} (S^i_k, T, b, \beta^i_k) \qquad (9.31)$$

enter the set of differential equations describing the evolution of the internal variables and the evolution of inelastic deformations. These quantities have to be determined by suitable experiments.

The constitutive law for the elastic-viscoplastic material under considerations has to be completed by the law for dissipation involved in internal processes, i.e. in this case

$$T \dot{\eta} = bh \underset{(b)}{\vartheta} + 2g\, B \underset{(\beta)}{\vartheta} \tag{9.32}$$

and by the law concerning the heat flux i.e., for instance, Fourier's law

$$q^i = - \bar{\lambda}\, T|^i \tag{9.33}$$

where $\bar{\lambda}$ represents the coefficient of heat conduction.

The transition to a rate insensitive elastic-plastic material cannot be achieved by assuming a quasi-static process with a deformation rate tending to zero. Elastic-plastic deformations require a separate consideration.

In our example we assume again the specific free enthalpy given in the form (9.1). Then also the balance equation (9.2) and the approximation (9.3) for the reversible deformations remain unchanged. The corresponding yield condition reads now

$$F = (t^i_k - \beta^i_k)\,(t^k_i - \beta^k_i) - k^2\,(b,\, T) = f^2 - k^2 = 0 \tag{9.34}$$

Concerning the plastic deformations we obtain according to (7.20)

$$\underset{(p)}{d}{}^{i}_{k} = \frac{\lambda}{\sqrt{\kappa^2}} \frac{\partial F}{\partial S^k_i} = \frac{\lambda}{\sqrt{\kappa^2}} 2 \, (t^i_k - \beta^i_k). \tag{9.35}$$

$\lambda$ can be determined from the consistency condition (7.21)

$$\dot{F} = \frac{\partial F}{\partial S^i_k} \overset{\nabla i}{S}_k + \frac{\partial F}{\partial T} \dot{T} + \frac{\partial F}{\partial b} \overset{\circ}{b} + \frac{\partial F}{\partial \beta^i_k} \overset{\nabla i}{\beta}_k$$

$$= 2 \, (t^i_k - \beta^i_k) \, (\overset{\nabla k}{t}_i - \overset{\nabla k}{\beta}_i) - \frac{\partial k^2}{\partial T} \dot{T} - \frac{\partial k^2}{\partial b} \dot{b} = 0, \tag{9.36}$$

when we express $\overset{\circ}{b}$ and $\overset{\nabla i}{\beta}_k$ by the respective evolution laws as it will be shown in the following. With respect to $\underset{(s)}{d}{}^i_k$ we suppose again

$$\underset{(s)}{d}{}^i_k = \varkappa \, (\overset{\nabla i}{t}_k - c\overset{\nabla i}{\beta}_k). \tag{9.37}$$

Correspondingly to (9.11), (9.12) we assume

$$\dot{b} = \frac{\delta}{\rho} (t^i_k - \beta^i_k) \, \underset{(s)}{d}{}^k_i + \frac{\mu}{\rho} (t^i_k - \beta^i_k) \, \underset{(p)}{d}{}^k_i - b \underset{(b)}{\vartheta} \tag{9.38}$$

$$\beta^i_k = \xi \, \underset{(s)}{d}{}^i_k + \nu \, \underset{(p)}{d}{}^i_k - \beta^i_k \underset{(\beta)}{\vartheta} \tag{9.39}$$

with side conditions (9.14) for $\underset{(b,\beta)}{\vartheta}$.

Analogously to viscoplastic deformations we conclude from the balance equations (9.14), (9.15)

$$\delta h = 1 \quad , \qquad \zeta = \frac{1}{2g\overset{\circ}{\rho}}$$
$$0 < h\mu = \xi \ll 1 \quad , \qquad \nu = \frac{\xi}{2g\overset{\circ}{\rho}} \tag{9.40}$$

and obtain

$$\underset{(h)}{\dot{w}} = \frac{1}{\overset{\circ}{\rho}} t^i_k \underset{(s)}{d^k_{\cdot i}} + \frac{\xi}{\overset{\circ}{\rho}} t^i_k \underset{(p)}{d^k_{\cdot i}} = \underset{(s)}{\overset{\circ}{w}} + \xi \underset{(p)}{\dot{w}} \tag{9.41}$$

$$\underset{(d)}{\dot{w}} = \frac{1}{\overset{\circ}{\rho}} (1 - \xi) t^i_k \underset{(p)}{d^k_{\cdot i}} = (1 - \xi) \underset{(p)}{\dot{w}} \geq 0 . \tag{9.42}$$

With the preceding results the evolution laws for the internal variables read (with $\zeta$ according to (9.23))

$$\dot{b} = \frac{x}{\overset{\circ}{\rho} h} (t^i_k - \beta^i_k) (\overset{\nabla}{t}{}^k_i - \overset{\nabla}{\beta}{}^k_i) + \frac{\mu}{\overset{\circ}{\rho}} \frac{\lambda}{\sqrt{k^2}} 2f^2 - b \underset{(b)}{\vartheta} \tag{9.43}$$

$$\overset{\nabla}{\beta}{}^i_k = x\zeta \, (\overset{\nabla}{t}{}^i_k - \overset{\nabla}{\beta}{}^i_k) + \xi\zeta \frac{\lambda}{\sqrt{k^2}} 2 \, (t^i_k - \beta^i_k) - \beta^i_k \underset{(\beta)}{\vartheta} \tag{9.44}$$

or finally

$$\overset{\nabla}{\beta}{}^i_k = \frac{1}{1+x\zeta} \left\{ x\zeta \, \overset{\nabla}{t}{}^i_k + \xi\zeta \frac{\lambda}{\sqrt{k^2}} 2 \, (t^i_k - \beta^i_k) - \beta^i_k \underset{(\beta)}{\vartheta} \right\} \tag{9.45}$$

$$\overset{\circ}{b} = \frac{1}{1+x\zeta} \frac{x}{\overset{\circ}{\rho} h} \left\{ (t^i_k - \beta^i_k) \overset{\nabla}{t}{}^k_i - \xi\zeta \cdot 2\sqrt{k^2}\,\lambda + (t^i_k - \beta^i_k) \beta^k_i \underset{(\beta)}{\vartheta} \right\} + 2 \frac{\xi}{\overset{\circ}{\rho} h} \sqrt{k^2} - b \underset{(b)}{\vartheta} . \tag{9.46}$$

Inserting these relations into the consistency condition (9.35) we obtain

$$\dot{F} = 0 = \frac{1}{1+x\zeta}\left[2 - \frac{x}{\mathring{\rho}\, h}\frac{\partial k^2}{\partial b}\right](t^i_k - \beta^i_k)\,\overset{\nabla k}{t}_i - \frac{2\xi\sqrt{k^2}}{1+x\zeta}\left[2 + \frac{1}{\mathring{\rho}\, h}\frac{\partial k^2}{\partial b}\right]\lambda$$

$$- \frac{\partial k^2}{\partial T}\dot{T} + \frac{1}{1+x\zeta}\left[2 - \frac{x}{\mathring{\rho}\, h}\frac{\partial k^2}{\partial b}\right](t^i_k - \beta^i_k)\,\beta^k_i\,\vartheta_{(\beta)} + \frac{\partial k^2}{\partial b}\, b\, \vartheta_{(b)} \; .$$

(9.47)

From this condition we can calculate $\lambda$ . Inserting this result into the relations (9.35), (9.45), and (9.46) we get the final form of the law for the plastic deformation rate and of the evolution laws for the internal variables. Furthermore we can derive from the consistency condition (9.47) the so-called loading condition on account of the requirement

$$\lambda > 0 \; . \qquad (9.48)$$

It results in

$$\frac{1}{1+\varkappa\zeta}\left[2 - \frac{\varkappa}{\mathring{\rho}\, h}\frac{\partial k^2}{\partial b}\right](t^i_k - \beta^i_k)\,\overset{\nabla k}{t}_i - \frac{\partial k^2}{\partial T}\dot{T}$$

$$+ \frac{1}{1+\varkappa\zeta}\left[2 - \frac{\varkappa}{\mathring{\rho}\, h}\frac{\partial k^2}{\partial b}\right](t^i_k - \beta^i_k)\,\beta^k_{i\,(\beta)} + \frac{\partial k^2}{\partial b}\, b\, \vartheta_{(b)} > 0. \qquad (9.49)$$

(9.34) and (9.49) form together the necessary and sufficient condition for the occurence of plastic deformations.

## 8.10. The thermo-mechanical process as initial boundary problem

As already stated in section 8.4 a thermo-mechanical process is based on the initial configuration of the body and the initial thermodynamical state of each material element. The process is governed by the history of the independent process variables on the one hand and by the general field equations as well as by the constitutive law of the material on the other hand. Mathematically the thermo-mechanical process represents an initial boundary problem.

The constitutive laws and the balance equations for energy and entropy are already discussed in many details. Let us still have a short look on the remaining material independent general field equations as specified in section 8.4.

The balance equation of mass is fulfilled already by the introduction of a body-fixed coordinate system. This procedure fails, however, if we have to deal with diffusion problems.

The compatibility conditions for the deformation rate can be written in the form

$$d^i_k\Big|^r_s + d^r_s\Big|^i_k - d^i_s\Big|^r_k - d^r_k\Big|^i_s \equiv 0. \tag{10.1}$$

They allow for eliminating of redundant quantities if necessary.

Using the weighted cauchy stress tensor the balance equation for momentum reads

$$S^i_k\Big|_i - S^i_k \varepsilon\Big|_i + \mathring{\rho}\,(f_k - a_k) = 0. \tag{10.2}$$

In this formula mean

$\varepsilon^r_r = \varepsilon$ bulk strain,

$f_k$ specific body forces,

$a_k$ acceleration against the space of observation in which $f_k$ is measured.

$f_k - a_k$ can be considered as resulting specific body force acting on the material element. With respect to the fact that the constitutive law in thermoplasticity defines a relation between the stress increments and the deformation rate we have to transform the balance equation for linear momentum into a balance equation for the stress rate. This leads (see, for instance, [5]) to

$$\overset{\nabla}{S}{}^i_k\Big|_i - \overset{\nabla}{S}{}^i_k \varepsilon\Big|_i - d^i_m\Big|_i S^m_k + \omega^m_{\cdot k}\Big|_i S^i_m + \mathring{\rho}\,(\overset{\nabla}{f}_k - \overset{\nabla}{a}_k) = 0. \tag{10.3}$$

The balance equation (10.3) can be transformed into several other versions.

The balance equation for the angular momentum of the material elements is within the frame of a classical continuum fulfilled by the symmetry of the Cauchy stress tensor.

The mathematical processing for the solution of the given initial boundary problem depends on the kind of the problem. This subject shall not be discussed here. In the following section only some hints on coupling phenomena will be given which can be observed in thermoplasticity.

## 8.11. Some coupling phenomena in thermoplasticity

We consider, for instance, a simple homogeneous deformation process in a finite specimen, e.g. a simple tension test with a cylindrical specimen or a simple torsion test with a thin-walled tube. At very low deformation rate (quasistatic process) the specimen behaves like in an isothermic process. At increasing deformation rate the process becomes non-isothermic because the heat produced by dissipation cannot be transferred completely to the surrounding. Owing to the increasing temperature during the process we obtain an interaction between hardening and softening which leads to a decreasing demand for mechanical work with increasing but still low deformation rate. At higher deformation rate the process runs nearly adiabatically. At the same time viscous effects become remarkable implying an increase of stresses with increasing deformation rate. In between we can find an optimum deformation rate minimizing the mechanical work for a given deformation. This is demonstrated in [2, 17] neglecting temperature differences inside the specimen. A more detailed analysis has to take into account also the inhomogenity of the hardening state due to the occuring temperature gradient and, at elevated temperature, annealing phenomena.

Bifurcation problems are in many cases very sensitive to material behaviour. This is particularly true when bifurcation is accompanied by strong localization of deformations [18]. In such cases it becomes very important to fit the constitutive law as near as possible to the real material behaviour. This requires also to take into account coupling effects between thermic and mechanical processes. The influence of such coupling phenomena concerns less the possibility and the very beginning of bifurcation [19] than the development of localization after bifurcation [20]. With respect to the latter the entropy production by inelastic work supports the localization of deformation. The effect, however, is rate sensitive in any case owing to the time dependent heat flux and to viscous effects. It needs more detailed investigations to clarify the interaction of the different coupling phenomena, even if annealing processes are neglected.

Solid phase transformations as, for instance, from austenite to martensite, are connected with volume changes. Therefore, in quenching processes these phase transformations can generate residual stresses

[24]. A more detailed investigation of this problem requires taking into account the enthalpy differences between the two phases and the energy dissipation in the dynamical process of phase transformation.

Residual stresses are also generated in arbitrary inhomogeneous inelastic deformations without phase transformations. In cyclic processes these residual stresses can give rise to low cycle fatigue. The generation of residual stresses depends strongly on the real material behaviour because residual stresses are traced back to small differences between comparable large strains. Therefore thermomechanical coupling phenomena have a remarkable influence on the generation of residual stresses.

The preceding considerations cannot exhaust the variety of coupling phenomena occuring in thermomechanical processes. They should be understood to be an illustration of some aspects of these coupling phenomena.

## REFERENCES


1. Raniecki, B. and Thermann, K., Infinitesimal thermoplasticity and kinematics of finite elastic-plastic deformations, *Mitt. Inst. Mech. Ruhr-Univ. Bochum*, Nr. 2, 1978
2. Lehmann, Th., Some aspects of non-isothermic large inelastic deformations, *Sol. Mech. Arch.* 3, 261, 1978
3. Zaremba, M.S., Sur une forme perfectionnée de la théorie de la relaxation, *Anz. Akad. Krakau, math.-naturw. Klasse* 1903, 594
4. Jaumann, G., Geschlossenes System physikalischer und chemischer Differentialgesetze, *Sitzungsber. Kais. Akad. Wiss. Wien, Abt. IIa*, 120, 385, 1911
5. Lehmann, Th. Formänderungen eines klassischen Kontinuums in vierdimensionaler Darstellung, *Proc. XI. Congr. Appl. Mech.*, München 1964 (ed. H. Görtler), Springer-Verlag Berlin/Heidelberg/New York 376, 1966
6. Fritsch, C. and Siegel, R., Kalt- und Warmfließkurven von Baustählen, *Inst. Forsch. Maschinenbau*, Karl-Marx-Stadt 1965
7. Lee, E.H., Elastic-plastic deformations at finite strains, *J. Appl. Mech.* 36, 1, 1969

8. Sidoroff, F., The geometrical concept of intermediate configuration and elastic-plastic finite strain, *Arch. Mech.* 25, 299, 1973

9. Mandel, J., Sur la décomposition d'une transformation élasto-plastique, *Comp. Rend. Acad. Sci.* Paris A 272, 276, 1971

10. Lee, E.H., Some comments on elastic-plastic analysis, *Int. J. Sol. Struct.* 17, 859, 1981

11. Sedov, L.I., *Introduction to the Mechanics of a continuous medium* (trans. from Russian), Addison-Wesley, London, 1965

12. Green, A.E., and Naghdi, P.M., A general theory of an elastic-plastic continuum, *Arch. Rat. Mech. Anal.* 18, 251, 1965

13. Lehmann, Th., On large elastic-plastic deformations, in: *Foundations of plasticity* (A. Sawczuk, ed.), Nordhoff Publ. Comp., Leyden, 571, 1973

14. Lehmann, Th. Some remarks on the decomposition of deformations and mechanical work, *Int. J. Eng. Sci.* 20, 281, 1982

15. Lehmann, Th., On the theory of large, non-isothermic, elastic-plastic and elastic-viscoplastic deformations, *Arch. Mech.* 29, 393, 1977

16. Lehmann, Th., On constitutive relations in thermoplasticity, in: *Three-dimensional constitutive relations and ductile fracture* (S. Nemat-Nasser, ed.) North-Holland Publ. Comp., Amsterdam 289, 1981

17. Bruhns, O. and Lehmann, Th., Optimum deformation rate in large inelastic deformations, in: *Metal forming plasticity* (H. Lippmann,

ed.) Springer-Verlag Berlin/Heidelberg/New York, 120, 1979

18. Rice, J., The localization of plastic deformation, in: *Theoretical and Applied Mechanics*, 14. IUTAM Congr. Delft, North-Holland Publ. Comp., Amsterdam 207, 1976

19. Bruhns, O. and Mielniczuk, J., Zur Theorie der Verzweigungen nicht-isothermer elastoplastischer Deformationen, *Ing. Arch.*, 46, 65, 1977

20. Litonski, J., Plastic flow of a tube under adiabatic torsion, *Bull. Acad. Polon. Sci.* 25, 1, 1977

21. Inoue, T., and Raniecki, B., Determination of thermal-hardening stress in steels by use of thermoplasticity, *J. Mech. Phys. Sol.*, 26, 187, 1978

22. Odquist, F.K.G., and Hult, J., *Kriechfestigkeit metallischer Werkstoffe*, Springer-Verlag Berlin 1962

23. Odquist, F.H.G., *Mathematical theory of creep and creeprupture*, Charendon Press, Oxford 1974

24. Rabotnov, Yu. N., *Creep problems in structural members*, translated from Russian, North-Holland Publ. Co., Amsterdam 1969

25. Perzyna, P., *Fundamental problems in viscoplasticity*, Advances of App. Mechanics, Academic Press Inc. New York, Vol. 9, 243, 1966

26. Penzy, R.K., and Marriott, *Design for creep*, Mc. Graw Hill, London 1971

27. Hult, J. (editor), *IUTAM second symposium on creep in structures*, Gothenburg, 1970, Springer-Verlag 1971

# BIFURCATION PROBLEMS IN PLASTICITY

O.T. BRUHNS
Department of Mechanical Engineering
University of Kassel, Kassel, FRG

## PREFACE

The text of the following chapters comprises six lectures given at the International Centre for Mechanical Sciences (CISM) in Udine, July 19 - 30, 1982, as part of the course "The Constitutive Law in Thermoplasticity" which has been coordinated by Professor Th. Lehmann.

The author feels indebted to Professor Lehmann for the invitation that induced him to prepare these lecture notes and Mrs. Deckart for the careful typing of the manuscript.

## 1. INTRODUCTION

Plastic buckling, i. e. the investigation of possible bifurcation of a structure when the stress in this structure exceeds the yield limit, has received a great deal of interest in the past. However, the progress in the subject has been very halting and at different levels as regards derivation and application of criteria for uniqueness and stability of elastic-plastic solids. One might only contemplate the classical confusion in column buckling regarding whether the 'tangent-modulus' or the 'reduced-modulus' theory gives an adequate description of the critical load, which lasted for more than half a century. In 1947 Shanley, by way of a simple model and careful experimentation, demonstrated the important distinction between uniqueness and stability in the range of elastic-plastic deformations. It was recognized that the tangent-modulus load is the lowest possible bifurcation load; the straight configuration loses its uniqueness at this load but not its stability. It took another decade until a continuum theory of bifurcation for solids exceeding the yield limit was laid down by Hill in a number of papers starting 1956. We therefore refer to the fundamental paper /13/. In some recent survey articles /9, 10, 12/ the authors have given a detailed exposé of the existing basic principles for uniqueness and stability including a bibliography of much of the enormous amount of work on plastic bifurcation problems. The state of affairs regarding structural buckling problems with emphasis on postbifurcation behaviour and imperfection sensitivity has been reviewed by Hutchinson /11/.

These lecture notes are devoted to bring about some little insight into this interesting field of research, the present author has been dealt with since ten years. Therefore these notes open with a recapitulation of the historical developement of the theory of bifurcation exhibiting the theory of plastic column buckling. Thereafter the classical concepts of uniqueness and stability of deformations of elastic-plastic solids established by Hill are explained in some extension. Both sections are followed by a number of illustrating examples. A few items finally are concerned with some deviations of this classical concept caused e. g. by non-associated constitutive laws in the plastic range.

## 2. PLASTIC COLUMN BUCKLING

In order to deal with something specific and at the same time as an introduction to a continuum theory, it is illustrative to consider a column hinged at both ends and subjected to an axial load as indicated in Fig. 2.1.

Only for simplicity we here consider a column of constant and symmetric cross-section, where the cartesian axes x, y, z are principal central axes.

Suppose that for some value of the compressive force F, buckling of the column takes place in the plane of least rigidity then w(x) denotes the displacement of the axis of the column.

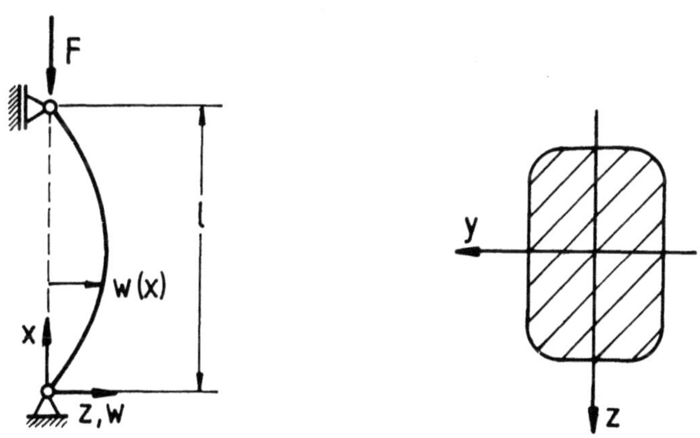


a) system b) cross-section

Fig. 2.1 Column in buckled configuration

The problem to calculate the critical value of F when the straight configuration of the column becomes unstable was first solved by Euler in 1744. Euler assumed a linear elastic material behaviour and therefore could show that the critical value of the load was

$$F_e = \pi^2 EI/l^2 \tag{2.1}$$

where I is the least moment of inertia of the section; here with respect to the z-axis, and E is Young's modulus.

However, if the column is loaded such that the stress $\sigma$ (= F/A in the unbuckled configuration, where A is the section area) exceeds the yield limit of the column material, plastic flow will occur. Thus the above analysis has to be modified since (2.1) includes the constitutive parameter E.

A typical stress/strain curve may be shown with Fig. 2.2a where the slope of the curve beyond the elastic limit $E_t = d\sigma/d\varepsilon$, the tangent modulus is still a function of stress.

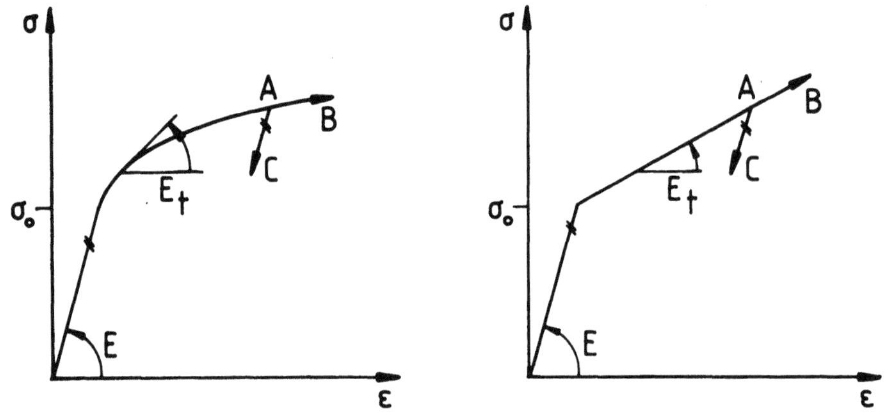


a) smooth curve b) two-segment curve

Fig. 2.2 Stress/strain curves

An often used simplification of this smooth stress/strain curve is that of a two-segment idealization (Fig. 2.2b) which in the analysis of elastic-plastic deformations usually is identified with the behaviour of a "linear hardening" material. This model seems to be generally favoured (since $E_t$ = const.), however, in problems of bifurcation the jump in the slope passing through the yield limit may cause some difficulties.

A first revised formula for the critical load was suggested in 1889 by Engesser /1/ who claimed that Euler's analysis was still applicable if only Young's modulus E was replaced by the tangent modulus $E_t$ in (2.1)

$$F_t = \pi^2 E_t I/l^2 \tag{2.2}$$

Shortly after the publication of Engesser's work this theory was criticized by Jasinski /3/ who observed that in the assumed buckling mode in one part of the section strain reversal will occur according to AC of the stress/strain diagram (Fig. 2.2) and consequently continued deformation will be purely elastic in that part. Only in the remainder of the section loading will continue and thus according to AB the tangent modulus remains valid. Jasinski also pointed out that this problem has been first realized and discussed by Considière in 1891 /4/ who introduced a reduced (or double) modulus $E_r$ with $E \geqq E_r \geqq E_t$ to give the lowest value of the critical load in the plastic range according to

$$F_r = \pi^2 E_r I/l^2 \tag{2.3}$$

where $E_r$ still is a function of geometric properties of the cross-section. Engesser accepted this critisism /2/ and moreover gave first suggestions to calculate this reduced modulus. Some examples of calculation later have been performed by v. Kármán /5/.

For the symmetric section under consideration (Fig. 2.3) the value of $E_r$ can be deduced in the following way. As so far buckling of the column is connected with the static stability criterion from which in our case we find

$$\delta F = 0 \tag{2.4}$$

During buckling the column experiences bending and therefore additional infinitesimal strains occur which still follow the hypothesis of plane sections and vanishing on the line n-n, separating the regions $A_1$ and $A_2$ in Fig. 2.3.

These additional strains induce infinitesimal incremental stresses $\delta\sigma(y)$ according to

$$\delta\sigma\,(y) = \begin{cases} E_t\;\delta\sigma\,(y) & \text{in the loading region } A_1 \\ E\;\;\delta\sigma\,(y) & \text{in the unloading region } A_2 \end{cases} \tag{2.5}$$

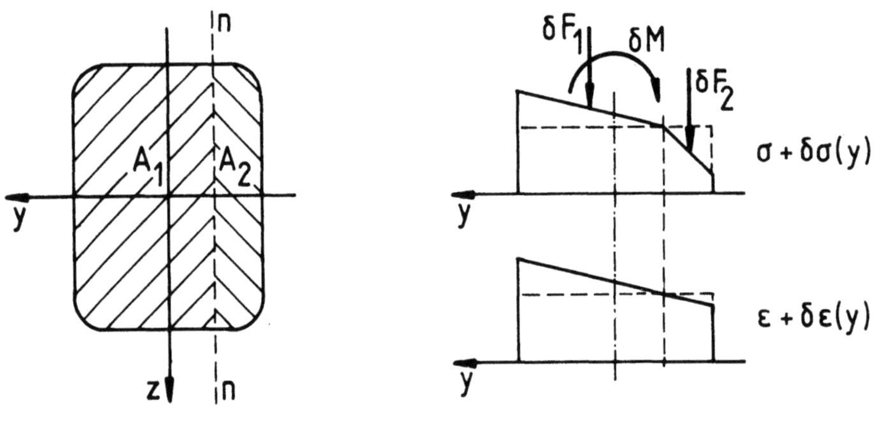


Fig. 2.3 cross-section, stress and strain profiles

Now, following (2.4) from Fig. 2.3 we find

$$\delta F_1 = \delta F_2 \tag{2.6}$$

and hence

$$S_1 E_t = S_2 E \tag{2.7}$$

where $S_1$, $S_2$ are the statical moments of the areas $A_1$, $A_2$ with respect to the separating line n-n. This equation determines the position of n-n, which obviously depends on the shape of the section and on the ratio $E_t/E$.

With this information the infinitesimal moment of the internal forces with respect to the z-axis can be calculated

$$\delta M = (E_t I_1 + E\, I_2)\delta\kappa = E_r\, I\delta\kappa \tag{2.8}$$

where in view of the corresponding result of Euler's theory and according to (2.3) we have introduced

$$E_r = (E_t I_1 + E\, I_2)/I \tag{2.9}$$

Here $I_1$, $I_2$ are respectively the moments of inertia of the areas $A_1$, $A_2$ with respect to the separating line and $\delta\kappa$ is the infinitesimal change of curvature.

For a rectangular cross-section, denoting h the height, b the width and mb the width of the loading zone $A_1$ ( $m < 1$), we find from (2.7)

$$m = (1 + \sqrt{E_t/E}\,)^{-1} \tag{2.10}$$

and further from (2.9) the reduced modulus

$$E_r = 4\, E_t\, m^2 \tag{2.11}$$

A second example refers to an idealized I section having two flanges each of area A/2 spaced a distance b apart by means of an infinitely rigid web of vanishingly small cross-sectional area. For this idealized I section we find /15/

$$m = (1 + E_t/E)^{-1}, \quad E_r = 2\, E_t m \tag{2.12}$$

The influence of these different cross-sections on the ratio $E_r/E_t$ has been depicted with Fig. 2.4. As $E_t/E$ decreases, the influence of the cross-section of the column on the ratio $E_r/E_t$ increases.

The reduced modulus theory was then accepted as the exact solution to the problem while the tangent modulus theory was considered as an approximative but convenient and conservative approach. Experiments that have been performed seemed to give results between both critical loads and deviations from the reduced modulus value were usually blamed on geometrical imperfections of the specimens and eccentricity of the loading device.

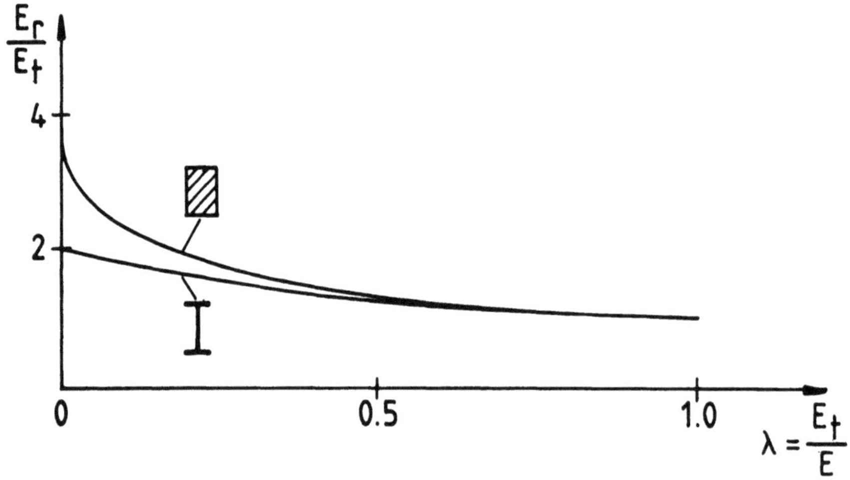


Fig. 2.4 Influence of the cross-section on the ratio $E_r/E_t$

For a long period no objections were raised to the validity of this approach, until 1946/47. Then Shanley /6, 7/ pointed out that quasistatic deflection indeed could begin at the tangent modulus value provided the axial load is simultaneously increased to permit continued plastic loading over the whole cross-section.

The reduced modulus solution is based on the use of the static stability criterion in a form which usually is applied to problems of stability of elastic systems. It is assumed that the column remains straight until stability breaks down, with the transition from straight to distorted state being achieved at an unchanged value of the axial load, i. e. $\delta F = 0$. Shanley's analysis represents a generalization of this problem. His procedure can be formulated as follows: What is the smallest value of the axial load at which a bifurcation of the equilibrium positions can occur, regardless of whether or not the transition to the bent position requires an increase of the axial load (i. e. $\delta F \geqq 0$)?

This problem has been solved by Shanley for a special case of an idealized column equivalent to the well-known two-spring model in Fig. 2.5 /9,11/.

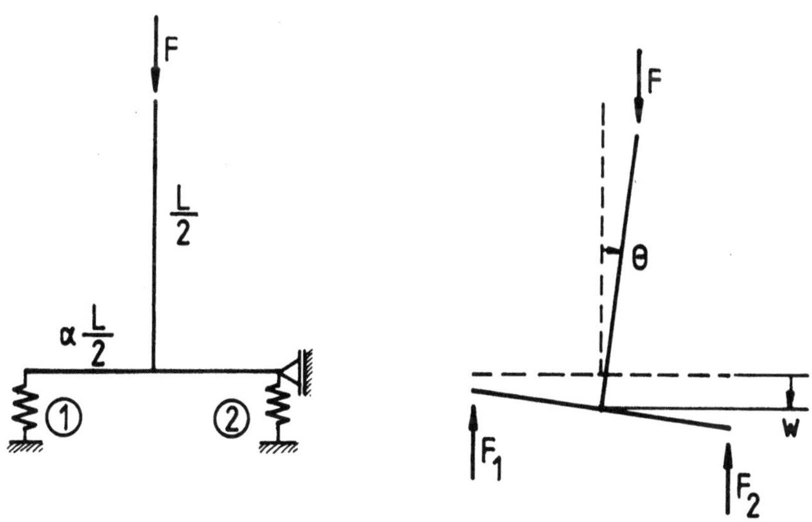


Fig. 2.5 Two-spring model

The model consists of a rigid T-piece (length L/2, with arms each of length $\alpha L/2$), whose join is constrained to move vertically as measured by w and rotate as measured by $\theta$. An increment in the compressive force in either of the two support springs depends on whether plastic loading or elastic unloading occurs. According to Fig. 2.2 we define

$$\delta F = \begin{cases} E_t \ \delta e & \text{during plastic loading} \\ E \ \delta e & \text{during unloading and loading in the elastic range} \end{cases} \tag{2.13}$$

where e here is the contraction of the spring. The tangent modulus $E_t$ is taken to be constant.

Equations of equilibrium and strain displacement are

$$\begin{aligned} F_1 + F_2 &= F \\ (F_1 - F_2)\,\alpha + F\,\theta &= 0 \end{aligned} \tag{2.14}$$

$$e_{\substack{2\\1}} = w \pm \frac{\alpha}{2} L\,\theta \tag{2.15}$$

where the subscripts denote springs Nos. 1 and 2 and taking into account only small values of the angle of rotation. For additional increments of deformation we find

$$\begin{aligned} \delta F_1 + \delta F_2 &= \delta F \\ (\delta F_1 - \delta F_2)\,\alpha + F\delta\theta + \delta F\theta &= 0 \end{aligned} \tag{2.16}$$

$$\delta e_{\substack{2\\1}} = \delta w \pm \frac{\alpha}{2} L\delta\theta \tag{2.17}$$

(i) elastic region

When the parameters of the model are such that buckling takes place in the elastic range the critical value of the axial force is found to be

$$F_e = E\alpha^2 L \tag{2.18}$$

(ii) elastic-plastic region

The lowest bifurcation load in the plastic range can be achieved assuming that both springs undergo continued loading beyond the elastic limit, what through (2.16) and (2.13) corresponds to an increase in the axial load. Similar calculations as before then give

$$F_t = E_t\alpha^2 L \tag{2.19}$$

the tangent modulus value of the axial load.

In the most general case spring No. 2 continues to load and No. 1 undergoes elastic unloading or neutral loading (i. e. $\delta e_1 = 0$) following bifurcation. With the bifurcation load denoted by $F_c$ as initial condition, equations (2.13), (2.16) and (2.17) give the differential equation

$$\frac{dF}{d\theta} = \frac{2F_t/(1+\lambda) - F}{\theta + \alpha(1-\lambda)/(1+\lambda)} \tag{2.20}$$

which can be integrated to give

$$F = \frac{2F_t\theta + F_c\alpha\,(1-\lambda)}{(1+\lambda)\theta + (1-\lambda)\alpha} \tag{2.21}$$

in the load-deflection (angle of rotation) plane, where $\lambda = E_t/E$.

The maximum of (2.21)

$$F_r = 2F_t/(1+\lambda) \tag{2.22}$$

is the reduced modulus value of the axial load since bifurcation at that level is related to a vanishing increment of F (i. e. $\delta F = 0$).

Some different load-deflection curves (2.21) are given with Fig. 2.6 with the bifurcation load in the range $F_t \leqq F_c \leqq F_r$. The tangent modulus load is the smallest load for which bifurcation is possible, as the compressive force increases deflection grows and tends to infinity as the reduced modulus value is approached. If we prevent deflections of the column up to some load $F_c > F_t$ and then release it, the column is deflected according to the dotted curve.

Thus buckling can occur for any load in the interval $(F_t, F_r)$. For loads $F > F_t$ deflections are accompanied by strain reversals and therefore Shanley's analysis, which here only for simplicity has been demonstrated for the very idealized column of Fig. 2.5, does not signify reversion of the original scheme of Engesser-v. Kármán. However, at the instant of bifurcation unloading is absent.

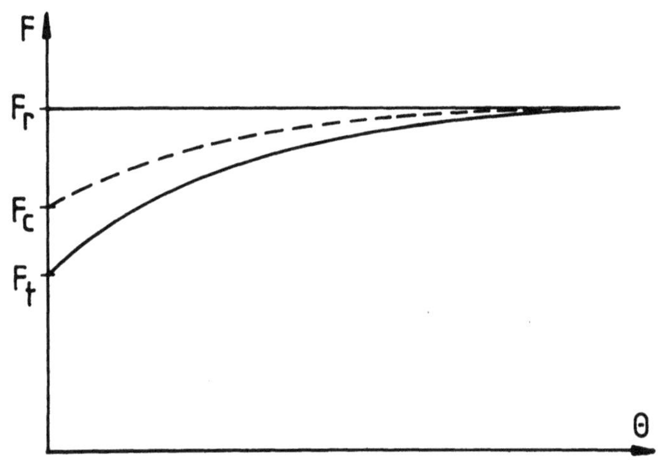


Fig. 2.6 Load-deflection curves

The great merit of Shanley's approach was that he realized the difference between a uniqueness problem and a stability problem /8/. In this sense the uniqueness problem determines, under prescribed increasing load, the magnitude of the ultimate load the column may withstand without the occurence of a bifurcation whereas in the stability problem the bifurcation of equilibrium under fixed external forces (dead load) is examined. It is evident that in this context the role of the boundary conditions is of great importance.

On the other hand, the inverse problem, a bar under axial tensile load has been the subject of many considerations on "tensile instability" in connection with the necking phenomenon which can be observed in ordinary tensile tests of ductile materials.

It is usually stated in elementary text-books /14/ that when the axial load reaches its maximum value, instability occurs with necking as a consequence. The critical condition for this event to occur is then derived in the following way.

At its maximum value the infinitesimal change of the total load $F = \sigma A$ vanishes and consequently during an infinitesimal time interval $\delta t$

$$\dot{F} = \dot{\sigma} A + \sigma \dot{A} = 0 \tag{2.23}$$

If elasticity effects are neglected and plastic deformation is assumed incompressible we find

$$\dot{V} = \dot{L} A + L \dot{A} = 0 \tag{2.24}$$

for a constant volume V of the column.

Introducing a logarithmic strain measure for the axial strains reads

$$\dot{\varepsilon} = \dot{L}/L \tag{2.25}$$

and thus by combination of (2.23) - (2.25) the critical condition becomes

$$\frac{d\sigma}{d\varepsilon} = \sigma \tag{2.26}$$

From this approach it is evident that (2.26) provides the ultimate nominal stress a bar may withstand when subjected to homogeneous deformation under a prescribed axial force. This condition is, however, neither necessary nor sufficient for the initiation of a neck in the tensile test. Especially since in the above analysis the form of the necking mode has not been taken into account.

As a third example, a thin-walled, closed-ended cylinder under hydrostatic pressure p and additional axial force T is examined to show, how "tensile instability" problems are ordinarily treated in the case of multiaxial stress states /30/.

For a cylindrical vessel of current magnitudes length L, mean radius R and thickness H (<< R) the principal Cauchy stresses are

$$\sigma_r \approx 0, \qquad \sigma_\varphi = pR/H, \qquad \sigma_z = \frac{1}{2}\sigma_\varphi + T/2\pi RH \tag{2.27}$$

in the radial, circumferential and axial directions, respectively.

Assuming now an incompressible elastic-plastic material behaviour satisfying normal $J_2$ flow theory in the plastic range the constitutive relations in principal directions become

$$\begin{aligned} \dot{\sigma}_\varphi &= 2G[2d_\varphi + d_z - \frac{3}{2\sigma_e^2}(1-\lambda)(\sigma_\varphi d_\varphi + \sigma_z d_z)\,\sigma_\varphi] \\ \dot{\sigma}_z &= 2G[2d_z + d_\varphi - \frac{3}{2\sigma_e^2}(1-\lambda)(\sigma_\varphi d_\varphi + \sigma_z d_z)\,\sigma_z] \end{aligned} \tag{2.28}$$

with $\lambda$ defined as before.

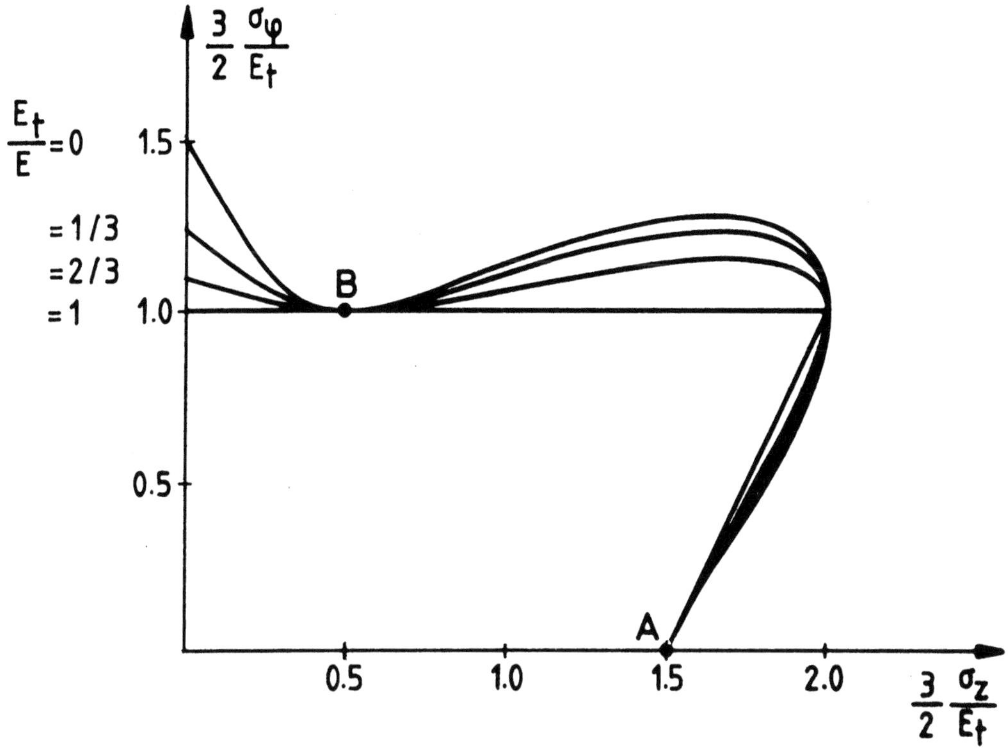


Fig. 2.7 Stress-interaction curves at load maximum for proportional as well as sequential loading

The principal strain rates herein are

$$d_\varphi = \frac{\dot{R}}{R}, \quad d_z = \frac{\dot{L}}{L}, \quad d_r = \frac{\dot{H}}{H} = -(d_\varphi + d_z) \tag{2.29}$$

and the effective stress $\sigma_e$ is

$$\sigma_e = (\frac{3}{2} J_2)^{1/2} = (\sigma_\varphi^2 - \sigma_\varphi \sigma_z + \sigma_z^2)^{1/2} \tag{2.30}$$

Due to the non-linearity of the problem different stress interaction curves can be achieved at which stability is lost - i. e., when either of the two or both loads reach a maximum. If for simplicity we restrict our further interest here to proportional loading (proportional stressing), the corresponding instability criterion becomes: $\dot{T} = 0$ or $\dot{p} = 0$ wherein $\sigma_z \sim \sigma_\varphi$. Considering the ratio $\alpha = 2\sigma_z/\sigma_\varphi - 1$ we obtain

$$\sigma_\varphi = \frac{2}{3} E_t , \quad \alpha \leqq 3,$$
$$\sigma_\varphi = \frac{2}{\alpha} E_t , \quad \alpha \geqq 3 \tag{2.31}$$

for the critical stresses at instability.

It can be verified that the result for $\alpha \leqq 3$ is connected with the condition $\dot{p} = 0$ whereas the second part of the result is due to condition $\dot{T} = 0$.

The regime of stable (proportional) loading has been depicted with Fig. 2.7, where the bounds of stability are given with the straight lines - independent of hardening ratio $E_t/E$ of the material. For sequential loading e. g. /30/ these results are relative to $E_t/E$ which varies from zero (ideal plastic and rigid-plastic, respectively) to unity (elastic material). As expected both results coincide for an elastic material. As special cases the maximum load values for simple tension (A) as well as simple internal pressure (B) are also included. In these points the different criteria coincide in their results.

Conditions of this kind are frequently adopted as a condition for a stability loss with accompanying bulging or necking of the shell. As in the preceding example it is evident that these conditions yield the maximum load the pressure vessel can withstand when deforming homogeneously. However, nothing can be said about bulging or necking of the shell since it is assumed that the shell keeps its perfect shape during expansion. Furthermore from these calculations it cannot be excluded that in the sense of Shanley there may exist bifurcations of the deformation with stable but deflected modes of deformation. To this end it seems to be obvious that a rigorous continuum theory of bifurcation is needed.

## 3. GENERAL BOUNDARY VALUE PROBLEM AND CONSTITUTIVE EQUATIONS

### 3.1 Formulation of the quasi-static boundary value problem

Consider a body of arbitrary shape to be in the state of motion. The corresponding boundary value problem will be formulated as a rate problem as follows.

At a generic instant t during finite elastic-plastic deformation, the current shape of the body, the deformation state, and the distribution of stresses are supposed to be determined, together with all necessary details regarding the history of plastic deformation. The interest is focused on the response of the body to an imposed infinitesimal change of external agencies.

A common type of boundary conditions for a mixed problem is that the rate of change of the nominal load vector $\dot{\tilde{t}}$, which is independent of any structural geometry changes, is prescribed on one part $S_T$ of the surface S of the body and rates of nominal body forces $\dot{\tilde{b}}$ (per unit volume) in its interior $\tau$. On the remainder $S_V$ of the surface the velocity $\tilde{v}$ is prescribed. As the current traction is then related to unit reference area it is convenient to introduce the first Piola-Kirchhoff (Lagrangean) stress tensor.

Referring to a fixed Cartesian coordinate system, the coordinates of a particle originally at $\tilde{X}$ in some reference configuration are denoted by $\tilde{x}$ in the current configuration at time t (Fig. 3.1).

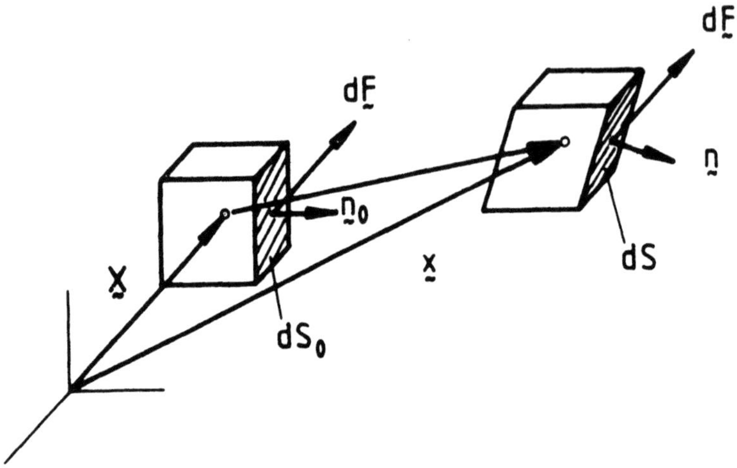


Fig. 3.1 Volume element in motion

Denoting the vector areas of a surface element $\tilde{n}_o dS_o$ and $\tilde{n}dS$, respectively, in the two configurations, where $\tilde{n}_o$ and $\tilde{n}$ are the outward unit normals to the surfaces, the current load vector $d\tilde{F}$ acting on these surfaces is

$$d\tilde{F} = \tilde{t}\, dS_o = \tilde{f}\, dS \tag{3.1}$$

where $\tilde{t}$ and $\tilde{f}$ are nominal and true tractions, respectively. Thus the nominal (first Piola-Kirchhoff) stress tensor $\tilde{S}$ and the Cauchy stress tensor $\tilde{\sigma}$ are defined through

$$\tilde{t} = \tilde{S}^T \tilde{n}_o \;, \quad \tilde{f} = \tilde{\sigma}\tilde{n} \;. \tag{3.2}$$

Since it is known that the vector areas of the surface elements in the two configurations are connected with each other, we finally find

$$\underset{\sim}{\sigma} = \frac{\rho}{\rho_o} \frac{\partial \underset{\sim}{x}}{\partial \underset{\sim}{X}} \underset{\sim}{S} \tag{3.3}$$

with $\rho$ and $\rho_o$ the mass densities in the current and the reference configurations, respectively.

Another stress measure can be defined introducing

$$\underset{\sim}{\tau} = \frac{\rho_o}{\rho} \underset{\sim}{\sigma} \tag{3.4}$$

which relates the true stress to the density of the reference configuration. This Kirchhoff stress is simply proportional to the true stress (equal when the material is incompressible). However, it has the later advantage that when deriving the constitutive relations the rate of mechanical work then is defined per unit initial volume /17/.

When formulating the rate problem it is convenient for the present purpose to let the reference configuration and the current configuration coincide at the generic instant. From the balance equations for the continued motion we thus get

$$\left.\begin{aligned} &\text{div}\, \dot{\underset{\sim}{S}}^T + \dot{\underset{\sim}{b}} = 0 \\ &\dot{\underset{\sim}{S}} - \dot{\underset{\sim}{S}}^T = \underset{\sim}{\sigma}\underset{\sim}{V}^T - \underset{\sim}{V}\underset{\sim}{\sigma} \end{aligned}\right\} \text{ in } \tau \tag{3.5}$$

and

$$\dot{\underset{\sim}{S}}^T \underset{\sim}{n} = \dot{\underset{\sim}{t}} \quad \text{on} \quad S_T \, , \qquad \underset{\sim}{v} = \underset{\sim}{v}^* \quad \text{on} \quad S_V \tag{3.6}$$

where $\underset{\sim}{V} = \partial \underset{\sim}{v}/\partial \underset{\sim}{x}$ is the gradiant of the velocity field and $\underset{\sim}{v}^*$ is the kinematical constraint prescribed on $S_V$.

With this choice of reference frame the Cauchy as well as the Kirchhoff and nominal stresses coincide at the considered instant, their material rates, however, from (3.3) and (3.4) differ through

$$\dot{\underset{\sim}{S}} = \dot{\underset{\sim}{\sigma}} - \underset{\sim}{V}\underset{\sim}{\sigma} + \underset{\sim}{\sigma} \mathrm{tr}\, \underset{\sim}{V}$$
$$\dot{\underset{\sim}{\tau}} = \dot{\underset{\sim}{\sigma}} + \underset{\sim}{\sigma} \mathrm{tr}\, \underset{\sim}{V} \tag{3.7}$$

To encompass a wide class of configuration - dependent loading in /19,22/ it is stipulated that the nominal traction and the body force rates are linear functions of the particle velocity. Furthermore it is assumed for the nominal traction rates to depend linearly on the gradiant of the velocity field

$$\dot{\underset{\sim}{b}} = \dot{\underset{\sim}{b}}_o + \underset{\sim}{h}\underset{\sim}{v} \qquad \text{in } \tau \tag{3.8}$$

$$\dot{\underset{\sim}{t}} = \dot{\underset{\sim}{t}}_o + \underset{\sim}{k}\underset{\sim}{v} + (\underset{\sim}{\theta}\underset{\sim}{V}^T)\, \underset{\sim}{n} \qquad \text{on } S_T \tag{3.9}$$

where $\dot{\underset{\sim}{b}}_o$, $\dot{\underset{\sim}{t}}_o$, $\underset{\sim}{h}$, $\underset{\sim}{k}$ and $\underset{\sim}{\theta}$ may be functions of the stress state and the position of particles. In many practical problems, hydrostatic pressure loading p is assigned on the body surface. The boundary conditions in this special case are given as

$$\underset{\sim}{t} = \underset{\sim}{S}^T \underset{\sim}{n} = - p\underset{\sim}{n} \tag{3.10}$$

and

$$\dot{\underset{\sim}{t}} = \dot{\underset{\sim}{S}}^T \underset{\sim}{n} = - \dot{p}\underset{\sim}{n} - p\,[\,\mathrm{tr}\, \underset{\sim}{V}\, \underset{\sim}{1} - \underset{\sim}{V}^T]\, \underset{\sim}{n} \tag{3.11}$$

during continued motion /19,21/.

### 3.2 Constitutive equations (associated plastic flow)

Since in the present context only the instantaneous motion of the body from the finitely deformed reference configuration is of interest we need not deal too much with the kinematics of the situation. As a measure of continued deformation the strain-rate tensor

$$\underset{\sim}{D} = \frac{1}{2}\,(\underset{\sim}{V} + \underset{\sim}{V}^T) \tag{3.12}$$

is adopted. Under the assumption that the total strain-rate can be split into two parts, namely one part $\underset{\sim}{D}_e$ which is related to the elastic (recoverable) deformations and another part $\underset{\sim}{D}_p$ related to the plastic deformations

$$\underset{\sim}{D} = \underset{\sim}{D}_e + \underset{\sim}{D}_p \tag{3.13}$$

the rate-constitutive relations for the isothermal, finite deformations of an elastic-plastic material characterized by a smooth yield surface and associated plastic deformations read /20/

$$\overset{\triangledown}{\underset{\sim}{\sigma}} = \underset{\sim}{L}\,\underset{\sim}{D} - \frac{\alpha}{g}\,(\underset{\sim}{\lambda}\cdot\underset{\sim}{D})\,\underset{\sim}{\lambda} \tag{3.14}$$

if the current stress is on the yield surface.

Herein g is a positive hardening function, $\overset{\triangledown}{\underset{\sim}{\sigma}}$ is any symmetric objective stress rate and

$$\alpha = \begin{cases} 1 \text{ if } \underset{\sim}{\lambda}\cdot\underset{\sim}{D} \geqq 0 \\ \\ 0 \text{ if } \underset{\sim}{\lambda}\cdot\underset{\sim}{D} < 0 \end{cases} \tag{3.15}$$

is an index classifying the regimes of plastic loading and unloading. When the stress lies within the yield surface the material behaves purely elastic and hence $\alpha = 0$. To obtain the inverse relations we suppose that

$$h = g - \underset{\sim}{\lambda} \cdot \underset{\sim}{\mu} > 0 \qquad (3.16)$$

where

$$\underset{\sim}{\mu} = \underset{\sim}{M}\,\underset{\sim}{\lambda} \quad , \qquad \underset{\sim}{M} = \underset{\sim}{L}^{-1} \qquad (3.17)$$

The unique inverse of (3.14) thus becomes

$$\underset{\sim}{D} = \underset{\sim}{M}\,\underset{\sim}{\overset{\triangledown}{\sigma}} + \frac{\alpha_1}{h}\,(\underset{\sim}{\mu} \cdot \underset{\sim}{\overset{\triangledown}{\sigma}})\,\underset{\sim}{\mu} \qquad (3.18)$$

with

$$\alpha_1 = \begin{cases} 1 \text{ if } \underset{\sim}{\mu} \cdot \underset{\sim}{\overset{\triangledown}{\sigma}} \geqq 0 \\ 0 \text{ if } \underset{\sim}{\mu} \cdot \underset{\sim}{\overset{\triangledown}{\sigma}} < 0 \end{cases} \qquad (3.19)$$

corresponding to (3.15). Here $\underset{\sim}{L}$ and $\underset{\sim}{M}$ are tensors of instantaneous elastic moduli and compliances with the symmetry property $\underset{\sim}{L} = \underset{\sim}{L}^T$ or equivalent $L_{ijkl} = L_{klij}$ except for the ordinary symmetries. $\underset{\sim}{\lambda}$ and its dual $\underset{\sim}{\mu}$ represent the normals to the hyperplane interfaces separating the domains of elastic and plastic behaviour in the strain-rate and stress-rate spaces, respectively, directed outwards from the elastic domain.

In /20/ Hill discussed the transformation of these constitutive relations in going from one choice of objective stress-rates to another. It has been mentioned that there is no reason for preferring one choice of stress-rates to another until some specific example of bifurcation is considered. This transformation therefore has been generalized in /24/. Two frequently used stress-rates that fall into the generalized form of /24/ are

(i) the rigid-body (Zaremba-Jaumann) derivative of Kirchhoff stress

$$\underset{\sim}{\overset{\triangledown}{\sigma}} = \underset{\sim}{\dot{\sigma}} - \underset{\sim}{W}\underset{\sim}{\sigma} + \underset{\sim}{\sigma}\underset{\sim}{W} + \underset{\sim}{\sigma}\,\mathrm{tr}(\underset{\sim}{D}) \qquad (3.20)$$

which by (3.7) can be expressed in terms of nominal stresses

$$\overset{\triangledown}{\underset{\sim}{\sigma}} = \dot{\underset{\sim}{S}} + \underset{\sim}{D}\underset{\sim}{\sigma} + \underset{\sim}{\sigma}\underset{\sim}{W} \tag{3.21}$$

(ii) the convected (Oldroyd) derivative of contravariant Kirchhoff stress (formed with respect to non-rigid local coordinates embedded in an element)

$$\overset{\triangledown}{\underset{\sim}{\sigma}} = \dot{\underset{\sim}{\sigma}} - \underset{\sim}{\sigma}\underset{\sim}{V}^T - \underset{\sim}{V}\underset{\sim}{\sigma} + \underset{\sim}{\sigma}tr(\underset{\sim}{D}) \tag{3.22}$$

which again expressed in terms of nominal stresses gives

$$\overset{\triangledown}{\underset{\sim}{\sigma}} = \dot{\underset{\sim}{S}} - \underset{\sim}{\sigma}\underset{\sim}{V}^T \tag{3.23}$$

where herein

$$\underset{\sim}{W} = \frac{1}{2}(\underset{\sim}{V} - \underset{\sim}{V}^T) \tag{3.24}$$

is the spin of the deformation.

It can be shown further that due to the present choice of reference system a strain-rate potential $W(\underset{\sim}{D})$ may be introduced

$$W(\underset{\sim}{D}) = \frac{1}{2}\overset{\triangledown}{\underset{\sim}{\sigma}}\cdot\underset{\sim}{D} \tag{3.25}$$

which according to (3.14) can be expressed as

$$W(\underset{\sim}{D}) = \frac{1}{2}\underset{\sim}{D}\cdot\underset{\sim}{L}\underset{\sim}{D} - \frac{1}{2}\frac{\alpha}{g}(\underset{\sim}{\lambda}\cdot\underset{\sim}{D})^2 \tag{3.26}$$

such that with

$$\overset{\triangledown}{\underset{\sim}{\sigma}} = \frac{\partial W(\underset{\sim}{D})}{\partial \underset{\sim}{D}} \tag{3.27}$$

(3.14) can be derived from (3.25). This potential is a continuous homogeneous function of degree two in the strain-rates, with continuous first derivative and sectionally continuous second derivative. The jump in the second derivative is due to the jump of $\alpha$ (3.15) when passing through the hyperplane $\underset{\sim}{\lambda}\cdot\underset{\sim}{D} = 0$. Such materials with strain-rate dependent coefficients will be named "non-linear solids" /19/.

We shall later need to refer to another "velocity-gradiant" potential

$$U(\underset{\sim}{V}) = \frac{1}{2}\,\dot{\underset{\sim}{S}} \cdot \underset{\sim}{V}^T \tag{3.28}$$

such that the nominal stress-rate can be expressed in terms of this potential

$$\dot{\underset{\sim}{S}} = \frac{\partial U(\underset{\sim}{V})}{\partial \underset{\sim}{V}^T} \quad . \tag{3.29}$$

Due to the aforementioned different definitions of objective stress-rates different relations between $U(\underset{\sim}{V})$ and $W(\underset{\sim}{D})$ can be achieved.

$$U(\underset{\sim}{V}) = W(\underset{\sim}{D}) + \frac{1}{2}\;\Sigma(\underset{\sim}{V}) \tag{3.30}$$

where $\Sigma$ is still a function of the stress state and the velocity field expressing the influence of geometry changes due to finite deformations. Introducing (3.7) into (3.20) and (3.22), respectively, for the Jaumann derivative we find

$$\Sigma(\underset{\sim}{V}) = \underset{\sim}{\sigma}\underset{\sim}{W} \cdot \underset{\sim}{W} - \underset{\sim}{\sigma}\underset{\sim}{D} \cdot \underset{\sim}{D} \tag{3.31}$$

whereas for the Oldroyd derivative

$$\Sigma(\underset{\sim}{V}) = \underset{\sim}{\sigma}\underset{\sim}{V}^T \cdot \underset{\sim}{V}^T \quad . \tag{3.32}$$

We now recall in mind the result of Shanley's analysis of the preceding section according to which in column buckling at the instant of bifurcation unloading is absent. In this same sense we introduce a second "linear solid" with the property that unloading is excluded through $\alpha = 1$. Thus

$$W_L(\underset{\sim}{D}) = \frac{1}{2}\,\underset{\sim}{D}\cdot\underset{\sim\sim}{L}\underset{\sim}{D} - \frac{1}{2}\,\frac{1}{g}\,(\underset{\sim}{\lambda}\cdot\underset{\sim}{D})^2 \tag{3.33}$$

and $U_L(\underset{\sim}{V})$ will be found as before ((3.31) and (3.32), respectively). This material according to Hill /16/ will be named a "linear comparison solid" for the "non-linear solid" (3.26) under consideration.

It has to be noted that most rate-constitutive relations of elastic-plastic solids with a smooth yield surface and associated plastic flow rule can be cast into the present more general form (3.14) to (3.19). This is especially true when as in classical plasticity a strictly convex yield function (potential function) $\Phi(\sigma,\underset{\sim}{\kappa})$ is introduced, where $\kappa$ (i. e. some set of 'internal' or 'hidden' variables) is a functional of the preceding plastic history and plastic flow occurs according to the 'normality rule',

$$\underset{\sim p}{D} = \begin{cases} \dot{\lambda}\,\dfrac{\partial\Phi}{\partial\underset{\sim}{\sigma}} \;, & \text{if } \left(\dfrac{\partial\Phi}{\partial\underset{\sim}{\sigma}}\cdot\underset{\sim}{\sigma}\right) \geqq 0 \;\text{ and }\; \Phi\,(\underset{\sim}{\sigma},\kappa) = 0 \\ 0 \;, & \text{otherwise} \end{cases} \tag{3.34}$$

i. e. associated with the plastic potential $\Phi$, where in (3.34) $\dot{\lambda}$ is a nonnegative scalar ('plastic multiplier'). In particular, when $\kappa$ is interpreted as one single scalar-valued internal variable, the plastic work, which is defined through

$$\dot{\kappa} = (\underset{\sim}{S}\cdot\underset{\sim p}{D}) \tag{3.35}$$

this model corresponds to the frequently used model of workhardening plasticity /23/. Comparing (3.34) with (3.18) we find

$$\underset{\sim}{\mu} = \frac{\partial\Phi}{\partial\underset{\sim}{\sigma}} \;, \qquad h = -\,\frac{\partial\Phi}{\partial\kappa}\,(\underset{\sim}{\sigma}\cdot\underset{\sim}{\mu}) \tag{3.36}$$

for this model. A yet more special case of this particular model is the well-known $J_2$ flow theory.

# 4. GENERAL BIFURCATION THEORY

## 4.1 Uniqueness criterion

In a series of papers (see e. g. /16/, /19/ or /9/) in the late fifties Hill has formulated a fundamental continuum theory for uniqueness and stability of solids with inelastic material properties. The uniqueness criterion derived here corresponds to this general theory.

Recapitulating, the boundary value problem is posed as follows; at a generic instant the current shape of a body (volume V and surface S), the internal stress distribution, its mechanical material properties together with the loading conditions described in the preceding section are considered known. The problem is to determine a stress-rate and a velocity field satisfying the rate relation (3.14), balance equations (3.5) and kinematical relation (3.12) in $\tau$ and boundary conditions (3.6) on $S_T$ and $S_v$, respectively. The velocity field is assumed to be continuous and to posess continuous first derivatives and at least sectionally continuous second derivatives. The interest is then focused on wether this ensuing quasi-static motion of the body is unique or not.

Now suppose that there could be two distinct solutions $(\cdot)^{(1)}$ and $(\cdot)^{(2)}$ of the problem and denote their difference by $\Delta(\cdot)$, e. g.

$$\Delta \underset{\sim}{v} = \underset{\sim}{v}^{(1)} - \underset{\sim}{v}^{(2)} \quad ; \quad \Delta \dot{\underset{\sim}{S}} = \dot{\underset{\sim}{S}}^{(1)} - \dot{\underset{\sim}{S}}^{(2)} \tag{4.1}$$

Note that herein $\Delta \underset{\sim}{v}$ and $\Delta \dot{\underset{\sim}{S}}$ do not necessarily correspond in the sense of being related by (3.14).

Since both solutions have to satisfy the same boundary conditions (3.6), it is evident then that

$$\Delta \dot{\underset{\sim}{t}} = 0 \quad \text{on } S_T , \qquad \Delta \underset{\sim}{v} = 0 \quad \text{on } S_V \tag{4.2}$$

and hence

$$\int \Delta \underset{\sim}{v} \cdot \Delta \dot{\underset{\sim}{t}} \, dS - \int \Delta \underset{\sim}{v} \cdot [\, \underset{\sim}{k} \Delta \underset{\sim}{v} + (\underset{\sim}{\theta} \Delta \underset{\sim}{V}^T) \, \underset{\sim}{n} ] \, dS_T = 0 \tag{4.3}$$

By use of divergence theorem the first integral of (4.3) becomes

$$\int \Delta \underset{\sim}{v} \cdot \Delta \dot{\underset{\sim}{t}} \, dS = \int (\Delta \underset{\sim}{V}^T \cdot \Delta \dot{\underset{\sim}{S}} + \Delta \underset{\sim}{v} \cdot \operatorname{div} \Delta \dot{\underset{\sim}{S}}^T) \, d\tau \tag{4.4}$$

such that with the local balance equation (3.5) and $\dot{\underset{\sim}{b}}$ given by (3.8) we find

$$\int [\Delta \underset{\sim}{V}^T \cdot \Delta \dot{\underset{\sim}{S}} - 2\psi_V (\Delta \underset{\sim}{v})] \, d\tau - 2 \int \psi_S (\Delta \underset{\sim}{v}) \, dS_T = 0 \tag{4.5}$$

where

$$\psi_V(\underset{\sim}{v}) = \frac{1}{2} \underset{\sim}{v} \cdot \underset{\sim}{h} \underset{\sim}{v}$$

$$\psi_S(\underset{\sim}{v}) = \frac{1}{2} \underset{\sim}{v} \cdot \underset{\sim}{k} \underset{\sim}{v} + \frac{1}{2} \underset{\sim}{v} \cdot (\underset{\sim}{\theta} \underset{\sim}{V}^T) \, \underset{\sim}{n} \tag{4.6}$$

It is evident now, that if (4.5) is valid for non-uniqueness of any solution then one <u>sufficient</u> condition for uniqueness is

$$\int [\, \Delta \underset{\sim}{V}^T \cdot \Delta \dot{\underset{\sim}{S}} - 2\psi_V (\Delta \underset{\sim}{v}) ] \, d\tau - 2 \int \psi_S (\Delta \underset{\sim}{v}) \, dS_T > 0 \tag{4.7}$$

This is true regardless of what the constitutive equations may be, because constitutive equations are not involved in (4.7).

In order to transform (4.7) into a form more convenient for applications, we suppose, instead of what is postulated to get (4.7), that $\underset{\sim}{v}^{(1)}$ and $\dot{\underset{\sim}{S}}^{(1)}$ as well as $\underset{\sim}{v}^{(2)}$ and $\dot{\underset{\sim}{S}}^{(2)}$ and thus $\Delta\underset{\sim}{v}$ and $\Delta\dot{\underset{\sim}{S}}$ are related by the constitutive equations of section 3.2. Due to the non-linearity of these equations through (3.15), different cases of loading and unloading have to be considered. Introducing $\nu_2 = \underset{\sim}{\lambda} \cdot \underset{\sim}{D}^{(2)}$, say, by use of (3.26), (3.28) and (3.30) we find

$$W(\Delta\underset{\sim}{D},\nu_2) = W_e(\Delta\underset{\sim}{D}) + W_p(\Delta\underset{\sim}{D},\nu_2) \tag{4.8}$$

where

$$W_e = \frac{1}{2}\Delta\underset{\sim}{D} \cdot \underset{\sim}{L}\Delta\underset{\sim}{D} \tag{4.9}$$

$$W_p = \begin{cases} -\frac{1}{2g}(\underset{\sim}{\lambda}\cdot\Delta\underset{\sim}{D})^2 & ;\ \nu_2 \geqq 0 \text{ and } \nu_2 \geqq -\underset{\sim}{\lambda}\cdot\Delta\underset{\sim}{D} \\ -\frac{1}{2g}(\underset{\sim}{\lambda}\cdot\Delta\underset{\sim}{D})(\nu_2+\underset{\sim}{\lambda}\cdot\Delta\underset{\sim}{D}) & ;\ -\underset{\sim}{\lambda}\cdot\Delta\underset{\sim}{D} \leqq \nu_2 < 0 \\ \frac{1}{2g}(\underset{\sim}{\lambda}\cdot\Delta\underset{\sim}{D})\,\nu_2 & ;\ -\underset{\sim}{\lambda}\cdot\Delta\underset{\sim}{D} > \nu_2 \geqq 0 \\ 0 & ;\ \nu_2 < 0 \text{ and } \nu_2 < -\underset{\sim}{\lambda}\cdot\Delta\underset{\sim}{D} \end{cases} \tag{4.10}$$

and hence the inequality (4.7) takes the form

$$\int [\,U(\Delta\underset{\sim}{V},\nu_2) - \psi_v\,]\,d\tau - \int\psi_s\,dS_T > 0 \tag{4.11}$$

where the velocity-gradient and strain-rate potentials are related through (3.30)

$$U(\Delta\underset{\sim}{V},\nu_2) = W(\Delta\underset{\sim}{D},\nu_2) + \frac{1}{2}\,\Sigma(\Delta\underset{\sim}{V}) \tag{4.12}$$

Suppose now further that $\underset{\sim}{v}^{(2)}$ is a known fundamental solution of the problem whose uniqueness is in question. Since the equality (4.5) must hold for any two distinct solutions it is evident that if for given $\nu_2$ and for <u>every</u> non-zero kinematically admissible velocity field $\underset{\sim}{v}$, say, vanishing on the boundary $S_v$ the inequality

$$R(\underset{\sim}{v},\nu_2) = \int[U(\underset{\sim}{V},\nu_2) - \psi_v(\underset{\sim}{v})]\, d\tau - \int\psi_s(\underset{\sim}{v})\, dS_T > 0 \tag{4.13}$$

holds, then the fundamental solution is unique.

This functional $R(\underset{\sim}{v},\nu_2)$, however, as consequence of the multi-valued form of the potential W has a complicated form and its direct application to estimate bifurcation stresses may lead to considerable difficulties. It is therefore preferable to replace this condition by a slightly weaker criterion introducing the linear comparison solid as defined in (3.33).

Comparing this expression with (4.8) and taking into account the four different cases of (4.10) it easily can be shown that

$$W_L(\Delta \underset{\sim}{D}) \leq W(\Delta \underset{\sim}{D},\nu_2) \tag{4.14}$$

the equality sign holding when both strain-rate fields call for additional loading. From (4.12) consequently

$$U_L(\Delta \underset{\sim}{V}) \leq U(\Delta \underset{\sim}{V},\nu_2) \tag{4.15}$$

and thus a weaker <u>sufficient</u> condition for uniqueness is

$$R_L(\underset{\sim}{v}) = \int[U_L(\underset{\sim}{V}) - \psi_v(\underset{\sim}{v})]\, d\tau - \int\psi_s(\underset{\sim}{v})\, dS_T > 0 \tag{4.16}$$

for <u>every</u> non-zero kinematically admissible velocity field vanishing on $S_v$.

Therefore, any solution of the rate boundary-value problem for the elastic-plastic solid is unique when the uniqueness of the analogous boundary-value problem for the comparison solid is assured. In other words, uniqueness of the velocity field cannot fail in the actual non-linear solid whenever the sufficient criterion (4.16) applied to its linear comparison solid is satisfied.

4.2 First bifurcation state

The advantage of transforming the uniqueness condition (4.7) to (4.16) is, of course, that $R_L(\underset{\sim}{v})$ is now a functional only of kinematically admissible velocity fields, with no stress-rate occuring explicitly. It has to be stressed, however, that (4.16) is only a sufficient - or even 'over-sufficient' /17/ - criterion and as such it is not the best tool to provide, say, any bifurcation state of the elastic-plastic solid. However, though not a necessary condition, (4.16) can be used as a device to determine the magnitudes of primary bifurcation stresses, which are ordinarily unknown.

Suppose that an elastic-plastic body is deformed such that he passes through a sequence of equilibrium states for which the inequality (4.16) holds. This sequence of states may be terminated at a configuration for which

$$R_L(\underset{\sim}{v}) \geqq 0 \qquad (4.17)$$

the equality holding for some velocity field, say $\bar{\underset{\sim}{v}}$. Such a state is the 'first bifurcation state' in the linear comparison solid.

In view of (4.17) the velocity field for which $R_L(\bar{\underset{\sim}{v}}) = 0$ must make the first variation of $R_L$ vanish.

$$\delta R_L = \delta[\int(U_L-\psi_V)\, d\tau - \int\psi_S\, dS_T] = 0 \qquad (4.18)$$

To prove this, the volume integral in (4.18) by use of (3.27) and (4.6) may be written as

$$\delta[\int(U_L-\psi_V)\,d\tau] = \int(\delta\bar{\underset{\sim}{v}}^T \cdot \dot{\underset{\sim}{S}} - \delta\bar{\underset{\sim}{v}} \cdot \underset{\approx}{h}\bar{v})\,d\tau \tag{4.19}$$

since $\psi_V$ has a quadratic form. Now, if the equilibrium conditions (3.5) are satisfied, then applying the divergence theorem to the right-hand side of the above equation and since $\delta\bar{\underset{\sim}{v}} = 0$ on $S_V$,

$$\delta R_L(\bar{\underset{\sim}{v}}) = \int[\delta\bar{\underset{\sim}{v}} \cdot \dot{\underset{\sim}{t}} - \delta\psi_S(\bar{\underset{\sim}{v}})]\,dS_T + \int\delta\bar{\underset{\sim}{v}} \cdot [\dot{\underset{\sim}{b}} - \underset{\approx}{h}\bar{v}]\,d\tau \tag{4.20}$$

Provided, the configuration dependent loading on $S_T$ (3.9) is conservative, which can be shown to be the case, when the coefficients $\theta_{ijkl}$ are antisymmetric with respect to interchanges of i and k, and j and l, respectively /19/,

$$\delta R_L(\bar{\underset{\sim}{v}}) = \int\delta\bar{\underset{\sim}{v}} \cdot [\,\dot{\underset{\sim}{t}} - \underset{\approx}{k}\bar{v} - (\underset{\approx}{\theta}\bar{V}^T)\,\underset{\sim}{n}\,]\,dS_T + \int\delta\bar{\underset{\sim}{v}} \cdot [\dot{\underset{\sim}{b}} - \underset{\approx}{h}\bar{v}\,]\,d\tau \tag{4.21}$$

As a consequence, to prove (4.18), the velocity field $\bar{\underset{\sim}{v}}$, being the solution of the variational equation (4.18), is at the same time the solution of a self-adjoint eigenvalue problem for the comparison solid, with homogeneous conditions

$$\begin{aligned}
\dot{\underset{\sim}{t}} &= \underset{\approx}{k}v + (\underset{\approx}{\theta}V^T)\,\underset{\sim}{n} \quad \text{on } S_T\,, \quad \underset{\sim}{v} = 0 \quad \text{on } S_V \\
\dot{\underset{\sim}{b}} &= \underset{\approx}{h}v \qquad\qquad\quad \text{in } \tau
\end{aligned} \tag{4.22}$$

Recall in mind now that $\underset{\sim}{v}^{(2)}$ is supposed to be a known 'fundamental solution' of the original problem (such as continuing uniform compression in a strut or externally pressurized body), i. e. a solution satisfying inhomogeneous boundary conditions, and suppose that the increment of this fundamental solution satisfies the condition that no unloading takes place throughout the region where the yield condition is satisfied, i. e. the material behaves like its linear comparison solid. Suppose further that $\bar{\underset{\sim}{v}}$ is the eigensolution of the corresponding homogeneous problem for the comparison solid (4.22), being normalized in some suitable fashion.

Since in passing from (4.12) to (4.13) $\Delta \underset{\sim}{v}$ is interpreted as $\underset{\sim}{v}$, it is clear then that through

$$\underset{\sim}{v}^{(1)} = \underset{\sim}{v}^{(2)} + a\bar{\underset{\sim}{v}} \tag{4.23}$$

where a is an arbitrary multiplier, solutions $\underset{\sim}{v}^{(1)}$ could be found where no unloading takes place at bifurcation, i. e. $\underset{\sim}{\lambda} \cdot \underset{\sim}{D}^{(1)} \geqq 0$. In other words, the first bifurcation state in an elastic-plastic solid can often be investigated in the sense of Shanley as if it were elastic solid with given tangent moduli as material coefficients. A necessary condition which has to be fulfilled, however, is that for the fundamental solution $\underset{\sim}{\lambda} \cdot \underset{\sim}{D}^{(2)} \geqq 0$ has to be guaranteed throughout that part of the body where the yield condition is satisfied.

## 4.3 Stability

Turning now to the question of stability, this concept is not given by any law of nature. A classical definition of stability in the dynamical sense claims that if a body, in equilibrium under dead loading on part of the boundary surface and rigid constraints on the remainder, is perturbed by any kind of agency, then the state is stable if the resulting motion, compatible with the constraints, stays vanishingly small (in some sense) when the perturbation itself does.

A sufficient condition for stability in this sense may be derived, starting from the requirement that during the initiated motion between time t = 0 and t = T the internal change of energy should exceed the work done by the external loads $\underset{\sim}{t}$ and the body forces $\underset{\sim}{b}$ on the displacements, i. e.

$$J = \int_o^T [\int \underset{\sim}{V}^T \cdot \underset{\sim}{S}\, d\tau - \int \underset{\sim}{v} \cdot \underset{\sim}{t}\, dS_T - \int \underset{\sim}{v} \cdot \underset{\sim}{b}\, d\tau ]\, dt > 0 \tag{4.24}$$

in terms of integrals over the fixed reference region $\tau$ with surface S /9/.

Due to the non-linearity of the constitutive relations of the underlying elastic-plastic material the value of J is strongly path-dependent. Concrete calculations of J therefore require a few more assumptions. In the present setting under rigid constraints and dead loading (i. e. when the inhomogeneous parts $\dot{\underset{\sim}{t}}_o$ and $\dot{\underset{\sim}{b}}_o$ of the surface and body force rates vanish), for small enough time intervals t a Taylor expansion of the integrand yields (don't mix up time t and traction $\underset{\sim}{t}$)

$$\int \underset{\sim}{V}^T \cdot [\, \underset{\sim}{S} + t\dot{\underset{\sim}{S}}\,]\, d\tau - \int \underset{\sim}{v} \cdot [\, \underset{\sim}{t} + t\dot{\underset{\sim}{t}}\,]\, dS_T - \int \underset{\sim}{v} \cdot [\, \underset{\sim}{b} + t\dot{\underset{\sim}{b}}\,]\, d\tau + 0\,(t^2) \tag{4.25}$$

Equilibrium of the given state balances out the first terms in the brackets. Thus from (4.24) we find

$$J \simeq \frac{1}{2}\, T^2 \, \{\int [\underset{\sim}{V}^T \cdot \dot{\underset{\sim}{S}} - 2\psi_V(\underset{\sim}{v})\,]\, d\tau - 2 \int \psi_S(\underset{\sim}{v})\, dS_T\} \tag{4.26}$$

for sufficiently small time intervals and with $\psi_V$ and $\psi_S$ as defined in (4.6). Therefore the following sufficient criterion for stability is obtained

$$\int [\, \underset{\sim}{V}^T \cdot \dot{\underset{\sim}{S}} - 2\psi_V(\underset{\sim}{v})\,]\, d\tau - 2 \int \psi_S(\underset{\sim}{v})\, dS_T > 0 \tag{4.27}$$

which is formally equivalent to the uniqueness criterion (4.7). As has been pointed out in deriving (4.7), the stress-rate and velocity-gradient (difference-)fields are not necessarily related by the constitutive equations (3.14) on account of different regimes of loading and unloading in the plastic zone.

Consequently (4.7) and (4.27) are distinct criteria. Nevertheless, there is a definite connexion in that (4.27) is always satisfied when (4.7) is (but not vice versa); for (4.7) reduces to (4.27) when one field of a pair vanishes identically. A partly plastic state in which the boundary value problem has a unique solution is therefore certainly stable; and even when the solution is not unique the state may still be stable /16/.

Coming back now to the simple problem of section 2, when $\dot{\underset{\sim}{b}} = 0$ and $\underset{\sim}{k} = 0$ as well as $\underset{\sim}{\theta} = 0$, for quasi-static motion (4.27) is equivalent to (in case of rigid constraints)

$$\int \underset{\sim}{V}^{T} \cdot \dot{\underset{\sim}{S}}\, d\tau = \int \underset{\sim}{v} \cdot \dot{\underset{\sim}{t}}\, dS_{T} > 0 \qquad (4.28)$$

and thus there always exists the possibility of stable bifurcations under changing external loads (when (4.28) is fulfilled but (4.7) fails) as once heuristically argued by Shanley.

# 5. EXAMPLES

The uniqueness criterion and the variational method for finding bifurcation states by minimizing expressions such as (4.16) have been applied to many problems in the last years. This method has been found advantageous especially in all those cases where one (or two) dimension(s) of a structure approximately can be taken as small compared with the others, i. e. in applications to columns, plates and shells. From the great number of works that could have been cited here we only want to refer to some typical examples of application /25, 29, 31, 33, 39, 41/ and particularly to the survey article in the booklet of Storåkers /10/. As usually, e. g. in shell theories, a velocity field $\bar{\underset{\sim}{v}}$ is introduced as explicitely linear function of thickness-coordinate z (say). Performing the integration procedure of (4.16) first in the z-direction reduces the originally three-dimensional problem to a two-dimensional one. The problem then is solved by introducing suitable kinematically admissible two-dimensional functions, satisfying the kinematical constraints of the shell, and minimizing the remaining functional (4.16).

In the present lecture notes, however, emphasis is laid on a second method for finding bifurcation states by searching for solutions of the governing equations (3.5) to (3.9) with the constitutive equations. This direct approach has been successful in exhibiting, especially for thick bodies under certain symmetrically distributed boundary conditions. Both symmetric modes of bifurcation (necking or bulging), as well as certain kinds of surface instabilities have been treated in numerous examples.

Returning now as a first example to the necking problem discussed in the second section this direct method runs as follows, with the results taken from the works of Cheng, Ariaratnam and Dubey /32/, Miles /37/, Hutchinson and Miles /36/ and the present author /28/.

We consider a specimen of isotropic incompressible elastic-plastic v. Mises material in accordance with constitutive relations (3.14), (3.18) and (3.34), respectively. In particular, the specimen is taken to be a circular cylinder of radius R and length L at a generic instant t during the process of deformation. The natural cylindrical coordinates r, θ, z are chosen with the origin at the centre of one end of the specimen. At time t the fundamental solution for the cylindrical solid is the state of uniaxial tension with a true stress denoted by σ. The relative axial displacement of the ends is prescribed to be uniform and is identified with a parameter C. Specifying the velocity field at this instant by u, v, w, the constitutive relation can be written as /30/

$$\overset{\nabla}{\underset{\sim}{\sigma}}{}' = 2G\,[\,\underset{\sim}{D} - \alpha\frac{\delta}{k}\,(\underset{\sim}{\sigma}' \cdot \underset{\sim}{D})\,\underset{\sim}{\sigma}'] \tag{5.1}$$

where $\underset{\sim}{\sigma}' = \underset{\sim}{\sigma} - \frac{1}{3}\,\mathrm{tr}\,(\underset{\sim}{\sigma})\,\underset{\sim}{1}$ is the deviatoric Cauchy stress,

$$\Phi(\underset{\sim}{\sigma},\kappa) = \underset{\sim}{\sigma}' \cdot \underset{\sim}{\sigma}' - k(\kappa) = 0 \tag{5.2}$$

is the v. Mises yield condition with isotropic workhardening k as function of the plastic work κ and from the definition of $E_t$ and (5.1)

$$\delta = 1 - E_t/E = 1 - \lambda \tag{5.3}$$

is a normalized hardening function.

Boundary conditions (3.6) are taken to be

$$\begin{aligned} &\text{(i) on } z = 0,\, L\ , \quad w = 0,\ \pm C, \quad \dot{\sigma}_{zr} = \dot{\sigma}_{z\theta} = 0 \\ &\text{(ii) on } r = R\ , \quad \dot{\sigma}_{rr} = \dot{\sigma}_{r\theta} = 0\ , \quad \dot{\sigma}_{rz} = \sigma\frac{\partial u}{\partial z} \end{aligned} \tag{5.4}$$

expressed in physical tensor components and where condition (ii) according to (3.7) expresses the requirement that the lateral surface of the specimen is traction free. The equations of continuing equilibrium (3.5) are

$$\operatorname{div} \dot{\underset{\sim}{\sigma}} = 0 \tag{5.5}$$

since the existing state of stress is uniform.

A solution involving homogeneous deformation is always

$$w = C\,\frac{z}{L}\ ,\quad u = -\frac{C}{2}\,\frac{r}{L}\ ,\quad v = 0 \tag{5.6}$$

the fundamental solution of the problem, regardless of which stress-rate is introduced in (5.1).

We now confine our attention to the linear comparison solid defined with $\alpha = 1$ everywhere in (5.1), and look for a non-zero solution (an eigenfield) satisfying the homogeneous boundary conditions with $w = 0$ on $z = 0, L$. If such an eigenfield exists, a bifurcation is obtained for the non-linear elastic-plastic solid by adding a small enough multiple of this eigenfield to the homogeneous solution (5.6) so that loading takes place everywhere. To keep the analysis as simple as possible, we look for an axisymmetric eigenfield, i. e. $v = 0$, as it seems plausible that such a mode should adequately describe necking.

If the Jaumann stress-rate (3.20) is used in (5.1), these equations with $\alpha = 1$ become

$$\begin{aligned}
\dot{\sigma}_{rr} &= 2G\,\{\frac{\partial u}{\partial r} - \frac{\delta}{6}\,[\,\frac{1}{r}\,\frac{\partial (ru)}{\partial r} - 2\,\frac{\partial w}{\partial z}\,]\,\} + \dot{q} \\
\dot{\sigma}_{\theta\theta} &= 2G\,\{\frac{u}{r} - \frac{\delta}{6}\,[\,\frac{1}{r}\,\frac{\partial (ru)}{\partial r} - 2\,\frac{\partial w}{\partial z}\,]\,\} + \dot{q} \\
\dot{\sigma}_{zz} &= 2G\,\{\frac{\partial w}{\partial z} + \frac{\delta}{3}\,[\,\frac{1}{r}\,\frac{\partial (ru)}{\partial r} - 2\,\frac{\partial w}{\partial z}\,]\,\} + \dot{q} \\
\dot{\sigma}_{rz} &= \frac{1}{2}\,\frac{\partial u}{\partial z}\,(2G + \sigma) - \frac{1}{2}\,\frac{\partial w}{\partial r}\,(2G - \sigma)
\end{aligned} \tag{5.7}$$

for the physical tensor components and with $\dot{\sigma}_{r\theta} = \dot{\sigma}_{\theta z} = 0$ as consequence of the assumed axisymmetry. $\dot{q}$ herein is the unknown hydrostatic pressure. The equilibrium equations (5.5) are

$$\frac{1}{r}\frac{\partial(r\dot{\sigma}_{rr})}{\partial r} + \frac{\partial\dot{\sigma}_{rz}}{\partial z} - \frac{1}{r}\dot{\sigma}_{\theta\theta} = 0$$
$$\frac{1}{r}\frac{\partial(r\dot{\sigma}_{zr})}{\partial r} + \frac{\partial\dot{\sigma}_{zz}}{\partial z} = 0 \tag{5.8}$$

Equations (5.7) and (5.8) together with the incompressibility of the material

$$\frac{\partial u}{\partial r} + \frac{u}{r} + \frac{\partial w}{\partial z} = 0 \tag{5.9}$$

comprise a system of seven equations for the seven unknowns, u, w, $\dot{\sigma}_{rr}$, $\dot{\sigma}_{\theta\theta}$, $\dot{\sigma}_{zz}$, $\dot{\sigma}_{zr}$ and q. The latter condition enables us to introduce a function $\Phi(r,z)$ such that with

$$u = -\frac{\partial\Phi}{\partial z} \quad \text{and} \quad w = \frac{1}{r}\frac{\partial(r\Phi)}{\partial r} \tag{5.10}$$

the incompressibility (5.9) is ensured.

Introducing (5.10) into (5.7), from (5.8) we get

$$L(\Phi)^{+}(1-\delta+\beta) + \Phi^{+++}(1+\beta) - \dot{q}'/G = 0$$
$$\frac{1}{r}[rL(\Phi)]'(1-\beta) + \frac{1}{r}(r\Phi)'^{++}(1-2\delta-\beta) + \dot{q}^{+}/G = 0 \tag{5.11}$$

two equations for the two functions $\Phi$ and $\dot{q}$, which by further manipulations can be reduced to give a single equation for

$$LL(\Phi)(1-\beta) + L(\Phi)^{++}(2-3\delta) + \Phi^{++++}(1+\beta) = 0 \tag{5.12}$$

where $\beta = \sigma/2G$ and the operator $L(\cdot)$ is defined by

$$L(\Phi) = \frac{1}{r}[(r\Phi)']' . \tag{5.13}$$

Here and in what follows we denote the derivatives $(\cdot)' = \partial(\cdot)/\partial r$, $(\cdot)^+ = \partial(\cdot)/\partial z$.

The homogeneous boundary conditions are

$$\text{(i)} \quad \text{on } z = 0, L \quad , \quad \Phi' = 0$$

$$L(\Phi)\ (1-\beta) - \Phi^{++}(1+\beta) = 0 \tag{5.14}$$

$$\text{(ii)} \quad \text{on } r = R \quad , \quad L(\Phi) - \Phi^{++} = 0$$

$$2\Phi'^{+} - \frac{\delta}{r}(r\Phi)'^{+} - \dot{q}/G = 0 \tag{5.15}$$

We now look for solutions of the eigenvalue problem in the form

$$\Phi(r,z) = F(r)\ \sin(mz)$$

$$\dot{q}(r,z) = G(r)\ \cos(mz) \tag{5.16}$$

where $m = n\pi/L$ and n being the number of 'necks' involved in the deformation mode, such that boundary conditions on $z = 0, L$ are satisfied automatically and ordinary differential equations are obtained.

From (5.12) we thus find

$$LL(F) + 2bm^2\ L(F) + cm^4F = 0 \tag{5.17}$$

where

$$2b = -\ (2-3\delta)/(1-\beta), \quad c = (1+\beta)/(1-\beta) \tag{5.18}$$

and furthermore the boundary conditions on $r = R$ give

$$L(F) + m^2 F = 0$$

$$[rL(F)]' (1-\beta) - (rF)' m^2 (3-3\delta-\beta) + 2m^2 F = 0 \tag{5.19}$$

The general solution to (5.17) satisfying boundedness condition on $r = 0$ is

$$F = c_1 J_1 (m\rho r) + c_2 J_1 (m\bar{\rho} r) \tag{5.20}$$

where $J_n$ is the Bessel function of order n of the first kind, here with complex argument $m\rho r$, $\rho^2 = b + i \sqrt{c-b^2}$ and the bar implies complex conjugation. Substitution of (5.20) into the boundary conditions (5.19) provides the eigenvalue equation for $\sigma_c$.

$$\mathscr{I}_m \{(1-\bar{\rho}^2) \bar{J}_1[\rho mR J_o (1-\beta)(1-2b+\rho^2) - 2J_1]\} = 0 \tag{5.21}$$

To obtain a relation between $\sigma_c$ and mR for small values of mR, we expand the eigenvalue equation in powers of $\rho$mR using the series representation of the Bessel functions. Thus from (5.21) we find that to the order $(mR)^6$ the critical stress compared with the stress at maximum load (2.26) becomes /36/

$$\frac{\sigma_c}{\sigma_m} = 1 + \frac{(mR)^2}{8} + \frac{(mR)^4}{576} \frac{1}{\lambda} \tag{5.22}$$

Next, in the limit $(mR) \to \infty$, i. e. for rather stubby specimens, again by power series expansion we find /40/

$$\sigma_c = \frac{G(1+3\lambda)}{1-\sqrt{\frac{1-\sigma_c/2G}{1+\sigma_c/2G}}} . \tag{5.23}$$

Some numerical calculations based on the exact eigenvalue expression (5.21) have been carried out for a hypothetical linear hardening material ($\delta$ = const.) to assess the range of accuracy of expansions (5.22)

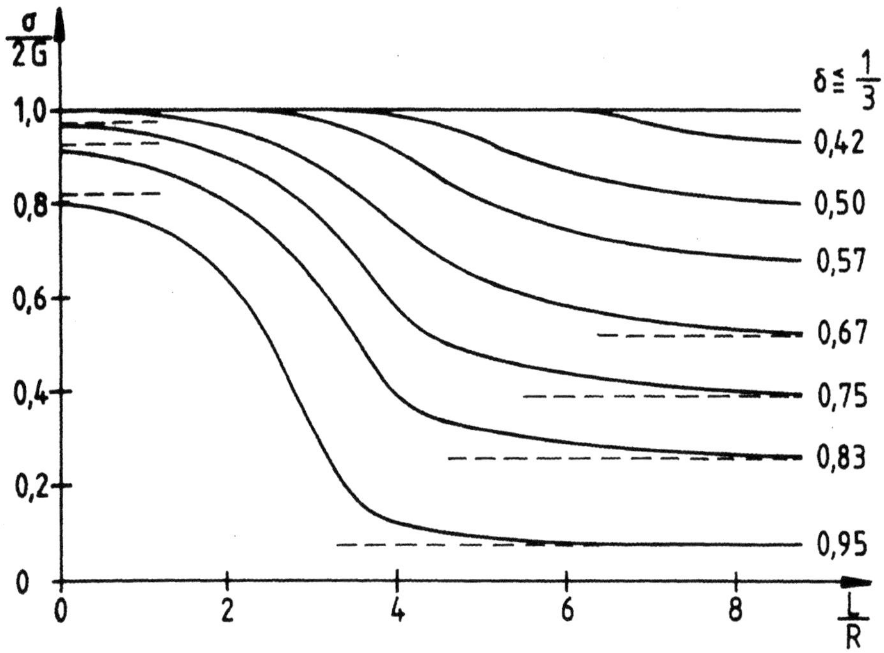


Fig. 5.1 Stresses at bifurcation in tension

and (5.23). Both approximations have been depicted in Fig. 5.1 together with the exact values as function of slenderness parameter L/R. The figure indicates that the simple formulae retain reasonable accurary for values of L/R > 4.0.

As a somewhat more complicated second example the necking (and bulging) of biaxially stretched elastic-plastic circular plates is discussed now. This problem has been treated by Needleman and Tvergaard /40/ and the present author /27/.

A circular plate of radius R and thickness H at a generic instant t is considered. As before the material is taken to be incompressible elastic-plastic v. Mises material and again cylindrical coordinates are introduced with the z-direction normal to the plate middle-surface. At the edge of the plate, a uniform radial displacement U is prescribed, while the tractions in circumferential direction and in the z-direction are assumed to vanish. The top and the bottom surfaces of the plate are taken to be traction-free. At time t the fundamental solution for this problem is

$$\sigma_{rr} = \sigma_{\theta\theta} = \sigma \tag{5.24}$$

for the non-vanishing components of true stresses. Thus, with the same notations as before, the boundary conditions for the plate problem are

$$\begin{aligned} &\text{(i) on } z = \pm H/2, \quad \dot{\sigma}_{zr} = \sigma \frac{\partial w}{\partial r}, \quad \dot{\sigma}_{z\theta} = \sigma \frac{\partial w}{\partial \theta}, \quad \dot{\sigma}_{zz} = 0 \\ &\text{(ii) on } r = R \quad , \quad u = \pm U \; , \quad \dot{\sigma}_{r\theta} = \sigma \frac{\partial u}{\partial \theta}, \quad \dot{\sigma}_{rz} = 0 \end{aligned} \tag{5.25}$$

The fundamental solution satisfying inhomogeneous boundary conditions is

$$u = U \frac{r}{R} \; , \quad w = - \; 2U \frac{z}{R} \; , \quad v = 0 \tag{5.26}$$

such that the loading condition is satisfied throughout the plastic region of the body.

Performing now the same procedure as before, i. e. looking for an axisymmetric eigenfield that satisfies the homogeneous boundary conditions with $u = 0$ on $r = R$, equations (5.5) and (5.7) to (5.13) of the first example can keep their validity, after replacing $\beta$ by $-\beta$, and vice versa, in (5.11) and (5.12), since $(5.7)_4$ for the plate problem reads

$$\dot{\sigma}_{rz} = \frac{1}{2} \frac{\partial u}{\partial z} (2G-\sigma) + \frac{1}{2} \frac{\partial w}{\partial r} (2G+\sigma) \quad . \tag{5.27}$$

The homogeneous boundary conditions (5.25) on the other hand become

$$\begin{aligned} &\text{(i) on } z = \pm H/2, \quad L\,(\Phi) - \Phi^{++} = 0 \\ &\qquad \frac{1}{r}\,(r\Phi)'^{+}\,(1-\beta) + \dot{q}/2G = 0 \end{aligned} \tag{5.28}$$

$$\text{(ii) on } r = R, \quad \Phi^{+} = 0$$

$$L(\Phi)(1+\beta) - \Phi^{++}(1-\beta) = 0 \tag{5.29}$$

Solutions to the foregoing eigenvalue problem can be found in the form /27/

$$\Phi(r,z) = J_1(\nu r)\, F(z)$$

$$\dot{q}(r,z) = J_o(\nu r)\, G(z) \tag{5.30}$$

where $\nu = j_n/R$ and $j_n$ is the nth positive root (taken in ascending order) of $J_1$. The mode (5.30) satisfies the boundary conditions on $r = R$ identically and introduced into (5.12) and (5.28) leads to ordinary differential equations. We thus find

$$F^{++++} + 2b\,\nu^2 F^{++} + c\,\nu^4 F = 0 \tag{5.31}$$

with parameters b and c defined in (5.18) and boundary conditions on $z = \pm H/2$ to be

$$F^{++} + \nu^2 F = 0$$

$$F^{+++}(1-\beta) - \nu^2 F^{+}(3-3\delta-\beta) = 0 \tag{5.32}$$

The solution to (5.31) can be written as

$$F(z) = c_1 \operatorname{sh}\rho z + c_2 \operatorname{ch}\rho z + c_3 \operatorname{sh}\bar{\rho} z + c_4 \operatorname{ch}\bar{\rho} z \tag{5.33}$$

where here $\rho^2 = \nu^2 (-b+i\sqrt{c-b^2})$ and again the bar implies complex conjugation. It is furthermore anticipated here that $c-b^2 > 0$.

The solution (5.33) has the form that with $c_1 = c_3 = 0$ antisymmetric modes of deformation (buckling) occur whereas for $c_2 = c_4 = 0$ these modes are symmetric (necking or bulging). Substitution of (5.33) into (5.32) then provides the eigenvalue equations for $\sigma_c$. Some numerical calculations have been carried out for a linear hardening material and are depicted with Fig. 5.2 for antisymmetric modes (Fig. 5.2a) and symmetric ones (Fig. 5.2b).

As has been pointed out by several authors /30, 35, 40, 42/ the straight line $\sigma = 2G$ in Fig. 5.2b marks points of accumulation of the solution. Whereas the value itself is not an eigenvalue, it can be shown that there exists an infinite sequence of solutions in the neighbourhood bounded from below by this value. For small values of $\nu H$, the eigenvalue equations are also expanded in powers of $\rho H$.

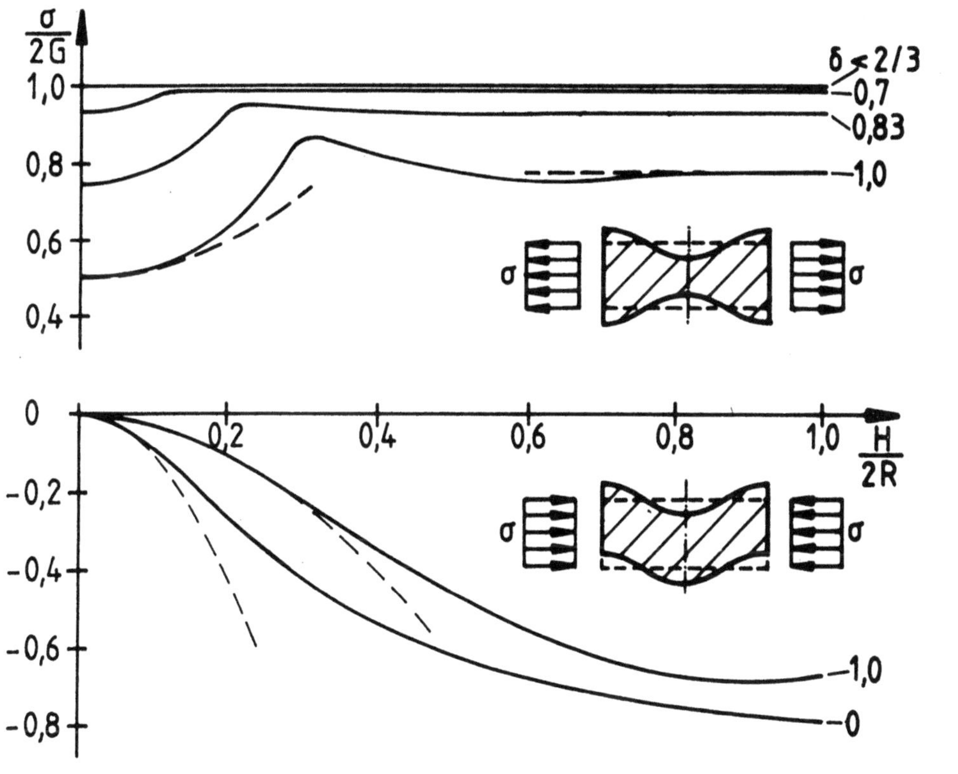


Fig. 5.2 Stresses at bifurcation in a circular plate

To the order $(\nu H)^4$ we find

$$\sigma_c = G[1+3\lambda]\{1+\frac{1}{12}(j_1\frac{H}{R})^2\} \tag{5.34}$$

for the necking mode of deformation and

$$\sigma_c \doteq -G[1+3\lambda]\frac{1}{12}(j_1\frac{H}{R})^2 \tag{5.35}$$

for the buckling mode. In the limit $(\nu H) \to \infty$ again (5.23) becomes valid for both modes. These approximations are also given with Fig. 5.2.

Problems of this kind are typical examples exhibiting what has been solved in the literature. In a similar manner, namely, rectangular elastic-plastic solids under uniaxial and biaxial tension (compression) and with the conditions of plane strain and plane stress have been treated e. g. by Ariaratnam and Dubey /26, 34/, Miles /38/, Hill and Hutchinson /35/ and Young /42/.

The foregoing procedure of solution, however, is restricted to those few simple cases, where

(i) the fundamental solution of the problem is characterized by a uniform stress state and moreover

(ii) analytic solutions can be found for the field equations of the eigenvalue problem.

For more practical, i. e. realistic cases these methods become useless and only numerical solutions of the governing eigenvalue problems can be achieved. This will be illustrated with a third example of bifurcation analysis, where now the third example of section 2 is treated as a thick-walled problem.

Let $(r, \varphi, z)$ be a cylindrical coordinate system, referred to the current configuration of a deformed thick-walled cylinder and $(r_o, \varphi_o, z_o)$ be the respective set of material coordinates, referred to the initial configuration of the body. Again this system is chosen with the origin at one end of the cylinder and the z- and $z_o$-axes, respectively, coinciding with the axis of symmetry. With similar material properties as before at time t the fundamental solution to this problem now consists of a non-uniform stress state of principal (true) stresses /43, 44/

$$\sigma_r = \sigma_r(r) \quad , \qquad \sigma_\varphi = \sigma_\varphi(r) \quad , \qquad \sigma_z = \sigma_z(r) \tag{5.36}$$

in the most general form wherein we have assumed plane and uniformly distributed deformations in the z-direction, i. e. as in $(2.29)_2$ it is assumed that

$$d_z = \frac{\dot{L}}{L} \tag{5.37}$$

Due to the radial equilibrium the stresses (5.36) are restricted to

$$\frac{\partial \sigma_r}{\partial r} - \frac{1}{r}(\sigma_\varphi - \sigma_r) = 0 \tag{5.38}$$

with the current configuration as reference configuration.

The boundary conditions for this problem are such that the external surface of the cylinder remains stress-free while the internal surface is subjected to pressure p. At the top of the cylinder, a uniform axial displacement W is prescribed, while the tractions in radial and circumferential directions are assumed to vanish at the end surfaces

$$\begin{aligned}
&\text{(i)} \quad \text{on } z = 0, L \quad , \quad w = 0, \pm W \quad , \quad \dot{S}_{zr} = \dot{S}_{z\varphi} = 0 \\
&\text{(ii)} \quad \text{on } r = b \quad , \quad \dot{S}_{rr} = \dot{S}_{r\varphi} = \dot{S}_{rz} = 0 \\
&\text{(iii)} \quad \text{on } r = a \quad , \quad \dot{S}_{rr} = -\dot{p} + p\frac{\partial u}{\partial r}, \\
&\qquad\qquad \dot{S}_{r\varphi} = p\frac{\partial u}{\partial \varphi} \, , \quad \dot{S}_{rz} = p\frac{\partial u}{\partial z}
\end{aligned} \tag{5.39}$$

expressed now in physical components of the nominal stress tensor and according to (3.11). The equations of continuing equilibrium are

$$\operatorname{div} \overset{\bullet}{\underset{\sim}{S}}{}^T = 0 \tag{5.40}$$

and the constitutive relations are given with (5.1) to (5.3) for the considered incompressible elastic-plastic material, where according to (3.21) the Jaumann stress rate is used

$$\overset{\bullet}{\underset{\sim}{S}}{}^T = \overset{\triangledown}{\underset{\sim}{\sigma}} - \underset{\sim}{\sigma}\underset{\sim}{D} + \underset{\sim}{W}\underset{\sim}{\sigma} \tag{5.41}$$

We now confine our interest here to the linear comparison solid and look for non-zero solutions (eigenfields) satisfying homogeneous boundary conditions with $w = 0$ on $z = 0, L$ and $\dot{S}_{rr} = p\partial u/\partial r$ on $r = a$. In addition to what has been stated before we furthermore denote the derivative $(\cdot)^* = \partial(\cdot)/\partial\varphi$ in the circumferential direction. Thus by virtue of (5.1), (5.40) and (5.41) the field equations of this problem in terms of principal stresses become

$$\begin{aligned}
&[\Delta u-2\frac{v^*}{r^3}]\,(1-\bar{\sigma}_r+\bar{\sigma}_z) + 2\dot{q}' - [\,Q(2\sigma_r-\sigma_\varphi-\sigma_z)]' \\
&- \frac{3}{r}\,Q(\sigma_r-\sigma_\varphi) + \frac{1}{r^2}\,(u^*-v')^*\,(\bar{\sigma}_\varphi-\bar{\sigma}_z) \\
&+ \frac{2}{r}\,(2u'+\overset{+}{w})(\bar{\sigma}_r-\bar{\sigma}_\varphi) &&= 0 \\
&[\,r\Delta(\frac{v}{r}) + 2\frac{u^*}{r}]\,(1-\bar{\sigma}_\varphi+\bar{\sigma}_z) + 2\dot{q}^* - [\,Q(2\sigma_\varphi-\sigma_r-\sigma_z)]^* \\
&+ (u^*+v'-2\frac{v}{r})(\bar{\sigma}_r-\bar{\sigma}_\varphi)' \\
&+ (v''-\frac{1}{r}v'-u'^*+\frac{1}{r}u^*)(\bar{\sigma}_r-\bar{\sigma}_z) &&= 0 \\
&[\,\Delta w+\frac{w}{r^2}\,]\,(1-\bar{\sigma}_z+\bar{\sigma}_\varphi) + 2\overset{\bullet +}{q} - [\,Q(2\sigma_z-\sigma_r-\sigma_\varphi)]^+ \\
&+ (\overset{+}{u}+w')(\bar{\sigma}_r-\bar{\sigma}_z)' + (w''+\frac{w'}{r}-u'^{+}+\frac{\overset{+}{u}}{r})(\bar{\sigma}_r-\bar{\sigma}_\varphi) &&= 0 \\
&u'+\frac{u}{r}+\frac{v^*}{r^2}+\overset{+}{w} = 0
\end{aligned} \tag{5.42}$$

where for abbreviation

$$Q = \frac{\delta}{\sigma_e^{\,2}} \left[ \sigma_r u' + \sigma_\varphi \left(\frac{u}{r} + \frac{v^*}{r^2}\right) + \sigma_z w^+ \right] \tag{5.43}$$

the operator in (5.42) is defined as

$$\Delta(\cdot) = \{\tfrac{1}{r}[r(\cdot)]'\}' + (\cdot)^{++} + \frac{1}{r^2}(\cdot)^{**} \tag{5.44}$$

and a bar above a quantity means division by 2G.

Homogeneous boundary conditions (5.39) on the other hand become

(i) on z = 0, L

$$\begin{aligned} & u^+ \,(1-\bar{\sigma}_r+\bar{\sigma}_z) + w' \,(1-\bar{\sigma}_r-\bar{\sigma}_z) = 0 \\ & v^+ \,(1-\bar{\sigma}_\varphi+\bar{\sigma}_z) + w^* \,(1-\bar{\sigma}_\varphi-\bar{\sigma}_z) = 0 \\ & w = 0 \end{aligned} \tag{5.45}$$

(ii) on r = a, b

$$\begin{aligned} & u' + \dot{q} - \frac{1}{2} Q \,(2\sigma_r-\sigma_\varphi-\sigma_z) = 0 \\ & u^* + v' - \frac{2}{r} v = 0 \\ & u^+ + w' = 0 \end{aligned} \tag{5.46}$$

Solutions to the equations governing the eigenvalue problem established so far can be written in the separated form

$$\begin{aligned} u &= u(r) \cos(\alpha z) \cos(m\varphi), \\ v &= v(r) \cos(\alpha z) \sin(m\varphi), \\ w &= w(r) \sin(\alpha z) \cos(m\varphi), \\ \dot{q} &= \dot{q}(r) \cos(\alpha z) \cos(m\varphi) \end{aligned} \tag{5.47}$$

where

$$\alpha = n\pi/L \quad , \quad n, m = 0, 1, 2, 3 \ldots \tag{5.48}$$

and 2n and m being the wave numbers of the solution in axial and circumferential directions, such that the boundary conditions on $z = 0, L$ are satisfied and ordinary differential equations in r are obtained. It must be emphasized here, that the coefficients of these ordinary differential equations are still very complicated functions of coordinate r.

As can be seen further, through the dependence of the slope of the effective stress-strain curve, the hardening function $\delta$ (5.3) is a function of r, since as defined in (2.30) the effective stress is

$$\sigma_e(r) = (\sigma_r^2 + \sigma_\varphi^2 + \sigma_z^2 - \sigma_r\sigma_\varphi - \sigma_r\sigma_z - \sigma_\varphi\sigma_z)^{1/2} \tag{5.49}$$

Introducing here a non-smooth effective stress-strain curve, as e. g. in connection with a linear hardening model, this causes a jump in the hardening function in passing through the yield limit. And contrary to those problems where the fundamental solution is characterized by a uniform stress state, this fact now causes difficulties for problems with non-uniform stress states in the fundamental solution, as has been mentioned in section 2. In these cases, where according to the stress state one part of the body may be yielded whereas the remainder is not, as consequence of what has been said before the hardening function jumps in passing the boundary separating the elastic and the plastic domains. Apart from the fact that discontinuities of the coefficients of equations (5.42) in general lead to numerical difficulties during the integration procedure what concerns the convergence of the solution, this furthermore can falsify the results and thus may lead to useless bifurcation stresses /44/. To demonstrate this phenomenon we consider the body in such a state that the elastic-plastic boundary coincides with the outer surface of the cylinder. This state which can be interpreted as fully plastic as well as plastic with an outer elastic zone (of vanishing thickness) is assumed to be relevant to some specific loading whose stability is in question. Provided now, that the fully plastic state is a critical one, the elastic-plastic state (connected with the same load) is not, since due to the jump in the δ-function this state is much stiffer than the other one. We thus result in an unreliable jump of the critical stresses.

To avoid such difficulties in problems of thick-body bifurcation with non-uniform stress states in the fundamental solution it seems to be convenient to introduce a smooth effective stress-strain curve of the form (say)

$$\varepsilon = \begin{cases} \dfrac{\sigma}{E} + \dfrac{\sigma_o}{B}\left(\dfrac{\sigma}{\sigma_o} - 1\right)^n & ; \ \sigma \geqq \sigma_o \\[2ex] \dfrac{\sigma}{E} & ; \ \sigma < \sigma_o \end{cases} \tag{5.50}$$

in the one dimensional case, where $\sigma_o$ is the yield stress, n and B are additional material parameters. Introducing this into the definition of the hardening function we finally arrive at

$$\delta = \begin{cases} 1 - \left\{1 + \dfrac{En}{B}\left(\dfrac{\sigma_e}{\sigma_o} - 1\right)^{n-1}\right\}^{-1} & ; \quad \sigma_e \geqq \sigma_o \\[2ex] 0 & ; \quad \sigma_e < \sigma_o \end{cases} \tag{5.51}$$

with the effective stress defined in (5.49).

This eigenvalue problem has been solved numerically by aid of a standard Runge-Kutta procedure. Solutions have been found while treating this problem as a boundary value problem with fixed values of the loading parameters $\lambda_i$ (i = 1, 2), where $\lambda_1$ is connected with the internal pressure and $\lambda_2$ with the axial load. A critical set of $\lambda_i$ is obtained when the determinant of the homogeneous equations (5.46) vanishes. Some results of the calculation based on the material data of an aluminium alloy /43/

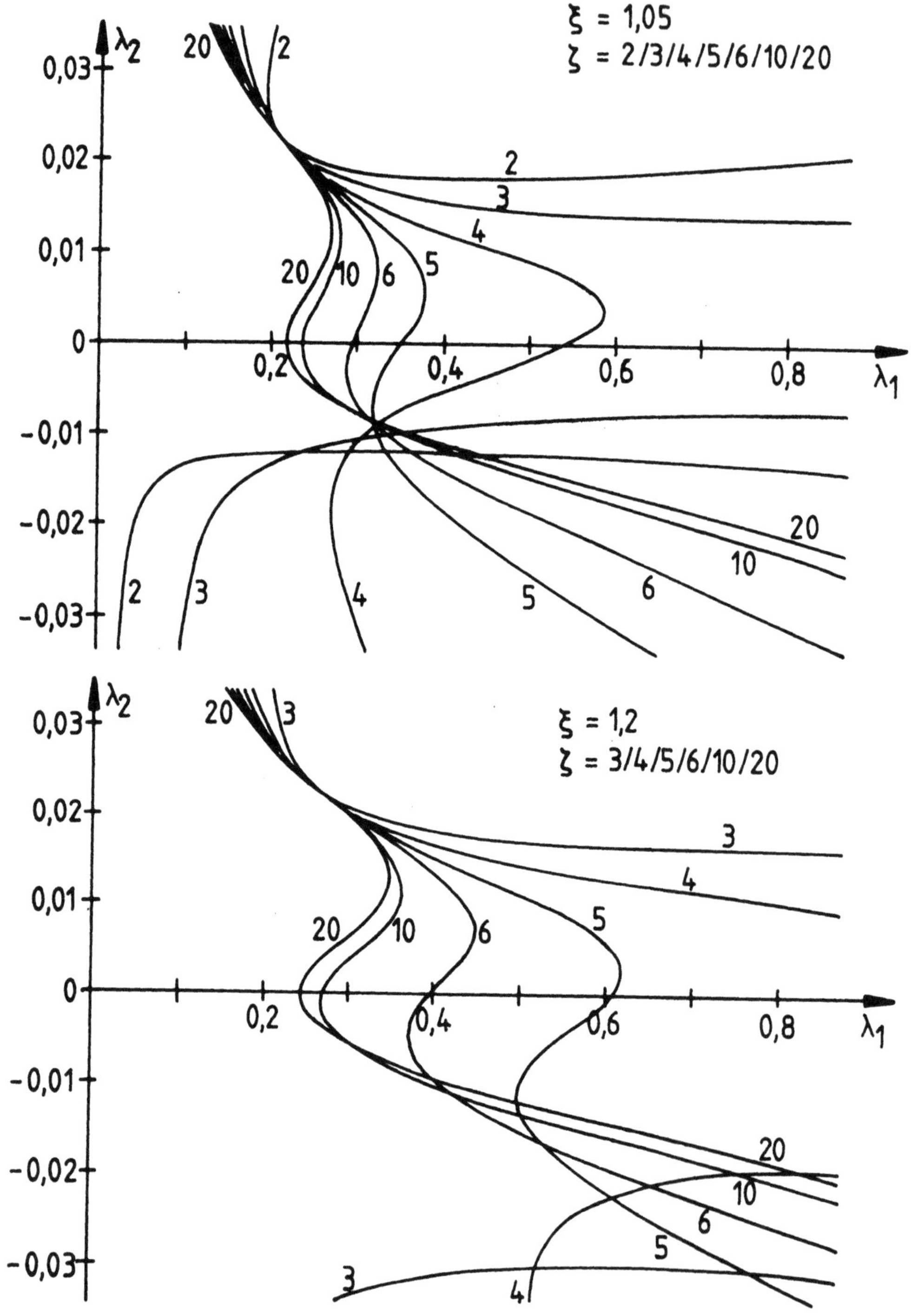


Fig. 5.3 Stresses at bifurcation in a thick-walled cylinder

$\sigma_o$ = 240 N/mm²

E = 59.500 "

B = 4.857,5 "

n = 2,80

have been depicted with Fig. 5.3 for different values of thickness and length ratios $\xi = b/a$ and $\zeta = L/R$, where here R is the mean radius of the cylinder.

## 6. THE INFLUENCE OF NON-ASSOCIATED PLASTIC FLOW LAWS

In deriving the expressions of section 4, i. e. the uniqueness criterion (4.13) and the related properties of the functionals R and $R_L$ we have assumed constitutive equations of the form (3.14) characterized by a smooth yield surface with associated plastic deformations where $\underset{\sim}{\lambda}$ and $\underset{\sim}{\mu}$, respectively, are the normals to the hyperplane interfaces (yield surfaces) separating the domains of elastic and plastic behaviour.

As has been pointed out in /24/ during the last two decades much effort has been made to investigate experimentally the behaviour of soil-like and rock-like materials in the plastic range and there is some evidence that they do not obey the normality rule. There are also some doubts whether this normality rule can adequately describe the behaviour of metals close to the region of localized deformations /53, 54/. Moreover, even though isothermal laws in metal plasticity are assumed to obey a normality rule, the corresponding adiabatic laws do not. Non-associated plastic flow laws are expected to describe better the behaviour of these types of materials or under the above mentioned conditions.

Similar to (3.14) the simplest generalization of the rate-constitutive relations of an elastic-plastic material at the yield point may be taken in the form

$$\underset{\sim}{\overset{\triangledown}{\sigma}} = \underset{\sim}{L}\,\underset{\sim}{D} - \frac{\alpha}{g}\,(\underset{\sim}{\lambda}^{f} \cdot \underset{\sim}{D})\,\underset{\sim}{\lambda}^{g} \tag{6.1}$$

with the same notations as in section three and where now $\underset{\sim}{\lambda}^{f}$ and $\underset{\sim}{\mu}^{f}$, respectively, are as before the normals to the hyperplane interfaces (yield surfaces) whereas $\underset{\sim}{\lambda}^g$ and its dual $\underset{\sim}{\mu}^g$ specify the direction of the plastic part of the strain rate. If

$$h = g - \underset{\sim}{\lambda}^f \cdot \underset{\sim}{\mu}^g > 0$$

where as before both duals $\underset{\sim}{\lambda}^f$ and $\underset{\sim}{\mu}^f$, and $\underset{\sim}{\lambda}^g$ and $\underset{\sim}{\mu}^g$, respectively, are related by (3.17) the unique inverse of (6.1) becomes

$$\underset{\sim}{D} = \underset{\sim}{M}\,\overset{\triangledown}{\underset{\sim}{\sigma}} + \frac{\alpha_1}{h}\,(\underset{\sim}{\mu}^f \cdot \overset{\triangledown}{\underset{\sim}{\sigma}})\,\underset{\sim}{\mu}^g \tag{6.2}$$

where furthermore $\alpha_1$ and $\alpha$ are defined as in section three (3.15), (3.19) when replacing $\underset{\sim}{\mu}$ and $\underset{\sim}{\lambda}$ by $\underset{\sim}{\mu}^f$ and $\underset{\sim}{\lambda}^f$.

The constitutive law of the form (6.1) has been introduced by Mandel and Mróz in the sixties (see e. g. /24, 47/) in order to incorporate in a uniform manner a number of previously proposed models for frictional granular materials for which the normality rule not necessarily holds. More recently this type of rate law has been discussed by Rice et. al. /53 to 56/ and has been motivated with reference to the dilatant, frictional response of brittle rocks and void-containing ductile metals. More specifically, Rudnicki and Rice /55/, in the analysis of localization phenomena in rock-type materials, used the simple generalization of the Prandtl-Reuss equation with

$$\begin{aligned} \underset{\sim}{\mu}^g &= \underset{\sim}{\sigma}'/2\tau_e + (\beta_o/3)\,\underset{\sim}{1} \\ \underset{\sim}{\mu}^f &= \underset{\sim}{\sigma}'/2\tau_e + (\nu_o/3)\,\underset{\sim}{1} \end{aligned} \tag{6.3}$$

where $\beta_o$ is the plastic dilatancy factor and $\nu_o$ is the internal friction coefficient measuring the pressure sensitivity at yield. Here

$$\tau_e^2 = \frac{1}{2}\underset{\sim}{\sigma}' \cdot \underset{\sim}{\sigma}' \tag{6.4}$$

and $\sigma'$ as in (5.1) is the deviatoric stress. Noteworthy is the fact that $\mu^f$ in (6.3) represents the normal to the yield surface associated with the classical Drucker-Prager yield condition for soils

$$\Phi(\sigma,\kappa) = \tau_e + \frac{\nu_o}{3} \operatorname{tr}(\sigma) - k(\kappa) = 0 \quad . \tag{6.5}$$

Introducing now as in section 4 two distinct solutions $(\cdot)^{(1)}$ and $(\cdot)^{(2)}$ of a problem and denoting their difference by $\Delta(\cdot)$ we as before arrive at (4.7) as a sufficient condition for uniqueness. This condition again may be transformed into a form more convenient for applications when the relations between $\Delta v$ and $\Delta\dot{S}$ according to the above mentioned constitutive equations are used. We therefore introduce $\nu_2 = \lambda^f \cdot D^{(2)}$ and thus find (4.8), (4.9) and (4.11), (4.12) and the uniqueness criterion (4.13) where now instead of (4.10)

$$W_p = \begin{cases} -(\lambda^f \cdot \Delta D)(\lambda^g \cdot \Delta D)/2g & ;\ \nu_2 \geqq 0 \text{ and } \nu_2 \geqq -\lambda^f \cdot \Delta D \\ -(\lambda^g \cdot \Delta D)(\nu_2 + \lambda^f \cdot \Delta D)/2g & ;\ -\lambda^f \cdot \Delta D \leqq \nu_2 < 0 \\ \nu_2(\lambda^g \cdot \Delta D)/2g & ;\ -\lambda^f \cdot \Delta D > \nu_2 \geqq 0 \\ 0 & ;\ \nu_2 < 0 \text{ and } \nu_2 < -\lambda^f \cdot \Delta D \end{cases} \tag{6.6}$$

However, due to the fact that now $\lambda^g \neq \lambda^f$ the property (4.14) loses its validity and consequently the linear solid defined with (3.33) is no longer a practicable "comparison" solid for materials that do not admit a potential. For this more general case, Raniecki /47/ recently has shown how this concept of "comparison" solids originally formulated by Hill for materials obeying normality can be extended to include constitutive equations of the non-associated form (6.1).

Consider therefore a one-parameter family of strain-rate potentials

$$W_L(D) = \frac{1}{2} D \cdot LD - \frac{1}{8rg} \left[ (\lambda^g + r\lambda^f) \cdot D \right]^2 \tag{6.7}$$

where r is assumed to be a positive parameter. From equations (4.8), (4.9) and (6.7) it follows that

$$\begin{aligned} \Delta &\equiv W(\underset{\sim}{\Delta D},\nu_2) - W_L(\underset{\sim}{\Delta D}) \\ &= W_p(\underset{\sim}{\Delta D},\nu_2) + [(\underset{\sim}{\lambda}^g + r\underset{\sim}{\lambda}^f)\cdot\underset{\sim}{D}]^2 / 8rg. \end{aligned} \tag{6.8}$$

Using now (6.6) for the four different cases of this expression we find

(i) $\nu_2 \geqq 0$ and $\nu_2 \geqq -\underset{\sim}{\lambda}^f\cdot\underset{\sim}{\Delta D}$ :

$$\begin{aligned} 8rg\Delta &= -(\underset{\sim}{\lambda}^f\cdot\underset{\sim}{\Delta D})(\underset{\sim}{\lambda}^g\cdot\underset{\sim}{\Delta D})4r + [(\underset{\sim}{\lambda}^g + r\underset{\sim}{\lambda}^f)\cdot\underset{\sim}{\Delta D}]^2 \\ &= [(\underset{\sim}{\lambda}^g - r\underset{\sim}{\lambda}^f)\cdot\underset{\sim}{\Delta D}]^2 \geqq 0 \end{aligned} \tag{6.9$_1$}$$

(ii) $-\underset{\sim}{\lambda}^f\cdot\underset{\sim}{\Delta D} \leqq \nu_2 < 0$ :

$$\begin{aligned} 8rg\Delta &= -(\underset{\sim}{\lambda}^g\cdot\underset{\sim}{\Delta D})(\nu_2 + \underset{\sim}{\lambda}^f\cdot\underset{\sim}{\Delta D})4r + [(\underset{\sim}{\lambda}^g + r\underset{\sim}{\lambda}^f)\cdot\underset{\sim}{\Delta D}]^2 \\ &= [2r\nu_2 - (\underset{\sim}{\lambda}^g - r\underset{\sim}{\lambda}^f)\cdot\underset{\sim}{\Delta D}]^2 - 4r^2\nu_2(\nu_2 + \underset{\sim}{\lambda}^f\cdot\underset{\sim}{\Delta D}) > 0 \end{aligned} \tag{6.9$_2$}$$

(iii) $-\underset{\sim}{\lambda}^f\cdot\underset{\sim}{\Delta D} > \nu_2 \geqq 0$ :

$$\begin{aligned} 8rg\Delta &= \nu_2(\underset{\sim}{\lambda}^g\cdot\underset{\sim}{\Delta D})\ 4r + [(\underset{\sim}{\lambda}^g + r\underset{\sim}{\lambda}^f)\cdot\underset{\sim}{\Delta D}]^2 \\ &= [2r\nu_2 + (\underset{\sim}{\lambda}^g + r\underset{\sim}{\lambda}^f)\cdot\underset{\sim}{\Delta D}]^2 - 4r^2\nu_2(\nu_2 + \underset{\sim}{\lambda}^f\cdot\underset{\sim}{\Delta D}) > 0 \end{aligned} \tag{6.9$_3$}$$

(iv) $\nu_2 < 0$ and $\nu_2 < -\underset{\sim}{\lambda}^f\cdot\underset{\sim}{\Delta D}$ :

$$8rg\Delta = [(\underset{\sim}{\lambda}^g + r\underset{\sim}{\lambda}^f)\cdot\underset{\sim}{\Delta D}]^2 \geqq 0 \tag{6.9$_4$}$$

Since in every case $\Delta$ is non-negative, inequality (4.14) now holds for the potential (6.7), so that

$$\overset{\triangledown}{\underset{\sim}{\sigma}} = \frac{\partial W_L(\underset{\sim}{D})}{\partial \underset{\sim}{D}} \quad , \qquad W_L(\underset{\sim}{D}) = \frac{1}{2} \overset{\triangledown}{\underset{\sim}{\sigma}} \cdot \underset{\sim}{D} \tag{6.10}$$

is the "linear comparison solid" of the underlying non-associated elastic-plastic solid (6.1). For materials that do not admit a potential expressions (4.14), (4.15) and finally the weaker uniqueness criterion (4.16) of section 4 still hold when replacing herein the comparison solid (3.33) by the more general one (6.7). It can be seen further that for $\underset{\sim}{\lambda}^f = \underset{\sim}{\lambda}^g$ and with $r = 1$, i. e. for materials obeying normality, (3.33) is a special case of (6.7). The basic practical meaning of the fictitious comparison body established here rests upon the fact that instability and bifurcation of an elastic-plastic body cannot precede primary bifurcation of the comparison body. Furthermore, the constitutive relation for the comparison body is not only linear but admits a potential which has the same structure as the corresponding potential for the elastic-plastic solid obeying normality when the tensor specifying the direction of the plastic strain rate is replaced by $(\underset{\sim}{\lambda}^g + r\underset{\sim}{\lambda}^f)/2\sqrt{r}$ . Therefore, the methods of analysis of bifurcation that have been developed so far may now also be used to determine lower bounds to bifurcation and instability states in solids with non-associated flow laws. The still undetermined parameter r in (6.7) can be chosen such that these lower bounds give most optimal ones /47/. However, due to the lack of variational principles, the first state for which the functional $R_L$ becomes positive-semidefinite is ordinarily not the bifurcation state of the underlying elastic-plastic solid. Therefore a second linear solid is introduced whose constitutive relations have the same form as an elastic-plastic solid in loading, i. e. with $\alpha = 1$

$$\overset{\triangledown}{\underset{\sim}{\sigma}} = \underset{\sim\sim}{L}\underset{}{D} - \frac{1}{g} (\underset{\sim}{\lambda}^f \cdot \underset{\sim}{D}) \, \underset{\sim}{\lambda}^g \quad . \tag{6.11}$$

The first eigenstate of this solid may be identified with some bifurcation state of the underlying elastic-plastic solid. It is therefore an upper bound to the primary bifurcation state.

For non-associated plastic flow laws ($\underset{\sim}{\lambda}^f \neq \underset{\sim}{\lambda}^g$), the search for the genuine primary bifurcation state is replaced by a search for upper and lower bounds. Only in the case when normality holds, i. e. when $\underset{\sim}{\lambda}^g = \underset{\sim}{\lambda}^f$ and $r = 1$, (6.10) and (6.11) coincide and therefore give exact solutions to the underlying bifurcation problem.

Applications of the foregoing theorems have been given with /58,59/. Needleman /46/ used a linear solid of the type (6.11) to analyse bifurcation states of a rectangular block subjected to uniaxial loading in plane strain. The material was taken to be incompressible and not necessarily obeying normality. For this specific case his results provide an upper bound to the primary bifurcation state.

In /58, 59/ the bifurcation of homogeneous cylindrical specimens of circular cross sections (with radius R and length L in the current configuration), uniformly stretched under axial loading $\sigma_z$ and with the lateral surface subjected to fluid pressure $\sigma$, have been analysed. At a generic instant the state of stress is given with

$$\sigma_{rr} = \sigma_{\theta\theta} = -\sigma , \qquad \sigma_{zz} = \sigma_z \tag{6.12}$$

in a cylindrical coordinate system. The bifurcation from this uniform fundamental solution is investigated.

Similar problems have been solved by Miles /21/, Hutchinson and Miles /36/ and the present author /28/. The investigations in /58, 59/, however, differ from those essentially in the constitutive relations of the considered material which is assumed to be a compressible, plastically dilatant and pressure sensitive material, with $\underset{\sim}{\mu}^f$ and $\underset{\sim}{\mu}^g$ given with (6.3). According to the previous statements two different linear solids have been introduced, a linear comparison solid (CS) and a second linear solid (SS), providing lower and upper bounds to the primary bifurcation stresses of the underlying solid.

Similar as in section 5 power series and asymptotic expansions of Bessel functions have been used to obtain relations between the critical stresses and slenderness ratio mR. As has been demonstrated before, two limiting cases of the solution of the eigenvalue equation can be investigated, the long wavelength limit with $(mR) \to 0$ and the short wavelength limit where $(mR) \to \infty$ .

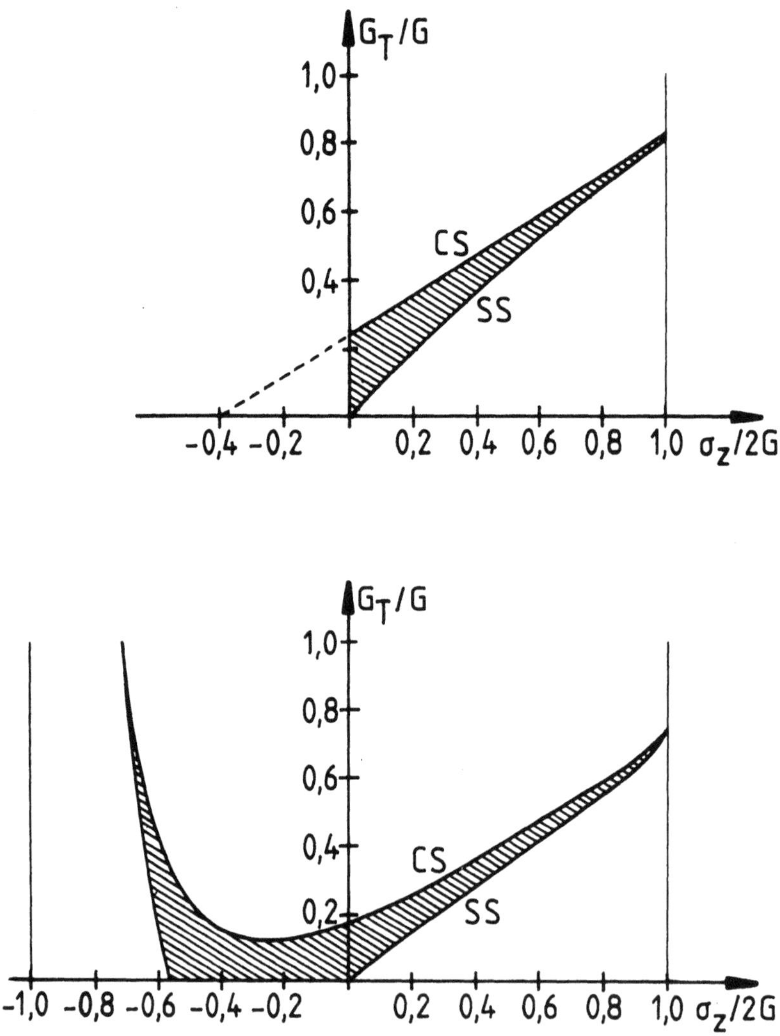


Fig. 6.1 Bounds to the critical stresses in bifurcation of slender (a) and stubby (b) specimens

Two typical results of the calculations of /59/ have been depicted with Fig. 6.1, where based on the properties $\nu = 0.3$; $\nu_o = \sqrt{3}/2$, $\beta_o = 0$, $\sigma = 0$ bounds are given to the critical stresses in bifurcation for slender specimens (long wavelength limit), Fig. 6.1a, and for stubby ones (short wavelength limit), Fig. 6.1b. For both cases the solutions for the comparison solid (CS) and for the second linear solid (SS) give the respective bounds to the critical region in the parameter space ($\sigma_z/2G$, $G_T/G$), where $G_T$ is the tangent shear modulus of the material measured in pure shear under hydrostatic pressure.

## 7. PLASTIC FLOW LOCALIZATION

When an initially uniform sheet is stretched so that the principal strain rates in the plane of the sheet are of opposite sign, localized necking is predicted along a line in the direction of zero extension in the plane, according to classical rigid-plastic flow theory with a smooth yield surface. Localized necking is not predicted by this theory, if both in-plane principal strain rates are positive, although even in this case localized necking is observed experimentally. This discrepancy - among others - has led to several investigations of localized necking in the last years.

Basically two different approaches have been used to explain this phenomenon analytically. In the first approach, originally suggested by Marciniak and Kuczynski /48/, the sheet is assumed to contain a small initial inhomogeneity, and it is found that necking results from a localization of the strains at this inhomogeneity. The second approach, introduced by Stören and Rice /49/, takes the sheet to be initially uniform and homogeneous. In this approach bifurcation results from a vertex-formation on the yield surface during straining. The question whether the explanation of the critical strains observed experimentally is that of vertex-formation or that of initial inhomogeneities has not yet been resolved. Some examples of application especially of the latter approach and some remarks concerning a comparison of both mechanisms have been given quite recently with the papers of Hutchinson and Tvergaard /50, 51/ and Tvergaard et. al. /52/. To the present author the vertex model seems to be quite attractive. Apart from the 'promising' influence on the results of specific bifurcation problems, this model has the advantage to correspond with the prediction of models of the elastic-plastic behaviour of polycrystalline metals based on the slip mechanism of single crystals. However, the experimental evidence of vertex-formation on yield surfaces seems to be still an open question.

The large majority of localized necking analyses are based on the Marciniak-Kuczynski-model (M-K-model) which has been generalized with the papers of Rice /53/ and Rice and Rudnicki /54/. A furthermore extension of this scheme to materials with non-associated flow laws, as e. g. to plastically dilatant pressure-sensitive rock-like or granular (or voided) materials, has been given by Rudnicki and Rice /55/ and Needleman and Rice /56/. With the notations of section 3.1 a homogeneous uniformly deformed solid is considered subjected to increments of deformation which could give rise to the homogeneous rate fields $\underset{\sim}{V}^{\circ}$, $\underset{\sim}{\dot{S}}^{\circ}$. We wish to determine if a bifurcation within a localized band, with orientation $\underset{\sim}{n}$ in the current configuration, is possible, in such a manner that the (difference) velocity field $\Delta\underset{\sim}{v}$ varies only across the band, i. e.

$$\Delta\underset{\sim}{v} = \underset{\sim}{f}(\underset{\sim}{n}\cdot\underset{\sim}{x}) \tag{7.1}$$

Note that in section 3.1 we have introduced the current state to coincide, instantaneously, with the reference state. The velocity gradient corresponding to (7.1) then is

$$\Delta\underset{\sim}{V} = \underset{\sim}{V} - \underset{\sim}{V}^{\circ} = \underset{\sim}{g}\underset{\sim}{n} \tag{7.2}$$

where $\underset{\sim}{g}$ is some function of $\underset{\sim}{n}\cdot\underset{\sim}{x}$ vanishing outside the band.

Equilibrium across the band at the instant of bifurcation requires that the nominal traction rates $\underset{\sim}{\dot{t}}$ on the band boundaries be continuous. Thus

$$\Delta\underset{\sim}{\dot{t}} = \underset{\sim}{\dot{t}} - \underset{\sim}{\dot{t}}^{\circ} = \Delta\underset{\sim}{\dot{S}}^{T}\underset{\sim}{n} = 0 \quad . \tag{7.3}$$

From the relation between nominal stress rates and Jaumann rates of Kirchhoff stress (3.21) and the constitutive law (3.14), which here is taken in the more general form

$$\overset{\triangledown}{\underset{\sim}{\sigma}} = \underset{\sim\sim}{\bar{L}D} \,, \tag{7.4}$$

both formulated inside and outside the band we find

$$\underset{\sim}{\Delta \dot{S}^T} = (\underset{\sim}{\bar{L}}-\underset{\sim}{\bar{L}}^\circ)\, \underset{\sim}{D}^\circ + \underset{\sim}{\bar{L}}\underset{\sim}{\Delta D} - \underset{\sim}{\sigma}\underset{\sim}{\Delta D} - \underset{\sim\sim}{\Delta W \sigma} \tag{7.5}$$

since $\underset{\sim}{\sigma} = \underset{\sim}{\sigma}^\circ$ and where $\underset{\sim}{\Delta D}$ and $\underset{\sim}{\Delta W}$ are defined through (3.12), (3.24) and (7.2). Thus from (7.5) the condition for continued equilibrium (7.3) becomes

$$(\underset{\sim\sim\sim}{n\bar{L}n}-\underset{\sim}{A})\, \underset{\sim}{g} = \underset{\sim}{n}\, (\underset{\sim}{\bar{L}}-\underset{\sim}{\bar{L}}^\circ)\, \underset{\sim}{D}^\circ \tag{7.6}$$

where herein

$$2\underset{\sim}{A} = \underset{\sim}{\sigma} + \underset{\sim\sim\sim}{\sigma n n} + (\underset{\sim}{n}\cdot\underset{\sim\sim}{\sigma n})\, \underset{\sim}{1} - \underset{\sim\sim\sim}{n \sigma n} \tag{7.7}$$

equivalent to $\Sigma$, e. g. in (3.30), is a term expressing the influence of geometry changes due to finite deformations. In /54/ it has been proved that even for materials which do not obey the normality rule the comparison theorem of section 4.2 holds. Therefore the lowest bifurcation point corresponds to continued plastic flow throughout the body and hence $\underset{\sim}{\bar{L}} = \underset{\sim}{\bar{L}}^\circ$. The condition for non-zero solutions of the localization problem thus becomes

$$\det\, (\underset{\sim\sim\sim}{nLn}-\underset{\sim}{A}) = 0 \tag{7.8}$$

provided the material is compressible. For an incompressible material the constitutive law (7.4) has to be replaced by

$$\overset{\nabla}{\underset{\sim}{\sigma}} = \underset{\sim\sim}{\bar{L}D} + \dot{q}\ \underset{\sim}{1} \qquad (7.9)$$

where now the additional unknown hydrostatic pressure difference $\Delta\dot{q}$ arising in (7.6)

$$(\underset{\sim\sim\sim}{nLn}-\underset{\sim}{A})\ \underset{\sim}{g} + \Delta\dot{q}\ \underset{\sim}{n} = 0 \qquad (7.10)$$

has to be eliminated through the condition of incompressibility

$$\mathrm{tr}\ (\Delta\underset{\sim}{D}) = \underset{\sim}{g} \cdot \underset{\sim}{n} = 0 \qquad (7.11)$$

This analysis is now carried out within the context of plane stress similar as performed by Hutchinson and Neale /57/.

We consider a thin flat sheet which is currently of uniform thickness and subjected to a homogeneous stress field with non-vanishing components (Fig. 7.1)

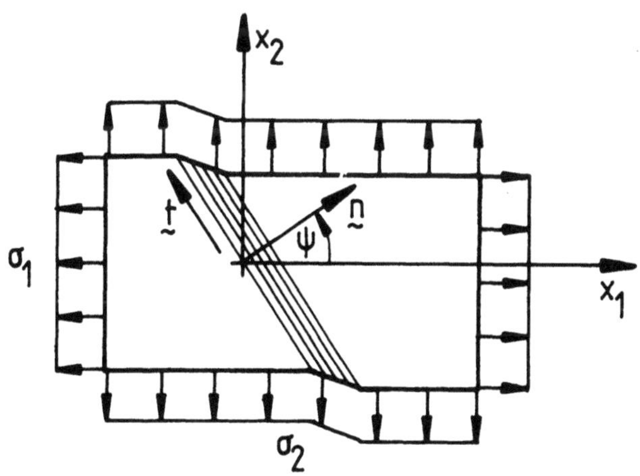


Fig. 7.1 Localization-band geometry

$$\sigma_{11} = \sigma_1 \quad , \quad \sigma_{22} = \sigma_2 \tag{7.12}$$

For simplicity we restrict our further interest here to a proportional loading (proportional stressing) of the sheet, i. e. we assume that

$$\alpha = \sigma_1/\sigma_2 = \text{const.} \tag{7.13}$$

holds during the whole loading process.

The material of the sheet is assumed to be incompressible and obeying $J_2$ flow theory according to relations (5.1) to (5.3). Due to the aforementioned assumption of plane stress, i. e. $\sigma_{33} = 0$, and the incompressibility condition, the hydrostatic pressure can be eliminated to give

$$\begin{aligned} \overset{\triangledown}{\sigma}_{11} &= \bar{L}_{1111}\, d_{11} + \bar{L}_{1122}\, d_{22} \\ \overset{\triangledown}{\sigma}_{22} &= \bar{L}_{2211}\, d_{11} + \bar{L}_{2222}\, d_{22} \\ \overset{\triangledown}{\sigma}_{12} &= \bar{L}_{1212}\, d_{12} \end{aligned} \tag{7.14}$$

equivalent to (2.28) where the instantaneous moduli are

$$\begin{aligned} \bar{L}_{1111} &= 4G - \frac{3}{2\sigma_e^{\,2}}\, \delta\, \sigma_1^2 \cdot 2G \\ \bar{L}_{2222} &= 4G - \frac{3}{2\sigma_e^{\,2}}\, \delta\, \sigma_2^2 \cdot 2G \\ \bar{L}_{1122} &= 2G - \frac{3}{2\sigma_e^{\,2}}\, \delta\, \sigma_1\sigma_2 \cdot 2G \\ \bar{L}_{1212} &= 2G \,, \end{aligned} \tag{7.15}$$

the effective stress is (2.30)

$$\sigma_e = (\sigma_1^2-\sigma_1\sigma_2+\sigma_2^2)^{1/2} \tag{7.16}$$

and $d_{ik}$ are in a cartesian coordinate system the components of the strain rate tensor $\underset{\sim}{D}$.

With the orientation of the band described by the components

$$n_1 = \cos\psi \ , \qquad n_2 = \sin\psi \tag{7.17}$$

of the normal to the band we find from (7.8)

$$\begin{aligned}
&[\bar{L}_{1111} + \rho^2 \bar{L}_{1212} - \sigma_1 - \tfrac{1}{2}\rho^2 (\sigma_1+\sigma_2)] \times \\
&[\bar{L}_{1212} + \rho^2 \bar{L}_{2222} - \tfrac{1}{2}(\sigma_1+\sigma_2) - \rho^2 \sigma_2] - \\
&- \rho^2 \{[\bar{L}_{2211} + \bar{L}_{1212}]^2 - \tfrac{1}{4}(\sigma_2-\sigma_1)^2\} = 0
\end{aligned} \tag{7.18}$$

where $\rho = \tan\psi$ and the critical state is a function of band orientation $\psi$. The pertinant critical state corresponds to that value of $\rho$ in (7.18) which minimizes the bifurcation stress. Some results of this computation are given with Fig. 7.2 in the stress plane ($\sigma_1$, $\sigma_2$) for different values of (constant) hardening ratio $\delta$ (5.3).

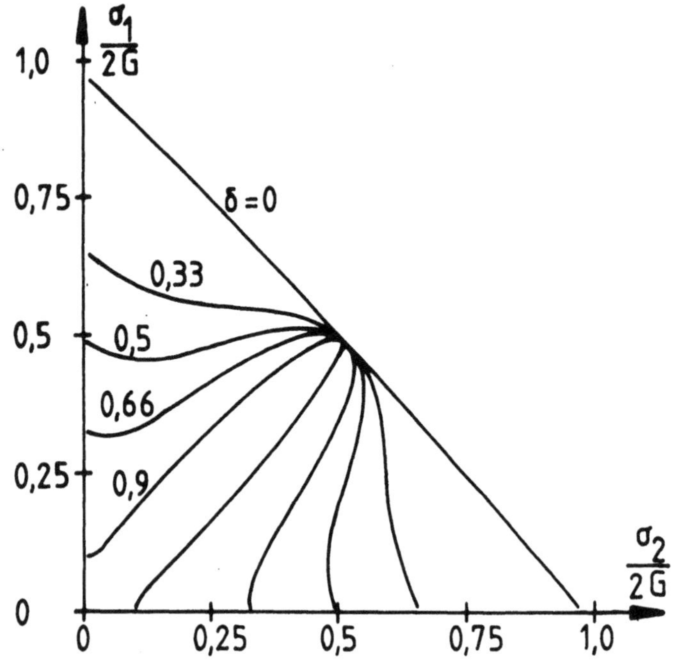


Fig. 7.2 Localization stresses for various values of (constant) hardening ratio

## REFERENCES


1. Engesser, F.: Über die Knickfestigkeit gerader Stäbe, Z. Architektur und Ingenieurwesen, 35, 455, 1889

2. Engesser, F.: Über Knickfragen, Schweizerische Bauzeitung, 26, 24, 1895

3. Jasinski, F.: Noch ein Wort zu den Knickfragen, Schweizerische Bauzeitung, 25, 172, 1895

4. Considière, A.: Résistance des pièces comprimées, Congrès International des Procédés de Construction, Paris, 1891, 371

5. v. Kármán, T.: Untersuchungen über Knickfestigkeit, Mitteilungen über Forschungsarbeiten VDI 81, 1910

6. Shanley, F. R.: The column paradox, J. Aero. Sci., 13, 678, 1946

7. Shanley, F. R.: Inelastic column theory, J. Aero. Sci., 14, 261, 1947

8. v. Kármán, T.: Comments on Shanley's paper, J. Aero. Sci., 14, 267, 1947

9. Sewell, M. J.: A survey of plastic buckling, in "Stability", SM Study No. 6, Leipholz, H. H. E., Ed., University of Waterloo, 1972, 85

10. Storåkers, B.: Uniqueness and stability at finite deformation of inelastic solids, Publ. 186, The Royal Institute of Technology, 1973

11. Hutchinson, J. W.: Plastic buckling, in "Advances in Applied Mechanics", Vol. 14, Yih, C.-S., Ed., Academic Press, 1974, 67

12. Storåkers, B.: On uniqueness and stability of elastic plastic deformation Arch. Mechanics, 27, 821, 1975

13. Hill, R.: Aspects of invariance in solid mechanics, in "Advances in Applied Mechanics" Vol. 18, Yih, C.-S., Ed., Academic Press, 1978, 1

14. Hornbogen, E.: Werkstoffe, Springer-Verlag, 1979

15. Pflüger, A.: Stabilitätsprobleme der Elastostatik, Springer-Verlag, 1975

16. Hill, R.: A general theory of uniqueness and stability in elastic-plastic solids, J. Mech. Phys. Solids, 6, 236, 1958

17. Hill, R.: Some basic principles in the mechanics of solids without a natural time J. Mech. Phys. Solids, 7, 209, 1959

18. Hill, R.: Bifurcation and uniqueness in non-linear mechanics of continua, in "Problems of Continuum Mechanics" N.I. Muskhelishvili Volume, Soc. Ind. Appl. Math, Philadelphia, 1961, 155

19. Hill, R.: Uniqueness criteria and extremum principles in selfadjoint problems of continuum mechanics, J. Mech. Phys. Solids, 10, 185, 1962

20. Hill, R.: On the classical constitutive relations for elastic/plastic solids, in "Recent Progress in Applied Mechanics", F.K.G. Odqvist Volume, Wiley, N.Y., 1967, 241

21. Miles, J. P.: Fluid-pressure eigenstates and bifurcation in tension specimens under lateral pressure, J. Mech. Phys. Solids, 21, 145, 1973

22. Storåkers, B.: On uniqueness and stability under configuration-dependent loading of solids with or without a natural time J. Mech. Phys. Solids, 25, 269, 1977

23. Bruhns, O. and Th. Lehmann: Optimum deformation rate in large inelastic deformations, in "Metal Forming Plasticity", Springer-Verlag, 1979, 120

24. Raniecki, B. and O. T. Bruhns: Bounds to bifurcation stresses in solids with non-associated plastic flow law at finite strain, J. Mech. Phys. Solids, 29, 153, 1981

25. Ariaratnam, S. T. and R. N. Dubey: Instability in an elastic-plastic cylindrical shell under axial compression, J. Appl. Mech., 36, 47, 1969

26. Ariaratnam, S. T. and R. N. Dubey: Some cases of bifurcation in elastic-plastic solids in plane strain, Quart. Appl. Math., 27, 349, 1969

27. Bruhns, O. und K. Thermann: Verzweigungslasten elastoplastischer Kreisplatten, Acta Mech., 20, 247, 1974

28. Bruhns, O. und. J. Mielnicznk: Zur Theorie der Verzweigungen nicht-isothermer elastoplastischer Deformationen, Ing.-Archiv, 46, 65, 1977

29. Bruhns, O.: Verzweigungslasten inelastischer Schalen, ZAMM, 57, 165, 1977

30. Bruhns, O.: On the stability of thin-walled pressure vessels, Nucl. Engng. Design, 58, 359, 1980

31. Chakrabarty, J.: Plastic buckling of cylindrical shells subjected to external fluid pressure, ZAMP, 24, 270, 1973

32. Cheng, S. Y., S. T. Ariaratnam and R. N. Dubey: Axisymmetric bifurcation in an elastic-plastic cylinder under axial load and lateral hydrostatic pressure, Quart. App. Math., 29, 41, 1971

33. Dubey, R. N.: Instabilities in elastic-plastic tubes, Int. J. Solids Struct., 5, 699, 1969

34. Dubey, R. N. and S. T. Ariaratnam: Bifurcation in elastic-plastic solids in plane stress, Quart. Appl. Math., 27, 381, 1969

35. Hill, R. and J. W. Hutchinson: Bifurcation phenomena in the plane tension test, J. Mech. Phys. Solids, 23, 239, 1975

36. Hutchinson, J. W. and J. P. Miles: Bifurcation analysis of the onset of necking in an elastic/plastic cylinder under uniaxial tension, J. Mech. Phys. Solids, 22, 61, 1974

37. Miles, J. P.: Bifurcation in plastic flow under uniaxial tension, J. Mech. Phys. Solids, 19, 89, 1971

38. Miles, J. P.: The initiation of necking in rectangular elastic/plastic specimens under uniaxial and biaxial tension, J. Mech. Phys. Solids, 23, 197, 1975

39. Neale, K. W.: A method for the estimation of plastic buckling loads, Int. J. Solids Struct., 10, 217, 1974

40. Needleman, A. and V. Tvergaard: Necking of biaxially stretched elastic-plastic circular plates, J. Mech. Phys. Solids, 25, 159, 1977

41. Storåkers, B.: On buckling of axisymmetric thin elastic-plastic shells, Int. J. Solids Struct., 11, 1329, 1975

42. Young, N. J. B.: Bifurcation phenomena in the plane compression test, J. Mech. Phys. Solids, 24, 77, 1976

43. Oeynhausen, H.: Verzweigungslasten elastoplastisch deformierter, dickwandiger Kreiszylinder unter Innendruck und Axialkraft, Mitt. Institut f. Mechanik, No. 29, RU Bochum, 1981

44. Bruhns, O.: Some remarks on the stability of extremely loaded pressure vessels, in "Mechanics of Inelastic Media and Structures" Mahrenholtz, O. and Sawczuk, A., Eds., Polish Academy of Sciences, 1982, 87

45. Maier, G. and T. Hueckel: Nonassociated and coupled flow rules of elastoplasticity for rock-like materials, Int. J. Rock Mech. Min. Sci., 16, 77, 1979

46. Needleman, A.: Non-normality and bifurcation in plane strain tension and compression, J. Mech. Phys. Solids, 27, 231, 1979

47. Raniecki, B.: Uniqueness criteria in solids with non-associated plastic flow laws at finite deformations, Bull. Acad. Polon. Sci., ser. sci. techn., 27, 391, 1979

48. Marciniak, Z. and K. Kuczynski: Limit strains in the process of stretch forming sheet metal, Int. J. Mech. Sci., 9, 609, 1967

49. Stören, S. and J. R. Rice; Localized necking in thin sheets, J. Mech. Phys. Solids, 23, 421, 1975

50. Hutchinson, J. W. and V. Tvergaard: Surface instabilities on statically strained plastic solids, Int. J. Mech. Sci., 22, 339, 1980

51. Hutchinson, J. W. and V. Tvergaard: Shear band formation in plane strain, Int. J. Solids Struct., 17, 451, 1981

52. Tvergaard, V., Needleman, A. and K. K. Lo: Flow localization in the plane strain tensile test, J. Mech. Phys. Solids, 29, 115, 1981

53. Rice, J. R.: The localization of plastic deformation, in "Theoretical and Applied Mechanics", Proc. 14th IUTAM Congr., Koiter, W. T., Ed., North-Holland, 1976, 207

54. Rice, J. R. and J. W. Rudnicki: A note on some features of the theory of localization of deformation, Int. J. Solids Struct., 16, 597, 1980

55. Rudnicki, J. W. and J. R. Rice: Conditions for the localization of deformation in pressure-sensitive dilatant materials, J. Mech. Phys. Solids, 23, 371, 1975

56. Needleman, A. and J. R. Rice: Limits to ductility set by plastic flow localization, in "Mechanics of Sheet Metal Forming" Koistinen, D. P. and Wang, N.-M., Eds., Plenum Publ., 1978, 237

57. Hutchinson, J. W. and K. W. Neale: Sheet-necking II. Time-independent behavior, in "Mechanics of Sheet Metal Forming" Koistinen, D. P. and Wang, N.-M., Eds., Plenum Publ., 1978, 127

58. Bruhns, O. und B. Raniecki: Ein Schrankenverfahren bei Verzweigungsproblemen inelastischer Formänderungen, ZAMM, 62, T111, 1982

59. Bruhns, O.: Bounds to the critical stresses in bifurcation of cylindrical specimens in the case of non-associated flow laws, in "Stability in the Mechanics of Continua", Springer-Verlag, 1982, 46

# CONSTITUTIVE LAW IN INELASTIC STRUCTURAL MECHANICS

J.F. BESSELING
Laboratory for Engineering Mechanics
of the Mechanical Engineering Department
University of Technology Delft

## 10.1. VARIATIONAL STATEMENTS FOR THE EQUATIONS OF MOTION AND OF HEAT CONDUCTION

The highly nonlinear field equations of the continuum theory for the inelastic behaviour of solids cannot serve as a basis for analysis in structural mechanics. The complex geometry of a structure entails a multitude of boundary value problems that even in the linearized elastic case necessitates an approximate representation by finite dimensional vectorspaces, amenable to analysis by means of computer algorithms.

The finite element method offers such a representation. It has been the natural follow-up of models, already used by Eiffel in the design of spectacular bridges and of the Paris'tower. However there is by no means one, clearly defined finite element method. Since the mathematicians rediscovered the finite element method as a special Ritz-Galerkin technique for the solution of boundary value problems there has been a tendency to stress the mathematical aspects of the method as an approximation procedure rather than its physical aspects as a method for setting up a model of a complex structure by considering this structure as a composition of relatively simple elements.

In particular for geometrically and physically nonlinear problems we have gone ahead with the physical modelling without bothering too much about the fact that no error estimates can be given, while mathematical convergence in most cases has been taken for granted. A direct application of the relevant physical laws to the finite dimensional approximation of the scalar-, vector- and tensorfields involved is then the surest way to obtain a valid mathematical model.

For the derivation of the finite element-equivalent to the non-scalar equations of mechanics the variational formulation of these equations is most appropriate. The expression itself, that is being considered, represents a scalar, and hence is frame invariant, while it encompasses the boundary conditions as well as the differential equations of the field. Thus for purely mechanical problems the principle of virtual work is employed to formulate the laws of balance of momentum and of moment of momentum. For thermal problems a principle of virtual heat incorporates the heatflow equation with the proper boundary conditions. It will be demonstrated that the variational statements of the two principles mentioned suffice for the derivation of the finite element equations in terms of the primary variables: stress, temperature, density, velocity and displacement.

In case of purely mechanical problems the principle of virtual work suffices. We shall employ this principle in the version which in [1] is attributed to Piola. The principle states as a necessary condition for any motion of a body B that

$$\int_{\partial B} t_i \delta u_i \, dA + \int_B \rho f_i \delta u_i \, dV = \int_B \rho \ddot{u}_i \delta u_i \, dV \tag{10.1}$$

holds for all rigid virtual displacements $\delta u_i$. Here $t_i$ are the cartesian components of the surface tractions and the components $\rho f_i$ represent the force density.

A rigid virtual displacement is characterized by the condition

$$\frac{1}{2}\left(\frac{\partial \delta u_i}{\partial x_j} + \frac{\partial \delta u_j}{\partial x_i}\right) = \delta \varepsilon_{ij} = 0. \tag{10.2}$$

in all points of B. Introduced by means a symmetric tensor of multipliers $t_{ij}$ condition (10.2) transforms (10.1) into

$$\int_{\partial B} t_i \delta u_i \, dA + \int_B \rho f_i \delta u_i \, dV = \int_B (\rho \ddot{u}_i \delta u_i + t_{ij} \delta \varepsilon_{ij}) \, dV$$

$$\forall \; \delta u_i \wedge \delta \varepsilon_{ij} = \frac{1}{2}\left(\frac{\partial \delta u_i}{\partial x_j} + \frac{\partial \delta u_j}{\partial x_i}\right). \tag{10.3}$$

By applying Green's transformation, we derive

$$t_i = t_{ij} n_j \quad \text{on} \quad \partial B \tag{10.4}$$

and

$$\frac{\partial t_{ij}}{\partial x_j} + \rho f_i = \rho \ddot{u}_i \quad \text{in} \quad B, \tag{10.5}$$

which identifies $t_{ij}$ as the stress tensor of Cauchy. Thus $t_{ij} \delta \varepsilon_{ij}$ in the right hand side of (10.3) may rightly be called the virtual work of deformation.

In cases where a non-mechanical energy flux is significant the principle of virtual work can be supplemented by a principle of virtual heat, equivalent to the virtual dissipation principle of Biot [2]. By the introduction of the entropyflux $\dot{h}_i$ the heatflux can be written as $q_i = T\dot{h}_i$, where T represents the absolute temperature ($T \geq 0$). The principal of virtual heat then states as a necessary condition for any heat flow

$$\int_{\partial B} T\delta h_i n_i \, dA + \int_B r_{ij}\dot{h}_i \delta h_j \, dV = 0 \qquad (10.6)$$

for all divergence-free virtual entropy displacements $\delta h_i$. Here $[r_{ij}]$ is the positive definite matrix of heatresistances, related to the matrix of heatconductivities by

$$[r_{ij}] = T[k_{ij}]^{-1}. \qquad (10.7)$$

If the condition

$$\frac{\partial \delta h_i}{\partial x_i} = 0 \qquad (10.8)$$

is introduced in (10.6) by means of a multiplier $\theta$ we have

$$\int_{\partial B} T\delta h_i n_i \, dA + \int_B \left( r_{ij}\dot{h}_i \delta h_j - \theta \frac{\partial \delta h_i}{\partial x_i} \right) dV = 0 \ \forall \ \delta h_i. \qquad (10.9)$$

Again by Green's transformation we obtain

$$\theta = T \quad \text{on} \quad \partial B, \qquad (10.10)$$

and

$$\frac{\partial \theta}{\partial x_i} = - r_{ij}\dot{h}_j \ \text{in } B,$$

or

$$k_{ij}\frac{\partial\theta}{\partial x_j} = -T\dot{h}_i = -q_i . \tag{10.11}$$

Hence the multiplier $\theta$ is the absolute temperature and equations (10.11) state Fourier's law of heat conduction.

If conditions (10.3) and (10.9) are both satisfied at any moment in a thermo-mechanical process, it is easily established that the energy equation and the entropy production law reduce to one scalar condition, locally to be fulfilled by the constitutive laws for $t_{ij}$ and T. After introduction of the internal energy per unit mass, e, the entropy and the entropy production per unit mass, s and $\sigma$, the energy equation and the entropy production law can be written as follows ($\frac{d}{dt}\int_B \rho\, dV = 0$)

$$\int_{\partial B}(t_i\dot{u}_i - T\dot{h}_i n_i)\,dA + \int_B \rho f_i \dot{u}_i\, dV =$$

$$= \int_B [\rho(\dot{e} - T\dot{s}) + \rho\ddot{u}_i\dot{u}_i + T\rho\dot{s}]\,dV , \tag{10.12}$$

$$\rho\dot{s} = -\frac{\partial\dot{h}_i}{\partial x_i} + \rho\sigma, \quad \sigma \geq 0. \tag{10.13}$$

Substitution of (10.11) and (10.13) into (10.12), and application of the divergence theorem, leads to

$$\int_B [\rho(\dot{e}-T\dot{s}) - t_{ij}\frac{1}{2}\left(\frac{\partial\dot{u}_i}{\partial x_j} + \frac{\partial\dot{u}_j}{\partial x_i}\right) - \frac{k_{ij}}{T}\frac{\partial T}{\partial x_i}\frac{\partial T}{\partial x_j} + T\rho\sigma]\,dV = 0,$$

or

$$T\rho\sigma = \frac{k_{ij}}{T}\frac{\partial T}{\partial x_i}\frac{\partial T}{\partial x_j} + t_{ij}d_{ij} - \rho(\dot{e} - T\dot{s}) \geq 0. \tag{10.14}$$

Here $d_{ij} = \frac{1}{2}\left(\frac{\partial\dot{u}_i}{\partial x_j} + \frac{\partial\dot{u}_j}{\partial x_i}\right)$ is the rate of deformation tensor.

We conclude that the variational conditions (10.3) and (10.9) are not only equivalent both to the laws of balance of momentum and of moment of momentum and to the heatconduction equation, but that the energy equation and the law of entropy production are covered as well, provided the constitutive equations for the stresses and the temperature satisfy locally condition (10.14).

Sofar we have employed the Eulerian description in terms of spatial coordinates. As functions of these coordinates the velocities and the mass density have to satisfy the law of conservation of mass:

$$\frac{d}{dt}\int_B \rho dV = \int_B \left(\frac{\partial \rho}{\partial t} + \dot{u}_i \frac{\partial \rho}{\partial x_i} + \rho \frac{\partial \dot{u}_i}{\partial x_i}\right) dV = 0, \qquad (10.15)$$

or locally

$$\frac{\partial \rho}{\partial t} + \dot{u}_i \frac{\partial \rho}{\partial x_i} + \rho \frac{\partial \dot{u}_i}{\partial x_i} = 0 \qquad (10.16)$$

As a basis for numerical analysis we shall now develop the finite element-equivalent of the thermo-mechanical equations. The essence of the finite element methods lies in the observation that solutions of the field equations of continuum theories can always be approximated locally to any desired degree of accuracy by interpolation polynomials of low degree, except for a neighbourhood of singularities in the field. Since knowledge of the state of a body at any time will be considered adequate in a phenomenological sense if in a finite number of subregions this local approximation is available, finite element descriptions of continuous fields aim at providing these local approximations.

We may distinguish between spatial finite elements and material finite elements. While the material finite elements are adjoining material bodies $B^e$, the spatial finite elements are fixed subdomains of space with matter flowing through their bounding surfaces. Like the lagrangian description in continuum theory the material finite elements are most appropriate for problems of deformation. For problems of unlimited flow the reference geometry of a piece of material looses its physical relevance for the actual state and like the eulerian description in continuum theory the spatial finite elements are then the proper tool for analysis.

## 10.2. SPATIAL FINITE ELEMENTS FOR PROBLEMS OF PLASTIC FLOW

Spatial finite elements have a fixed geometry and fill together a volume B within a so-called control surface in Euclidean pointspace. Matter flows through the control surface and through the bounding surfaces of the finite elements.

We introduce for each spatial finite element polynomial interpolation functions $\psi_N^e(\xi_k)$ such that in the domain of the finite element holds

$$x_i = \psi_N^e(\xi_k)\, x_{Ni}\, , \tag{10.17}$$

where $x_{Ni}$ are the fixed spatial coordinates of node N, which must be a node on the element under consideration. Outside the e-th element we have $\psi_N^e(\xi_k) = 0$. On the boundary $\partial B^c$ between adjacent elements p and q the functions (10.17) for these elements should be identical in order that the transformation from cartesian coordinates to local coordinates of the elements be continuous and single-valued

$$x_i^{\partial B^c} = \psi_N^p(\xi_k^{\partial B^c})\, x_{Ni} = \psi_M^q(\xi_\ell^{\partial B^c})\, x_{Mi}\, . \tag{10.18}$$

For (10.18) to be satisfied independent of the coordinates of the nodes it is necessary that in (10.18) $x_i^{\partial B^c}$ only depends on the coordinates of the nodes on $\partial B^c$, which the elements p and q have in common.

It is convenient to take for the local coordinates $\xi_k$ so-called natural coordinates. As examples we may consider triangular and quadrangular coordinates. Finite elements with straight edges are described by

triangular coordinates: quadrangular coordinates:

$$\begin{vmatrix} 1 \\ x \\ y \end{vmatrix} = \begin{bmatrix} 1 & 1 & 1 \\ x_1 & x_2 & x_3 \\ y_1 & y_2 & y_3 \end{bmatrix} \begin{vmatrix} \xi_1 \\ \xi_2 \\ \xi_3 \end{vmatrix}, \quad \begin{vmatrix} 1 \\ x \\ y \end{vmatrix} = \begin{bmatrix} 1 & 1 & 1 & 1 \\ x_1 & x_2 & x_3 & x_4 \\ y_1 & y_2 & y_3 & y_4 \end{bmatrix} \begin{vmatrix} \frac{1}{4}(1-\xi_1)(1+\xi_2) \\ \frac{1}{4}(1-\xi_1)(1-\xi_2) \\ \frac{1}{4}(1+\xi_1)(1-\xi_2) \\ \frac{1}{4}(1+\xi_1)(1+\xi_2) \end{vmatrix}, \quad (10.19)$$

$$0 \le \xi_i \le 1 \quad ,$$

$$-1 \le \xi_i \le 1.$$

Higher degree polynomials offer the possibility of finite elements with curved edges:

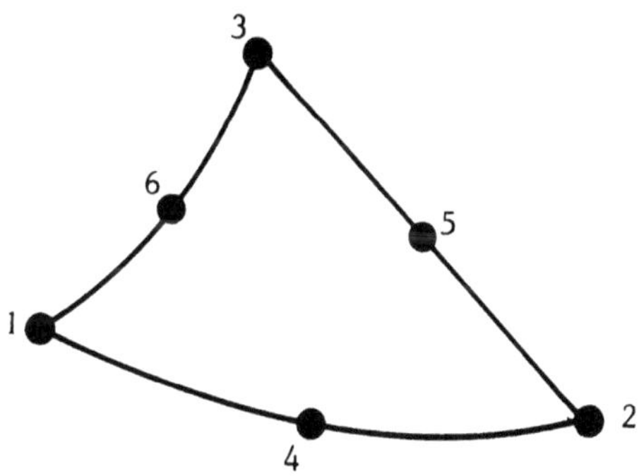


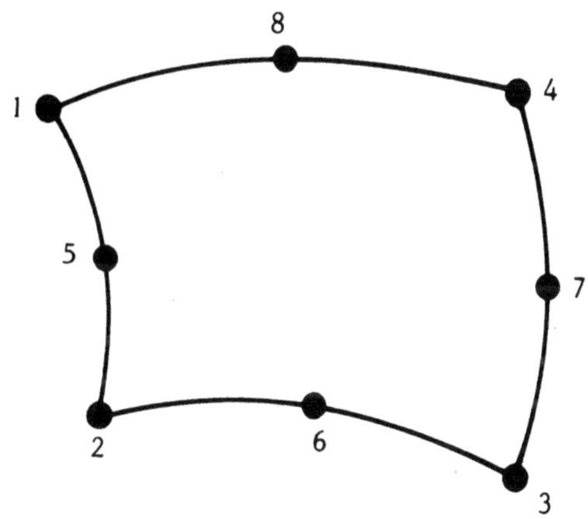

Triangular coordinates:

$$\begin{Bmatrix} 1 \\ x \\ y \end{Bmatrix} = \begin{bmatrix} 1 & 1 & 1 & 1 & 1 & 1 \\ x_1 & x_2 & x_3 & x_4 & x_5 & x_6 \\ y_1 & y_2 & y_3 & y_4 & y_5 & y_6 \end{bmatrix} \begin{Bmatrix} (2\xi_1 - 1)\xi_1 \\ (2\xi_2 - 1)\xi_2 \\ (2\xi_3 - 1)\xi_3 \\ 4\xi_1\xi_2 \\ 4\xi_2\xi_3 \\ 4\xi_3\xi_1 \end{Bmatrix}, \quad 0 \le \xi_i \le 1,$$

Quadrangular coordinates:

$$\begin{Bmatrix} 1 \\ x \\ y \end{Bmatrix} = \begin{bmatrix} 1 & 1 & 1 & 1 & 1 & 1 & 1 & 1 \\ x_1 & x_2 & x_3 & x_4 & x_5 & x_6 & x_7 & x_8 \\ y_1 & y_2 & y_3 & y_4 & y_5 & y_6 & y_7 & y_8 \end{bmatrix} \begin{Bmatrix} \frac{1}{4}(1-\xi_1)(1+\xi_2)(-\xi_1+\xi_2-1) \\ \frac{1}{4}(1-\xi_1)(1-\xi_2)(-\xi_1-\xi_2-1) \\ \frac{1}{4}(1+\xi_1)(1-\xi_2)(\xi_1-\xi_2-1) \\ \frac{1}{4}(1+\xi_1)(1+\xi_2)(\xi_1+\xi_2-1) \\ \frac{1}{2}(1-\xi_1)(1-\xi_2^2) \\ \frac{1}{2}(1-\xi_1^2)(1-\xi_2) \\ \frac{1}{2}(1+\xi_1)(1-\xi_2^2) \\ \frac{1}{2}(1-\xi_1^2)(1+\xi_2) \end{Bmatrix}, \quad -1 \le \xi_i \le 1 \tag{10.20}$$

The polynomials in (10.19) and (10.20) have the characteristic property

$$\sum_N \psi_N^e(\xi_k) = 1. \tag{10.21}$$

For the application of the principles of virtual work and of virtual heat all kinematically allowable rigid virtual displacements and all divergence-free virtual entropy displacements must be described. Furthermore a rate of divergence $\frac{\partial \dot{h}_i}{\partial x_i}$ and all components of the rate of deformation tensor should have the possibility of being unequal to zero independently of each other for each individual finite element. Otherwise no convergence by mesh refinement may be expected in general, since the absence of entropy divergence or of certain modes of deformation may then be induced not by the physics of the problem, but solely by the mathematical approximation of the fields.

The complete first degree polynomial in cartesian coordinates for velocity- and virtual displacement-components satisfies the requirements formulated above for any orientation of the axes.

$$\begin{vmatrix} \dot{u}_i \\ \delta u_i \\ \dot{h}_i \\ \delta h_i \end{vmatrix} = \begin{bmatrix} \alpha_{1i} & \alpha_{2i} & \alpha_{3i} & \alpha_{4i} \\ \delta\alpha_{1i} & \delta\alpha_{2i} & \delta\alpha_{3i} & \delta\alpha_{4i} \\ \beta_{1i} & \beta_{2i} & \beta_{3i} & \beta_{4i} \\ \delta\beta_{1i} & \delta\beta_{2i} & \delta\beta_{3i} & \delta\beta_{4i} \end{bmatrix} \begin{vmatrix} 1 \\ x \\ y \\ z \end{vmatrix} . \tag{10.22}$$

The three parameters $\alpha_{1i}$ give a description of the rigid translations, while the remaining nine parameters $\alpha$ take care of the three rigid rotations and the six components of the rate of deformation tensor. The parameters $\beta$ describe not only a divergence-free field, but also a field of arbitrary orientation with a divergence unequal to zero.

In our finite element model continuity of flow of matter and of entropy should be ensured. By choosing for the velocity- and the virtual displacement fields the same polynomial approximation as used in the transformation of local to cartesian coordinates we obtain not only continuity, but at the same time the first degree polynomials in (10.22) will be a subspace of the function space by which the fields are approximated. This follows immediately from the properties of the interpolation polynomials $\psi_N^e(\xi_k)$. Let us consider

$$\dot{u}_i = \psi_N^e(\xi_k)\ \dot{u}_{Ni} . \tag{10.23}$$

Now we take values for $\dot{u}_{Ni}$ in accordance with the linear functions (10.22).

$$\dot{u}_{Ni} = \begin{vmatrix} \alpha_{1i} & \alpha_{2i} & \alpha_{3i} & \alpha_{4i} \end{vmatrix} \begin{vmatrix} 1 \\ x_N \\ y_N \\ z_N \end{vmatrix} . \tag{10.24}$$

Substitution of (10.24) into (10.23) leads to

$$\dot{u}_i = \alpha_{1i} \sum_N \psi_N + \alpha_{2i} \psi_N x_N + \alpha_{3i} \psi_N y_N + \alpha_{4i} \psi_N z_N ,$$

or according to (10.18) and (10.21)

$$\dot{u}_i = \alpha_{1i} + \alpha_{2i} x + \alpha_{3i} y + \alpha_{4i} z . \tag{10.25}$$

This is the first degree polynomial in cartesian coordinates of (10.22).

In order that the interpolation polynomials in the natural coordinates for the velocities and the virtual displacements possess the properties required for convergence of solutions by mesh refinement it is essential that at least the polynomials used for the description of the geometry, (10.17), are represented. In case the same polynomials are used for geometry and for velocities and virtual displacements we speak of isoparametric elements.

For the derivation of the finite element equations by means of the principles of virtual work and of virtual heat we need not only a finite-dimensional representation of the velocity- and virtual displacement fields, but also the mass-density, the stresses and the temperature must be defined as functions of the coordinates before the integrations in the functionals (10.3) and (10.9) can be carried out. Since $t_{ij}$ and $\theta$ are multiplier fields, taking into account the conditions $\delta\varepsilon_{ij} = 0$ and $\frac{\partial \delta h_i}{\partial x_i} = 0$, for each finite element the number of parameters in the polynomial approximations of these fields should at least be equal to the number of deformation modes and divergence parameters. Therefore it is sufficient to choose for $t_{ij}$ and $\theta$ a representation that is obtained from $\delta\varepsilon_{ij} = \frac{1}{2}\left(\frac{\partial \delta u_i}{\partial x_j} + \frac{\partial \delta u_j}{\partial x_i}\right)$ and $\frac{\partial \delta h_i}{\partial x_i}$ by means of the polynomial approximations for $\delta u_i$ and $\delta h_i$. But this is no longer a representation by polynomials, except in the case that the transformation (10.17) is linear. For triangles, tetraheders, rectangles and rectangular parallelelopipeda we have this linearity of the transformation (10.17), but for arbitrary quadrangles and finite elements with curved boundaries the derivatives of the virtual displacements lead to complicated expressions in terms of the natural coordinates

$$\frac{\partial \delta u_i}{\partial x_j} = \frac{\partial \psi_N}{\partial \xi_k}\frac{\partial \xi_k}{\partial u_j}\delta u_{Ni} = J^{-1}_{jk}\frac{\partial \psi_N}{\partial \xi_k}\delta u_{Ni},$$

$$J = [\frac{\partial \psi_N}{\partial \xi_k} x_{Ni}] . \qquad (10.26)$$

The Jacobian matrix J has to be inverted. Only if J = constant, which implies a linear transformation, an explicite evaluation of $\frac{\partial \delta u_i}{\partial x_j}$ is useful as a basis for a polynomial representation of $t_{ij}$.

There are now two approaches for the derivation of the finite element equations from the conditions (10.3) and (10.9).

In the conventional approach the multiplier fields, being the stresses and the temperature, are expressed in terms of the displacement- or velocity fields by means of the constitutive equations of the material.

As a first example of this approach we consider the linearized elastic deformation problem. The constitutive equations in the isotropic case are

$$t_{ij} = [C\delta_{ij}\delta_{k\ell} + 2G(\delta_{ik}\delta_{j\ell} - \frac{1}{3}\delta_{ij}\delta_{k\ell})]\,\varepsilon_{k\ell}, \tag{10.27}$$

with the compression modulus C and the shear modulus G. The strain tensor $\varepsilon_{ij}$ in its linearized form is expressed in terms of the nodal displacements with the aid of the polynomial approximation for $u_i$ (10.23).

$$\varepsilon_{ij} = \frac{1}{2}\left(\psi_{N,\alpha}\frac{\partial\xi_\alpha}{\partial x_j}u_{Ni} + \psi_{N,\alpha}\frac{\partial\xi_\alpha}{\partial x_i}u_{Nj}\right). \tag{10.28}$$

Now we obtain from condition (10.3), in general by means of numerical integration

$$\int_{\partial B^t} t_i^c \delta u_i\, dA + \int_B \rho f_i^c \delta u_i\, dV = f_{Nj}^c \delta u_{Nj},$$

$$\int_B \rho\ddot{u}_i \delta u_i\, dV = \ddot{u}_{Mi} M_{MiNj} \delta u_{Nj},$$

$$\int_B t_{ij}\delta\varepsilon_{ij}\, dV = u_{Mi} K_{MiNj} \delta u_{Nj}, \tag{10.29}$$

The external forces, which are prescribed as a distribution of surface tractions $t_i^c$ on the part $\partial B^t$ of the external surface and as force densities $f_i^c$ in B, lead to a vector in a finite dimensional force space of which the components are known. The dual space of this force space is the space of unknown nodal displacements $u^c$. By denoting the nodal displacement components, which are given functions of time, by $u^o$, the principle of virtual work with the expressions (10.29) leads to the following finite element equations

$$K^{cc}u^c + M^{cc}\ddot{u}^c = f^c - K^{co}u^o - M^{co}\ddot{u}^o. \tag{10.30}$$

For given initial conditions $u^c$ and $\dot{u}^c$ at $t_o$ the system (10.30) can be solved. The solution describes small vibrations about a state of equilibrium. In the static case - $\ddot{u}^c = 0$, $\ddot{u}^o = 0$ - the solution for $u^c$ represents the elastic response to given forces $f^c$ and given nodal displacements $u^o$. Though the displacement field is continuous at the finite element boundaries, the derivatives of the displacements are not. Therefore the strains (10.28) and stresses (10.27) in the finite element solution will show discontinuities at the finite element boundaries, which are an indication of the approximative nature of this solution. However by reducing the mesh size we can obtain an approximation of the solution of the continuous field equations to any desired degree of accuracy, provided there are no singularities in the field.

A second example of the direct use of the constitutive equations in the principle of virtual work we find in the finite element equations for incompressible, viscous flow. In the constitutive equations

$$t_{ij} = - p\delta_{ij} + \mu \left( \frac{\partial \dot{u}_i}{\partial x_j} + \frac{\partial \dot{u}_j}{\partial x_i} \right) \tag{10.31}$$

the pressure p is physically not determined by the velocity field, but must follow from the equations of motion and their boundary conditions. We may approximate the incompressibility condition however by introducing a fictitious constitutive equation for p of the form

$$p = - \alpha\mu \frac{\partial \dot{u}_i}{\partial x_i} = - \alpha\mu \, \dot{u}_{Ni} \, \psi_{N,k} \frac{\partial \xi_k}{\partial x_i}, \quad \alpha >> 1. \tag{10.32}$$

Now the divergence of the velocity field is forced to remain small, because p remains bounded due to equations of motion and boundary conditions on $\partial B^t$. If inertia effects may be neglected the following expressions have to be evaluated in order to derive the finite element equations from condition (10.3).

$$\int_{\partial B^t} t_i^c \delta u_i \, dA + \int_B t_i^c \delta u_i \, dV = f_{Nj}^c \delta u_{Nj},$$

$$\int_B t_{ij} \delta \varepsilon_{ij} \, dV = u_{Mi} V_{MiNj} \delta u_{Nj} + \alpha \dot{u}_{Mi} P_{MiNj} \delta u_{Nj}. \qquad (10.33)$$

The introduction of the terms with $\alpha$ to take the incompressibility condition into account is also referred to as the penalty function approach. The larger the value of $\alpha$, the heavier counts the divergence of the velocity field in the equations. If the value of $\alpha$ is taken too large, then matrix of the system of equations becomes ill-conditioned due to a near-linear dependence of the nodal velocities.

Besides by the conventional approach it is possible to derive finite element equations without the use of the constitutive equations. But then the conditions $\delta\varepsilon_{ij} = 0$ and $\frac{\partial \delta h_i}{\partial x_i}$ from the virtual work and the virtual heat principles have to be enforced on the finite element level by the introduction of polynomial approximations for the multiplier fields $t_{ij}$ and $\theta$.

$$t_{ij} = t_{ijp}^{(e)} \phi_p^e(\xi_k),$$

$$\theta = \theta_q^{(e)} \phi_q^e(\xi_K). \qquad (10.34)$$

The number of parameters $t_{ijp}^{(e)}$ and $\theta_q^{(e)}$ must for each finite element correspond to the number of independent deformation parameters and divergence parameters, such as determined by the virtual displacement parameters.

For compressible flow the equation of conservation of mass for the massa density field has to be taken into account. Since the pressure p will be related to the density $\rho$ by a constitutive equation the polynomial approximation of the mass density field should contain the same number of parameters as the polynomial approximation for the pressure.

$$\rho = \rho_r^{(e)} \phi_r^e(\xi_k). \tag{10.35}$$

Local application of the equation (10.16) provides equations of the form

$$\rho_r + C_{rq\ell}\dot{u}_\ell \rho_q + D^\rho_{rq\ell}\dot{u}_\ell \rho_q = 0, \; D^\rho_{rq\ell} = 0 \quad \text{for} \quad r \neq q. \tag{10.36}$$

They are equal in number to the number of mass density parameters and constitute the discrete analogue of equation (10.16). For application of the principles of virtual work and of virtual heat appropriate boundary conditions must be given.

$$\begin{aligned} &\dot{u}_i^o \quad \text{on } \partial B^u \quad \text{and} \quad t_i^c \text{ on } \partial B^t, \\ &\dot{h}_i^o n_i \text{ on } \partial B^h \quad \text{and} \quad \theta^c \text{ on } \partial B^T. \end{aligned} \tag{10.37}$$

Substitution of the interpolation polynomials into (10.3) and (10.9) then leads to

$$\begin{aligned} &\delta u_k [t_i D^\sigma_{ik} + \rho_r (M_{rk\ell}\ddot{u}_\ell + N_{rk\ell m}\dot{u}_\ell \dot{u}_m) - f_k^c] = 0 \;\; \forall \;\; \delta u_k, \\ &\delta h_\ell [\theta_q D^\theta_{q\ell} - \frac{\theta_q}{k} R_{q\ell m}\dot{h}_m - \theta_\ell^c] = 0 \;\; \forall \;\; \delta h_\ell . \end{aligned} \tag{10.38}$$

The vectors $\delta\underline{u}$, $\dot{\underline{u}}$ and $\ddot{\underline{u}}$ and the vectors $\delta\underline{h}$ and $\dot{\underline{h}}$ in (10.38) are vectors in finite dimensional isomorphic vectorspaces, defined as the direct sum of the vectorspaces in the spatial nodes. The vectors $\underline{t}$, $\underline{\rho}$ and $\underline{\theta}$ are respectively vectors in the stress space, the mass density space, and the temperature space. These spaces are the direct sum of the corresponding subspaces belonging to the individual finite elements. The vectors $\underline{f}^c$ and $\underline{\theta}^c$ have prescribed components. They are vectors in the dual space of the virtual displacement space.

The finite element equations, resulting from (10.38), have to be supplemented by the constitutive laws for $\underline{t}$ and $\underline{\theta}$ in order to obtain a complete description of a physical model. These constitutive laws may be very complicated, such as is the case for elasto-visco-plastic material behaviour.

Relatively simply looking are the equations for incompressible viscous flow. With the constitutive equations (10.31) we obtain from (10.38)

$$- p_r D_{rk} + V_{k\ell}\dot{u}_\ell + (M_{k\ell}\ddot{u}_\ell + N_{k\ell m}\dot{u}_\ell\dot{u}_m) = f_k^c \tag{10.39a}$$

and from (10.36)

$$D^\rho_{q\ell}\dot{u}_\ell = 0. \tag{10.39b}$$

We shall not discuss here the problems connected with the solution of these equations, but we shall turn to the heat conduction problem for a simple illustration of the finite element representation of a physical problem.

Let us assume that the coupling between the heat conduction problem and the deformation problem may be neglected, as for solids is usually the case. Then we have from (10.38) the equations

$$\theta_q D^\theta_{q\ell} - \frac{\theta_q}{k} R_{q\ell m}\dot{h}_m = \theta^c_\ell , \tag{10.40}$$

based upon the polynomial approximations

$$\dot{h}_i = \psi_N^e(\xi_\alpha)\,\dot{h}_{Ni},$$

$$\theta = \theta_q^e \phi_q^e(\xi_\alpha). \tag{10.41}$$

The finite element description is completed by the local constitutive equations

$$e = c_v(\theta-\theta_o). \tag{10.42}$$

From (10.13) and (10.14) we now obtain

$$\dot{\theta} = \frac{k}{\rho c_v \theta}\frac{\partial\theta}{\partial x_i}\frac{\partial\theta}{\partial x_i} - \frac{\theta}{\rho c_v}\frac{\partial \dot{h}_i}{\partial x_i}. \tag{10.43}$$

Substitution of the polynomial representations (10.41) leads to

$$\dot{\theta}_q = \frac{\theta_q}{\rho c_v}\left[ - D^h_{q\ell}\dot{h}_\ell + \frac{k}{\theta^2_{(q)}} G_{qrs}\theta_r\theta_s\right]. \tag{10.44}$$

For a n-dimensional entropy velocity space and a m-dimensional temperature space the equations (10.40) and (10.44) form a system of n + m equations, that can be solved for given initial conditions. This system is nonlinear. This is due to the nonlinear relation between $\theta$ and s ($\theta = \theta_o \exp {}^s/c_v$), to the temperature dependence of $r_{ij}$ in (10.9), and to the positive definite expression for the entropy production connected with heat conduction. In fact the traditional description of the thermal properties of matter is such that the essential nonlinearity in the entropy production expression is neutralized in the partial differential equation for the temperature problem.

$$\theta = \frac{k}{\rho c_v}\frac{\partial^2\theta}{\partial x_i \partial x_i}. \tag{10.45}$$

Though the nonlinearity of the equations (10.40) and (10.44) presents no essential difficulty in a numerical integration algorithm, it is useful to observe that the nonlinearity has a negligible influence on the temperature distribution, provided the temperature differences remain small as compared to the absolute temperature $\theta$. Biot has already shown that linearization of the temperature-entropy problem leads to sufficient accuracy. The linearized version of the equations (10.40) and (10.44) is given by

$$\theta_q D^{\theta}_{q\ell} - \frac{\theta_o}{k} R_{\ell m} \dot{h}_m = \theta^c_\ell ,$$

$$\dot{\theta}_q = - \frac{\theta_o}{\rho c_v} D^h_{q\ell} \dot{h}_\ell = \frac{k}{\rho c_v} D^h_{q\ell} R^{-1}_{\ell m} (D^{\theta}_{rm} \theta_r - \theta^c_m). \tag{10.46}$$

As an example we consider a one-dimensional heat conduction problem. A wall of constant thickness is completely insulated at one side, while the other side from a certain time $t_o$ is exposed to a constant temperature increase $\Delta T$. Length and width of the wall are supposed to be sufficiently large in order to take into account for the larger part of the wall a dependence of the temperature on the thickness-coordinate only.

For the one-dimensional finite elements we choose the following polynomial approximations for the entropy velocity and for the temperature

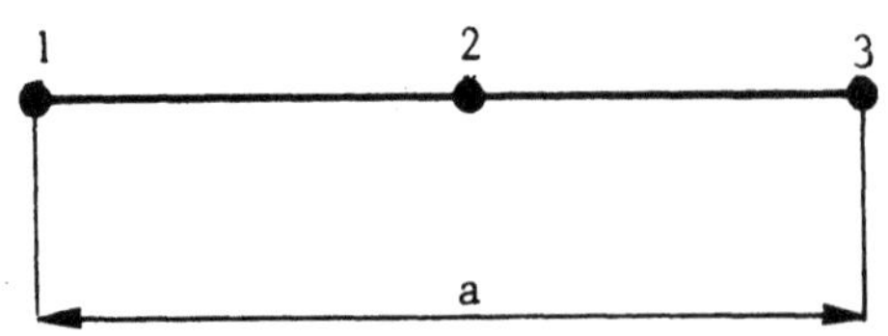

$$\begin{vmatrix} 1 \\ x \\ \dot{h} \end{vmatrix} \begin{bmatrix} 1 & 1 & 1 \\ x_1 & x_2 & x_3 \\ \dot{h}_1 & \dot{h}_2 & \dot{h}_3 \end{bmatrix} \begin{vmatrix} \frac{1}{2}\,\xi(\xi-1) \\ 1-\xi^2 \\ \frac{1}{2}\,\xi(\xi+1) \end{vmatrix}, \quad \theta = |\theta_1 \;\; \theta_3| \begin{vmatrix} \frac{1}{2}\,(1-\xi) \\ \frac{1}{2}\,(1+\xi) \end{vmatrix} \tag{10.47}$$

Notice that we took a linear distribution for $\theta$, because the divergence $\frac{\partial \dot{h}}{\partial x}$ is linear in x.

$$\frac{a}{2}\int_{-1}^{1} \dot{h}\delta h d\xi = \frac{a}{30}\,|\dot{h}_1 \;\; \dot{h}_2 \;\; \dot{h}_3| \begin{bmatrix} 4 & 2 & -1 \\ 2 & 16 & 2 \\ -1 & 2 & 4 \end{bmatrix} \begin{vmatrix} \delta h_1 \\ \delta h_2 \\ \delta h_3 \end{vmatrix} = \dot{h}^T R \delta h,$$

$$\int_{-1}^{1} \theta\,\frac{\partial \delta h}{\partial \xi}\,d\xi = \frac{1}{6}\,|\,\theta_1 \;\; \theta_3\,| \begin{bmatrix} -5 & 4 & 1 \\ -1 & -4 & 5 \end{bmatrix} \begin{vmatrix} \delta h_1 \\ \delta h_2 \\ \delta h_3 \end{vmatrix} = \theta^T D^\theta \delta h,$$

$$\begin{vmatrix} \dot{\theta}_1 \\ \dot{\theta}_3 \end{vmatrix} = -\frac{2\theta_o}{a\rho c_v} \begin{vmatrix} \left(\frac{\partial \dot{h}}{\partial \xi}\right)_{\xi=-1} \\ \left(\frac{\partial \dot{h}}{\partial \xi}\right)_{\xi=1} \end{vmatrix} = -\frac{\theta_o}{a\rho c_v} \begin{bmatrix} -3 & 4 & -1 \\ 1 & -4 & 3 \end{bmatrix} \begin{vmatrix} \dot{h}_1 \\ \dot{h}_2 \\ \dot{h}_3 \end{vmatrix} = \frac{-\theta_o}{\rho c_v}\,D^h \dot{h}. \tag{10.48}$$

If we take two finite elements over the wall thickness, then this temperature distribution as a function time and place is obtained from a solution of the following equations with given initial values for $|\theta^1_1 \;\; \theta^1_3 \;\; \theta^2_3 \;\; \theta^2_5|$

$$\frac{1}{6} \begin{bmatrix} -5 & -1 & 0 & 0 \\ 4 & -4 & 0 & 0 \\ 1 & 5 & -5 & -1 \\ 0 & 0 & 4 & -4 \end{bmatrix} \begin{vmatrix} \theta^1_1 \\ \theta^1_3 \\ \theta^2_3 \\ \theta^2_5 \end{vmatrix} - \frac{\theta_o a}{30k} \begin{bmatrix} 4 & 2 & -1 & 0 \\ 2 & 16 & 2 & 0 \\ -1 & 2 & 8 & 2 \\ 0 & 0 & 2 & 16 \end{bmatrix} \begin{vmatrix} \dot{h}_1 \\ \dot{h}_2 \\ \dot{h}_3 \\ \dot{h}_4 \end{vmatrix} = \begin{vmatrix} -\theta^c \\ 0 \\ 0 \\ 0 \end{vmatrix},$$

$$\begin{vmatrix} \dot{\theta}^1_1 \\ \dot{\theta}^1_3 \\ \dot{\theta}^2_3 \\ \dot{\theta}^2_5 \end{vmatrix} = - \frac{\theta_o}{a\rho c_v} \begin{bmatrix} -3 & 4 & -1 & 0 \\ 1 & -4 & 3 & 0 \\ 0 & 0 & -3 & 4 \\ 0 & 0 & 1 & -4 \end{bmatrix} \begin{vmatrix} \dot{h}_1 \\ \dot{h}_2 \\ \dot{h}_3 \\ \dot{h}_4 \end{vmatrix} . \qquad (10.49)$$

The first set of equations we can solve for the entropy velocities in terms of the temperatures. Substitution into the second set of equations then leads to a system of ordinary differential equations for the temperatures only.

If we take the case

$$t = 0 : \theta^c = \theta^c_o = \theta^1_1 = \theta^1_3 = \theta^2_3 = \theta^2_5 ,$$
$$t > 0 : \theta^c_o + \Delta\theta^c ,$$

then numerical integration produces temperature distributions as shown in Fig. 1. Notice that the average temperature of an element is approximated well, but that inevitably discontinuities occur.

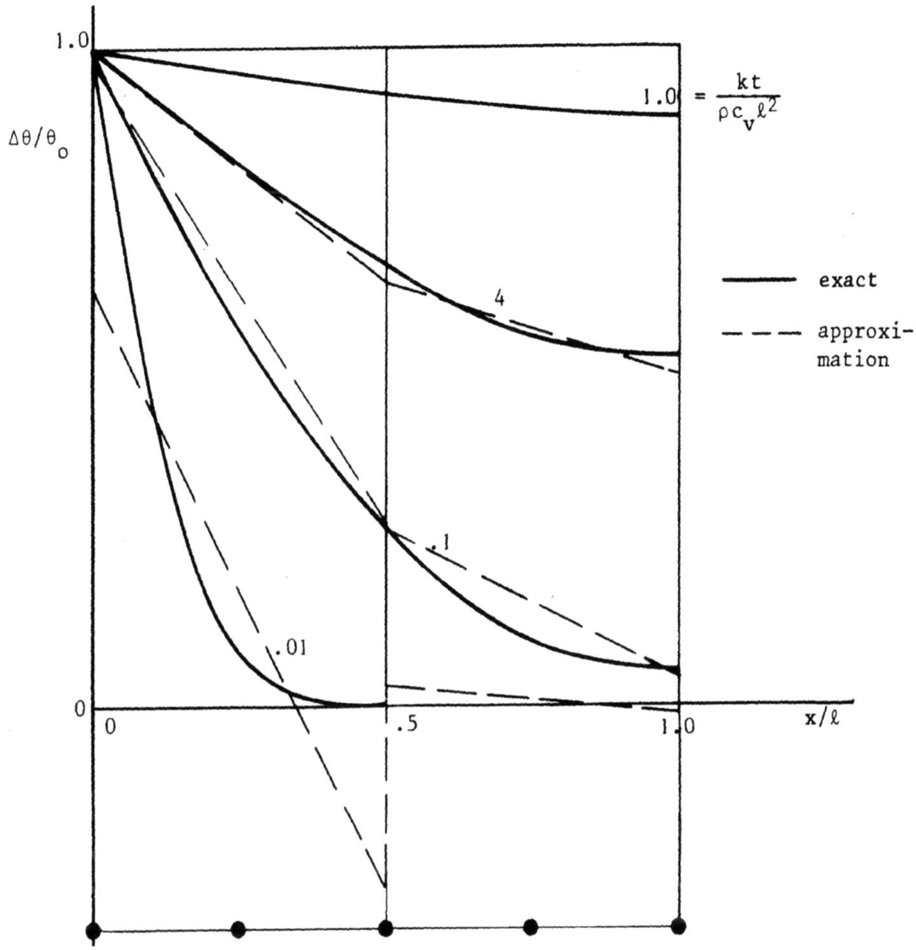


Fig. 1. Temperature distributions in wall.

In the case of plastic flow the local constitutive equations that have to supplement the equations from (10.38), are much more complicated than the equations (10.44), which were obtained by local application of the continuum equations (10.43). We shall give here a complete set of constitutive continuum equations for the thermo-visco-plastic deformation problem, details of which we shall discuss in the next paragraph.

With

$$s_{ij} = t_{ij} - \frac{1}{3} t_{nn}\delta_{ij},$$

$$\omega_{ij} = \frac{1}{2}\left(\frac{\partial \dot{u}_i}{\partial x_j} - \frac{\partial \dot{u}_j}{\partial x_i}\right), \quad d_{ij} = \frac{1}{2}\left(\frac{\partial \dot{u}_i}{\partial x_j} + \frac{\partial \dot{u}_j}{\partial x_i}\right), \tag{10.50}$$

we have

$$\dot{t}_{ij} = \frac{\partial t_{ij}}{\partial t} + \dot{u}_k \frac{\partial t_{ij}}{\partial x_k} = s_{kj}\omega_{ik} - s_{ik}\omega_{kj} + C(d_{nn} - \alpha\dot{\theta}) +$$

$$+ 2G\{\delta_{ik}\delta_{j\ell} - (1-h)\, Y^*_{ijk\ell}\}\,(d_{k\ell} - \frac{1}{3} d_{nn}\delta_{k\ell} - c_{k\ell}),$$

$$\rho c_v\dot{\theta} = \rho c_v\left(\frac{\partial\theta}{\partial t} + \dot{u}_k\frac{\partial\theta}{\partial x_k}\right) = - C\alpha\theta_o d_{nn} + \frac{k}{\theta}\frac{\partial\theta}{\partial x_i}\frac{\partial\theta}{\partial x_i} - \theta\frac{\partial \dot{h}_i}{\partial x_i} +$$

$$+ (1-h)\, s_{k\ell}(d_{k\ell} - c_{k\ell}) + s_{k\ell}c_{k\ell}, \tag{10.51}$$

where

$$c_{k\ell} = \frac{3}{2}\frac{\gamma(\theta)}{(\sigma^y)^2}\left(\frac{3J_2}{(\sigma^y)^2}\right)^{n-1} s_{k\ell},$$

$$Y^*_{ijk\ell} = \frac{1}{2} J_2^{-1} s_{ij}s_{k\ell} \quad \text{and} \quad 0 \le h \le 1,$$

$$\text{provided } 3J_2 - (\sigma^y)^2 = 0 \quad \text{and} \quad s_{ij}(d_{ij} - c_{ij}) \ge 0,$$

$$\text{otherwise } h = 1. \tag{10.52}$$

In the case of a hardening parameter $h > 0$ the yield stress $\sigma^y$ increases with the value of $J_2 = \frac{1}{2} s_{ij}s_{ij}$ in plastic deformation ($h \neq 1$), such that $(\sigma^y)^2 = (3J_2)_{max}$. Further the material is characterized by the material constants C, G, $\alpha$, $c_v$, k, and n, while $\gamma(T)$ represents the temperature dependence of the nonlinear viscous process that is usually denoted as creep. Notice that time independent plastic deformation ($h \neq 1$) occurs only if the deformation rate $d_{ij}$ exceeds the creeprate $c_{ij}$.

Local application of these constitutive equations with the aid of the polynomial approximations (10.34), together with the expressions for the velocities in each finite element, leads to the finite element constitutive equations, which together with the equations from (10.36) and (10.38) constitute the complete set of spatial finite element equations for problems of plastic flow.

## 10.3. CONSTITUTIVE RELATIONS FOR ARBITRARILY LARGE INELASTIC DEFORMATIONS

The industrially produced polycristalline metals, polymers and composite materials are conglomerations of differently orientated, anisotropic microbodies of a more or less well defined structure. But even if an accurate mathematical description of the processes on a microlevel would be possible, the complicated interaction of the microbodies would rule out a rigorous derivation of the macroscopic behaviour. Therefore the actual material is replaced by a continuum with at all times the same macroscopic massdensity as the actual material. Thus a mathematical description of the massflow can be given in terms of this continuum model, but we should realize that the so-called material points of the continuum define by no means the place of the actual material particles.

While it would be evidently inappropriate for fluids, for solids it is tempting to relate the geometric configuration of the material points of the continuum to the geometric configuration of the particles of the actual material. Then the thermo-dynamic state, in sofar as it is geometry dependent, will be determined by the deformation of the continuum with respect to a reference geometry of this continuum. For the reversible process of elastic deformation this approach is successful, but for plastic deformation and creep it must fail, even as a first approximation if the deformations become very large. The plate material will have lost its remembrance of the shape of the block from which the plate has been produced by a hot-rolling process.

Nevertheless subsequent elastic properties can be predicted fairly accurately, which in Ref. [3] prompted the idea of a generalization of Eckart's natural reference state [4] for a thermodynamic definition of stress. Stress is then by a thermodynamic equation of state uniquely related to a change of entropy and a deformation with respect to a geometric natural reference state. The description of the change of the natural reference state in the continuum then constitutes the theory of creep and plasticity.

Let us briefly recapitulate this theory of creep and plasticity.

Starting point is the assertion that at any moment the geometric configuration of the particles of the actual material, imbedded in a small neighbourhood of a material point of the continuum, differs from the configuration in a fixed natural reference state by an invertible transformation of line-elements.

$$da_\alpha = b_{\alpha i} dx_i \,, \quad \mathrm{Det}(b_{\alpha i}) > 0. \qquad (10.53)$$

With

$$b_{\alpha i} b_{\beta i} = C_{\alpha\beta} \qquad (10.54)$$

we have by the polar decomposition theorem

$$b_{\alpha i} = C^{\frac{1}{2}}_{\alpha\beta} R_{\beta i} \,, \qquad (10.55)$$

where

$$C^{\frac{1}{2}}_{\alpha\beta} = \sum_{p=1}^{3} \sqrt{C^{(p)}}\, n^{(p)}_{\alpha} n^{(p)}_{\beta}, \quad C^{(p)} > 0,$$

$$R_{\alpha i} R_{\beta i} = \delta_{\alpha\beta}.$$

The tensor components $C_{\alpha\beta}$ define a deformation, the components $R_{\alpha i}$ a rigid rotation. The latter determine the orientation of the material with respect to orthogonal triads, with fixed directions in space, attached to the material points of the continuum. Together with the entropy per unit mass, s, the tensor components $C_{\alpha\beta}$, independent of rigid rotations, are proper state variables in an internal energy function.

$$e = e(C_{\alpha\beta}, s). \tag{10.56}$$

The change of the natural reference state, characterized by $\dot{\overline{da}}_{\alpha}$, may be expressed in terms of quantities, determined by the local creep and plasticity process, as well as in terms of the rate of change of the tensor $b_{\alpha i}$ and the velocity field of the moving continuum:

$$\dot{\overline{da}}_{\alpha} = p_{\alpha i} dx_i = \left( \dot{b}_{\alpha i} + b_{\alpha j} \frac{\partial \dot{x}_j}{\partial x_i} \right) dx_i,$$

or

$$\dot{b}_{\alpha i} = p_{\alpha i} - b_{\alpha j} \frac{\partial \dot{x}_j}{\partial x_i}. \tag{10.57}$$

With the aid of (10.57) the rate of change of the state variables $C_{\alpha\beta}$ can be expressed by

$$\dot{C}_{\alpha\beta} = \dot{b}_{\alpha i} b_{\beta i} + b_{\alpha i} \dot{b}_{\beta i} = p_{\alpha i} b_{\beta i} + b_{\alpha i} p_{\beta i} - 2 b_{\alpha i} b_{\beta j} d_{ij}. \tag{10.58}$$

On the basis of the internal energy function (10.56) the expression (10.58) transforms the entropy production inequality (10.14) into

$$T\rho\sigma = \frac{k_{ij}}{T}\frac{\partial T}{\partial x_i}\frac{\partial T}{\partial x_j} + \left(t_{ij} + 2\rho\frac{\partial e}{\partial C_{\alpha\beta}} b_{\alpha i} b_{\beta j}\right) d_{ij}$$

$$- 2\rho\frac{\partial e}{\partial C_{\alpha\beta}} b_{\alpha i} p_{\beta i} - \rho\left(\frac{\partial e}{\partial s} - T\right)\dot{s} \geq 0. \qquad (10.59)$$

With the thermodynamic definition of stress and temperature

$$t_{ij} = - 2\rho\frac{\partial e}{\partial C_{\alpha\beta}} b_{\alpha i} b_{\beta j},$$

$$T = \frac{\partial e}{\partial s}, \qquad (10.60)$$

the entropy production is given by the sum of two positive definite terms

$$T\rho\sigma = \frac{k_{ij}}{T}\frac{\partial T}{\partial x_i}\frac{\partial T}{\partial x_j} + t_{ij} b^{-1}_{j\alpha} p_{\alpha i}. \qquad (10.61)$$

The second term represents the energy dissipation due to creep and plasticity.

The theory is completed by constitutive equations for the tensor components $p_{\alpha i}$.

First we make the obvious assumption that the natural reference state is characterized by an invariable massdensity $\rho_o$. In that case, as was shown in detail in [3], only the deviator components $s_{ij}$ of the stress tensor $t_{ij}$ contribute to the energy dissipation. With

$$t_{ij} = s_{ij} - p\delta_{ij}, \quad s_{ii} = 0, \qquad (10.62)$$

we then have

$$T\rho\sigma = \frac{k_{ij}}{T}\frac{\partial T}{\partial x_i} + s_{ij} b^{-1}_{j\alpha} p_{\alpha i}. \qquad (10.63)$$

For creep we introduce a stress and temperature dependent energy dissipation function $\pi$,

$$s_{ij} b^{-1}_{j\alpha} p_{\alpha i} = \pi(p, T, \frac{\rho_o}{\rho} s_{ij} R_{\alpha i} R_{\beta j}) \geq 0. \tag{10.64}$$

Notice that $s_{ij} R_{\alpha i} R_{\beta j}$ are the proper state variables, since also the rate of energy dissipation must be independent of rigid rotations of the material. However, the orthogonal transformations $R_{\alpha i}$ drop out again if the material is isotropic and consequently the rate of energy dissipation depends on stress through the invariants of the stress tensor.

Time independent plasticity can be considered as a limiting case of highly nonlinear creep. The yield surface then represents a surface of indeterminate rate of energy dissipation, which can be defined by

$$\phi = \phi(p, T, \frac{\rho_o}{\rho} s_{ij} R_{\alpha i} R_{\beta j}) = 0. \tag{10.65}$$

It was recently noted by v.d. Heijden [5], that the factor $\frac{\rho_o}{\rho}$ is needed in order to obtain the desired symmetry properties in the plasticity relations, if the constitutive equations for $p_{\alpha i}$ are based upon the normality condition for the tensor dual to $s_{ij}$

$$b^{-1}_{j\alpha} p_{\alpha i} = \lambda \frac{\partial \pi}{\partial s_{ij}} + \mu \frac{\partial \phi}{\partial s_{ij}}, \tag{10.66}$$

where $\mu \geq 0$ if $\phi = 0$ and $\dot{\phi} > 0$,

$\mu = 0$ if $\phi < 0$ or $\phi = 0$ and $\dot{\phi} < 0$.

For (10.57) and (10.58) we now have

$$\dot{b}_{\alpha i} = b_{\alpha k} \left( \lambda \frac{\partial \pi}{\partial s_{ik}} + \mu \frac{\partial \phi}{\partial s_{ik}} - \frac{\partial \dot{x}_k}{\partial x_i} \right)$$

$$\dot{C}_{\alpha\beta} = 2 b_{\alpha i} b_{\beta j} \left( \lambda \frac{\partial \pi}{\partial s_{ij}} + \mu \frac{\partial \phi}{\partial s_{ij}} - d_{ij} \right). \tag{10.67}$$

The material derivative of the Cauchy stress, $\dot{t}_{ij}$, can with the aid of

(10.67) be written as follows

$$\dot{t}_{ij} - t_{kj}\omega_{ik} + t_{ik}\omega_{kj} = \frac{\dot{\rho}}{\rho} t_{ij} + L_{ijmn}\left(d_{mn} - \lambda \frac{\partial\pi}{\partial s_{mn}} - \mu \frac{\partial\phi}{\partial s_{mn}}\right) + \frac{\partial t_{ij}}{\partial s}\dot{s}, \quad (10.68)$$

where

$$L_{ijmn} = 4\rho \frac{\partial^2 e}{\partial C_{\alpha\beta}\partial C_{\gamma\delta}} b_{\alpha i} b_{\beta j} b_{\gamma m} b_{\delta n} + \frac{1}{2}\left(t_{jn}\delta_{im} + t_{in}\delta_{jm} + t_{jm}\delta_{in} + t_{im}\delta_{jn}\right). \quad (10.69)$$

The scalar factor $\lambda$ is determined by the creep function

$$\lambda s_{ij} \frac{\partial\pi}{\partial s_{ij}} = \pi \Rightarrow \lambda = \left(s_{ij} \frac{\partial\pi}{\partial s_{ij}}\right)^{-1} \pi. \quad (10.70)$$

The calculation of the factor $\mu$ is more difficult. After putting

$$s^*_{\alpha\beta} = \frac{\rho_o}{\rho} s_{ij} R_{\alpha i} R_{\beta j}$$

we must require

$$\dot{\phi} = \frac{\partial\phi}{\partial s^*_{\alpha\beta}} \dot{s}^*_{\alpha\beta} + \frac{\partial\phi}{\partial p}\dot{p} + \frac{\partial\phi}{\partial T}\dot{T} \geq 0.$$

By observing that

$$R_{\alpha i}\dot{R}_{\beta i} = -\dot{R}_{\alpha i}R_{\beta i} = \Omega_{\alpha\beta} = -\Omega_{\beta\alpha},$$

$$R_{\alpha i}R_{\beta j}\Omega_{\alpha\beta} = \omega_{ij} = \dot{R}_{\alpha i}R_{\alpha j},$$

we can write

$$\frac{\partial\phi}{\partial s^*_{\alpha\beta}} \dot{s}^*_{\alpha\beta} = \frac{\partial\phi}{\partial s_{ij}}\left(\frac{\dot{\rho}}{\rho} s_{ij} + s_{kj}\omega_{ik} - s_{ik}\omega_{kj} + \frac{\partial s_{ij}}{\partial s}\dot{s} + \right.$$

$$\left. + L_{ijmn}\left(d_{mn} - \lambda \frac{\partial\pi}{\partial s_{mn}} - \mu \frac{\partial\phi}{\partial s_{mn}}\right) - \frac{\dot{\rho}}{\rho} s_{ij} - s_{kj}\omega_{ik} + s_{ik}\omega_{kj}\right).$$

Here we see that because of the factor $\frac{\rho_o}{\rho}$ the terms with $\frac{\dot{\rho}}{\rho} s_{ij}$ cancel and, as could be expected, the rigid rotations drop out since proper state variables were used in the yield function $\phi$. The factor $\mu$ can now be written as $(0 \leq h < 1)$

$$\mu = (1-h)\,\frac{\dfrac{\partial\phi}{\partial s_{ij}} L_{ijmn}\left(d_{mn} - \lambda\dfrac{\partial\pi}{\partial s_{mn}}\right) + \dfrac{\partial\phi}{\partial s_{ij}}\dfrac{\partial s_{ij}}{\partial s}\dot{s} + \dfrac{\partial\phi}{\partial p}\dot{p} + \dfrac{\partial\phi}{\partial T}\dot{T}}{\dfrac{\partial\phi}{\partial s_{k\ell}} L_{k\ell ab}\dfrac{\partial\phi}{\partial s_{ab}}}, \tag{10.71}$$

with the restriction $\mu \geq 0$.

Usually the pressure dependence, and to a lesser extent the temperature dependence of the yield function is neglected. Further for metals the stress deviator components $s_{ij}$ are independent of the entropy $s$. We then have

$$\dot{t}_{ij} - t_{kj}\omega_{ik} + t_{ik}\omega_{kj} = [L_{ijmn} - (1-h)\,Y_{ijmn} - t_{ij}\delta_{mn}]\left(d_{mn} - \lambda\frac{\partial\pi}{\partial s_{mn}}\right) + \frac{1}{3}\frac{\partial t_{kk}}{\partial s}\dot{s}, \tag{10.72}$$

with

$$Y_{ijmn} = \frac{L_{ijpq}\dfrac{\partial\phi}{\partial s_{pq}}\dfrac{\partial\phi}{\partial s_{rs}}L_{rsmn}}{\dfrac{\partial\phi}{\partial s_{k\ell}}L_{k\ell ab}\dfrac{\partial\phi}{\partial s_{ab}}}. \tag{10.73}$$

Note that no restrictions have been placed upon the amount of creep or plastic deformation, nor have the elastic deformations been assumed to be small.

But for metals the elastic deformations must remain small and then it is advantageous to replace the state variables $C_{\alpha\beta}$ by the elastic strain tensor $E'_{\alpha\beta}$.

$$E'_{\alpha\beta} = \frac{1}{2}(C^{-1}_{\alpha\beta} - \delta_{\alpha\beta}),\ e'_{\alpha\beta} = E'_{\alpha\beta} - \frac{1}{3}E_{\gamma\gamma}\delta_{\alpha\beta}. \qquad (10.74)$$

For an elastically isotropic metal the internal energy function depends on the deformation through the invariants

$$I_1 = E_{\alpha\alpha},\ I_2 = \frac{1}{2}e'_{\alpha\beta}e'_{\alpha\beta}.$$

In the case of small changes of entropy the internal energy can be given in the form

$$\rho_o e = \rho_o T_o s + \frac{1}{2}C\left(1 + \frac{C\alpha^2 T_o}{\rho_o c_v}\right)I_1^2 + 2GI_2 - \frac{C\alpha T_o}{c_v}I_1 s + \frac{\rho_o T_o}{2c_v}s^2. \qquad (10.75)$$

$T_o$ represents the temperature of the natural reference state. The remaining four constants have been expressed in terms of the commonly used physical constants: the coefficient of thermal expansion, $\alpha$, the specific heat at constant volume, $c_v$, the isothermal bulk modulus of elasticity, C, and the shear modulus, G.

If we take into account that the values of the moduli C and G are several orders of magnitude larger than the values the stresses may assume under normal conditions, while further for small values of $E'_{\alpha\beta}$ holds $|b_{\alpha i}-R_{\alpha i}| << 1$, then we derive for $L_{ijmn}$ according to (10.69) and (10.75):

$$L_{ijmn} = C\left(1 + \frac{C\alpha^2 T_o}{\rho_o c_v}\right)\delta_{ij}\delta_{mn} + 2G(\delta_{im}\delta_{jn} - \frac{1}{3}\delta_{ij}\delta_{mn}). \qquad (10.76)$$

In order to eliminate the entropy s from our equations we consider the results from (10.60) and (10.75).

$$T - T_o = -\frac{C\alpha T_o}{\rho_o c_v}I_1 + \frac{T_o s}{c_v},\ \frac{\partial t_{kk}}{\partial s} = -3\frac{C\alpha T_o}{c_v},$$

and obtain

$$\frac{\partial t_{kk}}{\partial s}\dot{s} = -3C\alpha\dot{T} - 3\frac{C^2\alpha^2T_o}{\rho_o c_v}d_{kk}. \qquad (10.77)$$

Next we introduce for creep the simplest energy dissipation function.

$$\pi = \gamma(T)\left(\frac{3J_2}{(\sigma^y)^2}\right)^n, \quad J_2 = \frac{1}{2}s_{ij}s_{ij}. \qquad (10.78)$$

This gives in accordance with (10.70)

$$\lambda\frac{\partial\pi}{\partial s'_{ij}} = c_{ij} = \frac{3}{2}\frac{\gamma(T)}{(\sigma^y)^2}\left(\frac{3J_2}{(\sigma^y)^2}\right)^{n-1} s_{ij}. \qquad (10.79)$$

With the yield function

$$\phi = 3J_2 - (\sigma^y)^2$$

time-independent plastic deformation is according to (10.73) determined by

$$Y^*_{ijmn} = \frac{1}{2G}Y_{ijmn} = \frac{1}{2}J_2^{-1} s_{ij}s_{k\ell}. \qquad (10.80)$$

If finally it is observed that in (10.72) the terms $t_{ij}\delta_{mn}$ may be neglected with respect to the moduli in $L_{ijmn}$, then (10.76), (10.77), (10.79) and (10.80) produce the constitutive equations for thermo-visco-plastic flow, given at the end of the previous paragraph.

It must be pointed out that only the simplest elasticity, creep and plasticity laws are incorporated in the equations (10.52). If isotropic stress dependent creep and isotropic plastic hardening is not a sufficiently accurate model of the material behaviour then the so-called fraction model with subelement stresses as internal state variables gives a way to generalize the theory for a more accurate representation [6]. In the past ten years numerous experiments [7] have demonstrated the success of this approach for plastic strains up to 2%.

## 10.4. MATERIAL FINITE ELEMENTS FOR PLASTIC DEFORMATION OF STRUCTURES

If the inelastic deformations remain small, then we can show that the elastic deformations $E'_{\alpha\beta}$ in (10.74) may be conceived as the difference between the total deformations and the inelastic deformations of the continuum, both taken with respect to the geometric reference state of the continuum.

We denote the cartesian coordinates of the material points of the continuum in its geometric reference state by $a^o_\alpha$, and we define the motion of the continuum by the three single-valued functions

$$x_i = x_i(a^o_1,\ a^o_2,\ a^o_3,\ t). \tag{10.81}$$

For line-elements in the deformed state and in the natural reference state (10.53) we have

$$dx_i = \frac{\partial x_i}{\partial a^o_\alpha}\, da^o_\alpha ,$$

$$da_\alpha = b_{\alpha i} dx_i = b_{\alpha i} \frac{\partial x_i}{\partial a^o_\beta}\, da^o_\beta . \tag{10.82}$$

In terms of the lagrangian strain tensor the total deformations are given by

$$E_{\alpha\beta} = \frac{1}{2}\left(\frac{\partial x_i}{\partial a^o_\alpha}\frac{\partial x_i}{\partial a^o_\beta} - \delta_{\alpha\beta}\right), \tag{10.83}$$

while the natural reference state is characterized by inelastic deformations

$$E''_{\alpha\beta} = \frac{1}{2}\left(b_{\gamma i}\frac{\partial x_i}{\partial a^o_\alpha} b_{\gamma j}\frac{\partial x_j}{\partial a^o_\beta} - \delta_{\alpha\beta}\right). \tag{10.84}$$

Provided

$$\left|b_{\alpha i}\frac{\partial x_i}{\partial a^o_\beta} - \delta_{\alpha\beta}\right| << 1 \tag{10.85}$$

we have

$$E_{\alpha\beta} - E''_{\alpha\beta} = \frac{1}{2}\,(b^{-1}_{\gamma k}b^{-1}_{\delta k} - \delta_{\gamma\delta})\; b_{\gamma i}\frac{\partial x_i}{\partial a^o_\alpha} b_{\delta j}\frac{\partial x_j}{\partial a^o_\alpha} \approx E'_{\alpha\beta}. \tag{10.86}$$

It is clear from (10.84) that conditions (10.85) imply that the inelastic deformations must remain small for (10.86) to be valid. It can be shown that for isotropic materials $|E''_{\alpha\beta}| << 1$ is also sufficient for (10.86) to hold. In the case of anisotropy there would still be the possibility of the difference between the rigid rotations of the continuum and those of the actual material growing large, such that for $|E''_{\alpha\beta}| << 1$ conditions (10.85) need not be fulfilled. This seems to be a rather rare possibility however.

For the formulation of the material finite element equations we shall now consider the subsidiary condition (10.2) of the principle of virtual work in terms of the lagrangian strain tensor. We derive

$$\int_B t_{ij}\delta\varepsilon_{ij}\,dV = \int_{B_o} \frac{\rho_o}{\rho}\frac{\partial a^o_\alpha}{\partial x_i} t_{ij} \frac{\partial a^o_\beta}{\partial x_j}\frac{\partial x_k}{\partial a^o_\alpha}\frac{\partial \delta u_k}{\partial a^o_\beta}\,dV_o . \tag{10.87}$$

After introduction of the symmetric pseudo-stress tensor

$$\sigma_{\alpha\beta} = \frac{\rho_o}{\rho}\frac{\partial a^o_\alpha}{\partial x_i} t_{ij} \frac{\partial a^o_\beta}{\partial x_j} , \tag{10.88}$$

and after observing that

$$\delta E_{\alpha\beta} = \frac{1}{2}\left(\frac{\partial x_k}{\partial a^o_\alpha}\frac{\partial \delta u_k}{\partial a^o_\beta} + \frac{\partial \delta u_k}{\partial a^o_\alpha}\frac{\partial x_k}{\partial a^o_\beta}\right),$$

we may write

$$\int_B t_{ij}\delta\varepsilon_{ij}\,dV = \int_{B_o} \sigma_{\alpha\beta}\delta E_{\alpha\beta}\,dV_o \tag{10.89}$$

We note that $\delta E_{\alpha\beta}$ is a first variation of a strain tensor, while $\delta\varepsilon_{ij}$ is not. As a consequence in elasticity theory the virtual work of deformation per unit undeformed volume, $\sigma_{\alpha\beta}\delta E_{\alpha\beta}$, can be considered as a first variation of an elastic potential and this leads to the powerful variational principle of minimum potential energy as an expression of stable elastic equilibrium [8].

For general deformation problems the principle of virtual work in the following form is an appropriate starting point.

$$\int_{\partial B} t_i \delta u_i\,dA + \int_{B_o} \rho_o(f_i - \ddot{u}_i)\delta u_i\,dV_o =$$

$$= \int_{B_o} \sigma_{\alpha\beta}\,\frac{1}{2}\left(\frac{\partial \delta u_\alpha}{\partial a^o_\beta} + \frac{\partial \delta u_\beta}{\partial a^o_\alpha} + \frac{\partial u_i}{\partial a^o_\alpha}\frac{\partial \delta u_i}{\partial a^o_\beta} + \frac{\partial \delta u_i}{\partial a^o_\alpha}\frac{\partial u_i}{\partial a^o_\beta}\right) dV_o\ \forall\ \delta \underline{u}. \tag{10.90}$$

A structure or body B is conceived as a set of continuously interconnected bodies $B^e$, which are called finite elements. We introduce for each finite element interpolation polynomials in terms of local coordinates, $\psi_N^e(\xi_k)$, such that in the reference configuration $B^o$ in the domain of the finite element for the cartesian coordinates holds

$$a_\alpha^o = \psi_N^e(\xi_k) a_{N\alpha}. \tag{10.91}$$

Here $a_{N\alpha}$ are the coordinates of nodal point N, which must be a node on the element under consideration. Further

$$a_\alpha^o = \sum_e \psi_N^e(\xi_k) a_{N\alpha}$$

must be continuous and single valued. Outside the e-th element we have $\psi_N^e(\xi_k) = 0$.

In contrast to the spatial elements the nodal velocities of the material finite elements can be integrated to give nodal displacements of material points. After going from the reference configuration $B^o$ to the deformed configuration B at time t we may write

$$x_i = a_i^o + u_i = \psi_N^e(\xi_k)\,(a_{Ni} + u_{Ni}) = \psi_N^e(\xi_k)\, x_{Ni}. \tag{10.92}$$

Thus we are led in a natural way to the concept of isoparametric elements, which ensures, as we have seen in paragraph 10.2, that all rigid virtual motions as well as all virtual deformation components will be represented if we consider for a finite element arbitrary virtual nodal displacements.

Substitution of the polynomial approximations for displacement fields into the variational condition (10.90) gives

$$\sum_e \left[ \int_{\partial B^e} t^c_\gamma \psi_N dA + \int_{B_o} \rho_o f^c_\gamma \psi_N dV_o - \ddot{u}_{M\gamma} \int_{B^e_o} \rho_o \psi_M \psi_N dV_o + \right.$$

$$\left. - \int_{B^e_o} \left\{ \sigma_{\alpha\beta} \psi_{N,i} \frac{\partial \xi_i}{\partial a^o_\beta} \left( \delta_{\alpha\gamma} + \psi_{M,j} \frac{\partial \xi_j}{\partial a^o_\alpha} u_{M\gamma} \right) \right\} dV_o \right] \delta u_{N\gamma} = 0 \; \forall \; \delta \underline{u}. \qquad (10.93)$$

This produces a set of equations, equal in number to the number of unknown nodal displacements. Before the integrals in (10.93) can be evaluated, however, constitutive equations for $\sigma_{\alpha\beta}$ must be given.

For small elastic deformations of an isotropic material we have

$$\sigma_{\alpha\beta} = C\{E_{kk} - \alpha(T - T_o)\}\, \delta_{\alpha\beta} + 2G(E_{\alpha\beta} - E''_{\alpha\beta} - \frac{1}{3} E_{kk} \delta_{\alpha\beta}). \qquad (10.94)$$

As it has already been pointed out the inelastic deformations $E''_{\alpha\beta}$ have to be small. Furthermore only $\dot{E}''_{\alpha\beta}$ is determined by constitutive equations and consequently the equations from (10.93) have to be differentiated with respect to time in order to obtain a set of equations by which for proper initial conditions a deformation process can be analysed.

A discrete model of a structure is obtained in a more direct way from the observation that the finite-dimensional displacement space of a material finite element defines a finite-dimensional virtual deformation space. Hence in the application of the principle of virtual work the subsidiary conditions (10.2) are for each finite element finite in number and can be introduced by means of a vector in the dual space of generalized stress variables.

The strain tensor components in the domain of a finite element are by (10.92) determined as functions of the nodal displacements. Writing for point p

$$E^e_{\alpha\beta p} = \frac{1}{2}\left(\frac{\partial u_\alpha}{\partial a^o_\beta} + \frac{\partial u_\beta}{\partial a^o_\alpha} + \frac{\partial u_i}{\partial a^o_\alpha}\frac{\partial u_i}{\partial a^o_\beta}\right)_p \qquad (10.95)$$

we introduce for $E_{\alpha\beta}$ in the domain of the finite element e interpolation polynomials

$$E^e_{\alpha\beta} = E^{(e)}_{\alpha\beta p}\,\phi^{(e)}_p(\xi_i). \qquad (10.96)$$

The number of linearly independent parameters $E^e_{\alpha\beta p}$ is taken to be equal to the dimension of the virtual deformation space, i.e. the number of nodal displacement components diminished by the number of rigid body degrees of freedom of the finite element. Using for the inelastic deformations $E''_{\alpha\beta}$ and for the thermal strains $\alpha(T-T_o)\delta_{\alpha\beta}$ the same interpolation polynomials as for $E_{\alpha\beta}$ we can express the stresses $\sigma_{\alpha\beta}$ by means of (10.94) in terms of these polynomials and the total, the inelastic, and the thermal strain parameters. After evaluation of the integrals in the variational statement, expressing the principle of virtual work.

$$\sum_e \left[\int_{\partial B^e} t^c_\gamma \psi_N dA + \int_{B^e_o} \rho_o f^c_\gamma \psi_N dV_o - \ddot{u}_{M\gamma}\int_{B^e_o} \rho_o \psi_M \psi_N dV_o\right]\delta u_{N\gamma} +$$

$$- \sum_e \int_{B^e_o} \sigma^{(e)}_{\alpha\beta}\,\delta E^{(e)}_{\alpha\beta}\, dV_o = 0 \;\forall\; \delta \underline{u},$$

we obtain

$$\delta u_k[f_k - M_{k\ell}\ddot{u}_\ell - \sigma_i D_{i,k}] = 0 \;\forall\; \delta u_k . \qquad (10.97)$$

The expressions (10.95) define the elements of a vector $\underline{\varepsilon}$ in the deformation space of the finite element:

$$\varepsilon_i = D_i(u_k),\; \delta\varepsilon_i = D_{i,k}\delta u_k, \qquad (10.98)$$

while the stress vector $\underline{\sigma}$ as dual vector of $\delta\underline{\varepsilon}$ follows from

$$\sum_e \int_{B_o^e} \sigma_{\alpha\beta}^{(e)} \delta E_{\alpha\beta}^{(e)} \, dV_o = (\varepsilon_i - \varepsilon_i'' - \varepsilon_i^T) \, S_{ij} \delta\varepsilon_j = \sigma_i \delta\varepsilon_j ,$$

or

$$\sigma_i = S_{ij}(\varepsilon_j - \varepsilon_j'' - \varepsilon_j^T). \tag{10.99}$$

The equations for $\dot{\varepsilon}_i''$ can be obtained by local application of the constitutive equations for creep and plasticity. In that case the approach based upon the deformationspace is comparable with the application of condition (10.93) with a finite number of integration points for a numerical evaluation of the integrals.

A consistent discrete model of a structure can also be obtained by a formulation of constitutive theory for creep and plasticity theory directly in terms of generalized strains and stresses for finite elements. This results in constitutive equations of the type

$$\dot{\sigma}_i = (S_{ij} - (1-h)Y_{ij})(\dot{\varepsilon}_j - c_j - \dot{\varepsilon}_j^T). \tag{10.100}$$

Here $c_j$ are the components of a creep vector and $Y_{ij} = Y_{ji}$ contains contributions of those elements for which stresses and rates of deformation are sufficiently high to produce plastic deformation [9]. In fact the relations (10.100) are completely analogous to the relations (10.52).

By differentiation with respect to time we obtain from the equations, resulting from (10.97), the following, highly nonlinear system

$$\{D_{i,k}(S_{ij} - (1-h)Y_{ij})D_{j,\ell} + \sigma_i D_{i,k\ell}\}\dot{u}_\ell + M_{k\ell}\ddot{u}_\ell =$$

$$= \dot{f}_k + D_{i,k}(S_{ij} - (1-h)Y_{ij})(c_j + \dot{\varepsilon}_j^T). \tag{10.101}$$

Most computerprograms deal only with the quasi-static case, i.e. problems where we may put $M_{k\ell}\ddot{u}_\ell = 0$. Then the incremental approach does not meet with serious numerical stability problems. In (10.101) we solve for $\Delta u_k = \dot{u}_k \Delta t$. Subsequently stress increments are calculated from (10.100), and new values for $Y_{ij}$ and $c_j$ can be evaluated. In this solution procedure a deviation from equilibrium can be compensated by applying the corrective forces

$$r_k = f_k - \sigma_i D_{i,k}. \qquad (10.102)$$

Of course the incremental procedure must fail as soon as the matrix of coefficients

$$D_{i,k}(S_{ij} - (1-h)Y_{ij})\, D_{j,\ell} + \sigma_i D_{i,k\ell}$$

becomes singular. We note that

$$\delta u_k D_{i,k}\, (S_{ij} - (1-h)Y_{ij})\, D_{j,\ell}\delta u_\ell > 0$$

is the condition for material stability, while

$$\delta u_k \sigma_i D_{i,k\ell}\delta u_\ell > 0$$

implies geometric stabilisation, and

$$\delta u_k \sigma_i D_{i,k\ell}\delta u_\ell < 0$$

geometric destabilisation due to existing stresses.

If the structure is only subject to creep, and if the geometric nonlinearity of $\sigma_i D_{i,k\ell}$ may be neglected, equations (10.100) and (10.101) can be reduced to a system of first order, nonlinear ordinary differential equations. With $Y_{ij} = 0$, $D_{i,k\ell} = 0$, $\ddot{u}_\ell = 0$, we solve equations (10.101) for $\dot{u}_\ell$ and calculate $\dot{\varepsilon}_i = D_{i,k}\, \dot{u}_k$. Substitution into (10.100) leads to

$$\dot{\sigma}_i = S_{ij}D_{j,k}[D_{m,k}S_{mn}D_{n,\ell}]^{-1}\, (D_{a,\ell}S_{ab}c_b + \dot{f}_\ell) - S_{ij}c_j \qquad (10.103)$$

The components of the creep vector $c_j$ are nonlinear functions of $\sigma_i$. The displacements are obtained by simultaneous integration of the equations (10.103) and of the equations

$$\dot{u}_k = [D_{i,k} S_{ij} D_{j,\ell}]^{-1} (D_{m,\ell} S_{mn} c_n + \dot{f}_\ell). \tag{10.104}$$

The integration of equations (10.103) and (10.104) for given initial conditions is a rather straightforward numerical procedure.

## 10.5. LIMIT ANALYSIS

Though at present computerprograms are available that make a complete elasto-plastic analysis of structures feasible, the ensuing lengthy calculations often can be circumvented by a judicious application of the so-called collapse and shakedown theorems of the theory of ideal plasticity. These theorems enable us to determine limit loads for a structure without going into the details of the whole deformation process prior to collapse.

In a fundamental article on general theorems for elastic-plastic solids Koiter has given an elegant development of the collapse and shakedown theorems [10] which we shall follow closely.

Limit analysis of structures is based upon the elastic-ideally plastic model of material behaviour. We introduce a vector of strain components $\varepsilon_i$ and a vector of stress components $\sigma_i$, such that $\sigma_i\dot{\varepsilon}_i$ represents the rate of work of deformation for the individual material elements, for which $\sigma_i$ and $\varepsilon_i$ are generalized stresses and generalized strains. A symmetric matrix of elasticity coefficient relates the stresses $\sigma_i$ to the elastic strains,

$$\sigma_i = S_{ij}\varepsilon'_j, \quad \varepsilon'_i = \varepsilon'_i - \varepsilon''_i, \tag{10.105}$$

while it is assumed that geometrical nonlinearities may be neglected:

$$\varepsilon_i = D_{ik} u_k . \tag{10.106}$$

A central role in the proof of the collapse and shakedown theorems plays the postulate for time independent plasticity, formulated in 1951 by Drucker [11]. It says: if a material element by an external agency is subjected to a stress cycle, starting and ending in the same state of initial stress $\sigma_i^*$, then the total work performed in such a complete cycle by the additional stresses $(\sigma_i - \sigma_i^*)$ is nonnegative. Hence for each material element holds

$$\oint_{\sigma_i} (\sigma_i - \sigma_i^*) \, d\varepsilon_i \geq 0. \tag{10.107}$$

We shall distinguish between a so-called "safe" state of stress, $\sigma_i^{(s)}$, which lies in the elastic domain, inside the yield surface, and a so-called "allowable" state of stress, $\sigma_i^{(a)}$, which is either in the elastic domain or on the yield surface. Then we can deduce from (10.107) the following inequality

$$(\sigma_i - \sigma_i^{(a)}) \dot{\varepsilon}_i'' \geq 0. \tag{10.108}$$

Since in geometric terms (10.108) implies that $\sigma_i^{(a)}$ lies on or at one side of a hyperplane in stress-space, which is perpendicular to the inelastic strainrate vector $\dot{\varepsilon}_i''$, it follows that in a regular point of the yield surface the normality condition

$$\dot{\varepsilon}_i'' = \mu \frac{\partial \phi}{\partial \sigma_i}, \quad \mu \geq 0,$$

must be satisfied as the hyperplane must be tangent to the yield surface.

Singular yield surfaces with edges or corners may be described by a number of intersecting yield surfaces. The inelastic strainrate at a singular point on the yield surface is then given by

$$\dot{\varepsilon}''_i = \sum_\alpha \mu_\alpha \frac{\partial \phi_\alpha}{\partial \sigma_i}, \quad \mu_\alpha \geq 0. \tag{10.109}$$

The fact that because of inequality (10.108) a state of stress on or inside the yield surface always lies on or at one side of a tangent hyperplane of the yield surface makes this yield surface convex. Further the fact that $\sigma_i^{(s)}$ lies by definition inside the yield surface rules out the possibility that $\sigma_i^{(s)}$ lies on the tangent hyperplane of the yield surface at $\sigma_i$. Hence we have the important strong inequality

$$(\sigma_i - \sigma_i^{(s)})\, \dot{\varepsilon}''_i > 0. \tag{10.110}$$

By means of the virtual work equation we now first prove the <u>auxiliary theorem</u>: if nonvanishing velocities occur in a structure under constant external loading, then for elastic-ideally plastic material behaviour the stressdistribution is constant.

Suppose $\dot{\sigma}_i \neq 0$ for $\dot{f}_k = 0$ and $\dot{u}_k \neq 0$. The stressrate must satisfy the equilibrium equations.

$$D_{ik}\dot{\sigma}_i = \dot{f}_k = 0.$$

Consequently we have

$$\text{I} : \dot{\sigma}_i D_{ik} \dot{u}_k = \dot{\sigma}_i \dot{\varepsilon}_i = 0.$$

But for

$$\varepsilon_i = \varepsilon'_i + \varepsilon''_i = S^{-1}_{ij}\sigma_j + \varepsilon''_i$$

it follows

$$\text{II} : \dot{\sigma}_i \dot{\varepsilon}_i = \dot{\sigma}_i S^{-1}_{ij} \dot{\sigma}_j + \dot{\sigma}_i \dot{\varepsilon}''_i = \dot{\sigma}_i S_{ij} \dot{\sigma}_j > 0 \;\; \forall \dot{\sigma}_i \neq 0.$$

Hence there is a contradiction between I and II if $\dot{\sigma}_i \neq 0 \Rightarrow \dot{\sigma}_i = 0$.

In the formulation of the collapse and shakedown theorems we speak of statically admissible stresses if they satisfy the equilibrium equations, $\sigma_i D_{ik} = f_k$, and of kinematically admissible strains if they satisfy the continuity equations, $\varepsilon_i = D_{ik} u_k$.

We shall now formulate the first collapse theorem, which was for the first time explicitely enunciated by Kazinczy (1914), and independently a few years later by Kist (1917). However many structural engineers must have been intuitively aware of its validity, when for a proof of the strength of their designs they relied on an approximate elastic analysis for the derivation of allowable stress distributions, in equilibrium with the external loads.

The first collapse theorem is expressed by the following statement: A structure will not collapse in the course of a given loading program, if at each stage of this loading program a "safe", statically admissible stress distribution can be found. On the other hand a structure cannot support a given external loading for which no "allowable", statically admissible stress distribution exists.

Hence according to this theorem the existence of a "safe" statically admissible stress distribution is a sufficient condition, the existence of an "allowable" statically admissible stress distribution a necessary condition for the capability of the structure of supporting the external loads.

For the proof of the sufficiency condition we assume that collapse takes place under constant loads $f_k$, though

$$D_{ik}\sigma_i^{(s)} = f_k, \quad \phi(\sigma_i^{(s)}) < 0.$$

By virtue of the auxiliary theorem, discussed above, the collapse mechanisme must be a purely plastic, kinematically admissible strain rate $\dot{\varepsilon}_i''$ with corresponding statically admissible constant stresses $\sigma_i$. From

$$\dot{\varepsilon}_i'' = D_{ik}\dot{u}_k \quad \text{and} \quad (\sigma_i - \sigma_i^{(s)})\, D_{ik} = f_k - f_k = 0$$

it follows

$$(\sigma_i - \sigma_i^{(s)})\, D_{ik}\dot{u}_k = (\sigma_i - \sigma_i^{(s)})\, \dot{\varepsilon}_i'' = 0.$$

But according to Drucker's postulate, (10.110),

$$(\sigma_i - \sigma_i^{(s)})\, \dot{\varepsilon}_i'' > 0 \quad \forall \quad \dot{\varepsilon}_i'' \neq 0 \;\Rightarrow\; \dot{\varepsilon}_i'' = 0 \;\Rightarrow\; \dot{u}_k = 0.$$

As to the necessary condition for the capability of supporting the external loads the necessity of the existence of an allowable, statically admissible stress distribution under a given load is self-evident.

Together with a proof of the first collapse theorem a clear ennunciation and proof of the second collapse theorem, both restricted to beams and frames, appeared in a paper by Greenberg and Prager in 1949. While the first collapse theorem provides a lower bound for the limit load of a structure, an upper bound for this limit load can be found by the <u>second collapse theorem</u>; which is expressed as follows: Collapse of a structure will take place if in the course of a loading program, for a given set of external forces, a kinematically admissible strainrate distribution can be found, which as a purely plastic strainrate distribution gives rise to a total energy-dissipation that is smaller than the rate of work of the external loads. On the other hand collapse of the structure cannot occur, if for all kinematically admissible plastic strainrate distributions the total energy dissipation exceeds the rate of work of the external loads.

To prove the sufficient condition for collapse suppose the structure does not collapse, though

$$f_k \overset{\bullet}{u}{}_k^{o} > \sigma_i \overset{\bullet}{\varepsilon}{}_i^{o''}, \quad \overset{\bullet}{\varepsilon}{}_i^{o''} = D_{ik} \overset{\bullet}{u}{}_k^{o}.$$

By the first collapse theorem there exists an "allowable", statically admissible stress distribution $\sigma_i^{(a)}$,

$$D_{ik}\sigma_i^{(a)} = f_k, \quad \phi(\sigma_i^{(a)}) \leq 0.$$

Hence

$$f_k \overset{\bullet}{u}{}_k^{o} = \sigma_i^{(a)} D_{ik} \overset{\bullet}{u}{}_k^{o} = \sigma_i^{(a)} \varepsilon_i^{o''},$$

or

$$0 > (\sigma_i - \sigma_i^{(a)}) \overset{\bullet}{\varepsilon}{}_i^{o''},$$

which contradicts inequality (10.108) ⇒ collapse will take place.

If collapse occurs under constant load, we have a purely plastic, kinematically admissible strainrate distribution and a statically admissible stress distribution. The ensuing equality

$$f_k \dot{u}_k = \sigma_i D_{ik} \dot{u}_k = \sigma_i \dot{\varepsilon}_i'',$$

proves that collapse cannot occur if for all kinematically admissible plastic strainrates holds

$$f_k \overset{\bullet}{u}{}_k^{o} < \sigma_i \varepsilon_i^{o''}.$$

The question can be raised whether the rate of energy dissipation,

$\sigma_i \dot{\varepsilon}_i''$, is uniquely determined, as it was tacitly assumed in the inequalities of the second collapse theorem, if for a given yield surface a certain plastic strainrate $\dot{\varepsilon}_i''$ is prescribed. The answer is in the affirmative, because, if the yield surface is strictly convex, the direction of the normal $\frac{\partial \phi}{\partial \sigma_i} = \frac{1}{\mu} \dot{\varepsilon}_i''$ determines a unique stress point on the yield surface, and, if this surface has straight generators or plane faces, all stress points on these generators or plane faces give rise to the same rate of energy dissipation, since they differ by vectors in stress space which are perpendicular to $\dot{\varepsilon}_i''$ in the same space.

The collapse theorems form the basis of the widely used limit analysis of structures. They should, however, be applied with caution. They have been proved under the limitations of geometric linearization. Thereby buckling and similar phenomena have been excluded. Furthermore the material has been endowed with the capacity of unlimited plastic deformation. But local excessive deformation even in rather ductile metals, may lead to failure, even though the overall deformations of the structure remains in the range, where geometric linearization, the neglect of the effect of changes in geometry on the conditions of equilibrium, is still justifiable.

If a structure is subjected to fluctuating loads the condition that in no stage of the loading process a collapse load is reached, is not sufficient to guarantee in the long run structural integrity. Most metals can sustain only a very limited number of cycles of plastic deformation, and in a loading program, in which cycles of plastic strainrates in alternating directions occur, so-called low-cycle fatigue may lead rapidly to failure. In addition to this type of failure, which is of special importance in pressure vessels, fluctuating loads may lead to cyclic plastic deformations that make the structure unserviceable due to excessive overall deformations.

A minimum requirement for permanent structural integrity in the presence of loads, which vary independently between certain limits, must be that plastic flow stops completely, if an initial period of such flow in the earlier stages of the loading program cannot be circumvented. Then during this initial period the structure will have to develop residual stresses such, that superimposed on the stresses according to elasticity theory at all subsequent times the resulting stress distribution does not violate the yield condition at any point. "The structure is then said to shake down to a state of permanent strains and corresponding residual stresses" (Koiter [10]).

The first shakedown theorem was given by Melan (1936). It furnishes a lower bound for the limits of the fluctuating loads at which ultimately a purely elastic response of the structure will be obtained. The theorem may be formulated as follows:

Let $\sigma_i^e$ represent the stresses, which according to elasticity theory would occur in the structure under a given set of loads. Suppose that a time independent distribution of statically admissible residual stresses $\sigma_i^r$ ($D_{ik}\sigma_i^r = 0$) can be found such that for all possible combinations of within certain limits fluctuating loads holds $\sigma_i^e + \sigma_i^r = \sigma_i^{(s)}$. Then plastic deformation cannot continue indefinitely and the structure will ultimately show a purely elastic response. The existence of residual stresses $\sigma_i^r$, such that $\sigma_i^e + \sigma_i^r = \sigma_i^{(a)}$, is the obvious necessary condition for "shake down".

For the proof of this theorem we consider the positive definite expression

$$A = \frac{1}{2}\,(\sigma_i-\sigma_i^{(s)})\;S_{ij}^{-1}\;(\sigma_j-\sigma_j^{(s)}) > 0 \;\;\forall\;\; \sigma_i^{(s)} \neq 0.$$

Since $\sigma_i$ are the actual stresses in the structure we have

$$S_{ij}^{-1}\,\dot{\sigma}_j = \dot{\varepsilon}_i' = \dot{\varepsilon}_i-\dot{\varepsilon}_i''.$$

In the "safe" stressdistribution $\sigma_i^{(s)} = \sigma_i^e+\sigma_i^r$, the stresses calculated on the basis of elasticity theory are the only time dependent stresses. They change with a change of load, where it should be remembered that inertia forces are being neglected. The time derivative of expression A is then given by

$$\dot{A} = (\sigma_i-\sigma_i^{(s)})\;S_{ij}^{-1}\;(\dot{\sigma}_j-\dot{\sigma}_j^{(s)}) = (\sigma_i-\sigma_i^{(s)})(\dot{\varepsilon}_i-\dot{\varepsilon}_i''-\dot{\varepsilon}_i^e).$$

But $(\sigma_i-\sigma_i^s)$ satisfies $D_{ik}(\sigma_i-\sigma_i^{(s)}) = 0$

and $\dot{\varepsilon}_i = D_{ik}\dot{u}_k$, $\quad \dot{\varepsilon}_i^e = D_{ik}\dot{u}_k^e$.

Hence

$$(\sigma_i-\sigma_i^s)(\dot{\varepsilon}_i-\dot{\varepsilon}_i^e) = (\sigma_i-\sigma_i^{(s)})\;D_{ik}(\dot{u}_k-\dot{u}_k^e) = 0.$$

According to Drucker's postulate:

$$(\sigma_i-\sigma_i^{(s)})\;\dot{\varepsilon}_i'' > 0,$$

which leads to the conclusion that the time derivative of A is negative

$$\dot{A} = -\;(\sigma_i-\sigma_i^{(s)})\;\dot{\varepsilon}_i'' < 0.$$

But from $A > 0$ and $\dot{A} < 0$ it follows that plastic deformation cannot continue unbounded.

It is not possible to give bounds for the local plastic strains in the shakedown state, and therefore the shakedown theorem, like the collapse theorems, has to be applied with caution in estimating the load carrying capacity of a structure. Some indication on the amount of plastic deformation prior to shakedown is obtained from an upper bound for the total energy dissipation, which can be derived as follows.

Let $n(\sigma_i^e+\sigma_i^r) = n\sigma_i^{(s)} = \sigma_i^{(a)}$, $n > 1$.

We then have in a loading program, which is accompanied by plastic deformation

$$(\sigma_i-\sigma_i^{(a)})\dot{\varepsilon}_i'' = (\sigma_i-n\sigma_i^{(s)})\,\dot{\varepsilon}_i'' \geq 0$$

and

$$-\dot{A} = (\sigma_i-\sigma_i^{(s)})\,\dot{\varepsilon}_i'' = \frac{\sigma_i-n\sigma_i^{(s)}}{n}\,\dot{\varepsilon}_i'' + \frac{n-1}{n}\,\sigma_i\dot{\varepsilon}_i'' \geq \frac{n-1}{n}\,\sigma_i\dot{\varepsilon}_i''.$$

Hence for the rate of energy dissipation holds

$$\sigma_i\dot{\varepsilon}_i'' \leq -\frac{n}{n-1}\,\dot{A}$$

and a bound for the total energy dissipation is found from

$$\int_o^t \sigma_i\dot{\varepsilon}_i''\,dt \leq \frac{n}{n-1}\,[A(o)-A(t)] \leq \frac{n}{n-1}\,\sigma_i^r S_{ij}^{-1}\sigma_j^r. \tag{10.111}$$

Apparently a factor of safety n with respect to shakedown, not much in excess of unity, is adequate to keep the energy dissipation within reasonable bounds.

It is obvious that the first collapse theorem can be considered as a special case of the first shakedown theorem, when the latter is applied to a loading program, in which all loads rise to certain constant values. Because of the complementary nature of the first and second collapse theorems a second, complementary shakedown theorem was sought for, and found by Koiter (1956).

In the second shakedown theorem the concept of an admissible plastic strainrate cycle $\varepsilon_i^{o''}(t)$ is introduced. It is defined by the property that the increments of the plastic strains in such a cycle for some time interval $t_c$ constitutive a kinematically admissible strain distribution

$$\Delta\varepsilon_i^{o''} = \int_o^{t_c} \varepsilon_i^{o''} dt, \quad \Delta\varepsilon_i^{o''} = D_{ik}\Delta u_k^o .$$

For $0 < t < t_c$ the plastic strainrates may give rise to residual stresses such that

$$D_{ik}\dot{\sigma}_i^o = D_{ik}S_{ij}(D_{j\ell}\dot{u}_\ell^o - \dot{\varepsilon}_j^{o''}) = 0,$$

from which it follows that

$$\dot{u}_k^o = [D_{ik}S_{ij}D_{j\ell}]^{-1} D_{a\ell}S_{ab}\dot{\varepsilon}_b^{o''} . \qquad (10.112)$$

The <u>second shakedown theorem</u> may now be expressed by the following statements: The structure will not shakedown, i.e. it will fail ultimately by cyclic plastic deformations, if any admissible plastic strainrate cycle $\dot{\varepsilon}_i^{o''}(t)$ can be found for which the work done by any external loads $f_k(t)$ within the prescribed limits exceeds the energy dissipation in the strainrate cycle. On the other hand the structure will shakedown if for all admissible plastic strainrate cycles and all external loads the energy dissipation is larger than the work done by the external loads.

For the proof of the first part of this theorem we assume that no further plastic deformation occurs, though we have found an admissible plastic strainrate cycle $\varepsilon_i^{o''}(t)$ such that

$$\int_0^{t} f_k \dot{u}_k^{o}\,dt > \int_0^{t_c} \sigma_i \dot{\varepsilon}_i^{o''}\,dt .$$

Then according to the first shakedown theorem there exists at any moment an allowable stress distribution

$$\sigma_i^{(a)} = \sigma_i^{e} + \sigma_i^{r}, \quad D_{ik}\sigma_i^{(a)} = f_k .$$

Hence

$$\int_0^{t_c} f_k \dot{u}_k^{o}\,dt = \int_0^{t_c} D_{ik}\sigma_i^{(a)} \dot{u}_k^{o}\,dt = \int_0^{t_c} \sigma_i^{(a)} (\dot{\varepsilon}_i^{o'} + \dot{\varepsilon}_i^{o''})\,dt ,$$

or

$$\int_0^{t_c} f_k \dot{u}_k\,dt = \int_0^{t_c} (\sigma_i^{e}\dot{\varepsilon}_i^{o'} + \sigma_i^{r}\dot{\varepsilon}_i^{o'} + \sigma_i^{(a)}\dot{\varepsilon}_i^{o''})\,dt .$$

But $\int_0^{t_c} \sigma_i^{e}\dot{\varepsilon}_i^{o'}\,dt = \int_0^{t_c} \varepsilon_i^{e}\dot{\sigma}_i^{o}\,dt = \int_0^{t_c} D_{ik} u_k^{e}\dot{\sigma}_i^{o}\,dt = 0,$

and

$$\int_0^{t_c} \sigma_i^{r}\dot{\varepsilon}_i^{o'}\,dt = \sigma_i^{r}\Delta\varepsilon_i^{o'} = 0 .$$

It follows that

$$\int_0^{t_c} f_k \dot{u}_k^{o}\,dt = \int_0^{t_c} \sigma_i^{(a)}\dot{\varepsilon}_i^{o''}\,dt$$

and

$$0 > \int_0^{t_c} (\sigma_i - \sigma_i^{(a)})\dot{\varepsilon}_i^{o''}\,dt ,$$

which contradicts the inequality (10.108). Hence there does not exist an allowable stress distribution $\sigma_i^{(a)}$ and the structure will not shakedown.

For the proof of the second part of the theorem we shall show that the energy dissipation in any loading program within prescribed limits is bounded if for all admissible plastic strainrate cycles holds

$$\int_0^{t_c} f_k \dot{u}_k^o \, dt < \int_0^{t_c} \sigma_i \dot{\varepsilon}_i^{o''} \, dt \, .$$

The actual rate of energy dissipation, $\sigma_i \dot{\varepsilon}_i''$, appears in the virtual work equation for the actual loading process.

$$f_k \dot{u}_k = \sigma_i D_{ik} \dot{u}_k = \sigma_i (\dot{\varepsilon}_i' + \dot{\varepsilon}_i'') \, .$$

The elastic strains $\varepsilon_i'$ can be considered as the sum of the strains according to elasticity theory, $\varepsilon_i^e$, and the residual elastic strains $\varepsilon_i^{r'} = S_{ij}^{-1} \sigma_j^r$.

If we integrate the virtual work equation over the interval of time $t_o < t < t_1$, where at $t_o$ we have the virgin state with residual stresses equal to zero, then we have

$$\Delta\varepsilon_i'' + \Delta\varepsilon_i^{r'} = D_{ik} (\Delta u_k - \Delta u_k^e)$$

and

$$\int_{t_o}^{t_1} f_k (\dot{u}_k - \dot{u}_k^e) \, dt = \int_{t_o}^{t_1} [(\sigma_i^r + \sigma_i^e) \dot{\varepsilon}_i^{r'} + \sigma_i \dot{\varepsilon}_i''] \, dt$$

$$= \int_{t_o}^{t_1} (\sigma_i^r \dot{\varepsilon}_i^{r'} + \sigma_i \dot{\varepsilon}_i'') \, dt \, ,$$

since

$$\sigma_i^e \dot{\varepsilon}_i^{r'} = \varepsilon_i^e \dot{\sigma}_i^r = D_{ik} u_k^e \dot{\sigma}_i^r = 0 .$$

We now construct an admissible plastic strainrate cycle as follows

$$t_o \leq t \leq t_1: \quad \dot{u}_k^o = \dot{u}_k - \dot{u}_k^e, \quad \dot{\varepsilon}_i^{o''} = \dot{\varepsilon}_i''; $$

$$t_1 \leq t \leq t_c: \quad \dot{u}_k^o = \frac{v_k}{t_c - t_1}, \quad \dot{\varepsilon}_i^{o''} = \frac{\Delta\varepsilon_i^{r'} + D_{ik} v_k}{t_c - t_1}.$$

The displacements $v_k$ must be determined such that $(\Delta\varepsilon_i^{r'} + D_{ik} v_k)$ can be realized as a plastic deformation with corresponding finite stresses producing an energy dissipation:

$$F(\Delta\varepsilon_i^{r'} + D_{ik} v_k).$$

For the given admissible plastic strainrate cycle the inequality

$$\int_o^{t_c} f_k \dot{u}_k^o \, dt < \int_o^{t_c} \sigma_i \dot{\varepsilon}_i^{o''} \, dt$$

turns into

$$\int_{t_o}^{t_c} f_k \dot{u}_k^o \, dt = \int_{t_o}^{t_1} (\sigma_i^r \dot{\varepsilon}_i^{r'} + \sigma_i \dot{\varepsilon}_i'') \, dt + \bar{f}_k v_k <$$

$$< \int_{t_o}^{t_1} \sigma_i \dot{\varepsilon}_i'' \, dt + F(\Delta\varepsilon_i^{r'} + D_{ik} v_k),$$

or with $k > 1$

$$(k-1) \int_{t_o}^{t_1} \sigma_i \dot{\varepsilon}_i'' \, dt \leq F(\Delta\varepsilon_i^{r'} + D_{ik} v_k) - k\bar{f}_k v_k - k \int_{t_o}^{t_1} \sigma_i^r \dot{\varepsilon}_i^{r'} \, dt.$$

It is now observed that F is a function which increases proportionally for a proportional increase of the residual stresses, resulting from the loading program, and the same is true for the term $\bar{f}_k v_k$, where we may take for $\bar{f}_k$ any time independent load vector within the prescribed limits.

Since the remaining term in the right hand side of the inequality is a quadratic expression in the residual stresses the whole expression in the right hand side obviously has an upper bound. Because $k > 1$ the total energy dissipation in the left hand side is then also bounded and plastic deformation cannot continue indefinitely.

In contrast to the collapse theorems the application of the shakedown theorems requires a complete elastic analysis, for all possible combinations of loads to determine $\sigma_i^e$ in the case of the first shakedown theorem, or for the calculation of $\dot{u}_k^o$ (10.112) in the case of the second shakedown theorem. The limit analysis itself, with respect to collapse as well as with respect to shakedown, generally leads to problems of nonlinear mathematical programming. Though the equilibrium equations and the continuity equations have been linearized, the yield criterion usually is still nonlinear. Replacing the yield criterion by a series of piecewize linear criteria we can formulate the limit analysis in terms of dual problems of linear programming algebra. The piecewize linear analysis for the model of beam behaviour, contained in the plastic hinge concept, has found widespread application in limit analysis of beam structures [9, 11].

REFERENCES


1. Truesdell, C. and R. Toupin, The classical field theories, p. 596, Encyclopedia of Physics, Vol. III/1, Springer Verlag, Berlin, Göttingen, Heidelberg (1960).

2. Biot, M.A., Variational lagrangian-thermodynamics of nonisothermal finite strain mechanics of porous solids and thermomolecular diffusion, Int. J. Solids of Structures, 13 (1977) 579-597.

3. Besseling, J.F., A thermodynamic approach to rheology, Proc. IUTAM-Symp. on irreversible aspects of continuum mechanics, Springer Verlag, Wien (1968) pp. 16-53.

4. Eckart, C., The thermodynamics of irreversible processes IV: The theory of elasticity and anelasticity, Phys. Rev. (2), 73, (1948) pp. 373-382.

5. Heijden, A.M.A. van der and J.F. Besseling, A large strain plasticity theory and the symmetry properties of its constitutive equations, to be published in Computers and Structures.

6. Besseling, J.F., A theory of elastic, plastic and creep deformations of an initially isotropic material. J. Appl. Mech. 25 (1958) 529-536.

7. Meijers, P. Experimental verification of constitutive equations for creep and plasticity based on overlay models. Lab. Eng. Mech., Delft (1981), Rep. nr. 704, 14 pages.

8. Koiter, W.T., The energy criterion of stability for continuous elastic bodies, I, II, Proc. Kon. Ned. Ak. Wet., Series B, 68 (1965) pp. 178-202.

9. Besseling, Plasticity and creep theory in engineering mechanics, Topics in continuum mechanics, Ed. by Zeman and Ziegler, Springer Verlag, Wien (1974), pp. 115-136.

10. Koiter, W.T., General theorems for elastic-plastic solids, Progress in Solid Mechanics, Vol. I. North-Holland Publ. Co., Amsterdam (1960) pp. 167-218.

11. Besseling, J.F., The force method and its application in plasticity problems, Computers and Structures, 8 (1978) pp. 323-330.

MIX
Papier aus verantwortungsvollen Quellen
Paper from responsible sources
FSC® C105338
FSC
www.fsc.org

If you have any concerns about our products,
you can contact us on
**ProductSafety@springernature.com**

In case Publisher is established outside the EU,
the EU authorized representative is:
**Springer Nature Customer Service Center GmbH**
**Europaplatz 3, 69115 Heidelberg, Germany**

Printed by Libri Plureos GmbH
in Hamburg, Germany